S0-CAH-374

Mihály Nógrádi

Stereoselective Synthesis

Distribution:
VCH Verlagsgesellschaft, Postfach 12 60/12 80, D-6940 Weinheim (Federal Republic of Germany)
USA und Canada: VCH Publishers, Suite 909, 220 East 23rd Street, New York NY 10010

ISBN 3-527-26467-1 (VCH Verlagsgesellschaft)
ISBN 0-89573-494-X (VCH Publishers)

Mihály Nógrádi

Stereoselective Synthesis

Prof. Dr. M. Nógrádi
Institute of Organic Chemistry
Technical Unversity
XI., Gellért tér 4
H-1521 Budapest

Editorial Director: Dr. Hans F. Ebel
Production Manager: Peter J. Biel

Deutsche Bibliothek Cataloguing-in-Publication Data

CIP-Kurztitelaufnahme der Deutschen Bibliothek

Nógrádi, Mihály:
Stereoselective synthesis / Mihály Nógrádi. –
Weinheim; New York: VCH, 1986.
ISBN 3-527-26467-1 (Weinheim)
ISBN 0-89573-494-X (New York)

Composition and Printing: Spandel-Druck, D-8500 Nürnberg
Binding: Großbuchbinderei Georg Gebhardt, D-8800 Ansbach
Cover Design: TWI, Herbert J. Weisbrod, D-6943 Birkenau
Printed in the Federal Republic of Germany

Preface

The present work is an attempt to review practical methods of stereoselective synthesis with emphasis on recent advances. It embraces a wide variety of subjects, such as hydrogenations over chiral catalysts, reductions with chiral hydride donors, stereoselective epoxidations, pericyclic reactions and the rapidly expanding field of "acyclic stereoselection".

This is a very broad topic and therefore several restrictions had to be imposed on the subjects to be covered.

First of all, enzymatic transformations were omitted because these have been excellently reviewed recently [1]. Neither are we going to discuss stereochemical aspects of reaction mechanisms in general. There is an almost endless number of examples in the literature for syntheses starting from an optically active compound, which are carried through a number of more or less selective transformations, to end up with a product in which the original stereogenic element is retained. No coverage will be given to such syntheses since this would have required the inclusion of such immense fields as transformations of steroids, terpenoids, carbohydrates and the like.

Although there is no specific date from which the literature has been processed, earlier methods giving poor stereochemical yields will not be discussed, unless they served as a starting point for more efficient processes. Most of these methods have been amply described in earlier works [2, 3].

The hectic activity in the field of stereoselective synthesis precludes a comprehensive treatment of even the literature of the last 15 years. Therefore methods with low stereochemical yields, but also the application of efficient methods to molecules of high complexity will generally be omitted.

Again for reasons of space, this book is somewhat biassed in favor of methods falling under the rather ill-defined term "asymmetric synthesis". Just by their sheer number diastereoselective methods could not have been covered with any claim to comprehensiveness. On the other hand, it would have been rather controversial not to deal at all with diastereoselectivity, since, as will be apparent from the introductory chapter, the underlying phenomenon of enantioselectivity is in fact diastereoselectivity.

In summary, we wished to serve the practically minded, synthetic organic chemist rather than the theoretician.

The author was induced to write this book, not only by the extremely rapid advancement and fascination of this field, but also by his conviction that the development of stereoselective synthetic methods has reached a turning point from which applications to practical problems have become a realistic proposition. Synthesis of natural amino acids, of non-racemic pharmaceuticals with fewer side-effects, of prostanoids and steroids, of insect hormones and pheromones are only the most rewarding fields in which such methods are of key importance.

The literature has been reviewed up to December 1984.

The author is indebted to Dr. C. Dyllick-Brenzinger for carefully revising the manuscript and to Mr. T. Goschi for his help in preparing the figures.

Budapest, August 1986

M. Nógrádi

---

[1] J. Rétey and J. A. Robinson, *Stereospecificity in Organic Chemistry and Enzymology*. Verlag Chemie, Weinheim, 1982.

[2] J. D. Morrison and H. S. Mosher, *Asymmetric Organic Reactions*. Prentice-Hall, Englewood Cliffs, N. J., 1971.

[3] Y. Izumi and A. Tai, *Stereodifferentiating Reactions*. Academic Press, New York, 1977.

Contents

List of Symbols and Abbreviations

Groups

Me	methyl
Et	ethyl
Pr	1-propyl
iPr	2-propyl
Bu	1-butyl
sBu	2-butyl
iBu	2-methylpropyl
tBu	1,1-dimethyl-l-propyl
Pent	1-pentyl
cPent	cyclopentyl
cP	cyclopentadienyl
Hex	1-hexyl
cHex	cyclohexyl
Ph	phenyl
Hept	1-heptyl
Bn	benzyl
Tol	*p*-tolyl
Oct	1-octyl
(−) Ment	(−) menthyl
BBN	9-borabicyclo[3.3.1]nonyl
Ac	acetyl
Bz	benzoyl
Tos	*p*-toluenesulfonyl
Tf	trifluoromethylsulfonyl
M	metal

Reagents and solvents

COD	1,5-cyclooctadiene
DBU	1,5-diazabicyclo[5.4.0]undec-5-ene
DME	1,2-dimethoxyethane
HMPA	hexamethylphosphoric amide
LDA	lithium diisopropylamide
LAH	lithium aluminum hydride
MCPBA	metachloroperbenzoic acid
MEM	(2-methoxyethoxy)-methyl
NBD	norbornadiene
PhMe	toluene
THF	tetrahydrofuran

Introduction

Ever since the stereoisomerism of organic molecules was discovered and the amazing stereoselectivity of living systems in synthesizing their products was recognized, chemists have been challenged to try their hands at preparing stereoisomers in a planned manner.

First their role was rather passive and confined mainly to the observation that in some reactions diastereomers were produced in unequal amounts. The reasons for such a selectivity remained obscure for a long time, and therefore it was left to chance which of the possible stereoisomers was obtained in excess. In fact, there was also not much practical demand for achieving stereoselectivity.

The birth of stereoselective synthesis probably dates back to 1890, when Emil Fischer recognized that the reaction of L-arabinose with hydrogen cyanide provided about 66% of one of the two possible diastereomers, namely, L-mannononitrile [1]. In this way asymmetric induction was discovered, and thus one of the cornerstones of diastereoselective synthesis laid down. This was followed at the turn of the century by the discovery of the partial kinetic resolution of racemic mandelic acid by esterification with (−)-menthol by Marckwald and McKenzie [2], the first example of a non-enzymatic enantioselective method.

During the next four decades stereoselective synthesis remained a marginal field of organic chemistry. After World War II, however, steroid hormones, manufactured industrially mostly by semi-synthesis, acquired enormous economic importance. This stimulated the interest of many of the leading organic chemists to search for practical methods for the preparation of a predetermined diastereomer of a compound. It was a logical development of this endeavor that in 1950 Barton was able propose a rationalization for a large number of hitherto unexplained examples of diastereoselection in the steroid and terpene field [3]. Barton's concepts were based on work by Hassel and Pitzer, who recognized that the stable conformation of cyclohexane derivatives was the chair form and the substituents preferred equatorial positions. Barton's ideas then became known as "conformational analysis", although nowadays this term is used in a somewhat different context. It was the adoption of Barton's concepts that first enabled, at least with compounds containing six-membered saturated rings, the planning of syntheses directed towards a given diastereomer.

However, methods for producing a required diastereomer of an acyclic compound remained in their infancy for a long time, although the rules of Cram [4] and Prelog [5] concerning nucleophilic addition to prochiral carbonyl groups were important milestones on the way towards efficient acyclic stereoselection. Furthermore, attempts at enantioselective synthesis using a wide variety of chiral aids (removable chiral groups, chiral catalysts, *etc.*) were almost invariably frustrated by low enantiomeric purity of the products. Characteristic of this situation is the book "Asymmetric Organic Reactions" by Morrison and Mosher [6] which reviewed the literature up to 1968. Here less than ten examples could be quoted in which products with more than 90% enantiomeric purity were obtained. The usual values were less than 20%.

Progress remained slow as long as steric hindrance alone was invoked to direct transformations towards a preselected stereoisomer. Perhaps influenced by the knowledge of how enzymes work, it slowly became clear that for high stereoselectivity it was necessary to immobilize the substrate in a suitable conformation. This fixation usually also involves the shielding of one of the molecular faces and thereby sets the stage for a stereoselective attack of the reagent. Two metals proved to be prominent aids to chemists in the realization of this concept, namely rhodium and lithium. Enantioselectivities which can be achieved in homogeneous hydrogenation using rhodium complexes of chiral biphosphines are really spectacular: by this method certain amino acids can now be prepared in almost total optical purity [7]. The process has also been realized on an industrial scale [8]. No less impressive are the results of methods in which lithium plays a key role. With selected combinations of substrates and reagents, electrophilic attack on lithium enolates may give rise to acylic products with total diastereselectivity, while that on chiral lithium enamides may provide almost total enantioselectivity [9]. Today one may venture to say that non-enzymatic stereoselective processes devised by organic chemists are almost as efficient as enzymatic systems. Both are characterized by achieving total stereoselectivity for a limited number of substrate—reagent pairs under strictly specified conditions, and both break down rapidly when either the optimal substrate—reagent combination or optimum conditions are abandoned.

References

[1] E. Fischer, Ber. dtsch. chem. Ges. **23,** 2611 (1890).
[2] W. Marckwald and A. McKenzie, Ber. dtsch. chem. Ges. **32,** 2130 (1899).
[3] D. H. R. Barton, Angew. Chem. **82,** 827 (1970).
[4] D. J. Cram and F. A. Abd Elhafez, J. Am. Chem. Soc. **74,** 5828 and 5851 (1952).
[5] V. Prelog, Helv. Chim. Acta **36,** 308 (1953).
[6] J. D. Morrison and H. S. Mosher, *Asymmetric Organic Reactions.* Prentice-Hall, Englewood Cliffs, N. J., 1971.
[7] B. Bosnich and M. D. Fryzuk, Topics in Stereochem. **12,** 119 (1980).
[8] B. D. Vineyard, W. S. Knowles and M. J. Sabacky, J. Mol. Catal. **19,** 159 (1983).
[9] P. A. Bartlett, Tetrahedron **36,** 2 (1980).

1 General Concepts of Stereoselective Synthesis

Although the present book is primarily oriented towards the practical aspects of stereoselective synthesis, it is necessary to describe briefly the basic principles of stereoisomerism, chemical selectivity in general and stereoselectivity in particular. Also, it is important to define the nomenclature and the system of notation to be used. The following introductory sections, however, should not be regarded as stereochemistry in a nutshell since only aspects important for our topic will be discussed.

1.1 Principles of Differentiating Molecules

The two main objectives of chemistry are the analysis and synthesis of molecules. The analysis of molecules is a rather abstract task, since there is no obvious, easily recognizable correlation between the outer appearance of a chemical substance and its internal properties which we generally call chemical structure. Differences in the structures of molecules are manifold, and it is possible to define a hierarchy of characteristics by which molecules can be distinguished. As we go down this ladder of hierarchic characteristics, molecules become more and more similar until we reach complete identity. Molecules which are identical in terms of higher ranking features may be distinguished by lower ranking ones.

(i) Molecules can differ in their *qualitative composition, i.e.* by the nature of elements they contain. Potassium carbonate and sodium carbonate, though both colorless crystalline solids, are of different qualitative composition.

(ii) Molecules of identical qualitative composition may differ in their *quantitative composition i.e.* by the ratio of the different elements they contain. Carbon monoxide and carbon dioxide, *e.g.*, differ in this respect.

(iii) Compounds of identical qualitative and quantitative composition may differ in their *molecular weight.* Acetylene, benzene, and cyclooctatetraene are examples for such a relationship.

(iv) Molecules which are found to be identical by criteria (i)−(iii), may be different due to the different *connectedness* of their atoms. Here we enter the domain of *isomerism,* and molecules which only differ by the sequence of their atoms are called *constitutional isomers.* Constitutional isomers, and of course the constitution of a single molecular species, can be fully characterized by enumerating each of their atoms and stating the nature and number of all the atoms connected to each particular atom by chemical bonds.

All this can be written down as a matrix, called a connectivity matrix, in a form suitable for computer processing. For some very simple molecules constitution can be defined in an abbreviated form. Thus isobutane is characterized by a carbon atom to which three other carbon atoms are bonded, while in n-butane there is no such carbon atom. It should be noted that the constitution of a molecule can always be adequately characterized without using words denoting directions such as “under” or “over”, “left” or “right”.

Molecules which are found to be identical by criteria (i)−(iv), but are nevertheless distinguishable are *stereoisomers*. Stereoisomers occupy two steps on our scale of differentation.

(v) *Diastereomers* are molecules of identical constitution but which can be differentiated by some scalar property, the most important being internuclear distances of a selected pair of groups (atoms) or in complex cases by the distances of several such pairs.

The following examples serve to illustrate how we can characterize diastereomers by internuclear distances. Thus diastereomers of 1,2-dibromoethene differ by the internuclear distance of the two bromo atoms. Conventional prefixes attached to names describing constitution enables one to identify diastereomers without taking recourse to formulas.

```
Br     Br          Br     H
  \   /              \   /
   C=C                C=C
  /   \              /   \
H      H           H      Br
```

(*Z*)-1,2-dibromoethene (*E*)-1,2-dibromoethene

The prefix *Z* in the above formula means that the distance between the selected pair of groups in this diastereomer is smaller than that in the *E* isomer. Rules for assigning the above and other conventional prefixes have been agreed upon by international conventions called the IUPAC Rules of Nomenclature. Most important from our point of view are the “Rules of Stereochemical Nomenclature” [1]. Any pair of diastereomers, however complex their constitution should be, can be adequately characterized by a set of statements referring to internuclear distances.

Thus the names in Fig. 1-1 give full information about the features differentiating the two diastereomeric stereoids. We only need to know that the conventional prefix β identifies groups which are on the same side as the reference group (smaller distance), while α is for groups which are on the opposite side (larger

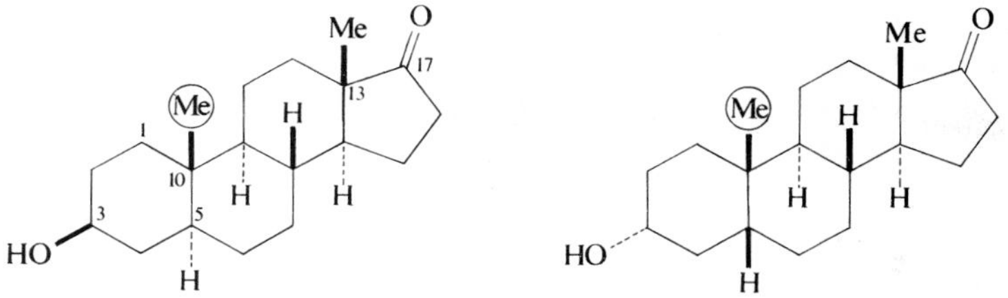

3β-Hydroxy-10β,13β-dimethyl-
-5α-gonan-17-one

3α-Hydroxy-10,β,13β-dimethyl-
-5β-gonan-17-one

Fig. 1-1. Characterization of diastereomers.

distance). In our case the C(10)–Me group serves as reference. With mobile molecules, in order to be able to establish larger-smaller relationships between internuclear distances, we have to fix the molecular framework in a certain arrangement. A classical conventional arrangement for linear molecules is the Fischer-projection shown in Fig. 1-2 for sorbitol and mannitol, both in a "sugar-short hand" and in a more detailed representation. The two diastereomers differ in the relationship of C(2)–OH to C(5)–OH (same side *vs.* opposite side). In carbohydrate chemistry the highest numbered chiral center is taken as the reference group (here C(5)) and special prefixes (like *gluco* and *manno*) are assigned to various relative arrangements of the four secondary hydroxyl groups in hexoses.

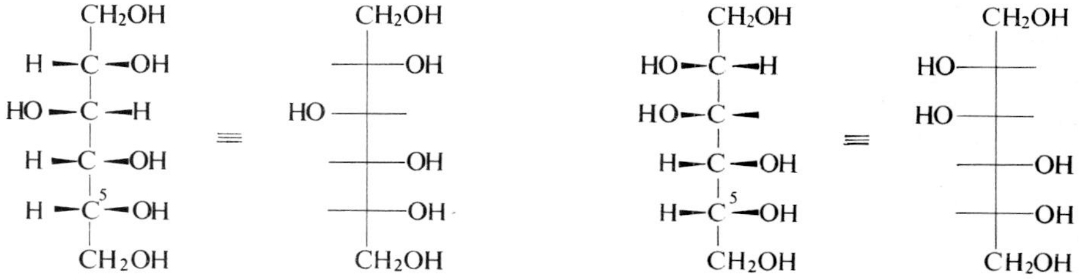

sorbitol
(*gluco* configuration)

mannitol
(*manno* configuration)

Fig. 1-2. The Fischer projection.

Differences in internuclear distances serve not only for the identification of diastereomers, more importantly they form the basis of differences in their physical and chemical characters.

(vi) *Enantiomers* are pairs of stereoisomers with the highest level of similarity. If their formulas are written down according to the same convention, internuclear distances for any given pair of atoms are identical. Enantiomers can, however, be distinguished by stating the *sequence* of selected groups following a certain convention. The conventional character of distinguishing enantiomers must be emphasized, because words such as clockwise – anticlockwise or right-handed –

left-handed are meaningless in themselves and come to life only by a world-wide agreement about their significance.

A typical statement describing the difference between dextrorotatory and levorotatory lactic acid is the following: when their formulas are depicted according to the same convention (hydrogen remote from the viewer) the sequence of the groups hydroxy, carboxy and methyl is anticlockwise for the dextrorotatory and clockwise for the levorotatory enantiomer.

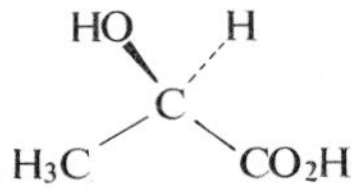

(−)-(*R*)-lactic acid 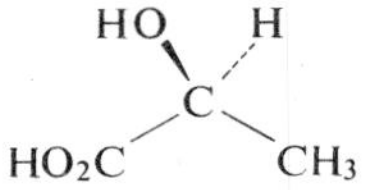 (+)-(*S*)-lactic acid

A system of nomenclature for the unambiguous characterization and distinction of enantiomers by pairs of simple prefixes (*R* and *S*, *P* and *M*) has been worked out by R. S. Cahn, C. K. Ingold, and V. Prelog (the s. c. C.I.P. convention) [2].

When represented following the same conventions, formulas of diastereomers are *not* mirror-images, while those of enantiomers are mirror-images. An object (*e. g.* a molecule) which is not identical with its mirror image is called *chiral,* otherwise it is achiral. Molecules forming enantiomers are chiral by definition, while chirality is not a condition for a diastereomeric relationship. Thus the diastereomeric 1,2-dibromoethenes are both achiral, while the gonanes shown above are both chiral. (−)-Tartaric acid and *meso*-tartaric acid are diastereomers, the former is chiral, the latter is achiral.

The last stage in our molecular identity – non-identity hierarchy is complete identity, which we are not interested in. Note that identity is a concept dependent on the depth of our insight. Thus molecules which we regard in our discussions as being identical may differ in their isotopic composition, electronic or nuclear quantum levels *etc.*

1.2 Characterization of Stereoisomers. Conformation and Configuration

Molecules can be characterized by a set of geometrical parameters. These are the van der Waals radii of the individual atoms (relevant to the concept of steric hindrance), equilibrium bond lengths between directly bonded atoms, equilibrium bond angles formed by the bonds of two atoms bonded to a common third atom and, finally, torsional angles describing the spatial relationship of the terminal atoms in a linear chain of four atoms.

The complete set of all possible torsional angles of a molecule defines its *conformation*. Certain well characterized conformations are called *conformers*. For practical purposes we usually disregard the torsional angle of bonds attached to a double bond (which we take as fixed at 0° and 180°, respectively) and those associated with groups rotating very fast, such as the methyl group. Molecular species having different conformations are, by definition, stereoisomers, since they are different entities with the same constitution, although such stereoisomers are usually inseparable due to their rapid interconversion. Examples for stereoisomers which differ in their conformation are the (*P*)-synclinal and antiperiplanar conformers of n-butane (inseparable) and (*R*)- and (*S*)-2,2′-diiodobiphenyl-6,6′-dicarboxylic acid [(*R*)- and (*S*)-(*1*)] (Fig. 1-3) (separable). The relationship of any two conformers may be either diastereomeric (as that of the two n-butanes) or enantiomeric (as that of the two biphenyls).

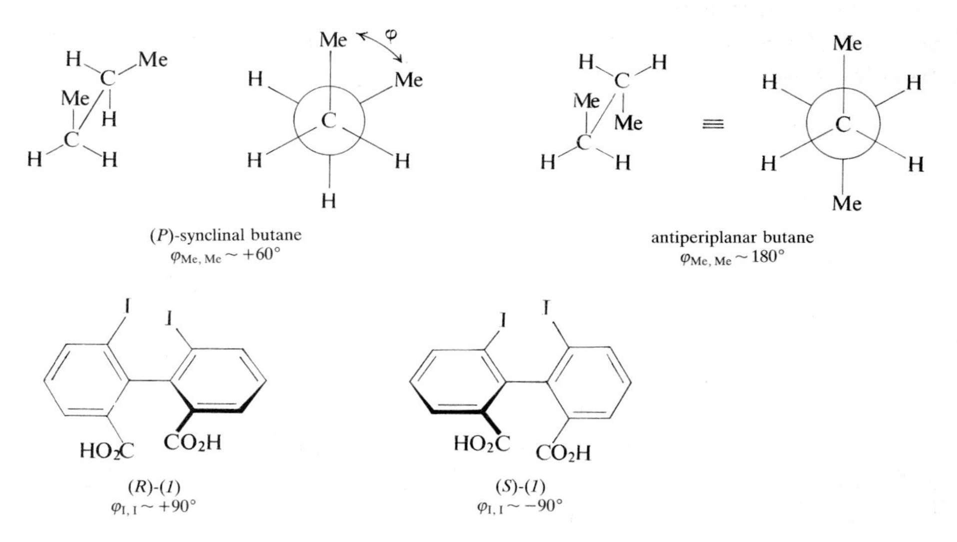

Fig. 1-3. Characterization of conformations.

While (apart from signs) conformation, *i. e.* describing stereoisomers by a set of torsional angles, is essentially a quantitative approach, *configuration* characterizes a molecule in a qualitative way. Configuration has a different role when describing diastereomers and enantiomers, but in both cases it essentially covers a system of conventions.

Diastereomers may be compared by their *relative configuration, i. e.* by differences in the intramolecular relationship of selected groups within each diastereomer. Examples for such an internal comparison were given in the preceding section, in which the distinguishing features were epitomized as conventional prefixes such as *E* and *Z*, α and β, *gluco* and *manno.*

Diastereomers which are interconvertible by rotation around a single bond can be described by the pertinent torsional angles or, more conveniently, by conventional names associated with specific ranges of torsional angles, such as synclinal for $\varphi = \pm 60° \pm 30°$, antiperiplanar for $180° \pm 30°$ *etc.*

Diastereomers which arise by different combinations of two or more chiral centers can be conveniently labelled by listing the configurational symbols *R* and *S* for each center.

CO_2H
R OH
HO *R*
CO_2H

(+)-(2*R*,3*R*)-tartaric acid

CO_2H
R OH
S OH
CO_2H

(2*R*,3*S*)-*meso*-tartaric acid

Fig. 1-4. Characterization of compounds with more than one chiral center.

In this book, in conformity with the IUPAC rules, the following nomenclature will be used to describe the relative configuration of diastereomers.

For geometrical isomers the *Z*−*E* notation will be used with some unavoidable exceptions, when, in order to embrace wider groups of compounds, we will be forced to fall back on the *cis-trans* notation. The prefixes *cis* and *trans* are useful to define the relative arrangement of groups attached to rings, for which purpose *Z* and *E* should not be used.

The names recommended by IUPAC (occasionally in their abbreviated form) will be applied to conformers generated by rotation around an A−B single bond. The relationships of the vicinal groups X and Y are denoted as in Fig. 1-5. Note that *syn* and *anti* are also used to describe the stereochemistry of addition and elimination.

Considerable confusion has been created recently in describing the relative configuration of two chiral centers in a linear molecule. Originally, the relative configuration of two groups which are on the same side in a Fischer projection (as in erythrose) was called *erythro,* while that of those on opposite sides (as in

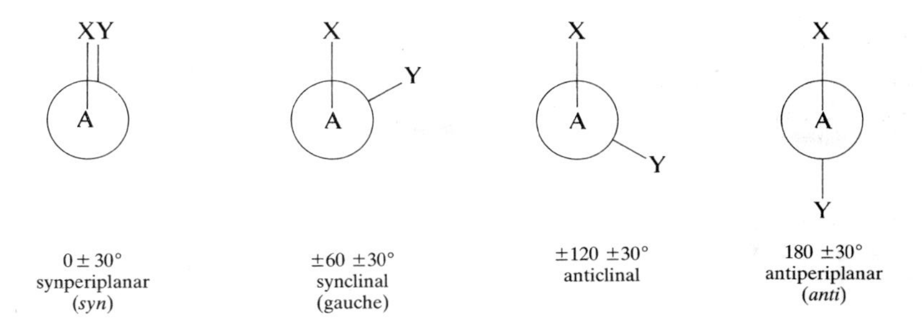

Fig. 1-5. Conventional names for the conformers of 1,2-disubstituted ethanes.

threose) *threo*. Apart from occasional difficulties in selecting the main chain, this system served well with but two chiral centers. The Fischer projection is undoubtedly an unnatural representation, and when the chain is drawn in the more realistic zig-zag form, *erythro* substituents end up on opposite sides whereas *threo* ones are on the same side. In Fig. 1-6 the (2*R*,3*R*) and the (2*R*,3*S*) diastereomers of 3-

(a)

(2*R*,3*R*)
erythro by the classical notation

threo by the "aldol" (Heathcock) notation

(b)

(2*R*,3*S*)
threo by the classical notation

erythro by the "aldol" (Heathcock) notation

Fig. 1-6. Characterization of relative configuration in aldol-type compounds.

hydroxy-2-methylbutanoic acid are shown in Fischer projection, in the zig-zag conformation and in a simplified representation of the latter. In order to put substituents back into their "customary" relationship, Heathcock suggested in 1981 [3] an inversion of the nomenclature, thereby causing bewildering confusion in the literature. Thus the (2*R*, 3*R*)-compound (a) previously called *erythro* now became *threo,* while its diastereomer was renamed *erythro.* Alas, Heathcock's suggestion gained rapid and widespread acceptance. The new problem was soon recognized and several suggestions for alternative ways to identify relative configuration in acyclic molecules have been devised [4, 5, 6]. The most comprehensive and consistent among them is the one by Prelog and Seebach [7] (*cf.* p. 27). To dissociate ourselves from both the old and new usage of *erythro* and *threo*, the prefixes *syn* and *anti* will be used in the sense shown in the following examples:*, **

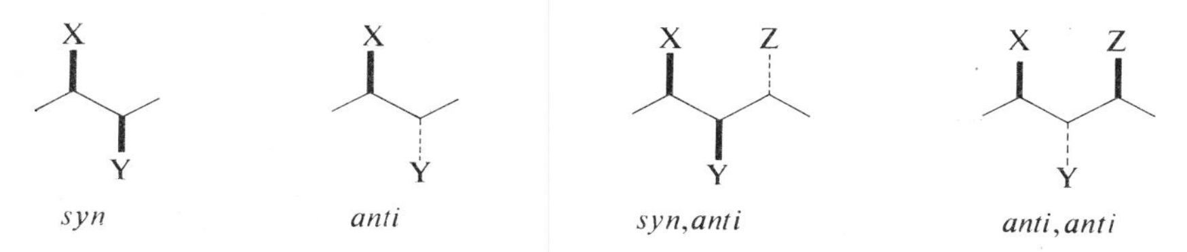

The steric disposition of groups with a geminal hydrogen atom will be indicated by heavy and broken lines, respectively. In the case of quaternary centers or when it is deemed necessary for a better presentation of the situation, traditional wedges will be used as well.

Enantiomers can be identified by quoting the sign of their optical rotation, by constructing or drawing a model representing the molecule or by reference to a certain convention. Fortunately nowadays only one system of convention, that of Cahn, Ingold and Prelog, also called the sequence rules, is in use. Its basic principles and application, at least for simple cases, is well known and need not be discussed here [2].

Well known optically active compounds containing more than one center of chirality, such as menthol, α-pinene, ephedrine *etc.* can be conveniently identified by their sign of rotation. Even this can be omitted for some compounds, such as the *Strychnos* alkaloids and steroids, which in nature only occur in one enantiomeric form.

---

* This is not a perfect solution either, since IUPAC recommended these words for the characterization of the mode of approach in addition reactions [8].

** In a recent review a repentant Heathcock turned to the *syn-anti* notation too [9].

1.3 Intramolecular Symmetry. Topicity and Prochirality

Topicity

Analysis of molecular symmetry is of fundamental importance for stereochemistry and therefore also for the thorough understanding of stereoselective reactions.

While for the discussion of stereoisomerism it is the global symmetry of a molecule which is relevant, from the point of view of stereoselectivity we also have to consider the symmetry relationships of certain subunits of the molecule, namely, those of groups and faces.

As a *group* we define in our context any subunit of a molecule; this can be as simple as a hydrogen atom or as complicated as a monosaccharide unit. Groups may be classified according to a hierarchical scheme similar to that used for molecules.

Thus groups, when regarded in isolation, may differ (i) in qualitative composition (*e. g.* Br and I, or CO_2H and COCl), (ii) in quantitative composition (*e. g.* CHO and CO_2H), in constitution (*e. g.* $-CH_2O(CH_2)_2OMe$ and $-CH_2OCH_2OEt$ or propyl and isopropyl) and (iii) in stereostructure (*e. g.* bornyl and isobornyl which are diastereomeric or (*R*)- and (*S*)-2-phenylethyl which are enantiomeric*).

Two or more groups in a molecule which are identical by the above criteria may have different relationships to each other.

(A) Groups can have the same or different connectedness with the rest of the molecule, in other words their *constitutional position* may be the same or different. Thus in picric acid nitro groups in the 2- and 6-position have the same connectedness [connected through two bonds with C(1)], while those in 2- and 4-position are constitutionally different [connected through two and four bonds to C(1)]. In a similar way bromine atoms in 2,3-dibromopropionic acid, and −HC= groups in crotonic acid are constitutionally different.

---

* Note that although often done for convenience, the assignment of *R* and *S* descriptors to groups is ambiguous since these are dependent on the nature of the fourth ligand. *E. g.*

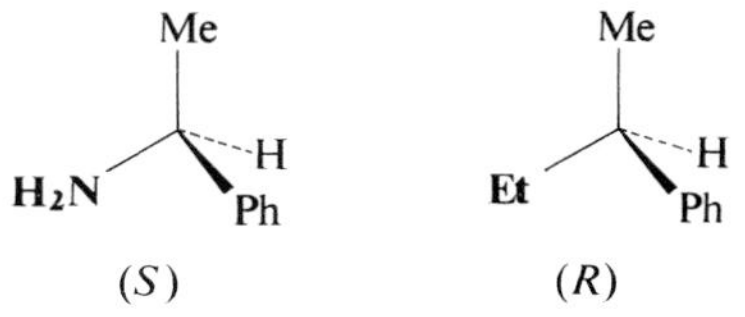

Me, H, H_2N, Ph — (*S*)

Me, H, Et, Ph — (*R*)

OH
O_2N 1 6 2 NO_2
4
NO_2

$BrCH_2—CH\mathbf{Br}—CO_2H$

$H_3C—\mathbf{HC}=\mathbf{C}H—CO_2H$

In discussing stereoselectivity we are only interested in identical groups of the same connectedness. Their relationship can be diastereotopic, enantiotopic or homotopic [10].

(B) *Diastereotopic* are groups which cannot be exchanged by any symmetry operation. Since in asymmetric molecules such as *2* or *3* in Fig. 1-7 symmetry elements cannot be present by definition, geminal groups in such molecules (set boldface) are always diastereotopic. Similarly to diastereomers, diastereotopic groups can be readily distinguished by their relationships (near-remote) to a reference group, *i. e.* in scalar terms. Chiral molecules with rotational symmetry (*e. g. 4*) and achiral molecules (*e. g. 5* and *6*) may also contain diastereotopic pairs of groups, but the symmetry element(s) must be unrelated to these groups. Thus geminal hydrogens within each CH_2 group (but not those at different carbons) of the cyclopentanone *4* (C_2 symmetry) are diastereotopic, one being near to, the other remote from the adjacent methyl group. Note that although *5* and *6* have a plane of symmetry, methyl and carboxyl groups resp. which lie in this plane cannot be exchanged by it. Although for conformationally mobile molecules such as *2* and *3* a given near-remote relationship between diastereotopic groups is only valid for a certain conformation, the molecular environment surrounding each member of the group is inherently different for each conformation. Although the magnitude of the difference (expressed as some physical parameter, *e. g.* magnetization) is depen-

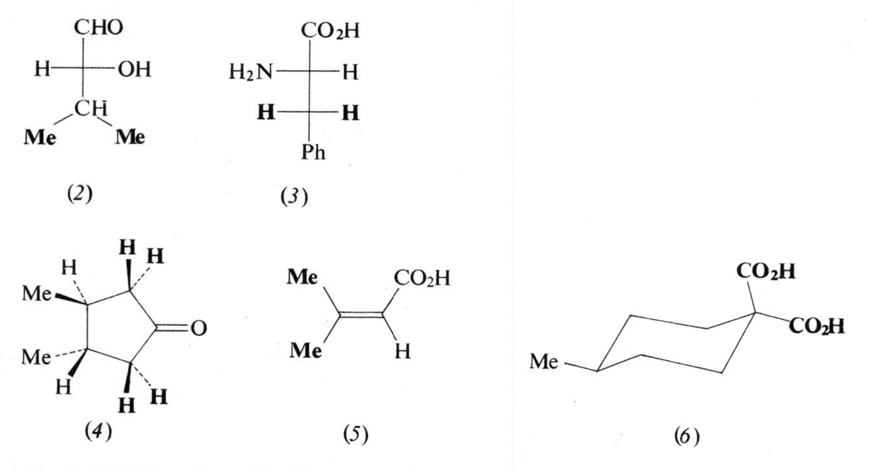

Fig. 1-7. Molecules with diastereotopic groups.

dent on conformational equilibria, the mere fact of this difference cannot be eliminated by fast rotation or ring inversion [11].

(C) *Enantiotopic* are groups which can be exchanged by a rotation-reflection axis, which is most often a plane ($\sigma = S_1$) or a center ($C_i = S_2$) of symmetry. Thus enantiotopic groups can only occur in achiral molecules.

Enantiotopic are CH_2CO_2H groups in citric acid (*7*) and subunits thereof (CH_2, CO_2H) or the phenyl or carboxy groups in *8*.

(D) *Homotopic* are groups which can be exchanged by a symmetry axis. It follows that any achiral or chiral molecule which has an axis of symmetry contains at least one set (usually a pair) of homotopic groups.

Compounds which contain a set of two, three and six homotopic hydrogens, respectively, are dichloromethane, chloromethane and benzene (Fig. 1-8).

Diastereo- and enantiotopic groups are called *heterotopic*.

The terms dia-, enantio- and homotopic express the relationship of one group to another and may therefore change with the partner. Thus, in *9* atoms H_A are homotopic and each has two kinds of enantiotopic relationships to atoms H_B, which also form a homotopic set. Neither is connected by any symmetry operation to the homotopic pair of atoms H_C which have a different connectedness. In *10* the H_A and H_B atoms form two enantiotopic sets, while any H_A is diastereotopic to any H_B.

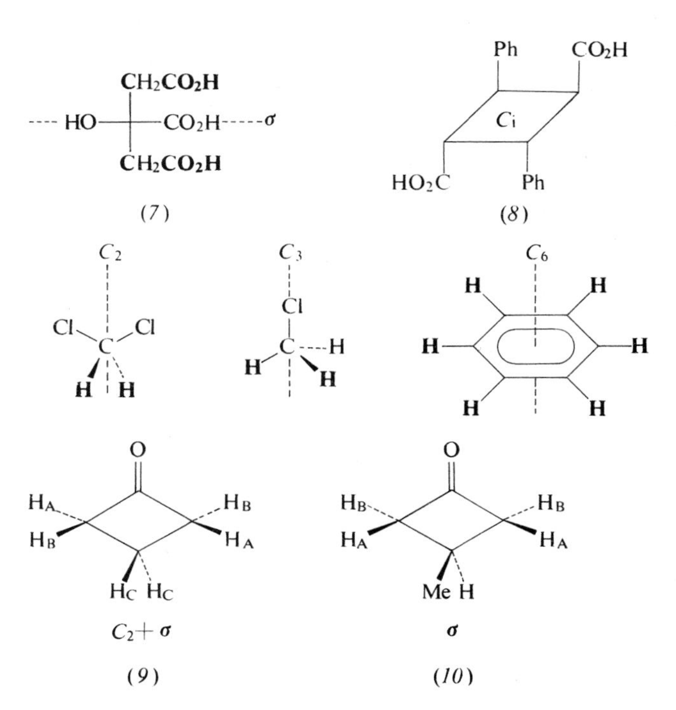

Fig. 1-8. Molecules with enantiotopic and homotopic groups.

One of the most frequent synthetic operations is the addition of a group to a tricoordinate center to form a tetracoordinate center. The tricoordinate center is usually a double bonded atom, the three valencies of which constitute a plane with two faces. The topicity notation explained above for groups can be conveniently extended to the symmetry relationships of such faces.

(i) *Diastereotopic* are two faces of any molecular plane which is no plane of symmetry and does not contain a coplanar axis of symmetry. Thus faces in asymmetric molecules (*e. g.* that of both C=O and C=C in *11*) are always diastereotopic, no matter how fast bond rotation may be.

A plane of symmetry perpendicular to the plane to be qualified is not incompatible with diastereotopicity, as can be seen with compound *10* which contains a carbonyl group with diastereotopic faces.

(ii) *Enantiotopic* are two faces of a molecular plane which is at the same time a molecular plane of symmetry but which does not contain a coplanar axis of symmetry. Enantiotopic are the faces of acetaldehyde, phenyl-methyl-sulfide and of the hypothetical molecule *12*, which contains a C_2 axis not coplanar with the $>$C=O planes.

Homotopic are two faces of a molecular plane which contains a coplanar axis of symmetry. Such faces can be found both in achiral molecules, such as acetone, isobutene, and in chiral ones as, *e. g.*, in *4*.

For conformationally mobile molecules symmetry relationships usually change with conformation. Here a practical standpoint can be adopted and conformational changes much faster than the process investigated should be disregarded. Thus for a low temperature NMR study the methyl groups and the faces of the carbonyl group of 2,2-dimethylcyclohexanone should be regarded as diastereotopic (*13*), while for a hydride transfer reaction at room temperature the s. c. *statistical symmetry* of the molecule, *i. e.* that of its most symmetrical (possibly non-populated) conformer (*14*, enantiotopic in the present case), should be invoked.

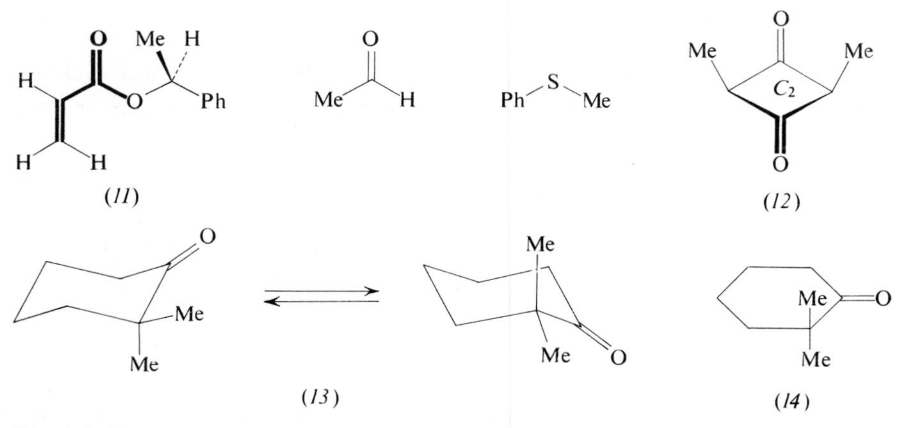

Fig. 1-9. Molecules with diatereotopic and enantiotopic faces.

Prochirality

Since, as will be discussed later, transformation of any one group of an enantiotopic pair of groups into another group or addition of a new group to a center with enantiotopic faces gives rise to a chiral compound, enantiotopic groups and faces are called *prochiral** [12]. The same is true for diasteretopic groups which do not coincide with a plane of symmetry and for diastereotopic faces which have no perpendicular plane of symmetry.

By application of the sequence rules, the labels pro-*R* and pro-*S* are given to geminal prochiral groups and prochiral faces. This is very convenient because a group or face involved in a given reaction can be identified without referring to a drawing. According to Prelog and Helmchen [13], a prochiral group is called pro-*R* (or briefly *Re*) when the sequence of ligands in decreasing order of priority (the other group is remote from the observer) is clockwise, and pro-*S* (*Si*), when it is anticlockwise (Fig. 1-10). The procedure for faces is even simpler: regarding the center to be qualified from the pro-*R* (*Re*) face the sequence of ligands in decreasing order of priority is clockwise, while antichlockwise from the pro-*S* (*Si*) face.

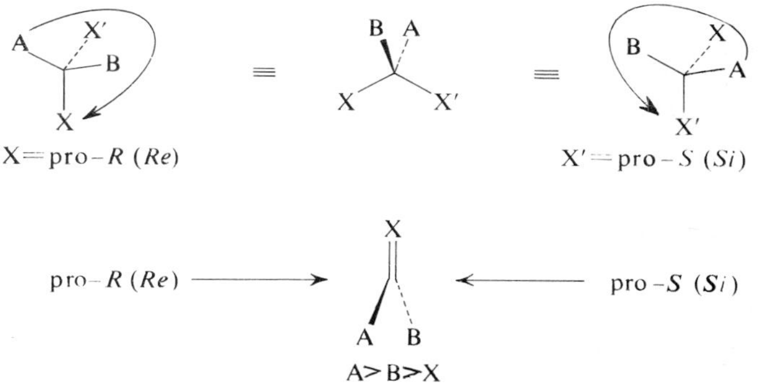

X = pro-*R* (*Re*) ≡ ≡ X′ = pro-*S* (*Si*)

pro-*R* (*Re*) → ← pro-*S* (*Si*)

A > B > X

Fig. 1-10. Prochirality.

Note that there is no correlation between the prochirality descriptor and the configuration of the product arising from transformation of the group or face concerned. For example:

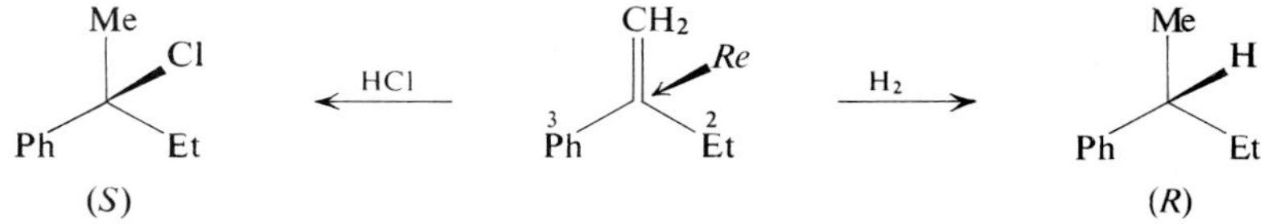

(*S*) ← HCl — CH_2 (*Re*) — H_2 → (*R*)

* In fact the concept of prochirality was proposed earlier than the topicity notation.

1.4 Selectivity in Chemistry

Two main modes of selectivity can be defined in chemistry, one concerns the substrates and the other the products of a reaction. These two modes are defined by the following schemes:
(i) Substrate selectivity:

$$A \xrightarrow[k_{AX}]{\text{reagent}} X\,; \quad B \xrightarrow[k_{BY}]{\text{reagent}} Y$$

$$k_{AX} \neq k_{BX}$$

(ii) Product selectivity:

$$A \xrightarrow{\text{reagent}} X + Y + Z + \ldots$$

$$[X] \neq [Y] \neq [Z] \ldots$$

Substrate selective is a *reagent* which transforms different substrates (*e. g.* A and B) under the same conditions at different rates to the products X and Y.

Product selective is a *reaction* or process in which more than one product can be formed but the products are formed in a ratio which differs from the statistically expected one. The latter is the ratio of the number of sites which can react (*e. g.* 2:2:1 for *o*-, *m*-, and *p*-nitrotoluene in the mononitration of toluene).

1.4.1 Substrate Selectivity

Substrate selectivity can be conveniently classified following the hierarchic scheme set up for the differentiation of molecules and described in Section 1.1.

Reactivity of constitutional isomers may differ markedly. Thus when a mixture of 1-hexene and 2,3-dimethyl-2-butene, 1 mole each, is subjected to catalytic hydrogenation, after the absorbtion of 1 mole of hydrogen almost all of the tetrasubstituted olefin will remain unchanged, while the monosubstituted one will be converted to hexane.

$$Me(CH_2)_3CH{=}CH_2 + Me_2C{=}CMe_2 \xrightarrow[\text{cat.}]{1\text{ mol } H_2} \begin{cases} \xrightarrow{k^1} Me(CH_2)_4Me \\ \xrightarrow{k^2} (Me)_2CH{-}CH(Me)_2 \end{cases} \qquad k^1 \gg k^2$$

Our real interest lies, however, in substrate selectivity exhibited by stereoisomers.

1.4.1.1 Substrate diastereoselectivity

Since diastereomers differ in their scalar properties it can be expected that their reactivity may also be different towards any type of reagent. In practice, these differences are the more significant the closer the reaction center is to the stereogenic units of the molecule.

Several modes of *substrate diastereoselectivity* can be distinguished:

1. Diastereomers can lead to the same product at different rates, *e.g.* $k_{ax}/k_{eq} = 3.2$ in 80% acetic acid for the following reaction [14]:

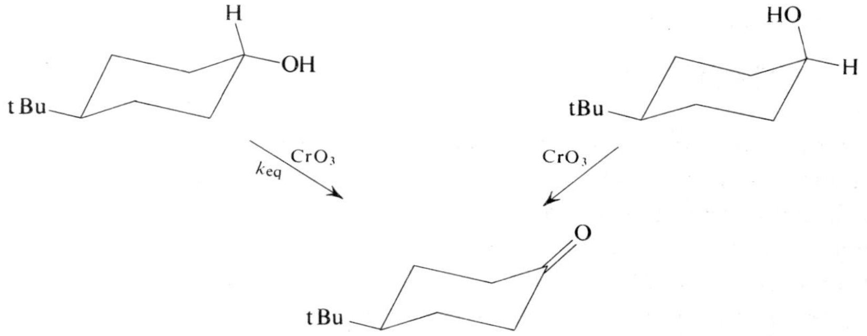

H, OH, tBu, CrO_3, k_{eq}, HO, H, tBu, CrO_3, O, tBu

2. Diastereomers can give rise to different diastereomers at different rates as exemplified by the debromination of diastereomeric 2,3-dibromobutanes [15]:

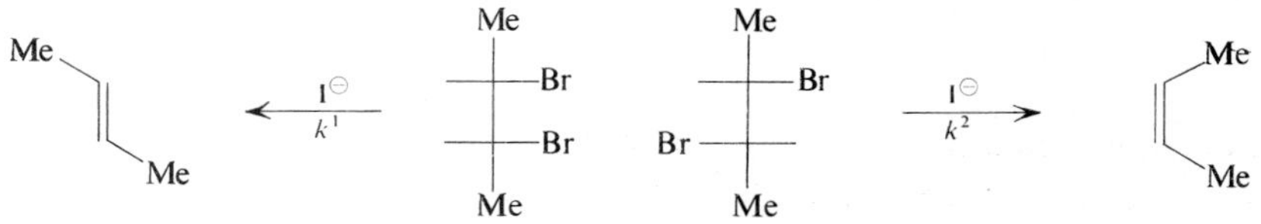

Me, Me, $I^{\ominus}$, k^1, Me, Br, Br, Me, Me, Br, Br, Me, $I^{\ominus}$, k^2, Me, Me

Rate differences between diastereomers may be so large that one of them does not react at all. Thus (±)-1,2-dichloro-1,2-diphenylethane is dehydrochlorinated in pyridine at 200°C, while the *meso*-diastereomer is inert under such conditions [16]:

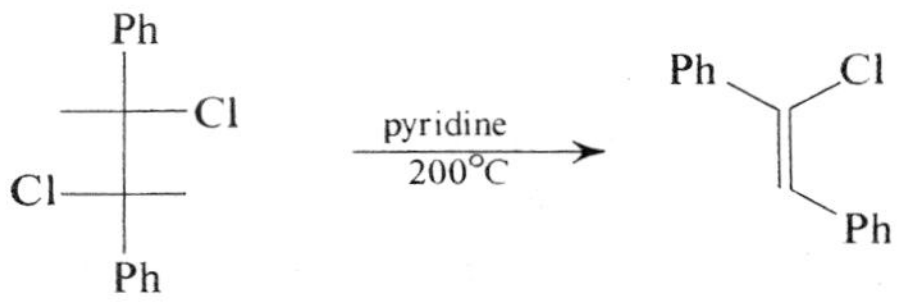

Ph, Cl, Cl, Ph, pyridine, 200°C, Ph, Cl, Ph

3. Diastereomers can even yield products with different constitutions, as shown by the classical example of the Beckmann-rearrangement of (*E*)- and (*Z*)-oximes:

(*E*) → PhNHCOMe

(*Z*) → PhCONHMe

4. Finally, different diastereomers can yield compounds which are not even isomers of each other [17]:

AcOH

AcOH

5. Since equilibria are in fact dynamic systems characterized by the ratio of the rates of the forward and reverse raction ($K = k/k^{-}$), equilibria involving diastereomers can be regarded as a special case of substrate diastereoselectivity. The simplest and most important type of such reactions is *epimerization*. One of the best known examples for the equilibration of diastereomers is the mutarotation in solution of α- or β-D-glucose, where the reagent is the solvent.

1.4.1.2 Substrate enantioselectivity. Kinetic resolution

Diastereomers can be relatively easily separated and therefore differences in their reactivity can be explored in separate experiments. With enantiomers the situation is just the opposite. Their separation is rather difficult and therefore conditions are sought under which they react at different rates, whereby ultimately their separation can be effected. Transformation of enantiomers at different rates is called in the present terminology *substrate enantioselectivity*.

Since enantiomers are the mirror images of each other, they can only be distinguished by effects or objects lacking reflectional symmetry. Such an effect is circularly polarized light and such an object is any non-racemic chiral reagent*.

This situation can be easily visualized by using two-dimensional chiral and achiral figures as suggested by V. Prelog [13]. A figure (*e. g.* a triangle) is chiral in two dimensions if it cannot be superimposed onto its mirror image by in-plane rotation and/or translation. Otherwise it is achiral. Molecules are symbolized by

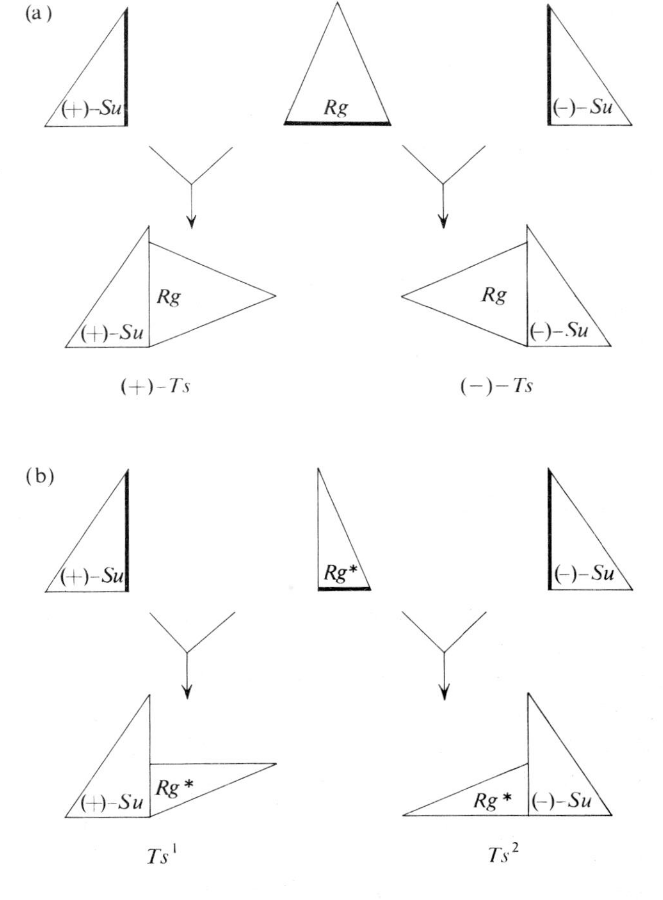

Fig. 1-11. Two-dimensional representation of the reaction of enantiomers with (a) achiral (*Rg*) and (b) chiral (*Rg**) reagents.

---

* A racemic mixture of chiral molecules behaves as if having reflectional symmetry. For the sake of brevity throughout this book the term "chiral reagent" denotes a non-racemic assembly of chiral molecules.

triangles, transition states, or interactions in general, by placing one side each of "substrate" and "reagent" in contact with each other. It is apparent from Fig. 1-11 (a) that interaction of enantiomeric chiral substrates [(+)-*Su* and (−)-*Su*] with an achiral reagent (*Rg*) produces "transition states" [(+)-*Ts* and (−)-*Ts*] which are related as mirror images. This remains true for any specific side ("functional group") of substrate and reagent brought into contact with another side.

When the same is done with a chiral reagent (*Rg**)* (Fig. 1-11 (b)) the resulting figures (Ts^1 and Ts^2) are no longer mirror images but diastereomeric and thus differ in their scalar properties.

Since the ground states of enantiomers are of the same free energy and, being diastereomeric, the transition states are of different free energy, it follows that in reactions involving chiral agents or effects enantiomers are transformed at different rates.

From a practical point of view we have to distinguish those cases when (i) the transformation rate of the less reactive enantiomer is virtually zero or (ii) when the two rates are comparable.

So far the first case has been restricted to enzymatic transformations and was first exploited by Pasteur. He subjected racemic ammonium tartrate to fermentation with the mold *Penicillium glaucum,* whereby the unnatural (−)-(2*S*,3*S*) enantiomer remained unchanged and could be isolated. A more recent example illustrates a case where it is the natural enantiomer which is required. Enzymic amidation of a racemic mixture of *N*-benzoyl amino acids (*15*) with aniline yields only the (*S*)-anilide, while the (*R*)-acid remains unchanged and is discarded [18].

$$\underset{\substack{(\pm)\text{—}(15)\\ \text{Ar} = 3,4\text{-(HO)}_2\text{C}_6\text{H}_3\text{-}}}{\text{ArCH}_2\text{CH(NHBz)CO}_2\text{H}} \xrightarrow[\text{papaine}]{\text{PhNH}_2} \underset{(S)}{\text{BzHN}\text{—}\overset{\text{CONHPh}}{\underset{\text{CH}_2\text{Ar}}{\text{C}}}\text{—H}} + (R)\text{-}(15)$$

In the second case, *i. e.* when the rates are comparable, one of the enantiomers is transformed faster, and therefore an excess of the less reactive substrate gradually builds up. This excess goes through a maximum and disappears on completion of the reaction (Fig. 1-12). If the reaction is interrupted before completion or if less than the necessary amount of reagent is applied, the result is a non-racemic mixture of the starting material and of a product in which an excess of the more reactive enantiomer of the substrate is incorporated. This way of partial separation of enantiomers is called *kinetic resolution.* The main types of kinetic resolution will be discussed below.

---

* Symbols marked with an asterisk represent chiral groups or non-racemic chiral compounds.

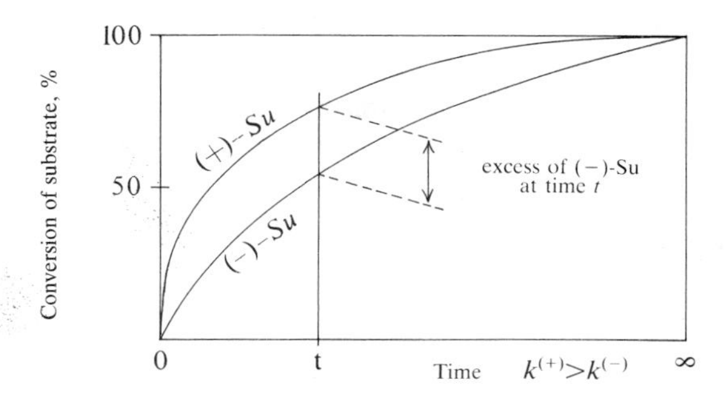

Fig. 1-12. Kinetic resolution as a function of time.

The chiral agent may be (i) a catalyst, a medium or a reagent donating or accepting a non-chiral entity (*e. g.* a hydride ion) or (ii) a reagent which is linked to the substrate in the course of the reaction.

In the first case the chirality of the substrate can either be retained or destroyed. If it is retained then the pair of enantiomeric substrates is transformed to a pair of enantiomeric products. When chirality is destroyed both enantiomers give rise to the same achiral product(s). In Fig. 1-13 examples for both situations are given.

The simplest chiral agent is circularly polarized light. On photodecomposition of the racemic diketone (±)-*16* to 55% conversion (Fig. 1-13 (a)) the remaining diketone is optically active (ee ~3.5%) [19].

In example (b) it is the medium which is chiral. When the racemic oxaziridine (±)-*17* was decomposed by heating for 1.25 min at 148° in cholesteryl benzoate, the unchanged starting material was enriched to 20% in the (–)-enantiomer [20].

Inclusion complex formation with cyclodextrin provides a chiral environment which is similar to a chiral solvent. Thus partial oxidation of (±)-*18* in the presence of α-cyclodextrin (c) leaves (+)-*18* in excess in the unchanged substrate [21]. Note that the product is achiral.

In the partial reduction to the achiral sulfide of (±)-methylphenylsulfoxide with (*S*)-*O*-ethyl-ethanthiophosphoric acid (d) it is the reagent which is chiral and leaves behind an excess of the (*S*)-sulfoxide [22]. Upon oxidation by the Sharpless method, discrimination between enantiomers of (±)-1-phenyl-2-pyrrolidino-ethanol (e) was exceptionally high [23].

The product may also be chiral and in the simplest case it is the enantiomer of the substrate itself. This process is called *deracemization* and can be brought about by equilibration in chiral media or using chiral catalysts. Thus, after heating 1-naphthylmethylsulfoxide in cholesterol *p*-nitrobenzoate the equilibrium mixture contained a 9% excess of the *R* enantiomer [20].

Partial deracemization of chiral sulfoxides can also be realized photochemically in the presence of a chiral sensitizer such as *N*-acetyl-2-(1-naphthyl)-ethylamine [24] or by acid catalysis using *e. g.* (+)-camphorsulfonic acid [25].

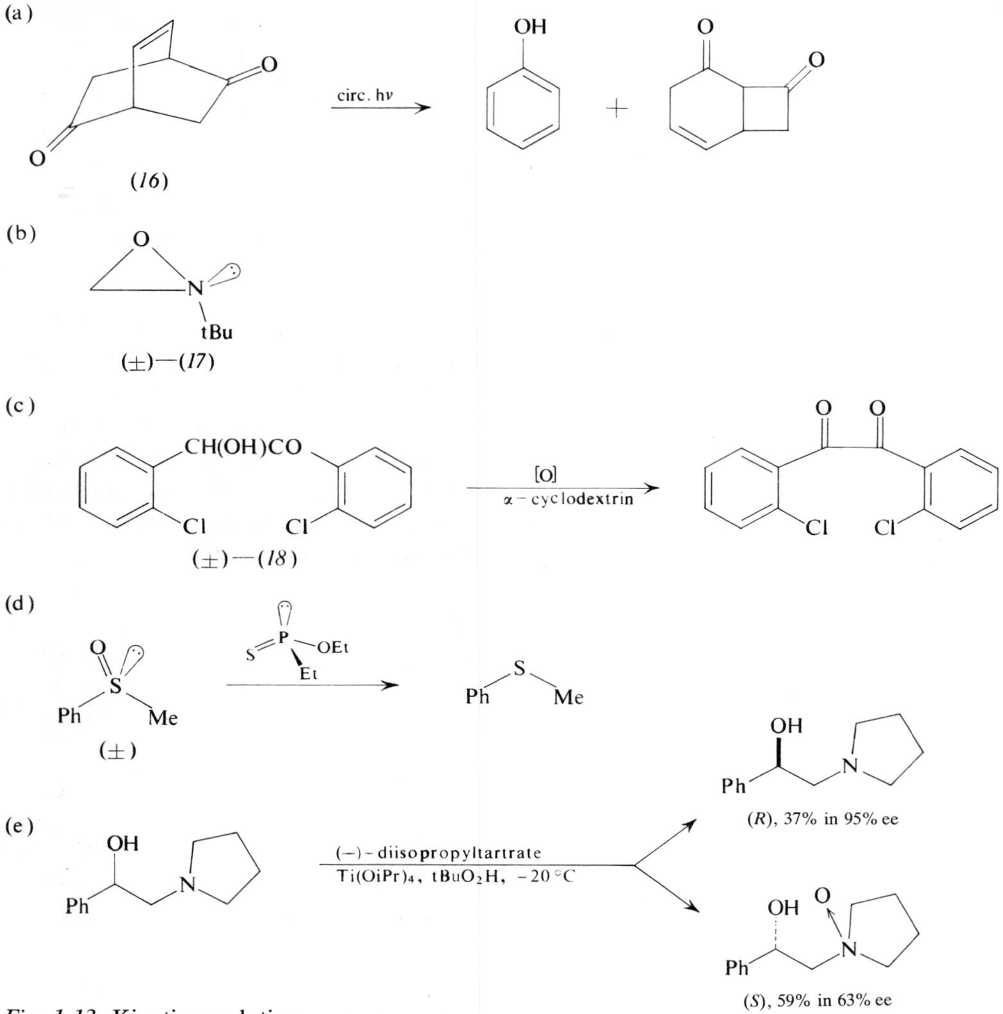

Fig. 1-13. Kinetic resolution.

When a transformation is enantioselective but does not affect the stereogenic element of the substrate, the configuration of the enantiomer in excess in the product is opposite to that in excess in the unchanged starting material. This is exemplified by the Meerwein-Ponndorf-Verley reduction of (±)-*19* with the chiral alcoholate (*S*)-*20*, in which the (*R*)-ketone was reduced 2.9 times faster than the (*S*)-ketone giving, when interrupted before completion, an excess of (*S*)-*19* and (*R*)-*21** [26].

* Owing to the C_2 symmetry of *21* CH(OH) is not a chiral center.

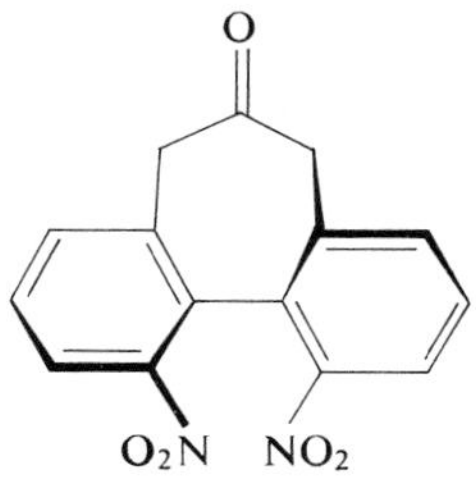

(S)-(19) 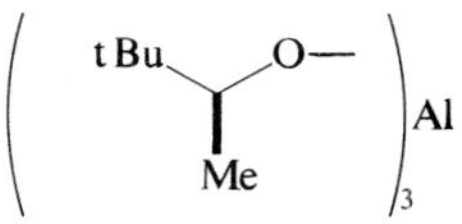 (S)-(20) 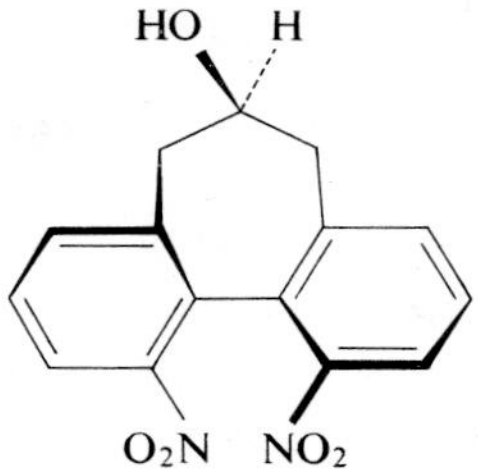 (R)-(21)

When the association of a racemic substrate with a chiral agent is not simply transient but leads to the formation of a compound in which the stereogenic elements of both are retained, the product is a mixture of two diastereomers. A classical example is the partial esterification of (±)-mandelic acid with (−)-menthol (Fig. 1-14). Actually this was the very reaction which led to the discovery of kinetic resolution by Marckwald and McKenzie in 1899 [27]. Incomplete reaction leaves an excess of (*S*)-mandelic acid, while total hydrolysis of the separated esters gives a mixture enriched in (*R*)-mandelic acid.

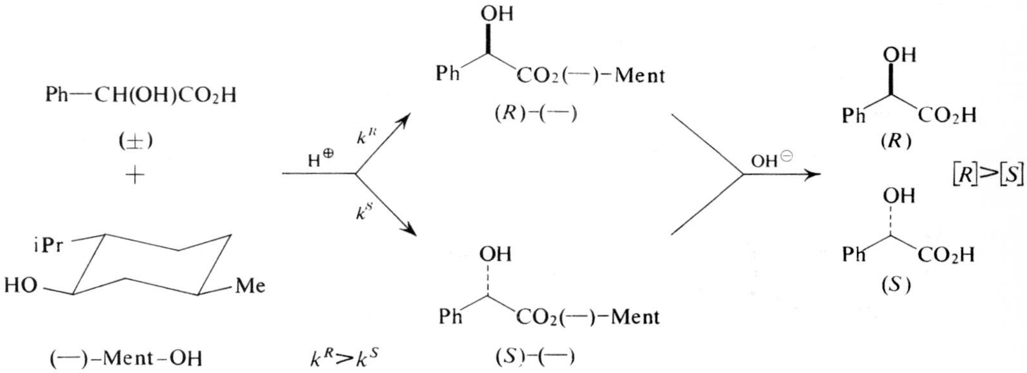

Ph—CH(OH)CO_2H (±) + (—)-Ment-OH; $H^{\oplus}$; k^R → (R)-(—) Ph–CH(OH)–CO_2(—)-Ment; k^S → (S)-(—) Ph–CH(OH)–CO_2(—)-Ment; $k^R > k^S$; $OH^{\ominus}$ → (R) Ph–CH(OH)–CO_2H, (S) Ph–CH(OH)–CO_2H; [R]>[S]

Fig. 1-14. The experiment of Marckwald and McKenzie.

Kinetic resolution is an inherently wasteful process for producing optically active compounds and can only compete with conventional resolution (of poor economy itself) when rate differences are extreme*. This has so far only been realized with enzymes, a field which lies outside the scope of this book.

Since different conformations of a compound can be regarded as unstable stereoisomers, their reactivity is governed by the same principles as those of stable stereoisomers. Differences in the reactivity of stereoisomeric conformers are, however, difficult to verify experimentally due to the fast equilibration of the substrates and, occasionally, also of the products.

* It can be calculated [28] that the enantiomeric purity of unreacted substrate at 50% conversion is 80% when the relative rate is 25, and about 93% when it is 100.

1.4.2 Product Selectivity

The main subject of our book is product stereoselectivity, *i. e.* the case when out of two or more possible stereoisomeric products, arising from a single substrate, one is formed preferentially.

When the number of products is more than two, selectivity can be best characterized by the percentage distribution of the products. When only two products (X and Y) are formed, selectivity may be characterized by giving the product ratio (r), or the percentage of the products or the excess (e) of the major product*. If $[X] > [Y]$, the following simple relationships hold:

$$r = \frac{[X]}{[Y]}, \quad e = \frac{[X]-[Y]}{[X]+[Y]} \cdot 100\% = \%\,X - \%\,Y = \frac{r-1}{r+1} \cdot 100\%$$

$$\text{and } X\,\% = \frac{100\,r}{r+1}$$

When the products are enantiomers or diastereomers the specific terms *enantiomeric excess* (ee) and *diastereomeric excess* (de) respectively may used.

When the final products of a diastereoselective reaction are the diastereomers formed themselves, selectivity will be characterized by their ratio or their percentage. When the diastereomeric products are later transformed by the removal of a chiral auxiliary group to a mixture of enantiomers, then diastereomeric excess values (de) will be given. The percentage of the major stereoisomer is often called the stereoselectivity of the reaction. In this case, however, the values should be standardized to

$$X\% + Y\% = 100\%$$

In the absence of intermolecular interactions between solute molecules the numerical value of enantiomeric excess is equal to optical purity (op) *i. e.* the ratio of observed specific rotation and the specific rotation of the pure enantiomer ($[\alpha]_{max}$):

$$\text{op} = \frac{[\alpha]}{[\alpha]_{max}} \cdot 100\%$$

For all practical purposes the two terms are equivalent. Nowadays, when high resolution NMR techniques and various chromatographic techniques find ever

---

* Reactions in which only a single stereoisomer can be detected are generally called stereospecific, but because of some controversy around this term we are not using it in this book.

increasing application to the determination of enantiomeric purity, it is more appropriate to use the term enantiomeric excess. It is not only for its simple relationship to optical rotation that the ee value rather than the percentage distribution should be given. Ee is in fact the upper limit of the amount of the major enantiomer which can be obtained by crystallization from a mixture of enantiomers forming a separate racemic solid phase (racemic compound), because for each mol of the minor component one mol of the major one is included in the crystals of the racemic phase. According to Jacques, unfortunately about 90% of the chiral compounds examined in this respect form such solid phases, while the rest form a conglomerate or a solid solution [29]. On crystallization conglomerate forming-enantiomers behave as different compounds, and therefore the upper limit for the isolable major enantiomer is its percentage and not its excess.

Product selectivity also comprises those reactions in which products of different qualitative or quantitative composition or of different molecular weight are formed, but these aspects are outside our interest.

Preferential formation of one constitutional isomer over other possible isomers is called *regioselectivity*. Regioselectivity arises from competition between groups (molecular sites) of different connectivity. An example is the chlorination of toluene which could lead, in principle, to four isomeric monochloro derivatives, of which practically only two are formed. Preference for *p*-chlorotoluene over the other isomers reflects genuine selectivity, since considering merely the number of competing sites a distribution of 1:2:2:3 would be expected for the *p, o, m,* and benzylic monochloro products.

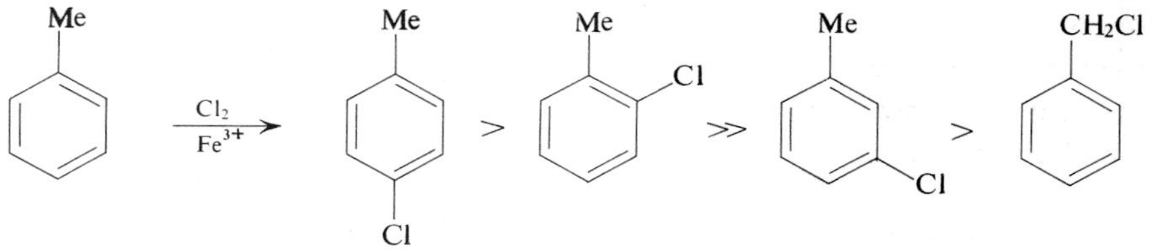

Me
Cl_2 / Fe^{3+} →
Me, Cl > Me, Cl ≫ Me, Cl > CH_2Cl

1.4.2.1 Formation of stereoisomers

What we are really interested in is product stereoselectivity, a prerequisite for which is the fact that the formation of at least two stereoisomers from the same substrate should be possible. Therefore we should first examine the symmetry requirements for the formation of stereoisomers.

1. Transformations of homotopic groups or addition to homotopic faces, even those in chiral molecules, gives rise to the same product. Thus bromination involving any of the homotopic methyl groups marked with ○ and ⊗, or hydride addition to any of the homotopic faces in acetone gives a single product (Fig. 1-15).

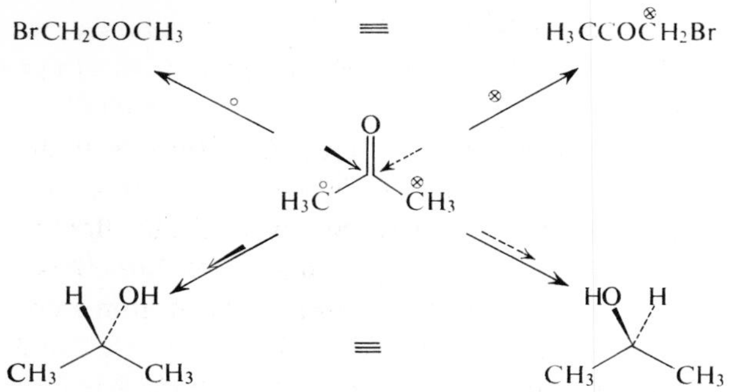

$BrCH_2COCH_3$ ≡ $H_3CCO\overset{\otimes}{C}H_2Br$

$H_3\overset{\circ}{C}$ $\overset{\otimes}{C}H_3$

H OH ≡ HO H

CH_3 CH_3 CH_3 CH_3

Fig. 1-15. Transformation of a compound with homotopic groups and faces.

An example for the transformation of homotopic faces in a chiral molecule is quoted on p. 20.

2. Transformation of enantiotopic groups or addition to enantiotopic faces usually produces stereoisomers*. If the reaction does not involve the permanent attachment of a chiral element to the substrate, the products are enantiomers, otherwise they are diastereomers.

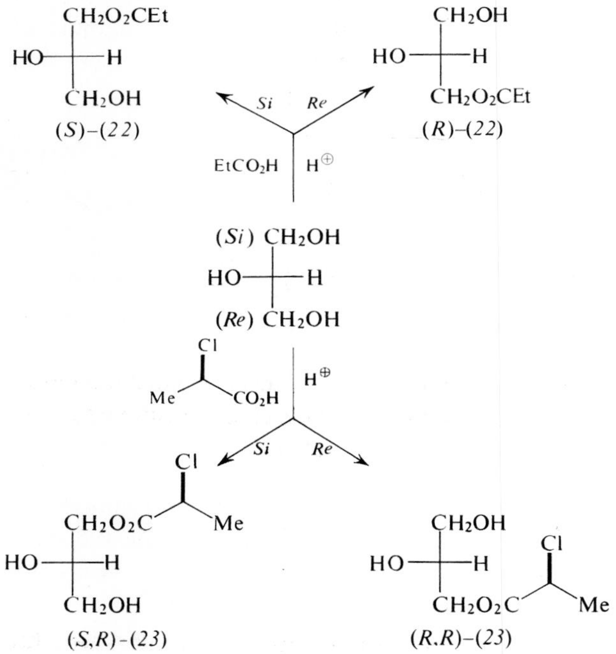

CH_2O_2CEt / HO—H / CH_2OH (*S*)-(*22*)

Si *Re*

$EtCO_2H$ $H^{\oplus}$

CH_2OH / HO—H / CH_2O_2CEt (*R*)-(*22*)

(*Si*) CH_2OH / HO—H / (*Re*) CH_2OH

Cl / Me CO_2H $H^{\oplus}$

Si *Re*

CH_2O_2C(Cl)Me / HO—H / CH_2OH (*S*,*R*)-(*23*)

CH_2OH / HO—H / CH_2O_2C(Cl)Me (*R*,*R*)-(*23*)

Fig. 1-16. Transformation of a compound with enantiotopic groups.

---

* For an exception *cf*. p. 25.

For example, mono-esterification of glycerol with propionic acid produces the enantiomeric esters (*R*)-*22* and (*S*)-*22*, whereas with (*R*)-2-chloropropionic acid the diastereomers (*R*,*R*)-*23* and (*R*,*S*)-*23* result (Fig. 1-16).

Addition to enantiotopic faces is more complex than transformation of enantiotopic groups and need not involve the formation of stereoisomers. Stereoisomers are only formed when the entering ligand is different from those already attached to the prochiral center. Also, changes at both ends of a double bond have to be considered and situations in which only one or both bridgehead atoms are prochiral should be distinguished. For example, in acetaldehyde and styrene only one of the bridgehead atoms is prochiral and addition of a new achiral ligand produces a pair of enantiomers.

H Re Ph Si $\xrightarrow{PhCO_3H}$ Re → Ph (*S*) H O; Si → Ph (*R*) H O

O Re Me H Si $\xrightarrow{EtMgBr}$ Re → HO Et Me H (*S*); Si → Et OH Me H (*R*)

If the incoming ligand contains a chiral element the products are diastereomeric.

Addition of a new ligand to a prochiral face fails to produce stereoisomers when, as a result of the reaction, two ligands become identical. *E. g.:*

H Me $\xrightarrow[H^{\oplus}]{H_2O}$ HO H Me Me

In non-terminal olefins usually both sp^2 carbons are prochiral centers and the stereochemical outcome of additions to such double bonds depends on the symmetry of both substrate and reagent. The complexity of this situation can be illustrated by the addition of oxygen, chlorocarbene and water to (*Z*)- and (*E*)-2-butene (Fig. 1-17).

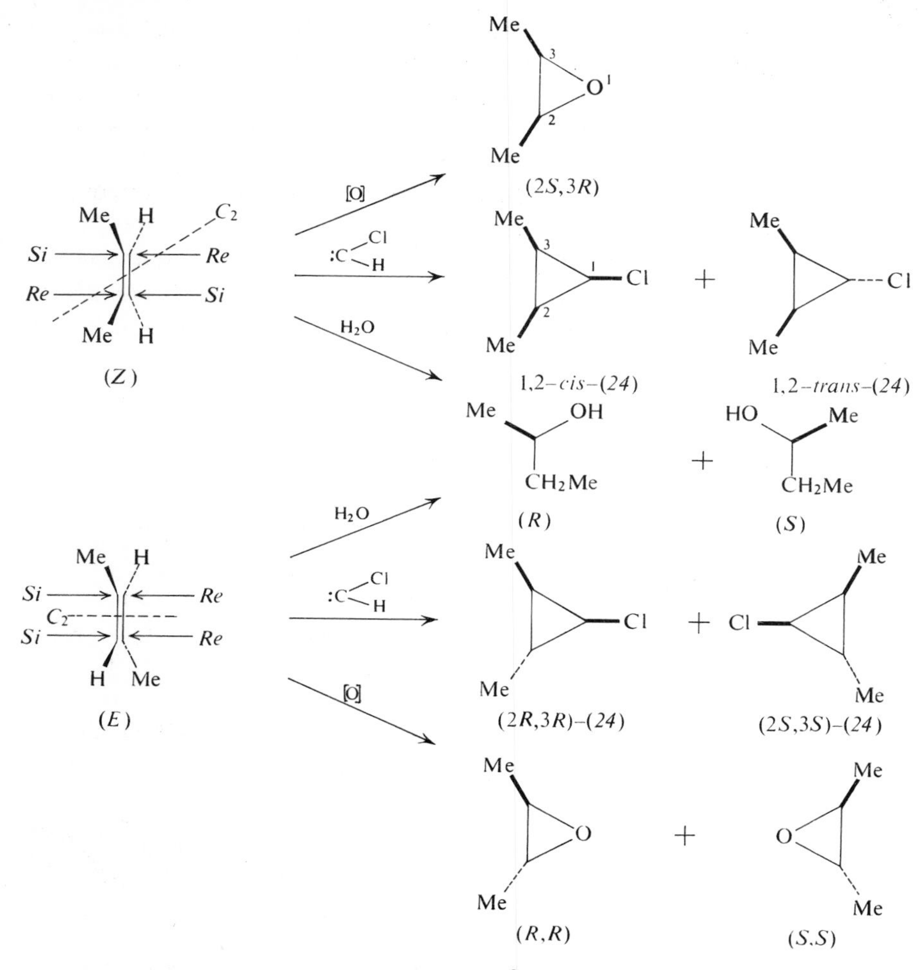

Fig. 1-17. Addition to olefins in which both sp^2 carbons are enantiotopic.

(*Z*)-2-Butene has a C_2 axis coplanar with the double bond and a mirror plane bisecting it. Therefore attachment of a point-like ligand, such as oxygen, from either side gives the same stereoisomer, namely, the achiral (2*R*,3*S*)-2,3-epoxybutane. Addition of chlorocarbene, which is of lower symmetry (C_s), gives rise to a pair of diastereomeric but still achiral chlorocyclopropanes (*24*). Finally, water is a reagent which forms different bonds (C–H and C–O) at the two prochiral centers and therefore these have to be considered separately. C–H bond formation does not create a chiral center, while attachment of OH from opposite sides yields (*R*)- and (*S*)-2-butanols.

Hydration of (*E*)-2-butene proceeds similarly but now there is a C_2 axis perpendicular to the plane of the double bond which renders the two centers, but not the two faces of the molecular plane, equivalent. Therefore both epoxidation and chlorocarbene addition lead to enantiomers. Note that C-1 in the chiral cyclopropanes (2*R*,3*R*)- and (2*S*,3*S*)-*24* is not a chiral center.

An important case is when in a reaction two planar trigonal atoms having enantiotopic faces are joined to form a pair of adjacent asymmetric centers. If the partners are different two pairs of diastereomers may be formed. In Fig. 1-18 (a) all four possible products are shown [30], while in (b) only two diastereomeric products are given. Heavy lined arrows indicate the preferred pathways. In example (a) one, in example (b) two pairs of atoms (one of them with enantiotopic faces) become linked, but in the present case this does not lead to additional stereoisomers because of the physical impossibility of *trans*-anellation.

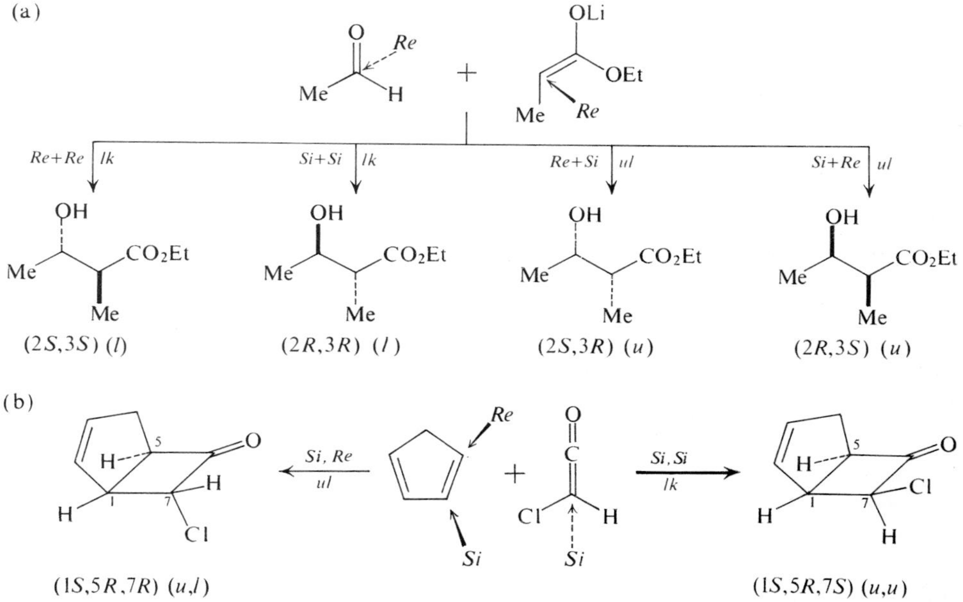

Fig. 1-18. Bond formation between two centers, both having enantiotopic faces. The Prelog-Seebach notation.

In order to characterize unambiguously both the mode of approach of the partners and the relative configuration of the products, Prelog and Seebach suggested a notation system based on the sequence rules [7].

If the approaching faces of the atoms to be linked carry the same descriptors, *i. e.* both are *Re* or both are *Si*, bond formation is labelled as "like" (*lk*), conversely if the corresponding descriptors are different (*Re* and *Si*) the label is "unlike" (*ul*). The relative configuration of the centers in the product, following each other in the order of their IUPAC approved numbering is labelled as like (*l*) for *R,R* and *S,S* pairs, and unlike (*u*) for *R,S* and *S,R* pairs. The system has been devised as to be applicable to reactions and products involving more than centers as well.

Advantages and disadvantages of the new notation are similar to those of the CIP system and have a common source: both are specific for individual compounds, but not for classes of compounds or types of reactions. Thus the generation of descriptors is unambiguous even in complicated situations, but may be reversed by changing the nature of substituents. Since in this book we intend to concentrate on the general aspects of stereoselective synthesis rather than on its specific applications, the Prelog-Seebach notation will only be used when a situation cannot be clearly characterized by simpler means.

More complex cases, such as addition to unsymmetrical, non-terminal olefins, attack by reagents which provide two new and different ligands (*e. g.* HOHal) or by reagents containing a chiral element can be, with some patience, analyzed along these lines.

3. Transformation of diastereotopic groups or the addition of a new ligand to a center with diastereotopic faces always gives rise to diastereomers. Depending on the symmetry of both substrate and reagent the products may be chiral or achiral. Examples for both situations are given in Fig. 1-19. The first two illustrate the transformation of diastereotopic groups, the second two those of diastereotopic faces.

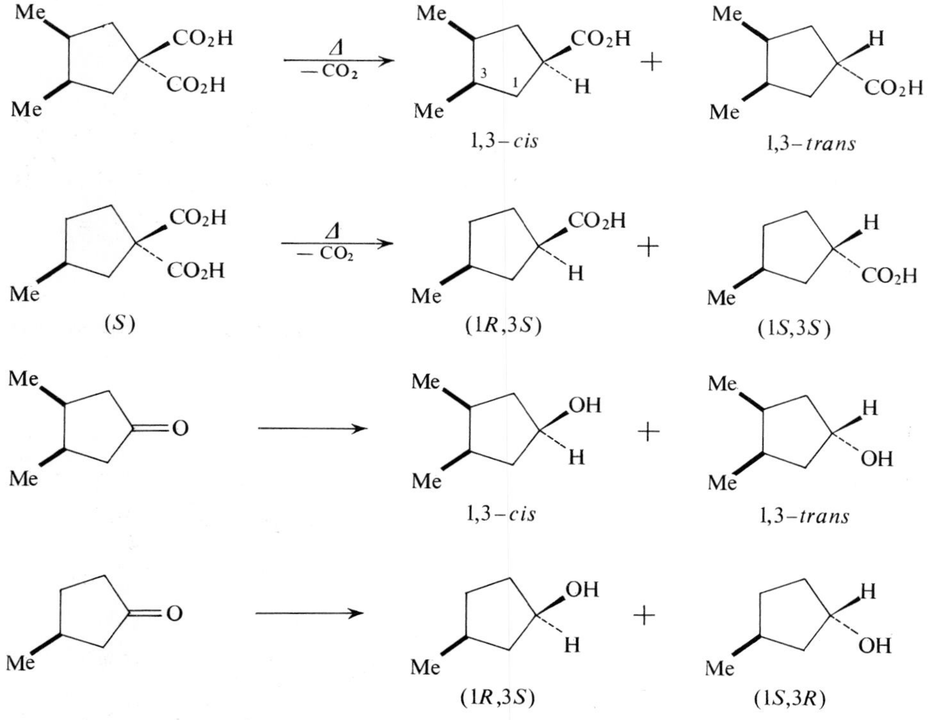

Fig. 1-19. Transformations of molecules with diastereotopic groups or faces.

The situation just described is not more complicated when the reagent is chiral, provided that it is a single pure enantiomer: but two stereoisomers are formed which are (except for very special situations)* both chiral.

The Prelog-Seebach notation can be conveniently applied to reactions in which a trigonal center with diastereotopic faces in a molecule containing a chiral center is transformed to a chiral tetrahedral center. *Re* + *R*, and *Si* + *S* combinations are labelled "like" (*lk*), while *Si* + *R* and *Re* + *S* combinations "unlike" (*ul*). Formation of (1*R*,3*S*)- and (1*S*,3*R*)-3-methylcyclopentanols is a *ul* (more precisely a 1,3-*ul*) addition of hydride anion giving rise to a *u* product. Note that *ul* processes need not yield *u* products.

Since the relationship of a given group or molecular face can be enantiotopic to one partner and diastereotopic to another, transformation of a single kind of group or face may lead to as many as four stereoisomers. Thus geminal hydrogens in *25* (Fig. 1-20) are diasteretopic, whereas those associated with different carbons form two enantiotopic pairs. Monobromination should give four stereoisomers, namely two racemic pairs of diastereomers [(2*R*,3*R*)- and (2*S*,3*S*)- further (2*S*,3*R*)- and (2*R*,3*S*)-2-bromo-3-methylglutaric acid]. In the reaction of norbornene (*26*) with ethene four stereoisomers could be expected, but in fact only approach from the less hindered *exo* face *i. e.* the formation of (1*R*,2*S*)- and (1*S*,2*R*)-*27* was observed (*cf.* Section 6.4.2).

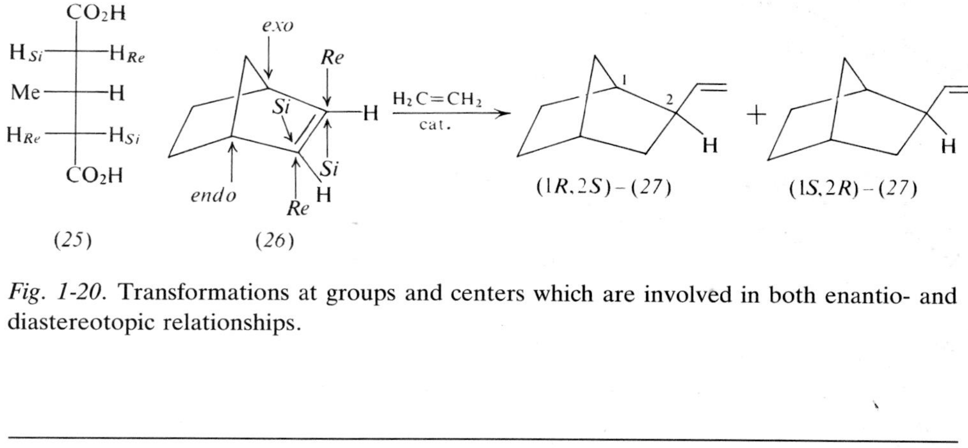

Fig. 1-20. Transformations at groups and centers which are involved in both enantio- and diastereotopic relationships.

---

* *E. g.*

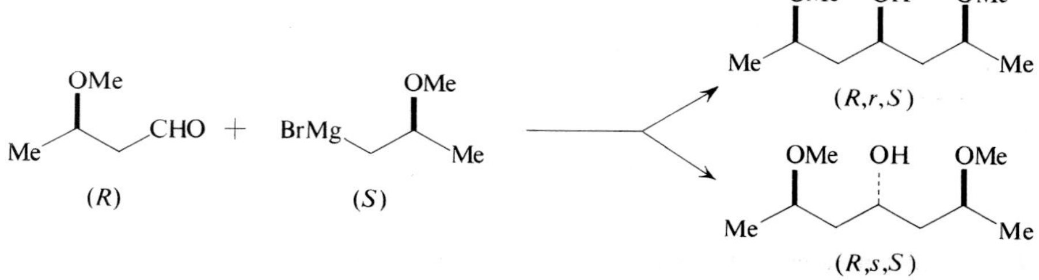

1.4.2.2 Conditions necessary for stereoselectivity

While it is rather difficult to predict the degree of stereoselectivity of a given transformation, it is very simple to define conditions which must be fulfilled that stereoselectivity should be possible at all.

The formation of diastereomers involves diastereomeric transition states, which, being geometrically different, are of different free energy. Therefore, whenever the formation of diastereomers is expected, it can also be anticipated that they are formed in unequal amounts. This applies equally to reactions of diastereotopic groups or faces with any reagent and to such reactions of enantiotopic groups or faces in which a reagent containing a chiral element forms a permanent bond with the substrate (*e. g.* the formation of *23* in Fig. 1-16).

Since the ground state free energy of diastereomers is different, the observed product ratio is either the result of unequal rates of formation (kinetically controlled selectivity) or subsequent equilibration (thermodynamically controlled selectivity) or of both.

Formation of a pair of enantiomers from a single substrate always involves the transformation of enantiotopic groups or faces. When such groups or faces interact with an achiral agent two different transition states arise which are related as mirror images. Consequently the free energies of activation associated with the two transition states and thus the rates of formation of the two products are equal. In other words, no selectivity can be expected in such reactions.

Interaction of enantiotopic groups of faces with a chiral agent (reagent, catalyst, solvent *etc.*) can be envisaged as the formation of a pair of transient compounds,

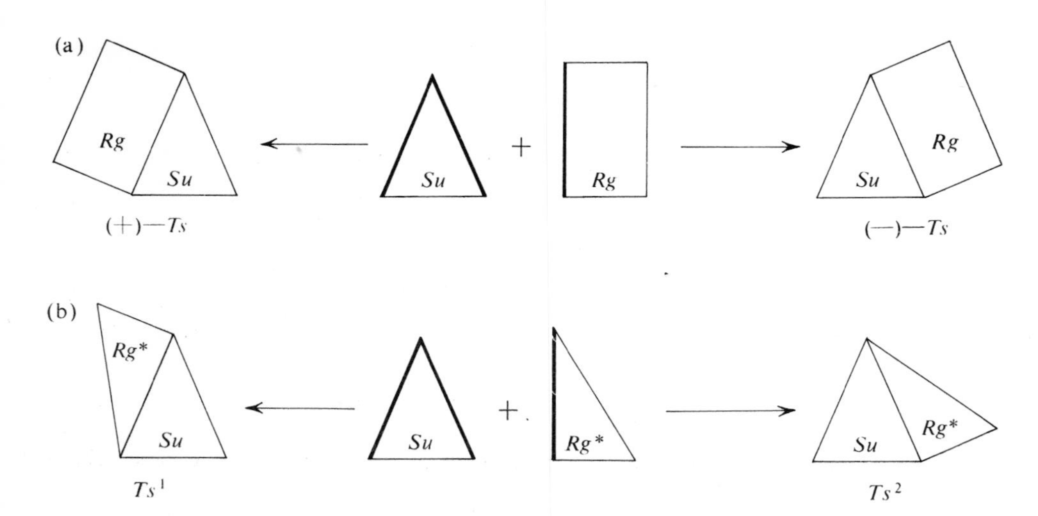

Fig. 1-21. Two-dimensional representation of the reaction of molecules with enantiotopic groups or faces with (a) achiral (*Rg*) and (b) chiral (*Rg**) reagents.

and we have just shown that such pairs are diastereomeric. Thus the same applies as to the formation of diastereomers, *i. e.* they are formed at unequal rates and their decomposition gives enantiomers in unequal amounts.

All this can be nicely illustrated by two-dimensional models. In Fig. 1-21 the substrate with enantiotopic groups is represented by a triangle having two equal sides (*Su*), the achiral reagent (*Rg*) by a rectangle, and the chiral reagent by a scalene triangle (*Rg**). It can be seen that any analogous association of *Rg* with the "enantiotopic" sides of *Su* leads to a pair of mirror image ("enantiomeric") figures (+)-*TS* and (−)-*TS* ("transition states"), while a similar operation with *Rg** gives TS^1 and TS^2, *i. e.* figures of different shapes (diastereomeric).

It follows from the above discussion that no reaction is genuinely enantioselective; there are only diastereoselective reactions which ultimately lead to a non-racemic mixture of enantiomers. The diastereomeric nature of the selectivity-determining interaction becomes increasingly apparent in the series of examples shown in Figs. 1-22 and 1-23.* In example (a) the diastereomeric interaction involves a chiral physical phenomenon, namely circularly polarized light [31]. In example (b) photoenergy transfer from a chiral sensitizer is the selectivity-determining step [32], while in example (c) it is the preferred abstraction of a proton by a chiral base from one of the enantiotopic methylene groups [33]. In example (a) (Fig. 1-23) acetophenone is reduced with the bulky complex hydride (*R*)-*28*

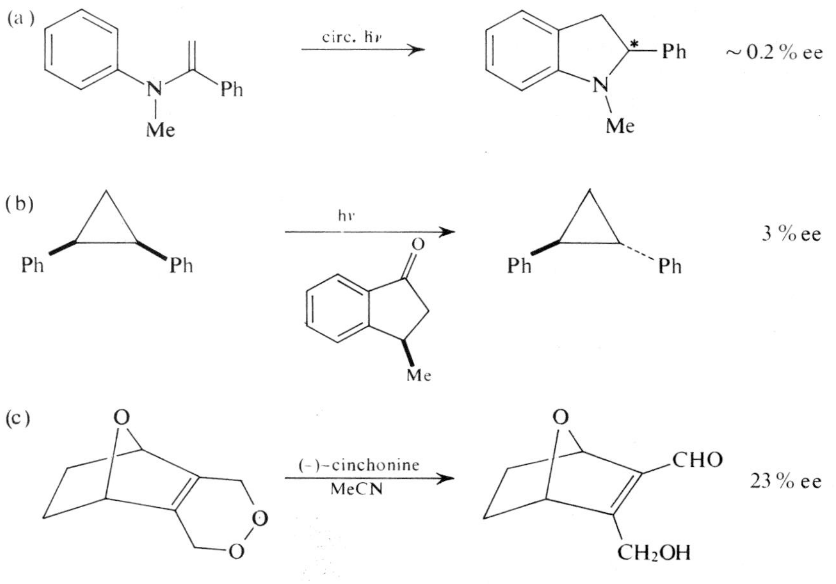

(a) circ. hν ~0.2 % ee

(b) hν 3 % ee

(c) (−)-cinchonine, MeCN 23 % ee

Fig. 1-22. Reactions involving enantiotopic groups or faces in the presence of chiral catalysts.

* An asterisk marks a chiral center. For mixtures of stereoisomers the major one is depicted.

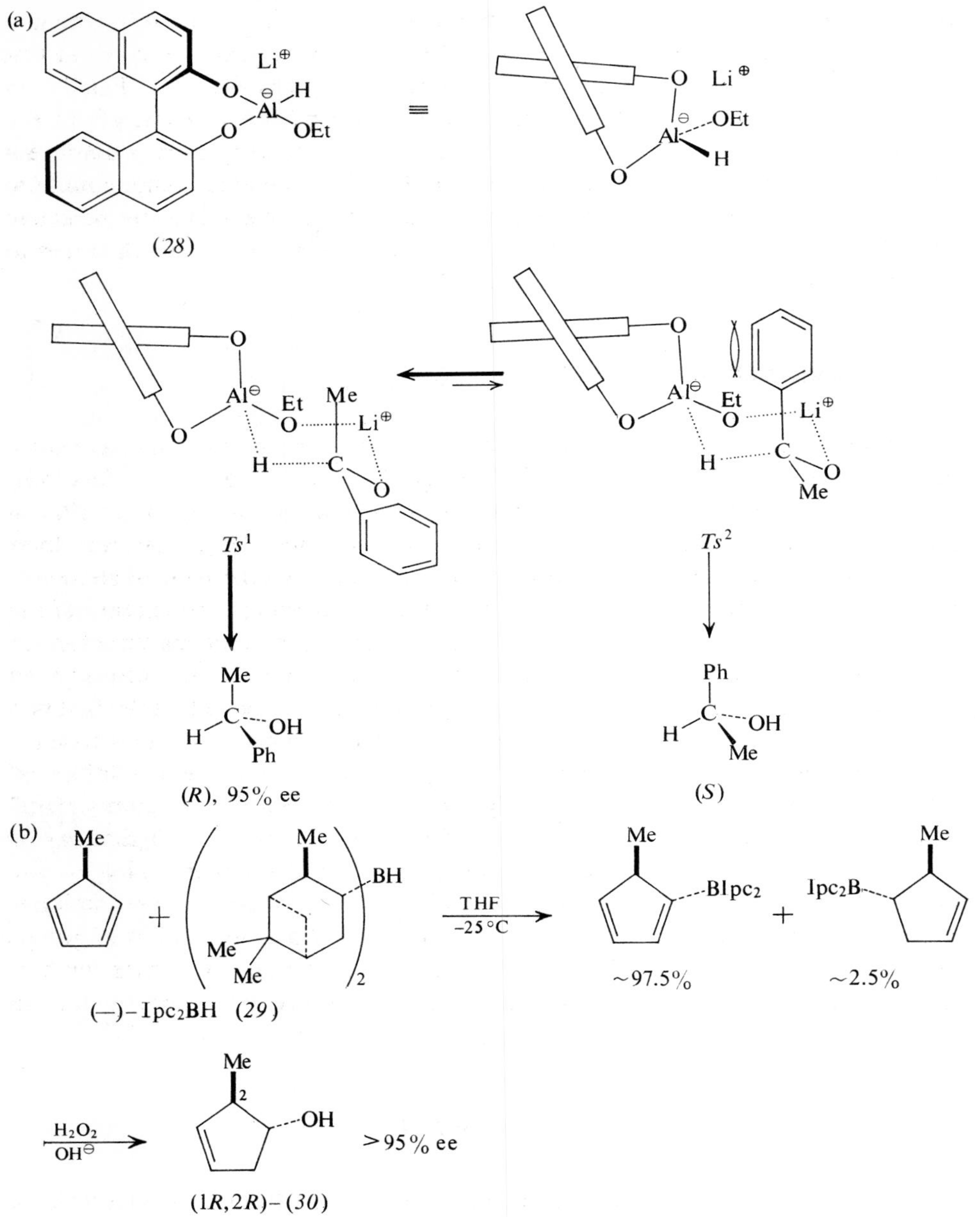

(a)

$Li^{\oplus}$ O Al H OEt O

(28)

O $Li^{\oplus}$ Al OEt H O

O Al Et Me O $Li^{\oplus}$ H C O

O Al Et O $Li^{\oplus}$ H C O Me

Ts^1 Ts^2

Me C OH H Ph

Ph C OH H Me

(*R*), 95% ee (*S*)

(b)

Me + Me BH Me Me 2 $\xrightarrow[-25\,°C]{THF}$ Me BIpc₂ + Ipc₂B Me

(−)-Ipc_2BH (29) ~97.5% ~2.5%

$\xrightarrow[OH^{\ominus}]{H_2O_2}$ Me 2 OH >95% ee

(1*R*,2*R*)-(30)

Fig. 1-23. Reactions of compounds with enantiotopic faces with chiral reagents.

prepared from (*R*)-2,2′-dihydroxy-1,1′-biphenyl. Here it is possible to formulate two diastereomeric transition states, one of which (TS^2) is unfavored. In fact at −100°C the (*R*)-1-phenylethanol was obtained in high purity [34]. Finally, in example (b) addition of (−)-diisopinocampheylborane [(−)-Ipc_2-BH, *29*] to the enantiotopic ethylene groups of the substrate gives a mixture of two diastereomers* which were not separated but oxidized directly to an enantiomeric mixture of alcohols (*30*). Here the selectivity-determining step is clearly diastereoselective hydroboration, since the product ratio is not altered by oxidation which serves to remove the remnants of the chiral carrier molecule.

1.4.2.3 Concept of stereodifferentiation

The above classification of stereoselective reactions according to the symmetry of products is useful inasmuch as it gives insight into the process but suffers from being not structured enough. Probably this prompted Izumi to suggest, in 1971, a new classification based on the symmetry of substrates [35], that was later explained in detail in a book [36]. Reactions giving unequal amounts of stereoisomers were called stereodifferentiating and prefixed according to the nature of the substrate as enantiomer- and diastereomer-, enantiotopos- and diastereotopos-, further enantioface- and diastereoface-differentiating reactions, according to whether stereoisomers, groups, or faces were differentiated. Note that the first two types cover substrate selective transformations, while the last four product selective ones. Izumi's classification is rather appealing because the conditions of selectivity can be defined very simply: enantio-differentiation requires chiral means, whereas diastereo-differentiation does not. Also very enlightening is Izumi's observation that the key for enantio-differentiation is provided by the environment (reagent, solvent, catalyst), while that for diastereo-differentiation is within the molecule (steric hindrance by, or electronic and other effects of groups around the reaction center). The Izumi notation is, unfortunately, rather cumbersome. This may be the reason why it did not gain wide acceptance and, yielding to convenience, we are not going to use it in this book either.

1.4.2.4 Methods for inducing stereoselectivity

In the foregoing sections the fundamentals of stereoselective reactions have been discussed, but little information has benn provided about the methodology of such reactions. For our topic, however, the various ways and means by which stereoselectivity can be generated is of primary importance.

---

* Note (i) that the process is totally diastereoselective giving only *trans* addition products and (ii) that oxidation proceeds with retention of configuration.

Enantioselectivity

Selectivity in the transformation of enantiotopic groups and faces can only be achieved, as explained before, by chiral means. These may be characterized as follows:

(i) *Photochemical transformations induced by "chiral", i. e. circularly polarized, light.* This is an impractical method giving only a few percent ee or less (see Fig. 1-13 (a)).

(ii) *Reactions conducted in a chiral solvent* (Fig. 1-24 (a)) [37]. Chiral solvents are, compared with their generally low efficiency, very expensive and thus do not qualify for practical stereoselective synthesis.

(iii) *Reactions carried out in the presence of a chiral additive.* Additives in this context are substances which emerge unchanged from a reaction and range from cosolvents (like DBB, *cf.* Fig. 1-22 (a)) through photosensitizers to catalysts. Enantioselective reactions carried out in the presence of chiral additives are illustrated by the photoisomerization of (*Z*)- to (*E*)-cyclooctene in the presence of

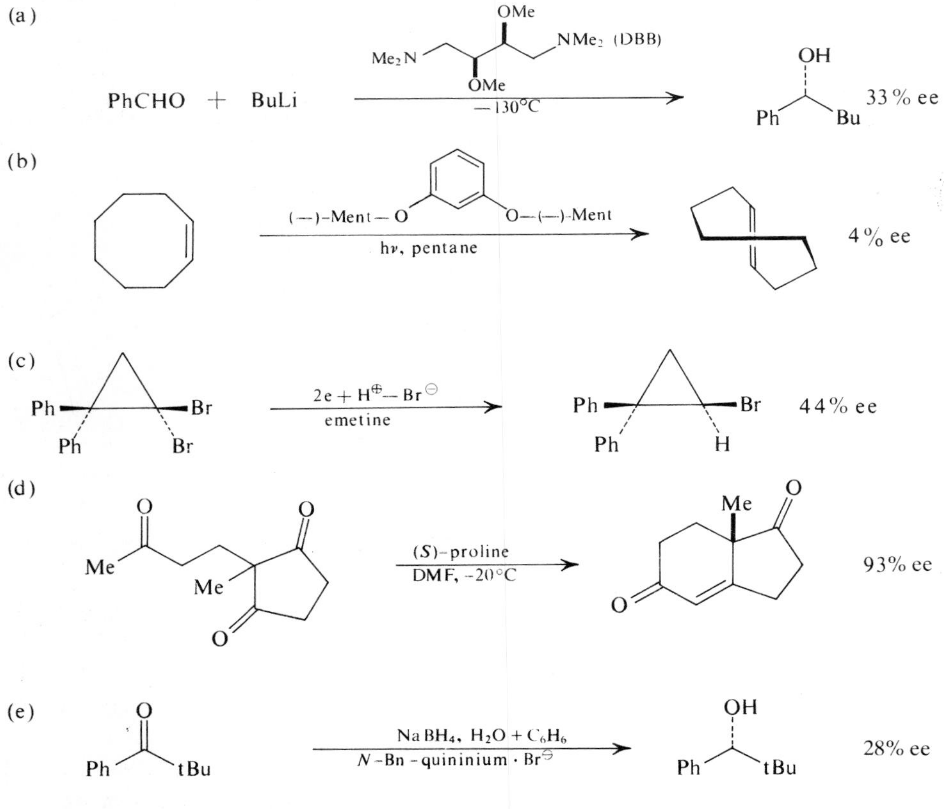

Fig. 1-24. Induction of enantioselectivity with the aid of chiral solvents or additives.

a chiral senzitizer (Fig. 1-24 (b)) [38], the electroreduction of a dibromocyclopropane in the presence of a chiral base (c) [39], base-catalyzed intramolecular aldol condensation (d) [40], two-phase reduction of an aromatic ketone aided by a chiral phase-transfer catalyst (e) [41], heterogeneous catalytic hydrogenation over Raney-nickel modified with a chiral acid Fig. 1-25 (a) [42] and, finally, homogeneous hydrogenation with rhodium (b) [43] and ruthenium catalysts (c) [44] prepared from chiral phosphines. The latter is one of the very few examples of a non-enzymatic enantioselective transformation affecting enantiotopic groups.

Methods (b) and (c) in Fig. 1-24 do not hold much promise for being developed into efficient procedures. The record of chiral phase transfer catalysis has also been poor so far, ee's were usually only a few percent [reaction (e) is rather an exception]. Heterogeneous catalysis by chirally modified catalysts still has a very limited scope, while homogeneous catalytic procedures have reached maturity, especially for the preparation of amino acids. Note that example (b) in Fig. 1-25 shows one of the rare instances when the selectivity of an enzymic process has been surpassed. Reduction with baker's yeast gives only 72% ee.

(iv) *Reaction of enantiotopic groups or faces with chiral reagents involving the transfer of an achiral species.* This approach has been mainly exploited for redox

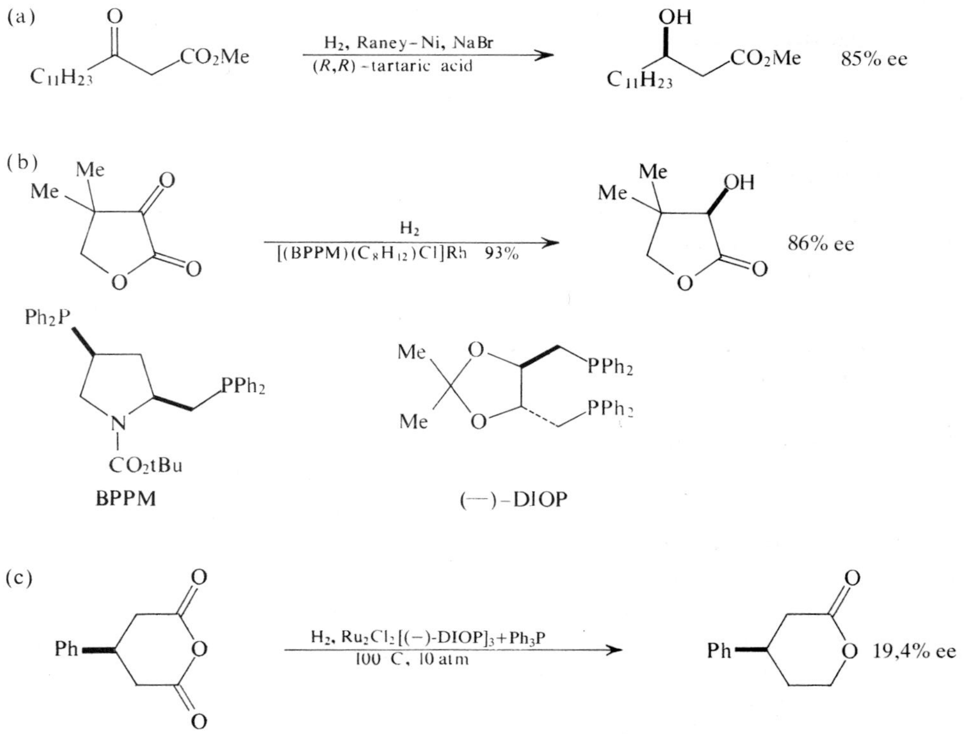

(a) $C_{11}H_{23}COCH_2CO_2Me$ —[H_2, Raney-Ni, NaBr; (*R*,*R*)-tartaric acid]→ $C_{11}H_{23}CH(OH)CH_2CO_2Me$ 85% ee

(b) —[H_2; $[(BPPM)(C_8H_{12})Cl]Rh$ 93%]→ 86% ee

BPPM

(−)-DIOP

(c) —[H_2, $Ru_2Cl_2[(-)\text{-DIOP}]_3 + Ph_3P$; 100 C, 10 atm]→ 19,4% ee

Fig. 1-25. Induction of enantioselectivity by chiral catalysts.

processes. The chiral complex hydride *28* (Fig. 1-23) is just one of the many chiral hydride ion donating reagents developed in the past two decades. Among them, those imitating the main reducing agent of nature, NADPH, such as *31* in Fig. 1-26 [45] are of high theoretical interest.

Rather often the hydride ion is transferred to the substrate from the chiral center itself, whereby the latter is destroyed. Such processes are called "self-immolative" [46], and one of them is shown in example (b). Oxygen transfer from chiral peracids is generally less efficient; however, selectivity is high with the azomethine in Fig. 1-26 (c)* [47].

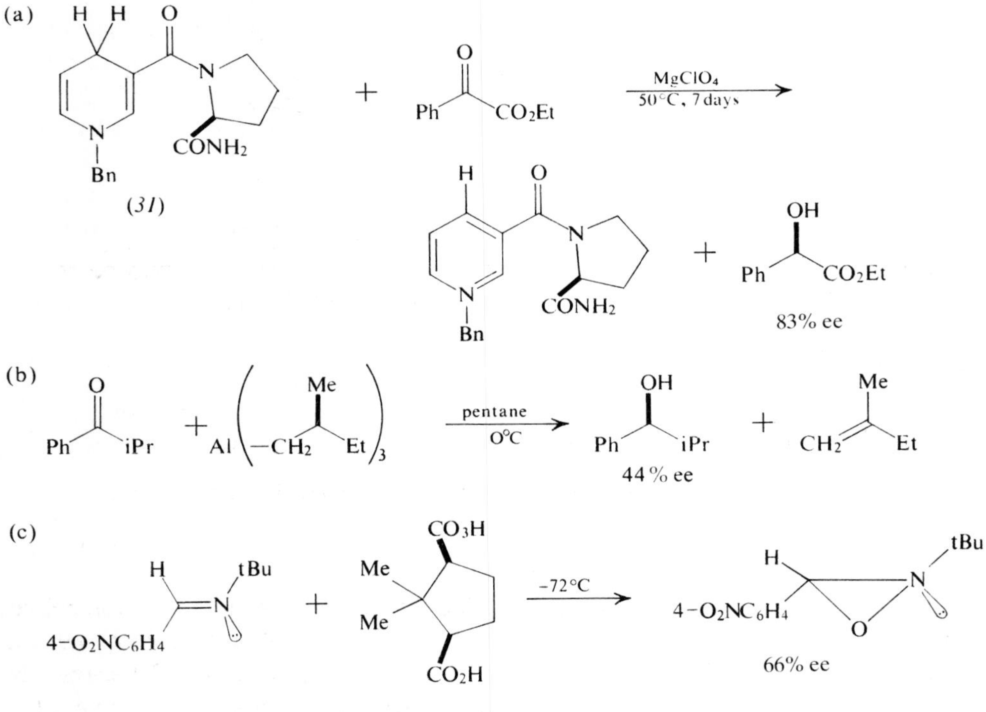

Fig. 1-26. Transfer of hydrogen or oxygen to substrates with enantiotopic faces.

(v) *The substrate with enantiotopic groups or faces is linked with a chiral auxiliary compound to form a derivative in which the former become diastereotopic.* Thus the necessary condition for stereoselectivity with any reagent (chiral or achiral) is met. After the required transformations have been carried out, the chiral auxiliary is removed to give rise to a non-racemic mixture of enantiomers. An ideal chiral

* Since the nitrogen atom in oxaziridines is configurationally stable the formation of *cis* and *trans* diastereomers is to be expected, but only one was formed, *i. e.* the reaction was totally diastereoselective.

auxiliary substance is one which (i) provides high asymmetry in the selectivity determining transition state, (ii) is recoverable, and (iii) is readily available in both enantiomeric forms in high optical purity and at low cost. Examples for recoverable (a) [48] and non-recoverable chiral auxiliaries (b) [49] are shown in Fig. 1-27.

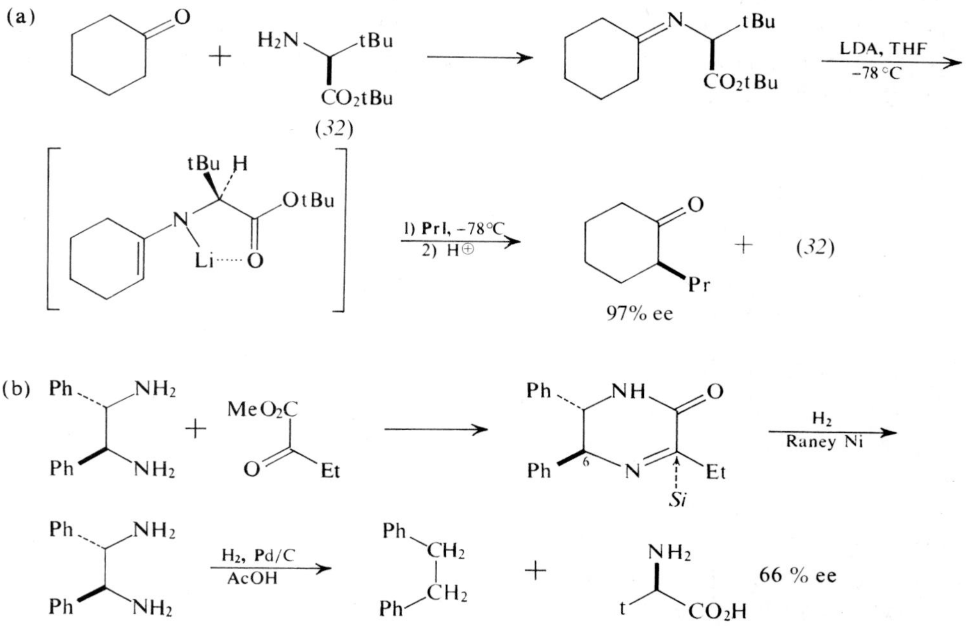

Fig. 1-27. Enantioselective synthesis with the aid of chiral auxiliary compounds.

The above examples require the following comments:

(A) It has to be noted that condensation of substrates with enantiotopic faces with a chiral auxilliary usually leads to the formation of *E-Z* isomers, although one on them may be highly preferred. This was avoided in example (a) because the original substrate had homotopic faces and in (b) because the chiral auxiliary had enantiotopic groups. It has to be borne in mind that no serious rationalization of stereoselectivity observed in a certain reaction is possible without knowing the configuration (*E* or Z) of the intermediate.

(B) In the stereoselective step the primary product is formed as a mixture of diastereomers which is then converted to a mixture of enantiomers by removing the chiral auxiliary.

(C) The enantiomeric excess in the end product does not necessarily reflect the ratio at which the diastereomers were formed for the following reasons: (a) equilibration of the diastereomers, (b) enrichment of one of the diastereomers during work up, (crystallization, chromatography *etc.*) and (c) racemization during the removal of the chiral auxiliary. An example for situation (a) is the addition of cyanide ion onto the Schiff-base *33*. The product (*34*) crystallizes out in 100% diastereomeric purity and can be converted to enantiomerically pure *35*.* This result, however, neither reflects the ratio of formation rates nor implies a much higher thermodynamic stability of *34* in solution. In fact, in chloroform a 2:1 equilibrium ratio of epimers was established [50].

Ph N=C Me Me O O Ph Me

(*33*)

NaCN
MeOH+AcOH

NH Ph Me CN Me O O Ph Me

(*34*)

Me NH_2 $PhCH_2$ CO_2H

(*35*)

100 % ee

(D) It is assumed that as chiral auxiliaries enantiomerically pure substances are used. In practice this is not always very easy to realize and therefore optically impure reagents are often employed, resulting in correspondingly lower ee values.** In this case namely, for reasons of symmetry, each of the diastereomeric intermediates is contaminated by exactly as much of its enantiomer, as was the chiral auxiliary substance used.

In the extreme case, with a racemic chiral auxiliary, the end product will also be racemic, regardless of whether diastereomers are separated or not. On the other hand, the ratio of diastereomers obtained before the removal of the chiral auxiliary is equal to the enantiomeric purity of the end product obtained with an enantiomerically pure auxiliary. Thus, if one wishes to examine the performance of a chiral auxiliary, it is sufficient to employ it in the racemic form and determine in some way the ratio of the diastereomeric intermediates [51]. This also has the advantage that results are not influenced by possible racemization in the decomposition step.

Diastereoselectivity

It has been demonstrated in the preceding section that enantioselectivity can always be traced back to diastereoselectivity. The selective preparation of diastereomers for their own sake is an even more important task and has been the

---

* Such epimerizations associated with the crystallization of one of the epimers are called *second order asymmetric transformations* in the older literature.

** In this book, in order that data should be comparable, whenever possible ee values were corrected for enantiomerically pure reagents.

$$ee_{corr} = \frac{ee_{obs}}{ee_{reagent}}$$

subject of innumerable studies, mainly in the natural product field. Therefore, it would be very useful if it were possible to formulate some general rules controlling diastereoselectivity. Owing to the diversity of organic reactions, chances for an all-embracing correlation are poor, but several rules of more or less limited scope are known and will be discussed in due course. The factors controlling diastereoselection can be identified as being of stereoelectronic and purely steric nature.

For example, the well-known fact that addition of bromine to fumaric acid gives exclusively (2*R*,3*S*)-2,3-dibromosuccinic acid [52] can be well explained by the mechanism of the reaction (stereoelectronic control), while attack at the *Si* face in example (b) of Fig. 1-27 is rationalized (less convincingly) by shielding of the *Re* face by the phenyl group at C-6 (steric hindrance). In the following reaction the preferred formation of the thermodynamically less favored axial alcohol [53] can be ascribed both to steric hindrance by the axial 3-methyl group and to the bulkiness of the reagent. However, this explanation also implies the stereoelectronic requirement that hydride transfer should occur in a plane perpendicular to the ring and not laterally.

Me Me O 3 5 Me $\xrightarrow[\text{THF, 0°C}]{\text{Li[(tBuO)}_3\text{AlH]}}$ Me Me OH H Me (96%) + Me Me H OH Me (4%)

Diastereoselectivity was first recognized by Emil Fischer [54], who pointed out that the ratio of diastereomers arising by the formation of a new asymmetric center in a molecule was biassed by those already present. No successful attempt was made to predict even qualitatively the direction of this bias until the seminal papers by Prelog [55] and by Cram and Elhafez [56] were published in the early fifties. The rules of Cram and Prelog based on these papers and their further developments are closely connected with the addition reactions of carbonyl compounds and will be discussed in Section 3.2.1.

Among the very few quantitative approaches to the prediction of diastereoselectivity the one by Ruch and Ugi developed in 1969 on the basis of group theory should be mentioned [57]. This is a semi-empirical method operating with so-called substituent constants and should be, in principle, amenable to any type of reaction. Surprisingly few applications [58], some of them unsuccessful [59, 60], have been reported so far, indicating that there is still ample room for activity in this field. Stereoselectivity in nucleophilic addition to ketones was predicted by computational methods by Wipke and Gund [61] (*cf.* Section 3.2.1) and by Hirota *et al.* [62], who also modelled the oxidation of chiral sulfides.

The traditional term for diastereoselectivity experienced in chiral molecules is *asymmetric induction*. This term is useful in a way that it points to the source of selectivity, but also somewhat narrow because it excludes diasteretopicity in achiral molecules. Clearly the structural features which may induce stereoselectivity in the chiral and achiral substrates shown in Fig. 1-19 are essentially the same.

It is much more controversial to take a stand on another traditional term, *i. e. asymmetric synthesis.* In Marckwald's formulation [63] "asymmetric syntheses are those reactions which produce optically active substances from symmetrically constituted compounds with the intermediate use of optically active material but with the exclusion of all analytical processes". The examples in Figs. 1-22 to 1-25 all correspond to this definition, although what Marckwald probably had in mind was the use of chiral auxiliary compounds as exemplified in Fig. 1-27. In their book "Asymmetric Organic Reactions" [64] Morrison and Mosher extended the definition to transformations of any kind of substrate containing prochiral groups or faces. This was a logical development, since, from the point of view of the selectivity-determining step, it was irrelevant whether the presence of the inducing group in the molecule was transient or permanent. The scope of this definition is, however, rather wide and points towards a third usage of the term, *i. e.* for synthetic sequences in general starting from and arriving at a chiral compound. In view of the above mentioned ambiguities, whenever possible we are going to replace "asymmetric synthesis" by the more specific terms enantioselective and diastereoselective synthesis, respectively.

In order to assess asymmetric induction in reactions involving an achiral reagent we need not use an enantiomerically pure substrate since modern analytical methods (most conveniently GLC, HPLC or NMR) permit the determination of the ratio of diastereomeric racemates ($[RR]+[SS]/[RS]+[SR]$). In the absence of intermolecular interactions between substrate molecules (a resonable assumption in dilute solution), rates of formation for enantiomers must be equal and therefore

$$\frac{[RR]+[SS]}{[RS]+[SR]}=\frac{[RR]}{[RS]}=\frac{[SS]}{[SR]}\ .$$

Transfer of hydride anion to racemic 2-methylcyclohexanone (Fig. 1-28), a typical example for this situation, is illustrated also in two dimensions (b). Diastereotopic faces of the chiral substrate are represented by different sides of a scalene triangle, the achiral hydride donor (*HD*), *e. g.* sodium borohydride, by a quadrangle. It is apparent that the transition states TS_{RR} and TS_{SS}, as well as TS_{SR} and TS_{RS}, are mirror images, which is equivalent to equal rates of formation of the enantiomeric products. In the lower part of the figure reaction with a chiral hydride donor (HD^*) is depicted, the reactive side of the reagent being marked with a heavy line. Transition state symbols are all of different shape, *i. e.* the relationship of any two is diastereomeric. Consequently, even the rates of formation of enantiomeric products are different, in other words the product is optically active. This phenomenon is called *double asymmetric induction* and has considerably synthetic

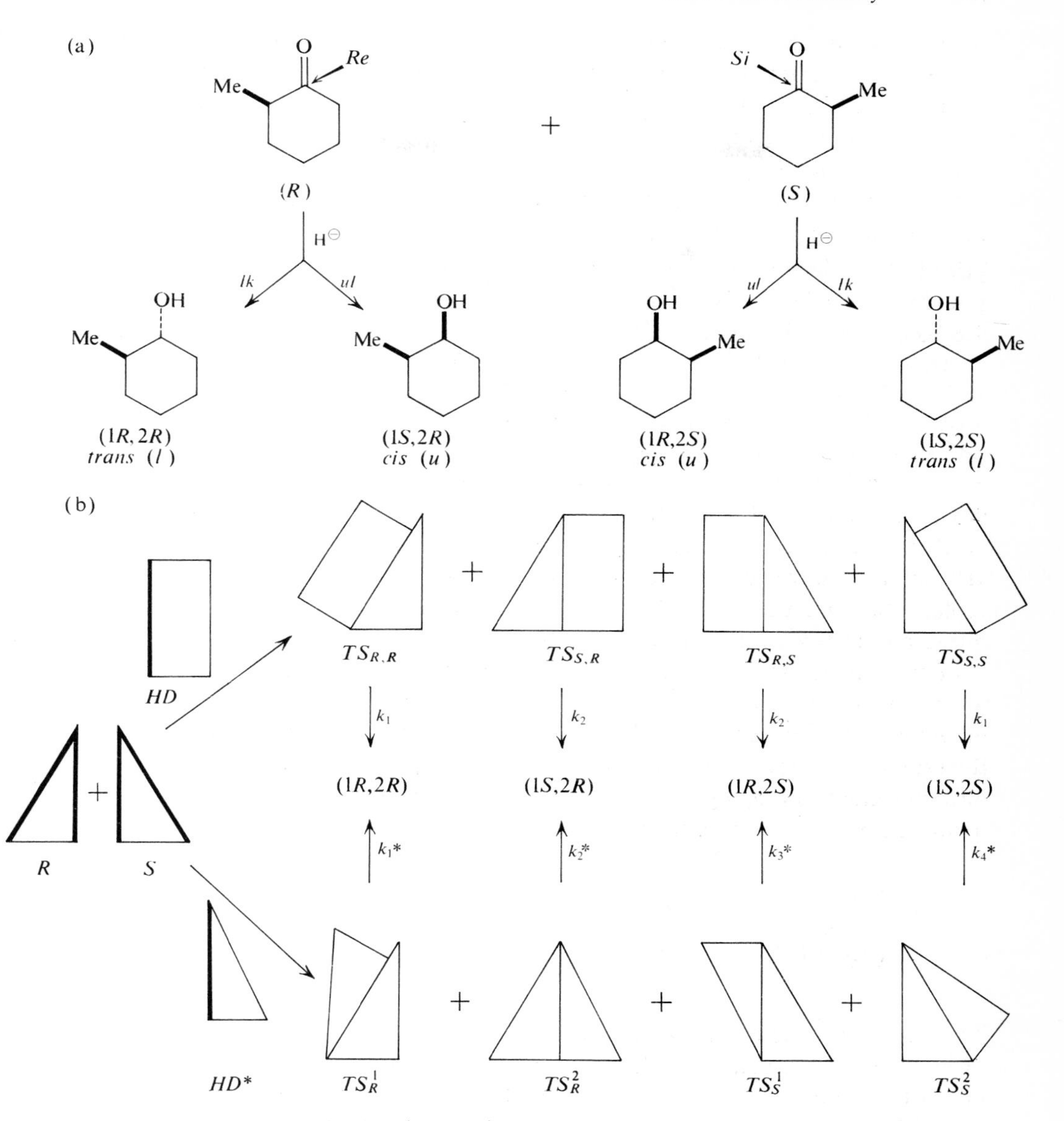

Fig. 1-28. Asymmetric induction (see text).

utility. In practice, usually a certain stereoisomer is needed and very often it is difficult to predict which would be formed in excess with a given reagent. For reasons of symmetry, when one enantiomer of the product is in excess with a certain chiral reagent, the same excess of its antipode must be obtained using the antipode of the same reagent. Guetté and Horeau [65] have shown that in the reaction of a racemic substrate with a chiral reagent the optical purity of a given diastereomer is inversely proportional to its yield:

$$\frac{ee_{D_1}}{ee_{D_2}} = \frac{[D_2]}{[D_1]}$$

Thus while reduction of (+)-2-methylcyclohexanone with $NaBH_4$ in methanol gave a 2:1 ratio of *trans*- and *cis*-alcohols [66], reduction with (−)-diisopinocampheylborane produced 92% of the *cis*-alcohol containing the 1*S*,2*S* enantiomer in 1.8% ee and 8% of the *trans*-alcohol with 13.3% ee for the 1*R*,2*S* enantiomer [65].

Another important feature of reactions with double asymmetric induction is that, when the substrate is a pure enantiomer, reaction with one or the other enantiomer of a reagent gives different diastereomeric ratios. This can be easily

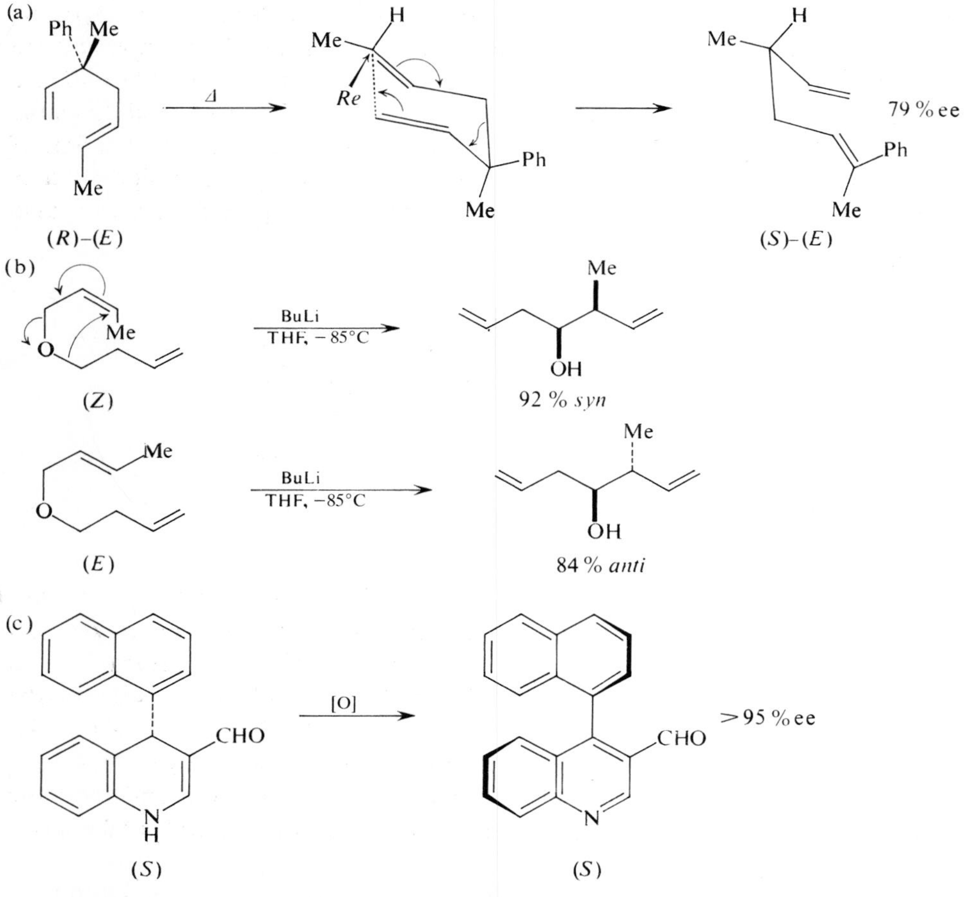

Fig. 1-29. Stereoselective synthesis by intramolecular transfer of chirality.

recognized by exchanging the roles of substrate and reagent in Fig. 1-28 (b). Taking *HD** as substrate, it is apparant that $k_1^*/k_2^* \neq k_4^*/k_3^*$. In this respect it is irrelevant whether the transition states relax by dissociation of the chiral reagent (to give enantiomers) or the reagent remains attached to the substrate to give diastereomers. Examples for the latter are given in Section 5.2.1.

Finally, the technique of chirality transfer* should be mentioned, in which the stereoselective formation of a new chiral center is linked to the elimination of another.

Intermolecular versions of this approach are usually of poor selectivity [*cf.* example (b) in Fig. 1-26], while intramolecular chirality transfer reactions are mainly sigmatropic rearrangements which usually proceed with high stereoselectivity. In all the examples in Fig. 1-29 transposition of a double bond is involved. Due to the concertedness of the reactions, formation of the new double bond is totally stereoselective. Example (a) [67] involves a six-membered cyclic transition state, while (b) contains a five-membered one [68]. In example (c), chirality of a center is transferred to an axially chiral element [69]. Enantioselectivity depends on preference for one of two competing diastereomeric transition states. In example (a), the more favored one of them, *i. e.* attack at the *Re* face, is shown. Attack at the *Si* face would force the phenyl group into an axial position, which is less favored. Example (b) is out of line, since here not a chiral center is sacrificed for the creation of another one, but a double bond of a given diastereomeric configuration is stereoselectively converted to a pair of chiral centers. The products are of course racemic.

1.4.3 Stereoselective Synthetic Strategies

1.4.3.1 Enantioconvergent synthesis

The synthesis of optically active compounds containing several chiral centers from an achiral starting material poses serious problems of economy, even at a laboratory scale. Apart from less than quantitative chemical yields, at least half of the material is lost when generating the first chiral center, while losses in the creation of the other chiral centers are rather unpredictable owing to the diastereoselective character of the transformations. At least one highly enantioselective step or resolution of enantiomers at some stage is unavoidable, and, since the more material involved the higher the costs due to labor, chemicals and energy,

* Self-immolative asymmetric synthesis in Mislow's terminology [46].

this operation has to be carried out at an as early stage as possible. Enantioselective synthesis, even if it produces some of the unwanted enantiomer, is certainly the more efficient approach, since the maximum yield of resolution cannot be higher than 50%. If the chiral compound obtained either by resolution or enantioselection contains a single element of chirality it may be possible to racemize the unwanted isomer, whereas the racemization of diastereomers is only very rarely possible. An example in which the resistance against simultaneous inversion of two chiral centers could be outwitted was provided by Trost *et al.* [70] and is shown in Fig. 1-30. The racemic mixture of alcohols (*33*) was converted to diastereomeric urethanes, which were separated. The useful one was hydrolyzed, while the other converted, by equilibration, to a mixture rich (2:1) in the useful diastereomer. Note that the inversion of configuration at C-1 is formal, *i. e.* it covers no actual transformation.

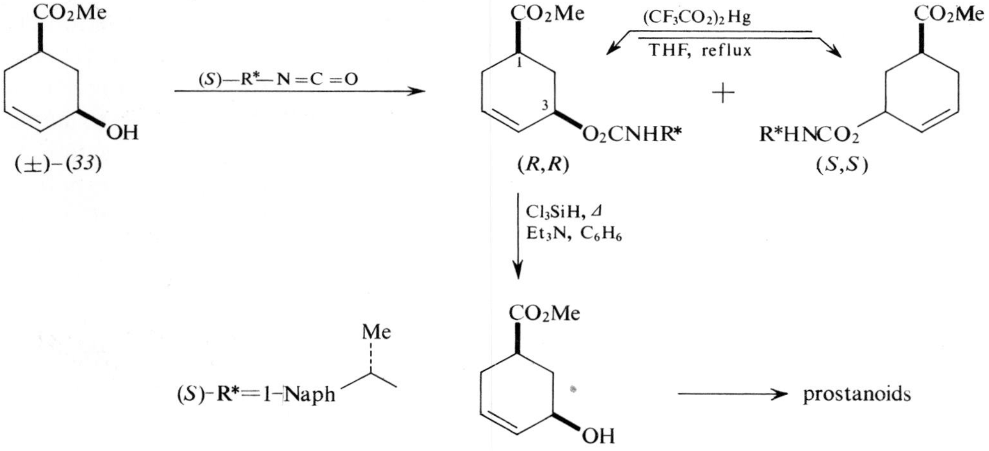

CO_2Me OH (±)-(*33*) (*S*)—R*—N=C=O CO_2Me 1 3 O_2CNHR^* (*R*,*R*) $(CF_3CO_2)_2Hg$ THF, reflux + CO_2Me R^*HNCO_2 (*S*,*S*)

Cl_3SiH, Δ Et_3N, C_6H_6

Me (*S*)-R*=1-Naph CO_2Me OH → prostanoids

Fig. 1-30. Utilization of both enantiomers of a chiral substrate by racemization of the unwanted product.

In 1975 Fischli *et al.* [71] proposed an ingeniously simple scheme by which a starting material with enantiotopic groups can be converted completely to a predetermined enantiomer. The original scheme is shown in Fig. 1-31 (a). Its essence is the reversal in the order of transformations. First, a bifunctional prochiral starting material is reacted with one equivalent of a chiral auxiliary (C*) to give a pair of diastereomers (D_1 and D_2). These must be separated, and then the unreacted group (X) is transformed to group B in diastereomer D_1 and to A in D_2. Removal of the chiral auxiliary group and transformation of the liberated functional groups to groups A and B, respectively, gives in both cases, the same enantiomer. Interchanging the order of transformations at both branches provides an antipodal end product.

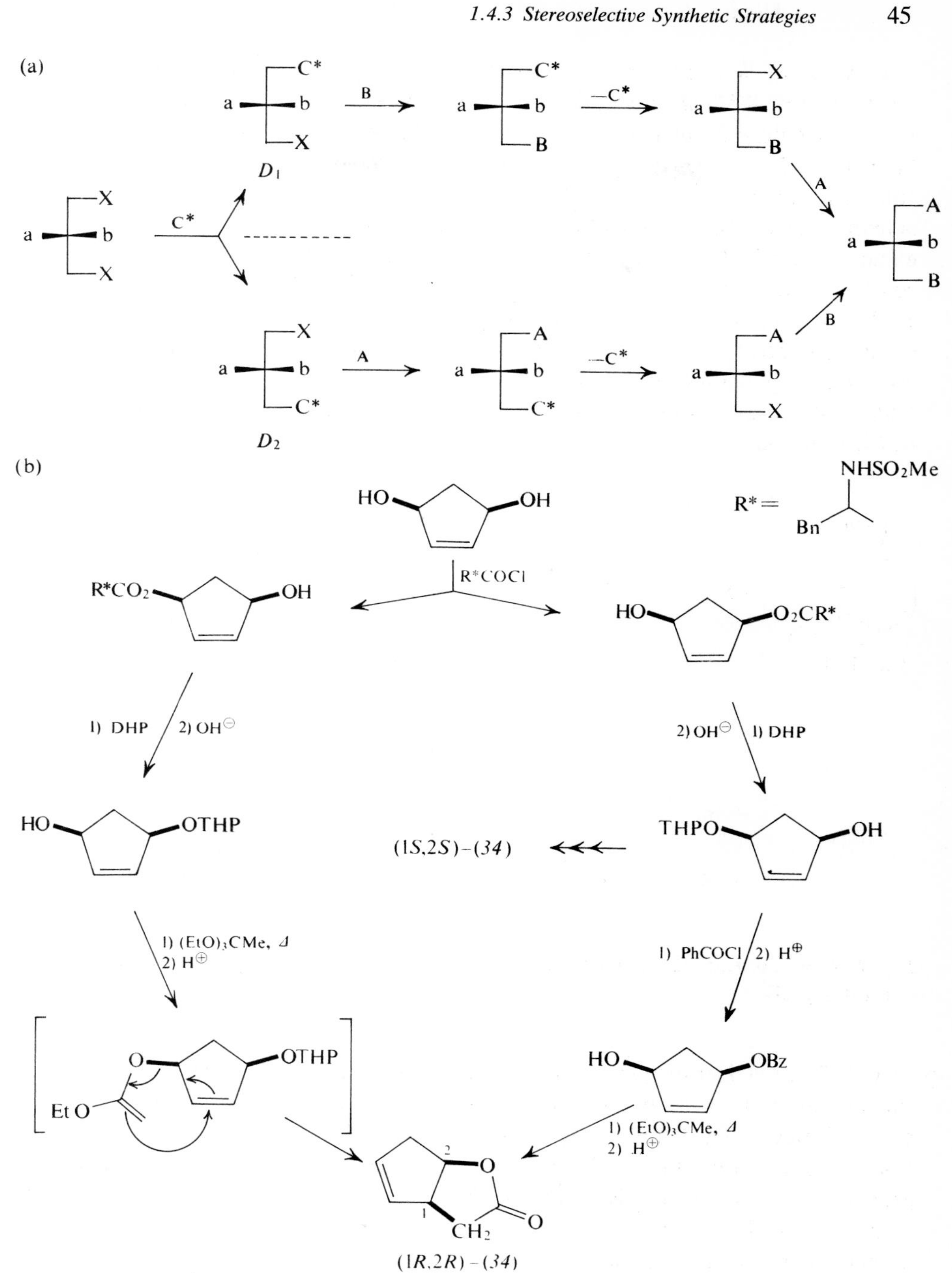

Fig. 1-31. Enantioconvergent synthesis according to the Fischli scheme.

In practice the scheme is fraught with considerable difficulties, and in the original paper no perfect example for its realization was presented, but a version of it in which the unwanted diastereomer was reverted to the starting material and two enantiomeric end products were synthesized by reversing the order of operations.

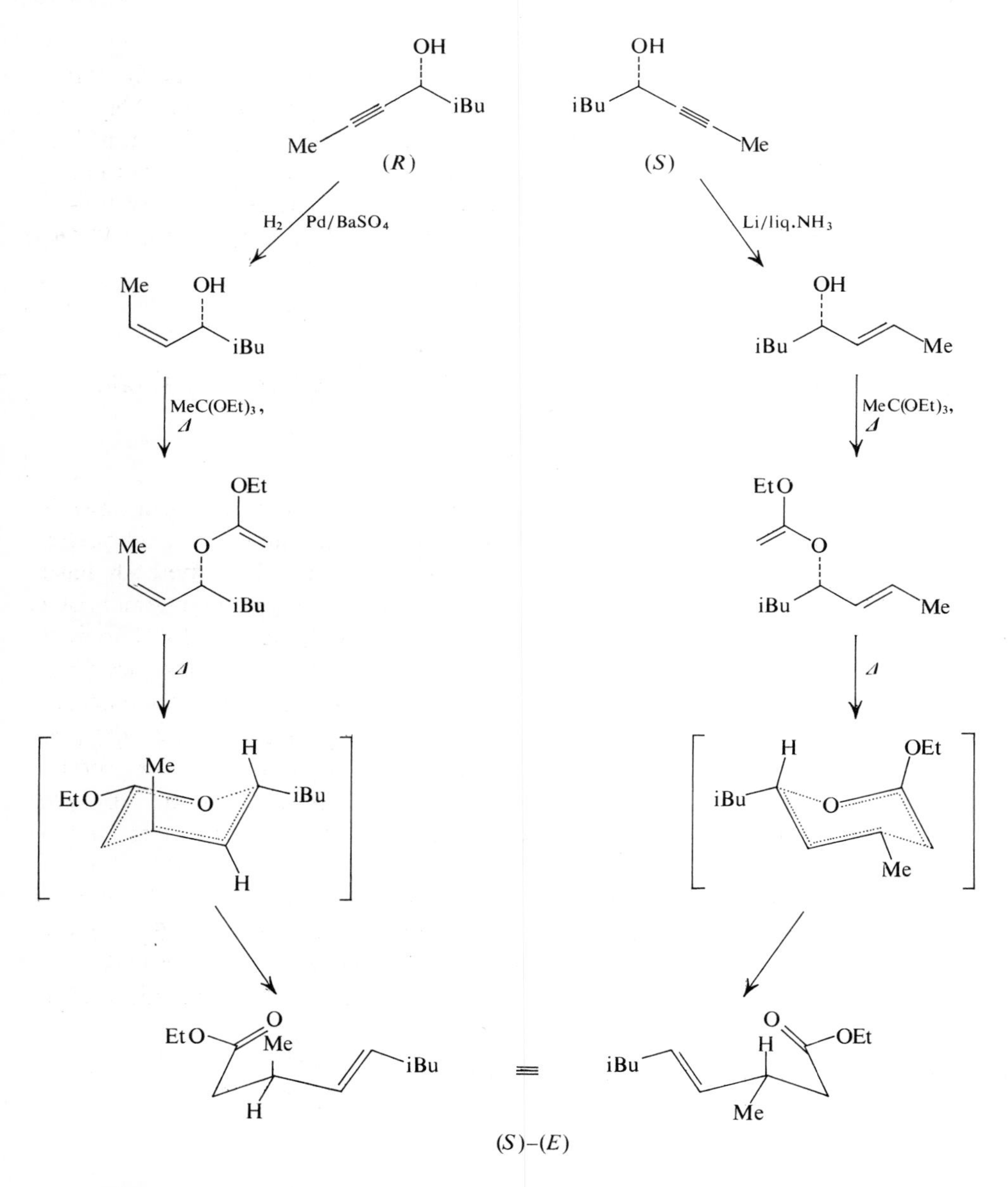

Fig. 1-32. Enantioconvergent synthesis according to the scheme of Cohen and Saucy.

Apart from the problem that synthetic operations cannot be interchanged at will, it is not always easy to establish the absolute configuration of D_1 and D_2 which is essential for the choice of the correct order of transformations.

A relatively simple enantioconvergent scheme, somewhat different from the original proposition, was reported by Terashima and Yamada [72] and is shown in Fig. 1-31 (b). Note that any of the intermediates can be readily converted to any enantiomer of the end product *(34)*.

The synthetic sequence of Cohen *et al.* (Fig. 1-32) [73], which leads from a pair of enantiomeric alcohols, obtained by resolution, to a single enantiomer corresponding to the side chain fragment of tocopherol, features two important steps: (i) Diastereoselective reduction of a triple bond to either a (*Z*)- or an (*E*)-olefin. It is the configuration of the olefinic bond which is traded-in in the next step for a chiral center. (ii) Chirality transfer in a concerted sigmatropic process, *i. e.* the Claisen-rearrangement. An interchange of the reduction methods would give the (*R*)-(*E*) isomer from both starting materials.

1.4.3.2 Selective preparation of both enantiomers from a single substrate

Often there is a need for both enantiomers of a compound, or only one is required but it cannot be foreseen which will emerge as the major product from a stereoselective reaction. Since resolution is not always practical, methods have been devised by which a chiral substrate can be converted to both enantiomers of a product or which can be conducted in a way that the configuration of the product could be controlled. Both approaches can best be illustrated by examples (Fig. 1-33). The scheme in example (a) is based on asymmetric induction, separation of diastereomers, and finally destruction of the inducing center. Example (b) [74] resembles the Fischli scheme inasmuch as the configuration of the major product can be controlled by the order in which the different alcohols are applied. Diastereoselectivity in this reaction is moderate, (+)- and (−)-esters are formed in a ratio of 2:1.

Meyers and his coworkers have described several methods by which both enantiomers of an end-product can be prepared. In example (c) [75] a chiral oxazoline is first alkylated (for details *cf.* Fig. 6-16) with MeI and then with BuI. Alkylation with BuI first and then with Me_2SO_4 gave (*R*)-2-methylhexanoic acid in 70% ee.

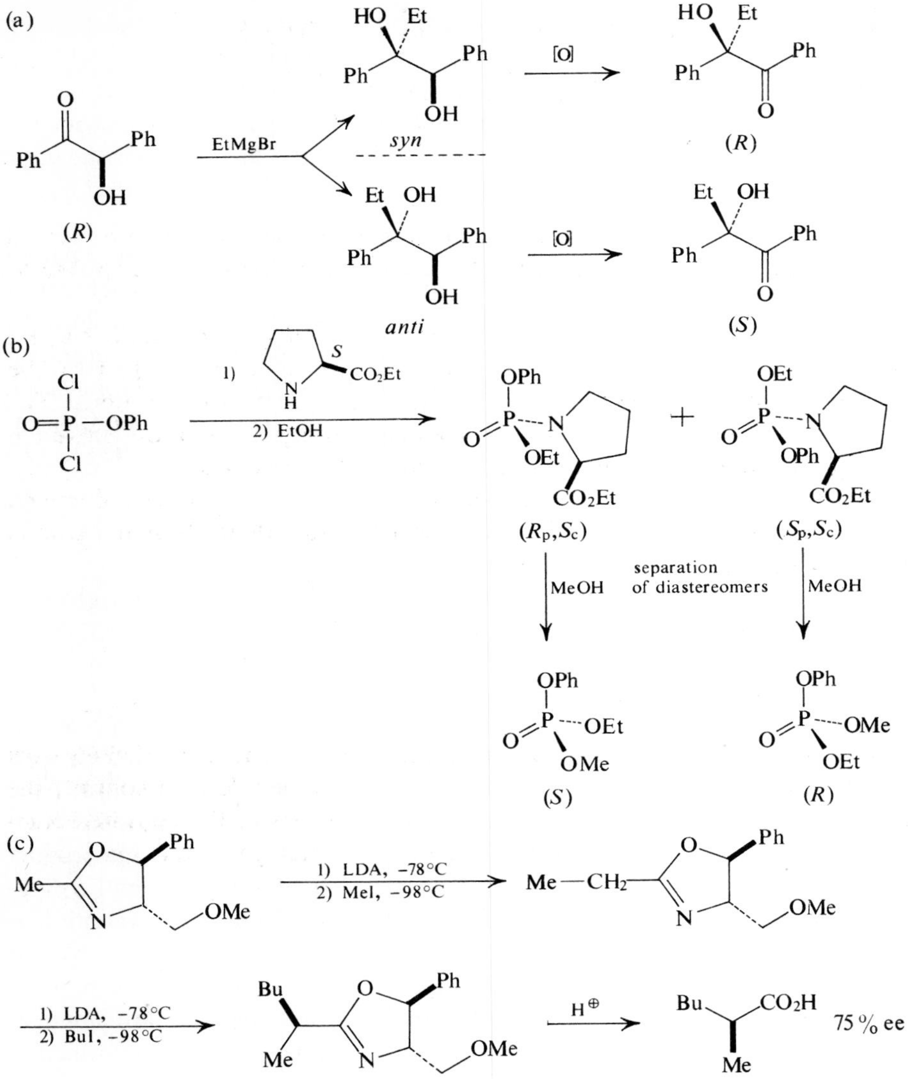

(a)

(R) EtMgBr syn anti [O] (R) (S)

(b)

1) CO_2Et 2) EtOH

(R_p,S_c) + (S_p,S_c)

MeOH separation of diastereomers MeOH

(S) (R)

(c)

1) LDA, −78 °C 2) MeI, −98 °C

1) LDA, −78 °C 2) BuI, −98 °C

$H^{\oplus}$ 75 % ee

Fig. 1-33. Selective preparation of both enantiomers from a single chiral substrate.

1.5 Kinetics and Thermodynamics of Stereoselective Reactions

Before concluding this introductory chapter it is appropriate to restate some well known facts and concepts of reaction kinetics and thermodynamics as applied to stereoselective reactions.

In reactions giving rise to more than one product, the product ratio may be controlled either by the relative rates of formation (kinetic control) or by the equilibrium constant of the products (thermodynamic control) or both. The contribution of the two effects depends on the relative magnitudes of the activation free energies for the formation and equilibration of the products.

When equilibrium is reached, the ratio of products, in our case of stereoisomers, is determined by the equilibrium constant K, which in turn depends on the ground state free energy difference of the products:

$$A \longrightarrow X \rightleftharpoons Y; \quad K = [Y]/[X]$$

$$-\mathrm{R}T \ln K = G_Y^\circ - G_X^\circ = \Delta G_{YX}$$

At the other extreme, when the equilibration between products is infinitely slow as compared with their rate of formation, *i. e.* under pure kinetic control, the product ratio is determined by the relative concentrations of the activated complexes, which are related to their free energies by an analogous equation:

$$Y \xleftarrow{Y^{\ddagger}} A \xrightarrow{X^{\ddagger}} X$$

$$-\mathrm{R}T \ln \frac{[Y]}{[X]} = \ln \frac{100 + e_Y}{100 - e_Y} = G_Y^{\ddagger} - G_X^{\ddagger} =$$

$$= \Delta G_Y^{\ddagger} - \Delta G_X^{\ddagger} = \Delta \Delta G_{XY}^{\ddagger}$$

Since both X and Y are derived from the same substrate, free energies can be replaced by free energies of activation ($\Delta G^{\ddagger} = G^{\ddagger} - G^\circ$) and concentrations can be expressed as the excess of the major product (e) (ee or de).

Thus the correlation of the free energy with both thermodynamically and kinetically controlled selectivity can be expressed by a common graph (Fig. 1-34). This graph should be kept in mind whenever constructing stereochemical models for the explanation of preference for a certain product. Free energy differences at low selectivities are very small and fall into the range characteristic for solvation, conformational changes and other unaccountables. Consequently the predictive

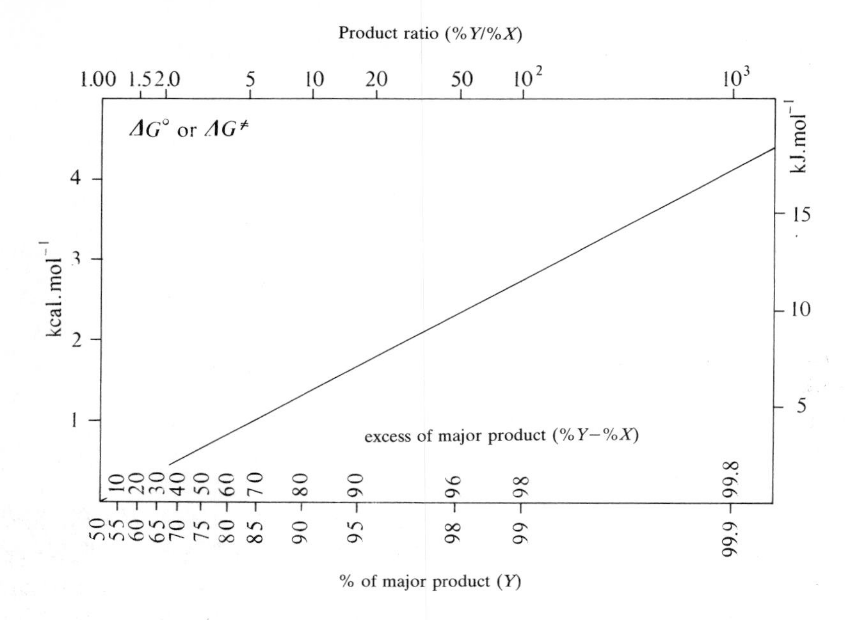

Fig. 1-34. Relationship between free energies ($\Delta G°$ for equilibria and $\Delta\Delta G^{\ddagger}$ for kinetically controlled reactions) and product distribution at 25°C.

power of stereochemical models constructed for reactions of low selectivity is very limited, and since they usually lack any physicochemical or spectroscopic evidence except the observed product distribution itself, they may not point to the real motives of selectivity. Therefore in this book stereochemical models will be employed only in a few justified cases, mostly in conjunction with reactions of high selectivity.

The above relationships are based on the assumption that one single substrate is transformed to two or more products. This holds for conformationally rigid substrates but it is an oversimplification for those which can exist in several conformations. Such systems behave as a mixture of substrates and in situations when, for stereoelectronic or other reasons, one of the products can only be derived from a certain conformer and the other from another conformer, the question about the influence of the conformational equilibrium on product distribution inevitably arises. By simple argumentation Curtin and Hammett demonstrated [76] that for such systems product distribution only depends on the difference of the free energies in the transition state, *i. e.*:

$$\ln([Y]/[X]) = -RT(G^{\ddagger}_{Y} - G^{\ddagger}_{X}) = -RT\Delta G^{\ddagger}_{YX}.$$

The energy diagram for this situation shown in Fig. 1-35 and the Curtin-Hammett principle itself invites several comments:

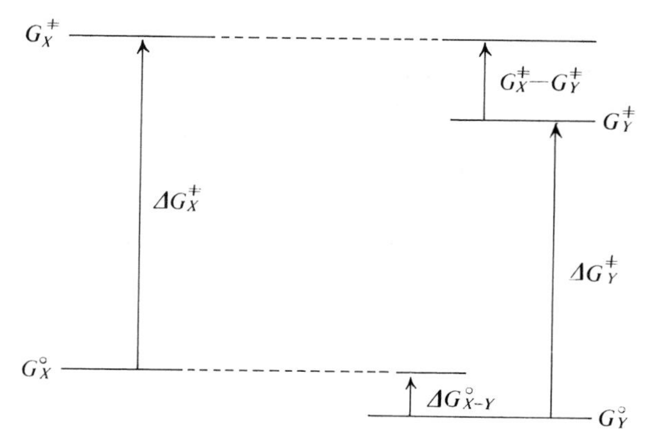

Fig. 1-35. The Curtin-Hammett principle.

(i) Product distribution and ground state stability are, in principle, unrelated. A less stable conformer may lead to a transition state of smaller energy. This is not an unlikely case since free energies of activation for chemical transformations are generally much larger than free energy differences between conformers (if $\Delta G^{\circ}_{XY} > 11$ kJ. mol^{-1} the minor conformer is practically undetectable). A typical example is the rhodium-chiral phosphine catalyzed enantioselective hydrogenation of 2-acylaminoacrylic acids (*cf.* Section 2.1.1.5) which proceeds *via* an undetectable minor intermediate.

(ii) If one has good reason to assume that the free energies of activation for the competing products ($\Delta G^{\ddagger}_X$ vs. $\Delta G^{\ddagger}_Y$) are very close to each other, then the product distribution can be predicted on the basis of conformer distribution. Thus, in the hydrogenation of some macrocyclic exo-methylene ketones the ratio of *cis* and

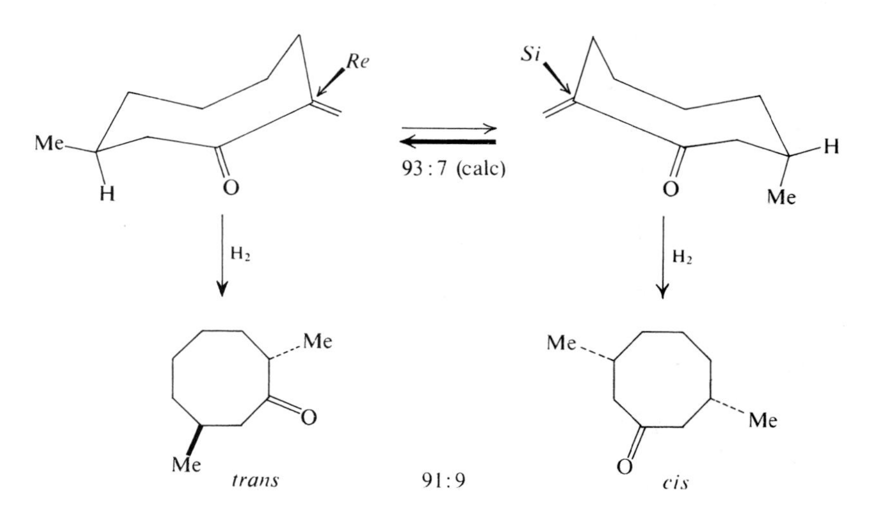

Fig. 1-36. Example for the control of product ratio by substrate conformation.

trans products was predicted by Still and Galynker [77] on the basis of the calculated conformational distribution of the substrate (Fig. 1-36).

(iii) The same as in (ii) applies when ΔG°_{XY} is in the order of ΔG^{+}. In this case it is unlikely that the lower stability of the minor conformer could be compensated for by a smaller free energy of activation.

2 Stereoselective Catalytic Reductions

2.1 Stereoselective Homogeneous Hydrogenations with Rhodium-Phosphine Catalysts

The development of stereoselective homogeneous phase catalytic hydrogenation methods is one of the glorious chapters of synthetic organic chemistry. The role played by rhodium complexes was overwhelming, and therefore most of this chapter (Sections 2.1.1.1 to 2.1.1.5) will be devoted to them. The "rhodium-phosphine story" is briefly as follows.

In 1966 Wilkinson and his coworkers discovered that chlorotris(triphenylphosphine)rhodium, a complex soluble in apolar solvents (*e. g.* benzene), can be used as an efficient hydrogenation catalyst [1]. Not much earlier, methods for the preparation of optically active phosphines with phosphorus as the chiral center were developed by Horner and Mislow [2–4]. The idea of replacing triphenylphosphine in the Wilkinson catalyst by a chiral phosphine, notably by methylpropylphenylphosphine (*1*, Fig. 2-1), was reported independently by Knowles and Sabacky at the Monsanto laboratories in the USA [5, 6] and by Horner and his coworkers in Mainz, Germany [7]. Enantioselectivities achieved with 2-phenylacrylic acid and 2-phenyl-1-propene as substrates were low (15 and 8% ee resp.) but promising. In 1969 Abley and McQuillin improved the record to more than 50% ee with a rhodium catalyst prepared by reduction of a Rh(III)complex in a chiral amide solvent [8], but this method could not be developed further [9]. Morrison *et al.* [10] were the first to recognize that phosphines achiral at phosphorus but carrying a chiral substituent (*e. g. 5*) may also serve as ligands for enantioselective hydrogenation.

Meanwhile, much progress was made with *P*-chiral phosphines as well. Replacement of propyl by *o*-anisyl (*2*, PAMP) enhanced enantioselectivity to 50–60%, while an exchange of phenyl for cyclohexyl led to a ligand (*3*, CAMP) which was already eligible for practical application since it gave 80–88% ee [11]. The search

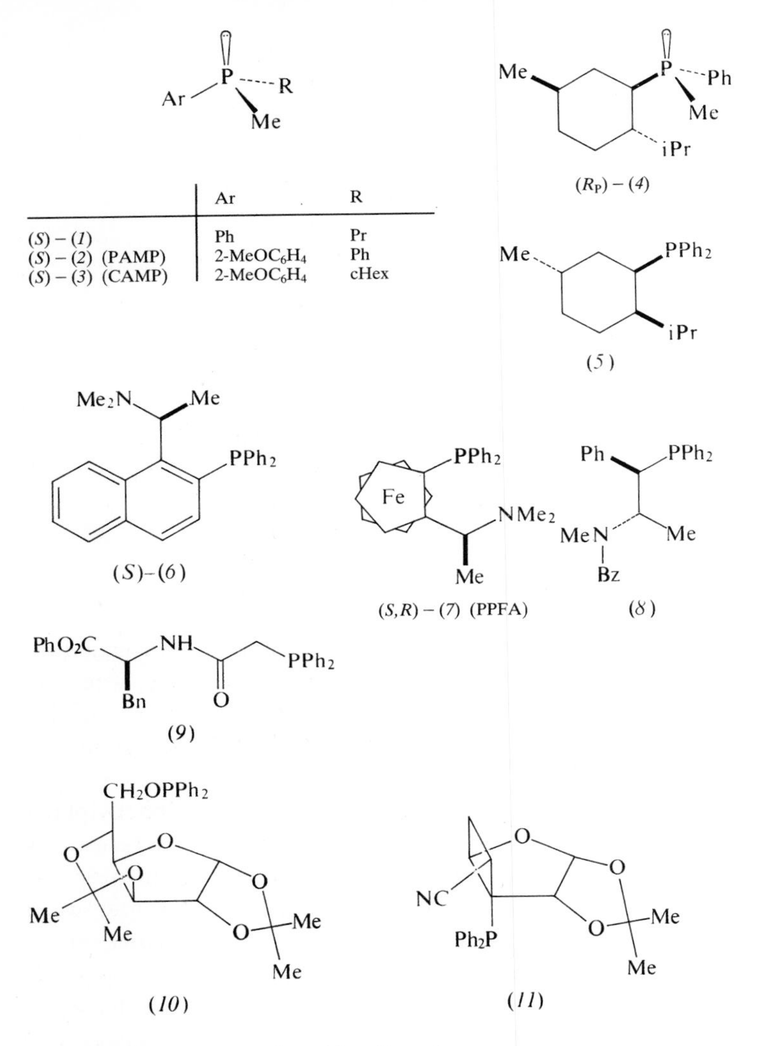

	Ar	R
(S) – (1)	Ph	Pr
(S) – (2) (PAMP)	2-$MeOC_6H_4$	Ph
(S) – (3) (CAMP)	2-$MeOC_6H_4$	cHex

Fig. 2-1. Chiral monophosphine ligands.

for better catalysts became more efficient when it was recognized that the best substrates for rhodium catalyzed enantioselective hydrogenations were 2-acylaminocinnamic acids. The best among *P*-chiral ligands proved to be a diphosphine (*12*, DIPAMP, Fig. 2-2) [12].

The discovery that biphosphines are more selective catalysts than monophosphines can be credited to Kagan and his group [13, 14], who developed one of the most efficient ligands (−)-DIOP (*34*) which could be easily prepared from natural tartaric acid in five steps. By this time the period of fundamental discoveries was more or less over, and researchers turned to working out the details. The literature became flooded with papers on new ligands, only very few of which proved to be

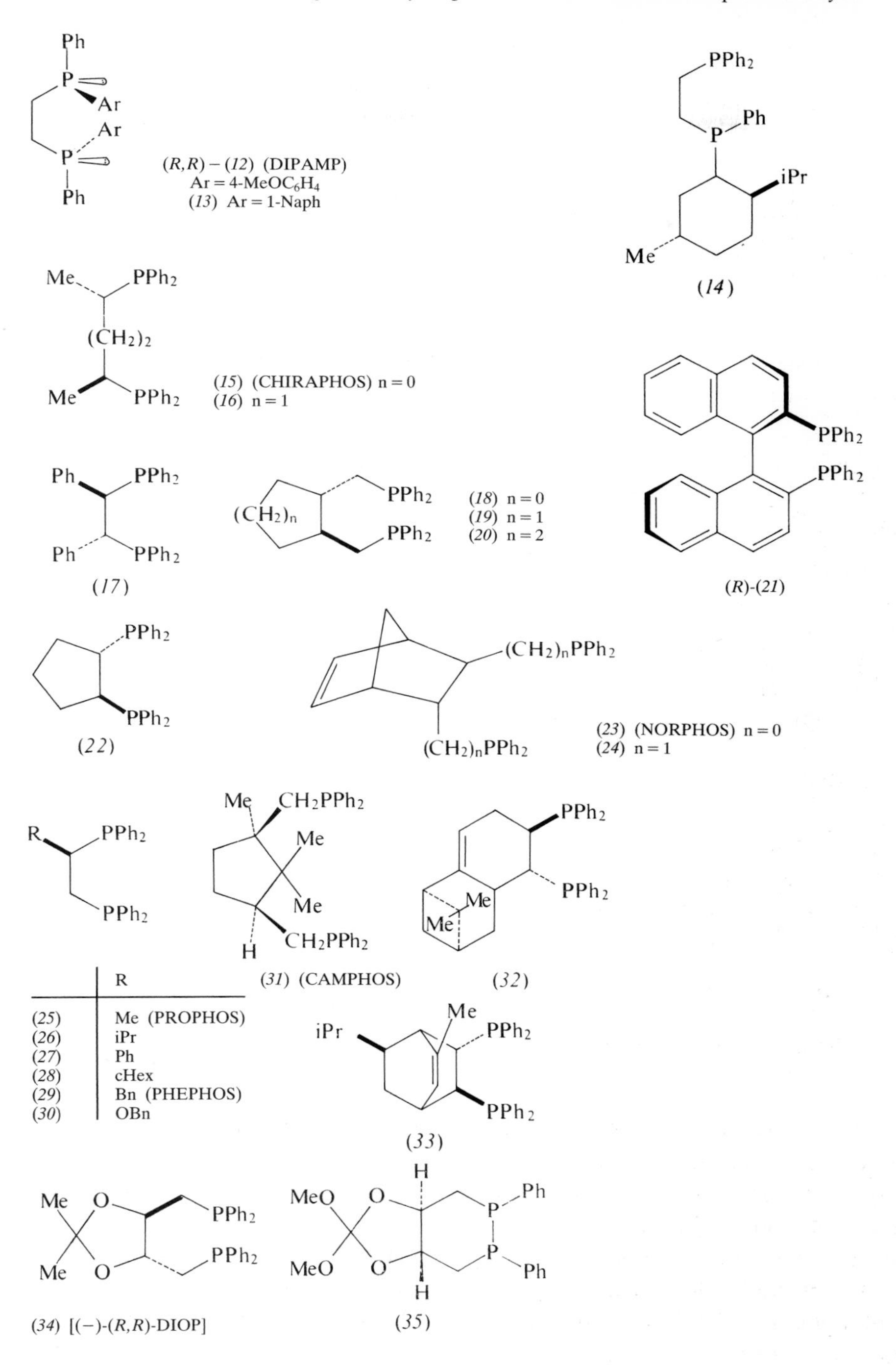

Fig. 2-2. Chiral diphosphine ligands.

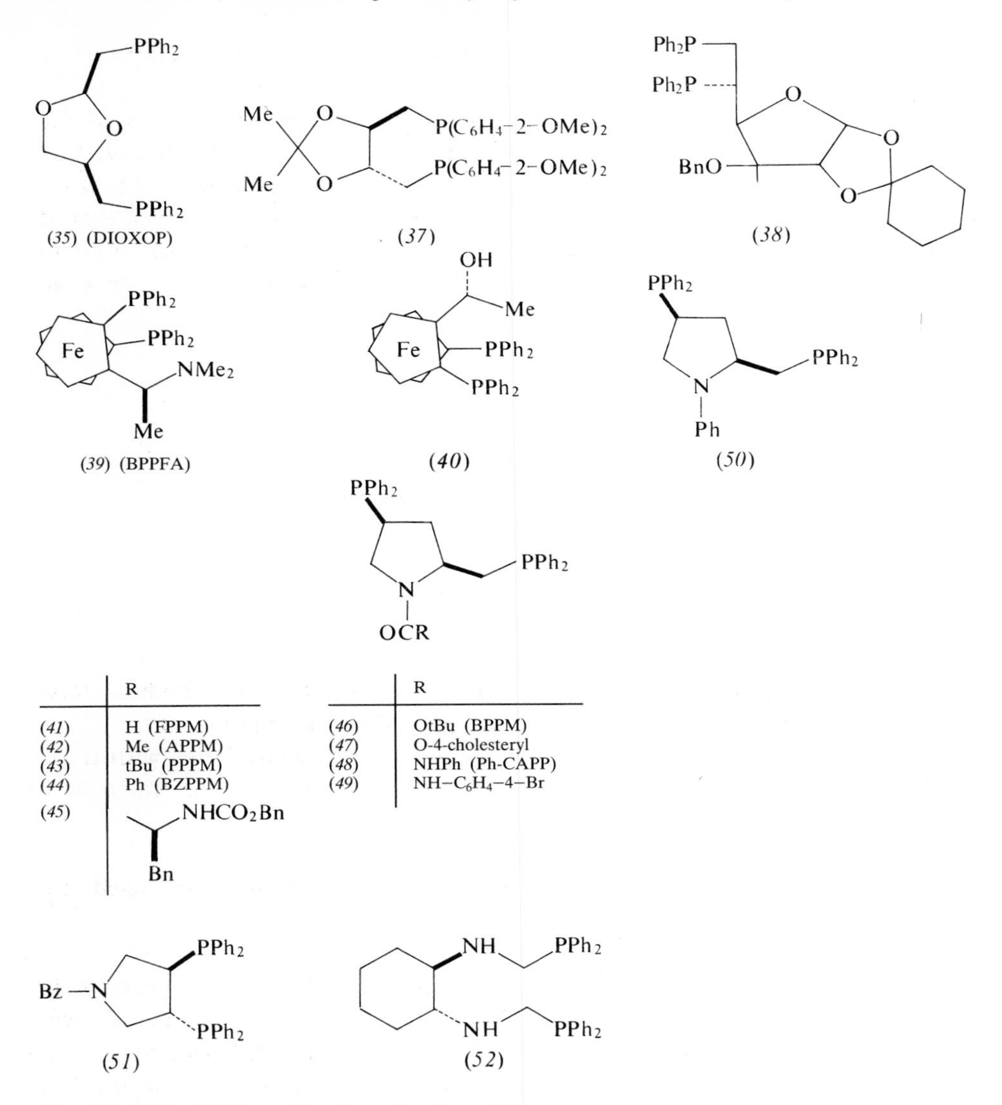

	R
(41)	H (FPPM)
(42)	Me (APPM)
(43)	tBu (PPPM)
(44)	Ph (BZPPM)
(45)	(structure: NHCO2Bn, Bn)

	R
(46)	OtBu (BPPM)
(47)	O-4-cholesteryl
(48)	NHPh (Ph-CAPP)
(49)	NH–C_6H_4–4–Br

Fig. 2-2. Chiral diphosphine ligands. (contd.)

superior to existing ones. Also, intensive research began on the mechansim of the reaction, which shed light on many important details of the process but still failed to explain the very high selectivities achieved with certain rhodium-chiral phosphine catalyst systems. One major deficiency of the method, which may be inherent to highly specialized systems in general, is that the spectrum of substrates which can be hydrogenated with high selectivity, not only with a given catalyst but also with the whole class of chiral rhodium-phosphine catalysts, is very limited. The monopolistic position of rhodium(I) as complexing metal is also surprising, in view of the enormous amount of research into organic transition metal complexes in recent decades.

It is not easy to separate essential from inessential when faced with the bewildering number of papers published on this topic. As a compromise, we first present chiral phosphorus(III) containing compounds recommended as ligands for rhodium catalyzed hydrogenations. This is followed by a description of the most important experimental aspects and the range of substrates amenable to enantioselective hydrogenation. Finally, we summarize the present state of knowledge concerning the mechansim of the reaction.

2.1.1 Hydrogenation of Olefinic Bonds

2.1.1.1 Chiral phosphorus containing ligands

Chiral phosphorus containing compounds which are suitable for rhodium catalyzed hydrogenation can be classified according to the following aspects:

(i) the nature of the ligands attached to phosphorus: phosphines with three carbons as ligands, phosphinites with two carbons and one oxygen, and aminophosphines with two carbons and one nitrogen;

(ii) the number of phosphorus atoms;

(iii) the source of chirality, which may be either the phosphorus atom itself or a chiral substituent attached to it.

Chiral ligands may also be incorporated into a polymer.

In Fig. 2-1 monophosphines, in Fig. 2-2 diphosphines, in Fig. 2-3 phosphinites, in Fig. 2-4 aminophosphines, and in Fig. 2-5 some polymeric ligands are shown. Altogether, the preparation and testing of well over a hundred different ligands has been reported [15]. Most of them, however, have been applied only with a few standard substrates, most often with (*Z*)-2-acetamido- and (*Z*)-2-benzamidocinnamic acid, their methyl esters, 2-acetaminoacrylic acid, and itaconic acid. Comparision of the selectivity of different chiral ligands is not easy, since both the set of test substrates and the experimental conditions (solvent, preparation of catalyst, temperature and pressure) are widely divergent. Also, a decline of ee with increasing conversion has occasionally been observed. In Table 2−1 ee values and prevailing configurations obtained with selected substrates and rhodium(I) catalysts prepared using the ligands shown in Figs. 2-1−2-5 are listed. Ee values under optimum conditions are quoted, not only because data obtained under identical conditions were rarely available, but also because the main purpose of this tabel is to give information about the practical usefulness of the individual ligands.

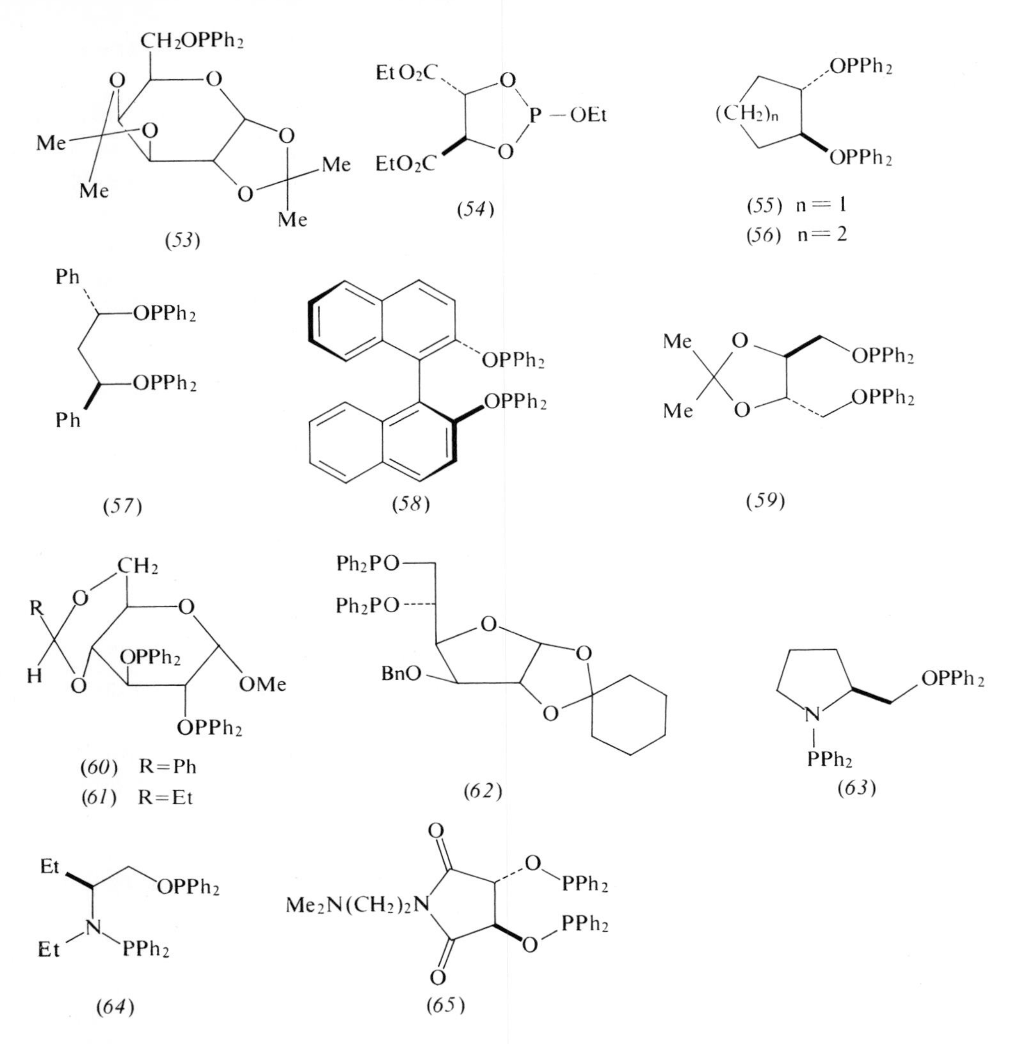

Fig. 2-3. Chiral phosphinite ligands.

Namely, being highly specialized systems, the performace of rhodium-chiral phosphine catalysts is very sensitive to reaction conditions.

Inspection of Tables 2-1, 2-3, 2-4, and 2-5 (see later), as well as of Figs. 2-18 to 2-22 makes it clear that very few of the ligands tested gave satisfactory selectivities. It is generally accepted that ee values in excess of 80% are necessary to secure by recrystallization acceptable yields of an optically pure end-product (*cf.* Section 1.4.2). In addition, several practical requirements have also to be considered, such as ease of preparation, availability and optical purity of the starting material, chemical stability *etc.*

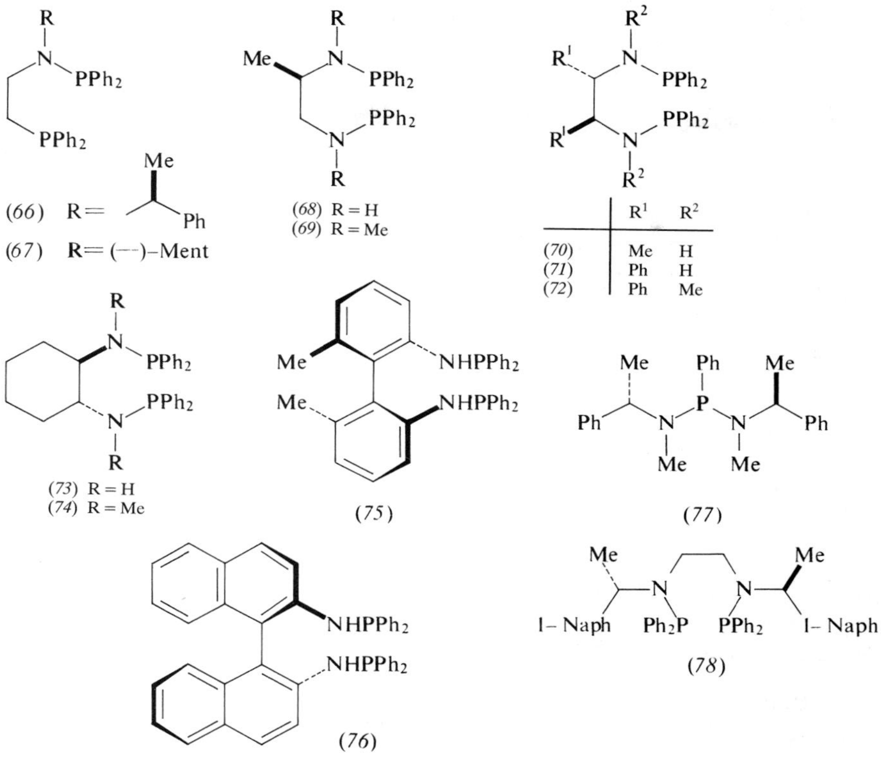

Fig. 2-4. Chiral aminophosphine ligands.

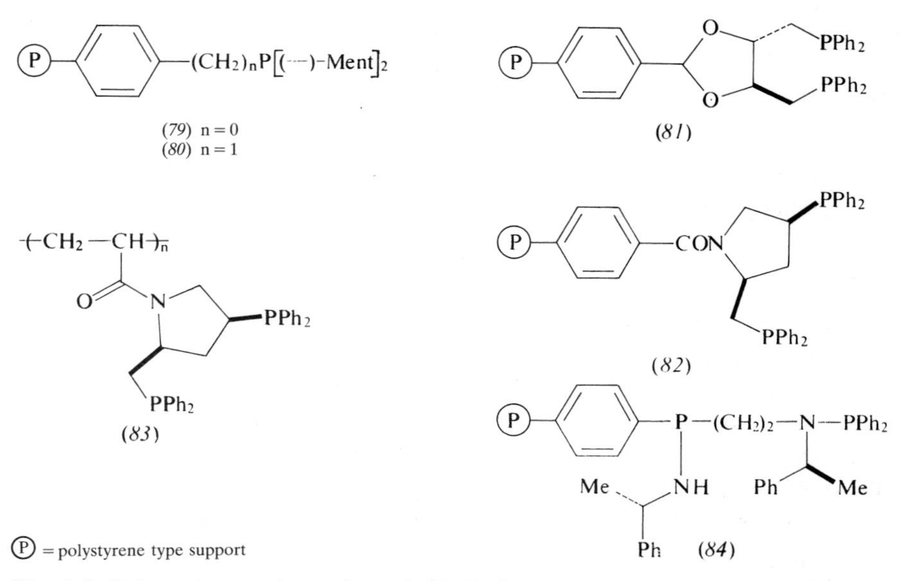

Fig. 2-5. Polymeric or polymer-bound chiral phosphorus containing ligands.

Table 2-1. Enantioselectivities in the Hydrogenation of Standard Substrates with Chiral Rhodium-Phosphine, -Phosphinite, and -Aminophosphine Catalysts under Optimized Conditions. (AcNH–CA = (*Z*)-2-acetamidocinnamic acid, BzNH–CA = (*Z*)-2-benzamidocinnamic acid, AcNH–CE = methyl (*Z*)-2-acetamidocinnamate, AcNH–AA = 2-acetamidoacrylic acid, ITA = itaconic acid.)

Ligand[a]	Substrates										References	
	AcNH–CA % ee	(Conf.)	BzNH–CA % ee	(Conf.)	AcNH–CE % ee	(Conf.)	AcNH–AA % ee	(Conf.)	ITA % ee	(Conf.)	Preparation	Application
Monophosphines												
2	55	(*S*)[b]	58	(*S*)[b]							11	11
3[f]	89	(*S*)	89	(*S*)	60	(*S*)	63	(*S*)			11	11
4	80	(*R*)									24	24
6									43	(*R*)	25	25
7	84	(*R*)					58	(*R*)			26	26
8	77	(*S*)									27	27
10	67	(*R*)									28	28
11	86	(*S*)									29	29
Diphosphines												
12	96	(*S*)	94	(*S*)	96	(*S*)	93.5	(*S*)			30	6, 30, 31
13	76	(*S*)									32	32
(+) *14*	62	(*R*)	44	(*R*)	66	(*R*)					33	33
(−) *14*	60	(*S*)	85	(*S*)	58	(*S*)					33	33
15	89	(*R*)	99	(*R*)			91	(*R*)			34, 35[c]	34, 35[c]
16	98	(*R*)	90	(*R*)			93	(*R*)			36	36
17							61	(*R*)	24	(*R*)	37	37, 38
18	86	(*R*)	91	(*R*)	44	(*R*)	72	(*R*)			39	40–42
19	68	(*R*)					72	(*R*)			43	42
20	35	(*S*)	8	(*S*)	1	(*R*)	40	(*S*)			44	42, 44
21	84	(*S*)	100	(*S*)	93	(*S*)					45–47	45, 47
22			100								48	48
23	96	(*S*)	89	(*S*)	88	(*S*)	90	(*S*)	63	(*R*)	49	49
24[d]	53	(*R*)									50	50
25	91	(*S*)	93	(*S*)			90	(*S*)			36	36, 51, 52
26	94	(*S*)	95	(*S*)	95	(*S*)	88	(*S*)			53	53
27	82	(*S*)	84	(*S*)	88	(*S*)	82	(*S*)			32, 37, 54	37, 54
28	84	(*S*)	90	(*S*)							37	37, 55
29	99	(*S*)					84.5	(*S*)			53	53
30							91	(*S*)			56	56
31	17				22				11	(*R*)	57, 58	57, 58
32									75	(*S*)	59	59
33									41	(*R*)	59	59
34	81	(*R*)	70	(*R*)	69	(*R*)	73	(*R*)			60, 106	58
35	71	(*R*)					17	(*R*)			61	61
36	90	(*S*)			53	(*S*)	86	(*S*)			62	63–65
37	31	(*S*)	71	(*S*)	66	(*S*)	8	(*S*)	37	(*S*)	66	66
38	54	(*S*)			41	(*S*)					67	67
39	67	(*S*)	55	(*S*)	80	(*S*)	57	(*S*)			68	69, 70
42							50[e]	(*R*)			71	71

Table 2-1. Continued

Ligand[a]	Substrates					References	
	AcNH–CA % ee (Conf.)	BzNH–CA % ee (Conf.)	AcNH–CE % ee (Conf.)	AcNH–AA % ee (Conf.)	ITA % ee (Conf.)	Preparation	Application
43			59[e] (*R*)	87 (*S*)		71	71
44	23 (*R*)			89 (*S*)		72	73, 74
46	91 (*R*)	84 (*R*)	15 (*R*)	95 (*R*)	93.6 (*S*)	75	76–79
48	95.4 (*S*)		95.6 (*S*)		95.4	80	80
51	99 (*S*)					81	81
52	73 (*S*)	85 (*S*)				82	82
Phosphinites							
53	67 (*R*)				49 (*S*)	83	83
54	26 (*R*)					84	84
56	68 (*S*)			79 (*S*)		85	85
57	96 (*R*)	93 (*R*)	79 (*R*)	77 (*R*)	93.5	86	86
58	76		9	6		87	87
59	15 (*R*)			14 (*R*)		88	88
60	75 (*S*)		65 (*S*)	80 (*S*)		89, 90	90, 91
61	80 (*S*)		10 (*S*)			92	92
62	40 (*R*)		24 (*R*)			67	67
63	50 (*S*)			80 (*S*)	20 (*R*)	93	93, 94
64	23 (*S*)			55 (*S*)	10 (*R*)	93	93, 94
65	36 (*S*)	29 (*S*)		26 (*S*)	65 (*R*)	88	88, 95
Aminophosphines							
66	84 (*R*)			77 (*R*)		96	96
67	93 (*S*)			83 (*S*)		96	96
68	28 (*R*)	21 (*R*)				97	97
69	70 (*S*)	48 (*S*)				97	97
70	94 (*S*)	32 (*S*)				97	97, 98
71	94 (*S*)			89 (*S*)	71 (*R*)	97	38, 97
72	68 (*R*)			86 (*R*)	6 (*S*)	97	97
73	49 (*R*)	41 (*R*)		24 (*R*)		97	97
74	88 (*S*)	92 (*S*)		91 (*S*)		97	97, 99
75	81 (*R*)					100	100
76	91 (*R*)	89 (*R*)	69 (*R*)	86 (*R*)	80 (*R*)	101, 102	101, 102
77			20 (*R*)			103	103
78	92 (*R*)			88 (*R*)		103	104, 104

[a] No data were available for any of the standard substrates for catalysts with the following ligands: *1* [6], *5* [16], *9* [17, 18], *40* [19], *45* [20], *47* [21], *50* [22] and *55* [23].

[b] Results for the 4-hydroxy-3-methoxyphenyl analogue.

[c] Refers to (*R,R*)-(*15*).

[d] The (+)-enantiomer.

[e] The *N*-tert-butoxycarbonyl derivative was used.

[f] Configuration not given in ref. [87].

At present, CAMP (*3*), PPFA (*7*), DIPAMP (*12*), CHIRAPHOS (*15*), NORPHOS (*23*), PHEPHOS (*29*), (+)- and (−)-DIOP (*34*), DIOXOP (*36*), BPPFA (*39*), some members of the PPM family [*e. g.* FPPM (*41*), PPPM (*43*), BZPPM (*44*), BPPM (*46*) and *48*], and *73* can be seriously considered as ligands for enantioselective hydrogenations. The syntheses of the most important ones will be briefly described below.

The synthesis of (*R*,R)-DIPAMP (*12*) [30] is shown in Fig. 2-6. Unfortunately, on esterification with (−)-menthol and separation of the diastereomers, the required ester, *i. e.* the one with an *R* configuration at phosphorus, was the minor product. Note that both introduction of the anisyl group and deoxygenation proceeded with inversion of configuration. The same sequence of reactions with the major (S_P) menthyl ester afforded (*S,S*)-DIPAMP which, however, gave a catalyst yielding the unnatural (*R*)-amino acids.*

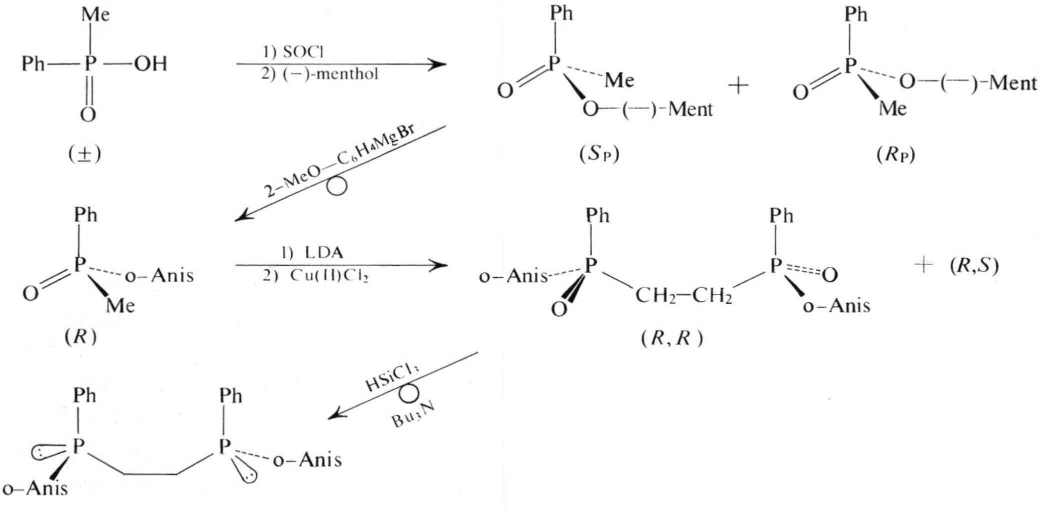

Ph—P(=O)(Me)—OH (±)
1) SOCl, 2) (−)-menthol
(S_P) + (R_P) [O—(−)-Ment esters]
2-MeO—C_6H_4MgBr
(*R*)
1) LDA, 2) Cu(II)Cl_2
o−Anis, CH_2−CH_2, (*R*,*R*) + (*R*,*S*)
$HSiCl_3$, Bu_3N
(*R*,*R*)-DIPAMP (*12*)

Fig. 2-6. Synthesis of (*R*,*R*)-DIPAMP (*12*).

The synthesis of NORPHOS (*23*) [49] (Fig. 2-7) was based on the Diels-Alder synthesis, which required the activation of the dienophile by conversion to the phosphine oxide. Resolution was performed with dibenzoyl (2*R*,3*R*)-tartaric acid and both enantiomers could be obtained pure.

(*R*)-PHEPHOS (*29*) was prepared from (*S*)-phenylalanine in four steps in good overall yield (54%) [53] (Fig. 2-8).

* When the starting material is *o*-anisylmethylphosphinic acid it is the major (S_P) diastereomer which leads to (*R, R*)-DIPAMP [6].

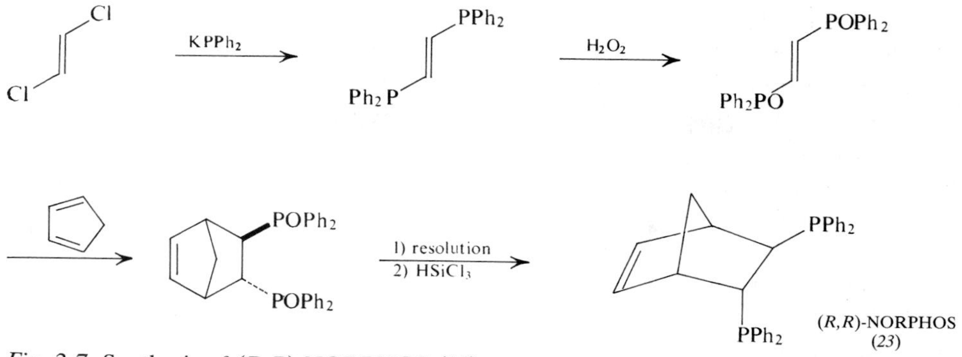

Fig. 2-7. Synthesis of (*R,R*)-NORPHOS (*23*).

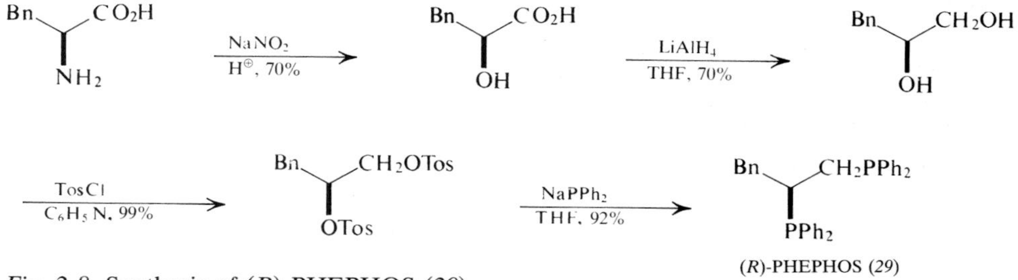

Fig. 2-8. Synthesis of (*R*)-PHEPHOS (*29*).

(−)-DIOP (*34*), probably the most successful chiral phosphine ligand so far, can be prepared from an inexpensive starting material of unlimited availability, namely, (+)-(2*R*,3*R*)-tartaric acid. Although the unnatural (−)-tartaric acid is about 50 times more expensive than its enantiomer, it is also available and can be converted to (+)-DIOP. DIOP was first synthesized by Dang and Kagan in 1971 [51]; an improved procedure giving 49% overall yield was published later (Fig. 2-9 (a)) [105]. Červinka and Gajewski have described an alternative preparation of (+)-DIOP from L-mannitol (Fig. 2-9 (b)) [106].

DIOXOP (*36*), which seems to be the best ligand derived from a sugar, was prepared in four steps (*cf.* Fig. 2-10) from laevoglucosan (*85*), which is itself readily available by pyrolysis of starch.

The chiral ferrocenylphosphine *38* was prepared in 58% yield by the transformation of the ferrocenylamine (*S*)-*86*, the latter being obtained by resolution (*cf.* Fig. 2-11) [107].

A whole array of highly efficient chiral auxiliaries, among them biphosphines, can be prepared from hydroxyproline. Its conversion to BPPM (*46*) [75, 76] and other *N*-acylpyrrolidines [71], as well as to the urea *48* [80], is shown in Fig. 2-12.

One of the reasons why rhodium-phosphines are extremely active is that they are soluble in the solvents used for hydrogenation. Since they are also very sensitive to air, their solubility makes it practically impossible to recover them unharmed. Therefore, many attempts have been made to combine the activity and selectivity

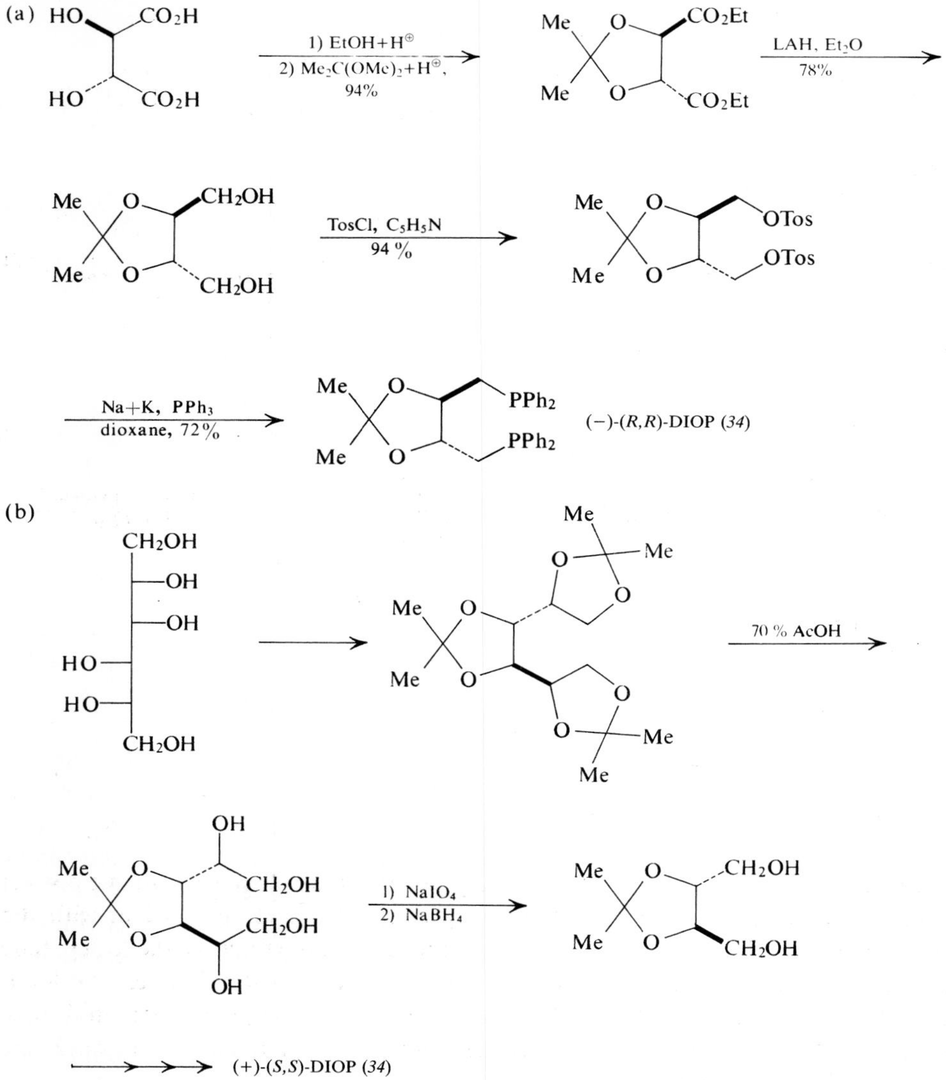

(a) HO, CO_2H / HO, CO_2H —— 1) EtOH+$H^{\oplus}$ 2) $Me_2C(OMe)_2$+$H^{\oplus}$, 94% ——> Me, Me, O, O, CO_2Et, CO_2Et —— LAH, Et_2O 78% ——>

Me, Me, O, O, CH_2OH, CH_2OH —— TosCl, C_5H_5N 94 % ——> Me, Me, O, O, OTos, OTos

—— Na+K, PPh_3 dioxane, 72% ——> Me, Me, O, O, PPh_2, PPh_2 (−)-(*R*,*R*)-DIOP (*34*)

(b) CH_2OH, OH, OH, HO, HO, CH_2OH ——> Me, Me, O, O, Me, Me, Me, Me —— 70 % AcOH ——>

Me, Me, O, O, OH, CH_2OH, CH_2OH, OH —— 1) $NaIO_4$ 2) $NaBH_4$ ——> Me, Me, O, O, CH_2OH, CH_2OH

——>——>——> (+)-(*S*,*S*)-DIOP (*34*)

Fig. 2-9. Synthesis of (−)-(*R*,*R*)- and (+)-(*S*,*S*)-DIOP (*34*).

of soluble catalysts with the ease of recovery of heterogeneous phase catalysts. Di-(−)-menthylphosphine linked to polystyrene (*79* and *80*) [108], (−)-DIOP linked to polystyrene (*81*) [109, 110] and to Amberlite [111], as well as several copolymers of DIOP-substituted styrene- and acrylic-type monomers [112−114], polymers incorporating the chiral phosphine moiety of BPPM (*82* and *83*) [74, 113, 115], as well as aminophosphine side chains (*84*) [27] have been synthesized and tested (Fig. 2-5). Although much interesting information was gained from the study of

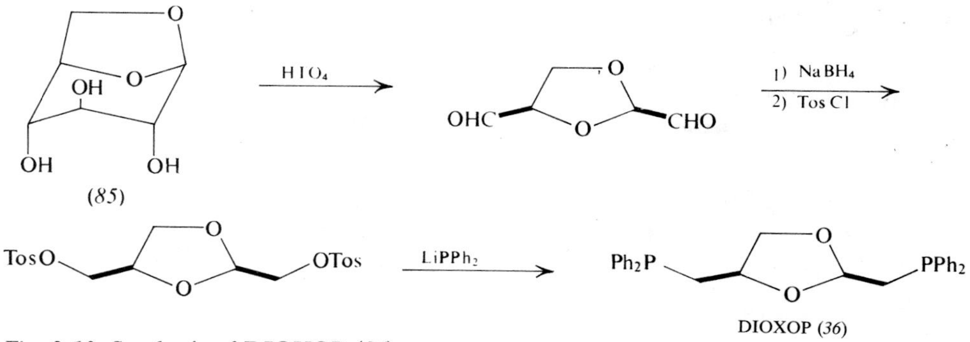

(*85*) DIOXOP (*36*)

Fig. 2-10. Synthesis of DIOXOP (*36*).

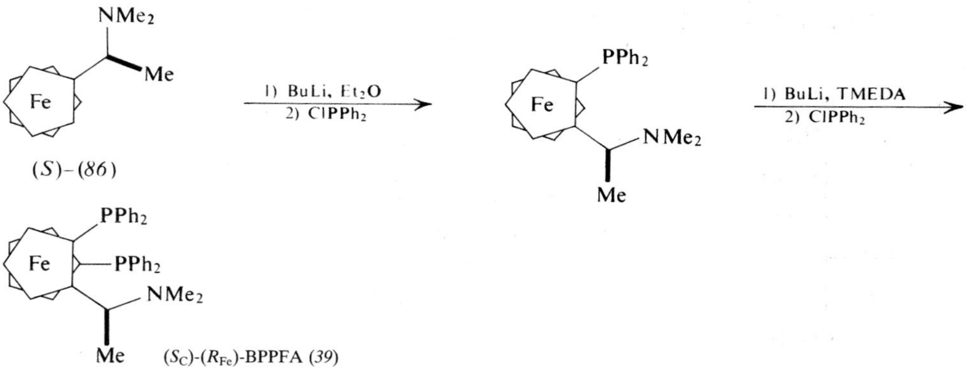

(*S*)-(*86*) (S_C)-(R_{Fe})-BPPFA (*39*)

Fig. 2-11. Synthesis of BPPFA (*39*).

such polymer supported catalysts, from a practical point of view they have not yet fulfilled expectations. Thus selectivities were usually much lower than with the corresponding soluble catalysts, and while not so sensitive to air as the latter, they lose much of their activity when recycled or stored [113]. After 5 cycles, rhodium-DIOP and rhodium-BPPM catalysts adsorbed onto charcoal pretreated with Cr(III)-acetate and triethylamine retained 90% of their activity, but were less selective than the parent catalysts [116].

2.1.1.2 Rhodium-phosphine catalyst systems

The classical Wilkonson catalyst, *i. e.* [$Rh(Ph_3P)_3Cl$], has a square planar structure. By analogy, rhodium monophosphine catalysts were formulated as *87* (Fig. 2-13), where P* represented the chiral monophosphine. *87* may be in equilibrium with the dimer *88* in which the P*−Rh ratio is only 2:1. On addition of hydrogen *87* is supposed to be transformed to a coordinatively saturated (18 electron) octa-

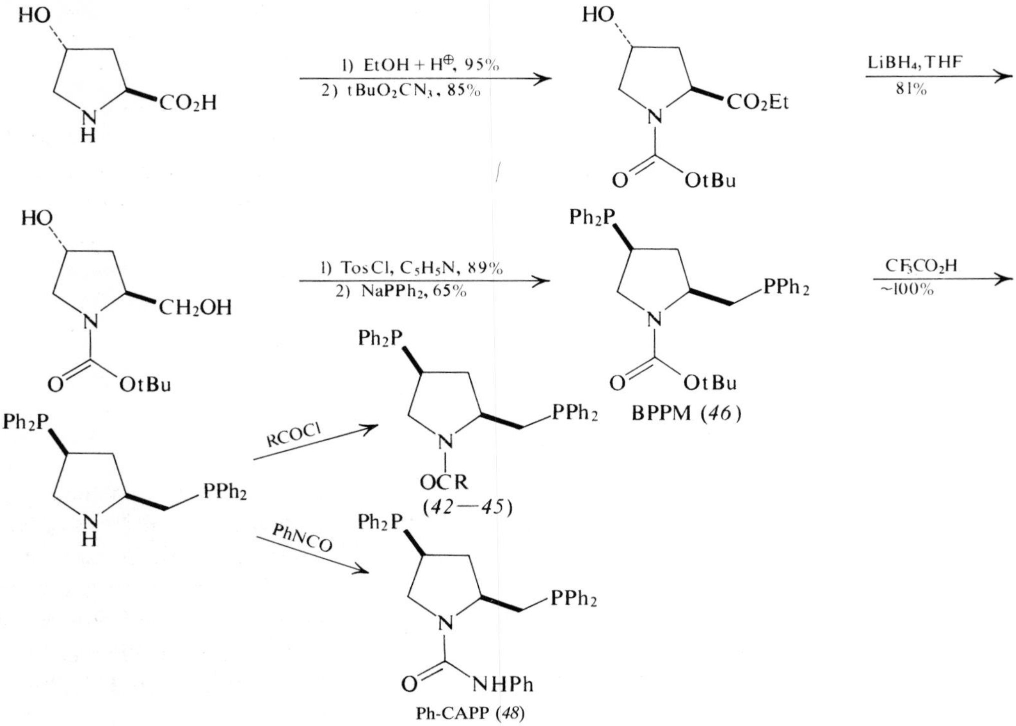

Fig. 2-12. Synthesis of chiral diphosphines from (2*S*,4*R*)-hydroxyproline.

hedral species such as *89*. This is in equilibrium with a solvated species (*90*) and the substrate-catalyst complex (*91*).

From a practical point of view rhodium-phosphine complex catalysts are applied in two forms. The catalyst is either prepared *in situ* in the hydrogenation solvent, or in an isolable ionic form.

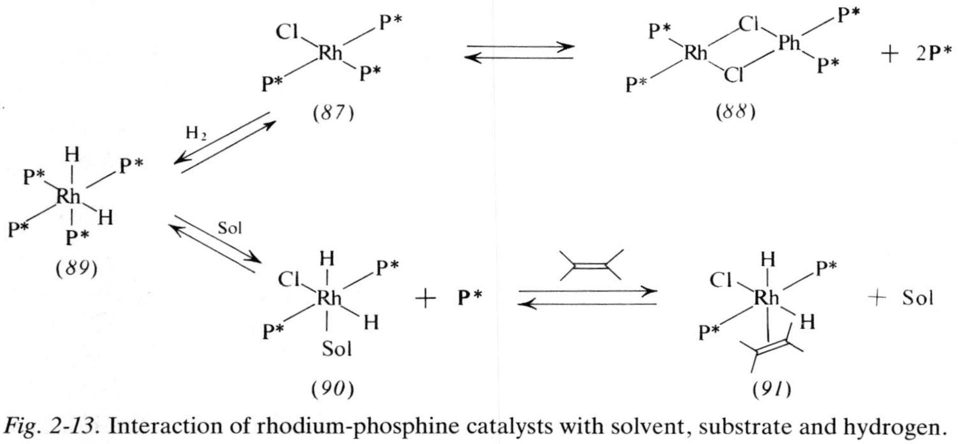

Fig. 2-13. Interaction of rhodium-phosphine catalysts with solvent, substrate and hydrogen.

Neutral *"in situ"* catalysts are prepared by mixing, with the rigorous exclusion of air, a rhodium-olefin complex, such as $[RhCl(cyclooctene)_2]_2$, $[RhCl(1,5\text{-}cyclooctadiene)]_2$, $[RhCl(norbornadiene)]_2$, with an excess of the diphosphine. There is an optimum for the Rh:P ratio which varies with the ligand and other parameters but is generally around 1:3. On hydrogenation the olefinic ligand becomes saturated, loses its affinity to the metal and is displaced by the solvent and/or the diphosphine.

The isolable cationic catalysts, *e. g.* rhodium(1,5-cyclooctadiene)(−)-DIOP tetrafluoroborate (*92*) [117], are made by mixing one of the above mentioned neutral diene complexes with a slight excess of the phosphine and adding an equivalent of

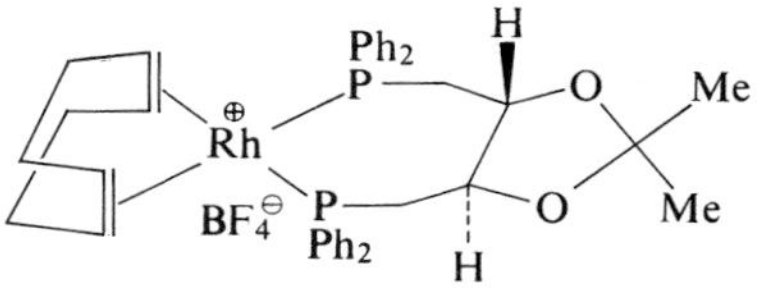

Rh⊕ BF$_4^{\ominus}$ P Ph$_2$ P Ph$_2$ H O O Me Me H (*92*)

$NaBF_4$. This catalyst is air sensitive, too. In such complexes both ligands must be *cis* coordinated. It should be noted that not every ligand is suitable for the preparation of ionic complexes.

Differences in rate and enantioselectivity between ionic and neutral complexes of the same ligand have often been recorded [*e. g.* 54, 118]. Thus the *in situ* catalyst prepared from *27* is much less rective than the ionic complex and also shomewhat less selective towards certain, but not all, substrates [119].

Catalyst preparation is only one of the many experimental variables which have to be optimized for maximum selectivity. As will be seen later, the search for useful general correlations is a frustrating undertaking.

Correlating ligand structure and enantioselectivity is an arduous task too. So far, two points hae become firmly established: (i) chirality at phosphorus is not necessary for enantioselectivity, and (ii) diphosphines are more efficient ligands than monophosphines. Since, except for CAMP (*3*) and PPFA (*7*), no monophosphine ligand can qualify for practical application, we shall restrict the present discussion to diphosphines.

The majority (>80%) of diphosphines shown in Figs. 2-2 to 2-5 form five- or seven-membered chelate rings with rhodium, and very efficient catalysts can be found in both groups. The conformation of the chelate ring is of crucial importance, since this is the lever by which the chirality of the ligand is transmitted to the substrate.

A five-membered chelate complex of rhodium, even with the achiral bis-diphenylphosphinoethane, is intrinsically chiral, although mirror image conformations of C_2 symmetry undergo fast interconversion (Fig. 2-14 (a)) [120]. In the case of a chiral ligand, one of the conformations may be highly preferred, and thus a stable chiral environment is provided for complexation with the prochiral ligand.

Very few chiral phosphines forming six-membered chelates have been described. Comparison of analogues forming five- and six-membered chelates is somewhat

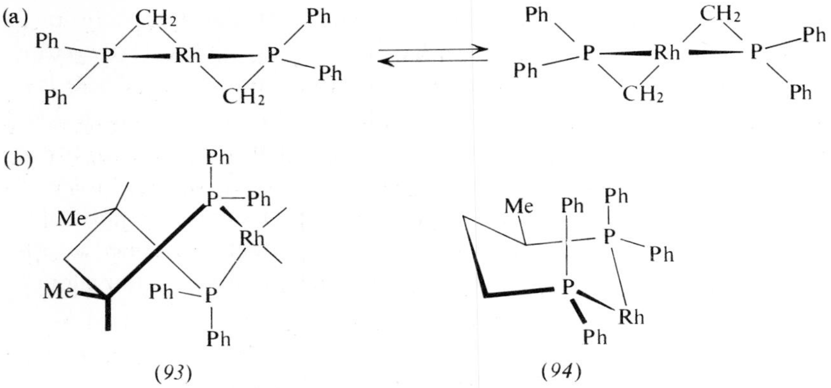

Fig. 2-14. Conformation of rhodium-phosphine complexes forming a six-membered chelate ring.

bewildering. CHIRAPHOS (*15*) and its homologue (*16*) both give a highly selective catalyst (ee ≥ 89%), while the homologue of PROPHOS (*25*) is a poor ligand. For an explanation it was assumed [120] that for high enantioselectivity the chelate ring must take up a twist-chair conformation (Fig. 2-14, *93*), which provides an antisymmetrical orientation of the phenyl rings. The PROPHOS homologue should be in a chair conformation with mirror image symmetry for the phenyl rings (*94*).

Diphosphines forming seven-membered chelates constitute by far the most populous group and the majority of highly efficient catalysts belongs here. Selective ligands are generally cyclic compounds, notable exceptions being the binaphthyls *21, 58* and *75*, as well as the aminophosphines *69–71*.

According to an analysis by Brown and his coworkers [120], two chiral conformers should be considered for the seven-membered chelate ring, the chair (*C*) and the twist-boat form (*TB*) (Fig. 2-15 (a)). (One or both of the atoms marked with a dot are substituted.) The most important difference between the two conforma-

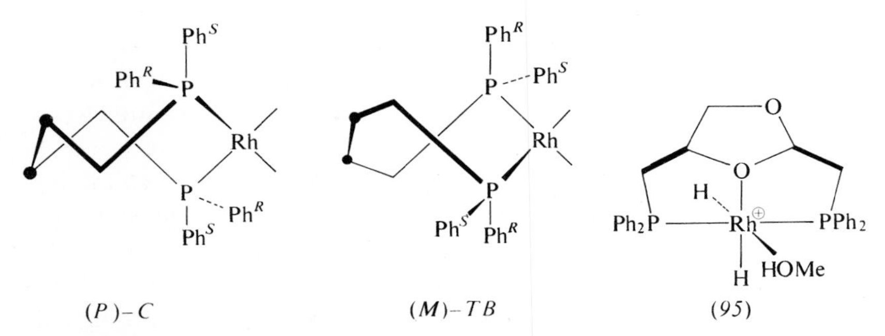

Fig. 2-15. Conformation of rhodium-phosphine complexes forming seven- and eight-membered chelate rings.

tions is that the quasi-axial and quasi-equatorial dispositions for the pro-*R* and pro-*S* phenyl groups are interchanged, and the helicity of the CH_2–P–Rh–P–CH_2 sequence is inverted (*P* in *C* and *M* in *TB*). Since the latter is the very feature of the complex which controls enantioselectivity, prediction of the configuration of the product in excess may become feasible. This requires that both the prevailing conformer of the ring and the mode of substrate binding should be known. According to the empirical rule suggested by Brown, conformer (*P*)–*C* should give (*S*)-amino acids from (*Z*)-enamides, while conformer (*M*)-*TB* provides the *R* enantiomers. Note that the conformational equilibrium $C \rightleftarrows TB$ is controlled by the constitution and relative configuration of the ligand, and helicity by its absolute configuration. Thus, for the antipode of the ligand the enantiomers of the conformers shown [*i. e.* (*M*)-*C* and (*P*)-*TB*] should be considered.

Inspection of models suggested that (−)-DIOP and its carbocyclic analogues (*18–20*) preferred the (*M*)-*TB* conformer. In fact, all give (*R*)-amino acids in excess, and for the homologues selectivity decreases with ring size (*cf.* Table 2-1). The dramatic inversion of the prevailing configuration on *N*-methylation with the aminophosphine pairs *68–69, 71–72,* and *73–74* can also be rationalized by a change of conformation. While experiments with the DIOP analogues *96* seemed to support Brown's interpretation, no explanation can be given for the reversal of the stereochemical course of the reaction in the case of the *o*-methoxy analogue (Table 2-2).

Eight-membered and larger rings seem to be too flexible for effective transfer of their chirality to the substrate. The poor performance of CHIRAPHOS (*31*) confirms this proposition, and the efficient ligand DIOXOP (*36*) was shown [121] to form a hydrogenated complex containing not a single eight-membered ring, but two anellated five-membered rings (*95*). Diphosphine ligands capable of forming

Table 2-2. Enantioselective Hydrogenation of (*Z*)-2-Acetamidocinnannic Acid with the Rhodium Complexes of DIOP Analogues.

Me O PAr_2
Me O PAr_2
(*96*)

Ar	% ee	(Conf.)	Ref.
2–Me–C_6H_4	27	(*R*)	127
2,5–Me–C_6H_3	44	(*R*)	127
3–Me–C_6H_4	87	(*R*)	127
2,2′–C_6H_4–C_6H_4	28	(*R*)	127
1–Naph	0		128
2–Naph	83	(*R*)	128
2–MeO–C_6H_4	93	(*S*)	120

nine- or even ten-membered rings all belong to the phosphinite or aminophosphine group, and due to lack of experimental data relevant to the structure of the complexes their behavior cannot be analyzed. Certainly a separation of the phosphorus centers by 5–7 atoms is not incompatible with high stereoselectivity. For example, *57*, the phosphinite analogue of CAMPHOS (*31*), is a very poor catalyst for amino acid precursors but gives 93.5% ee with itaconic acid.

2.1.1.3 Influence of experimental conditions

In the field of rhodium-phosphine catalyzed hydrogenations any attempt to establish correlations between experimental parameters and enantioselectivity is a controversial undertaking, and the final conclusion from all what follows here is that each catalyst-substrate system has to be optimized individually.

There is, however, one feature common to all of the catalysts, namely that, though not pyrophoric, all are sensitive to oxygen, since phosphines are readily oxidized to phosphine oxides.

Solvent effects

Alcohols (mainly methanol), benzene, benzene-alcohol mixtures, tetrahydrofuran, and even aqueous alcohols can be used as solvents. The catalyst prepared from a cholesteryl-phosphine (*47*) is active even in cyclohexane [21].

Solvent effects are rather unpredictable. For example, no or insignificant changes were observed in the hydrogenation of pulegone with Rh-DIPAMP (DMF or MeOH) [129] and of the standard substrates with Rh-PROPHOS (*25*) (THF and EtOH) [51]. Moderate (< 15%) and substrate dependent effects were recorded, *e. g.* for Rh-CHIRAPHOS (*15*) (THF and EtOH) [34, 35], but for many systems the effect was dramatic.

Since it has been shown that solvent molecules are integral parts of the catalysts, it is the absence rather than the presence of a solvent effect that is surprising. In Table 2-3 selected examples for significant solvent effects are presented, and it seems that no general conclusions can be deduced from these data. Even with the same catalyst, the optimum solvent may change with the substrate. Inversion of the solvent effect with the aminophosphines *77* and *78* is especially intriguing and may reflect profound differences between the structure of monophosphine and diphosphine complexes.

Pressure effects

Since one of the important steps in the hydrogenation process is the uptake of dihydrogen by the square planar Rh(I)-phosphine complexes, whereby they are transformed to octahedral hydrido Rh(III) complexes, it can be anticipated that

Table 2-3. Solvent Effect on Enantioselectivity in Rhodium-Phosphine Catalyzed Hydrogenations. (B = benzene, E = ethanol, M = methanol, T = tetrahydrofuran. Numbers are % ee (conf.) values for phosphine ligands.)

Substrate	DIOP	BPPM	*73*	*77*	*78*
CO_2H Ph NHAc			B:41 (*R*) E:72 (*R*) [99]		
CO_2Me Ph NHAc				B:20 (*R*) M: 4 (*R*) [123]	B: 9 (*R*) M:66 (*R*) [123]
AcNH CO_2Me Ph		B:33 (*S*) M:47 (*S*) [122]			
AcNH CO_2Me Me		B:37 (*R*) M: 3 (*S*) [122]			
CO_2H CH_2CO_2H		B + M (2:1):50 (*S*) M:71 (*S*) [73]			
O Me Me O O		B:87 (*R*) E:32 (*R*) T:81 (*R*) [124, 125]			
PhCOMe	B:80 M:53 [126]				

hydrogen pressure should be a crucial parameter. In heterogeneous catalytic hydrogenation of substrates with a single reducible group, the only effect of enhanced hydrogen pressure is a higher rate of reduction. Acceleration of hydrogen absorption under higher pressure is of course a welcome effect in rhodium-phosphine catalysis, but it is, unfortunately, often accompanied by loss of selectivity. However, it was found that the latter can be partly compensated for, although at the price of lower rates [130], by adding a base, usually triethylamine, to the system. Since a pressure increase may often result even in an inversion of the preferred configuration, the pressure effect is clearly a manifestation of competition between two different mechanisms. Some characteristic data are given in Fig. 2-16.

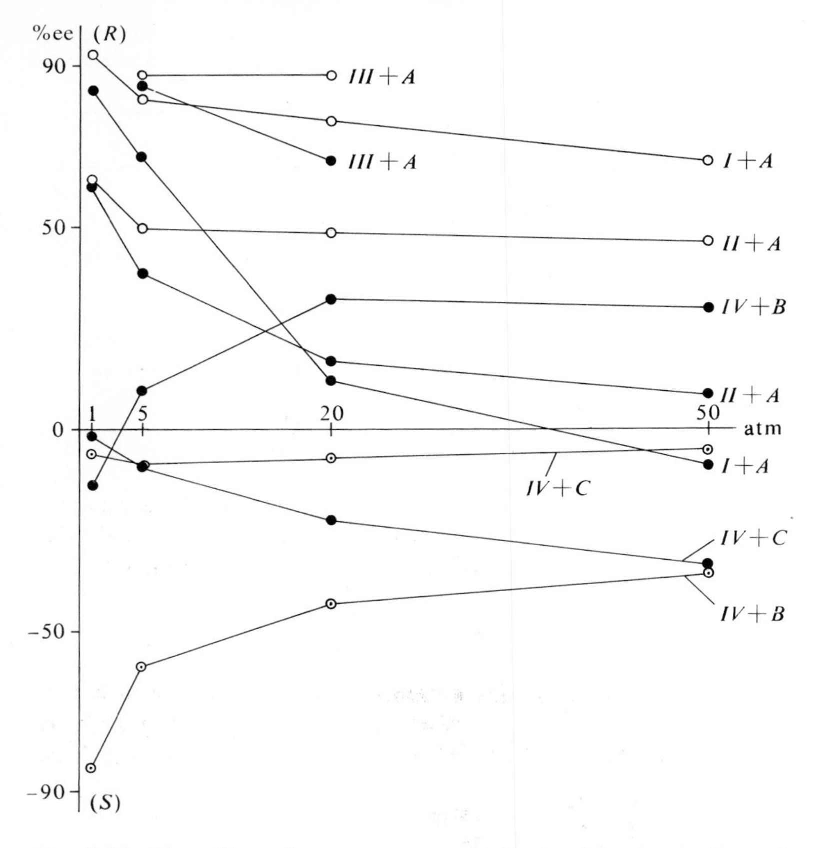

Fig. 2-16. The effect of pressure on enantioselectivity in rhodium-phosphine catalyzed hydrogenations. ● Without base. ○ Et_3N added. ⊙ 1-Phenylethylamine added. Catalysts: *I* = BPPM, *II* = (−)-DIOP, *III* = DIPAMP, *IV* = DIOXOP. Substrates: *A* = (*Z*)-2-benzamidocinnamic acid, *B* = 2-acetamidoacrylic acid, *C* = methyl ester of *B*.

The pressure effect depends slightly on whether the catalyst is ionic or neutral [79]; in Fig. 2-16 data on neutral catalysts are shown. Contrary to expectation, added base has a stabilizing effet even when the substrate is an ester. For an interpretation of the pressure effect, *cf.* Section 2.1.1.5.

Effect of temperature

Very few data are available on this topic, and these show that selectivity often increases on lowering the temperature [90, 100, 131]. Since the latter is usually accompanied by a sharp decrease in rate, this measure is seldom practicable. A study by Sinou [64] revealed that higher temperature may result in an increased selectivity too, *cf.* Fig. 2-17.

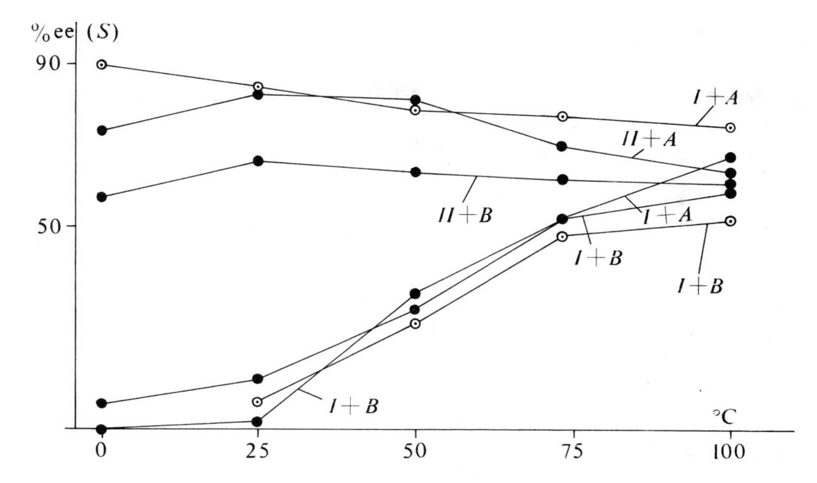

Fig. 2-17. Effect of temperature on enantioselectivity in rhodium-phosphine catalyzed hydrogenation. ● Without base. ⊙ 1-Phenylethylamine added. Catalyst: *I* = DIOXOP, *II* = (+)-DIOP. Substrate: *A* = (*Z*)-2-acetamidocinnamic acid, *B* = methyl ester of *A*.

Increase of selectivity on elevation of the temperature was also reported by Ojima *et al.* [79]. An interpretation of this phenomenon was proposed by Halpern [131] (*cf.* Section 2.1.1.5), but not for the opposite effect. Certainly both originate from a competition of at least two pathways.

Effect of added base

Attenuation of the pressure effect by added base has already been mentioned, and from Fig. 2-16 it is also apparent that this is accompanied by a net increase in enantioselectivity. Triethylamine and other bases obviously act by converting the carboxylic substrates to the anions. How many moles of amine are optimal has again to be determined experimentally. Thus, *e. g.* with 2-acylaminocinnamic and acrylic acids, 2 moles of triethylamine for 1 mole of substrate seems to be best, while with dicarboxylic acids 1 mole suffices for maximum selectivity and 2 moles may even be detrimental [79]. The behavior of 2-benzyloxycarbonylacrylic acid was remarkable inasmuch as enantioselectivity sharply decreased or even inverted in the presence of triethylamine [71]. Another exception was the hydrogenation of 2-acetamido-3-(2-indolyl)-acrylic acid to *N*-acetyltryptophane with a DIOP-based catalyst, in which ee dropped from 86% to 49% when triethylamine was added [133].

Amine addition is, not unexpectedly, ineffectual with esters.

Different catalysts respond to base addition in various ways. While DIOXOP is a poor catalyst in the absence of triethylamine [62], no improvement was observed on addition of the same to *48*.

2.1.1.4 Substrate dependence of enantioselectivity

Although rhodium-phosphine catalyzed hydrogenation is one of the most elegant among methods to prepare chiral compounds in high optical purity, the range of substrates to which it can be successfully applied is quite narrow. This becomes apparent from Tables 2-4 and 2-5 and Figs. 2-18 and 2-19, in which the most important results obtained with substrates, other than those shown already in Table 2-1 are compiled. This survey contains a few simple prochiral olefins, some additional enamines and a series of 2,3-unsaturated carboxylic acids. Ketones will be discussed in Section 2.1.2.

L*	%ee (conf.)	ref.
(−)-DIOP	24.5 (*S*)	23
CPPM	25 (*S*)	21
(*32*)	31 (*R*)	59
(*33*)	18 (*R*)	59
(*47*)	25 (*S*)	21
(*55*)	60 (*R*)	23
(*56*)	33 (*R*)	85

L*	% ee (conf.)	ref.
(−)-DIOP	63 (*S*)	23
(R_P) − (*4*)	27 (*S*)	134
(*5*)	28 (*S*)	16
(*33*)	12 (*S*)	59
(*38*)	36 (*R*)	67
(*55*)	63 (*S*)	23

Fig. 2-18. Enantioselective hydrogenation of olefins with chiral rhodium-phosphine catalysts (I).

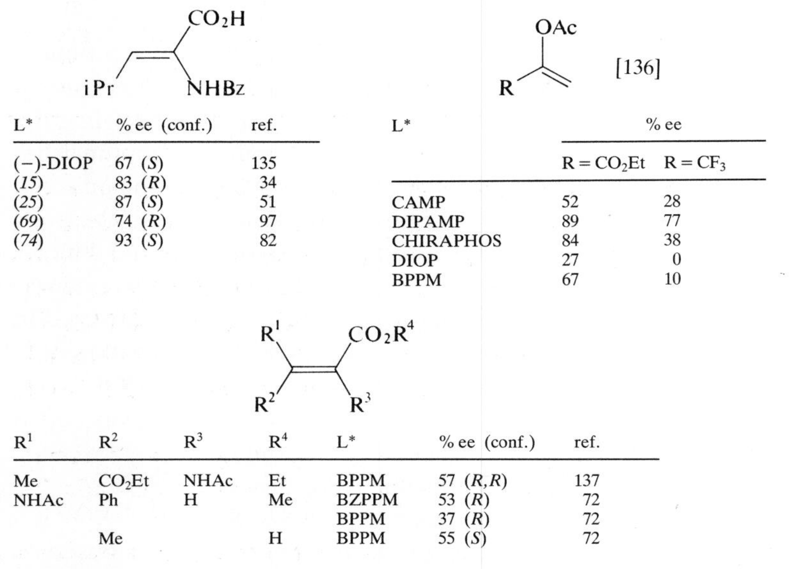

L*	% ee (conf.)	ref.
(−)-DIOP	67 (*S*)	135
(*15*)	83 (*R*)	34
(*25*)	87 (*S*)	51
(*69*)	74 (*R*)	97
(*74*)	93 (*S*)	82

L*	% ee	
	R = CO_2Et	R = CF_3
CAMP	52	28
DIPAMP	89	77
CHIRAPHOS	84	38
DIOP	27	0
BPPM	67	10

R^1	R^2	R^3	R^4	L*	% ee (conf.)	ref.
Me	CO_2Et	NHAc	Et	BPPM	57 (*R*,*R*)	137
NHAc	Ph	H	Me	BZPPM	53 (*R*)	72
				BPPM	37 (*R*)	72
	Me		H	BPPM	55 (*S*)	72

Fig. 2-19. Enantioselective hydrogenation of olefins with rhodium-phosphine catalysts (II).

Table 2-4. Enantioselective hydrogenation of α,β-unsaturated carboxylic acids and esters with chiral rhodium-phosphine catalysts. (For atropic acid see Fig. 2-17.)

$$R^1R^2C{=}C(R^3)CO_2R^4 \xrightarrow[Rh-L^*]{H_2} R^1R^2CH{-}{}^{*}CH(R^3)CO_2R^4$$

R^1	R^2	R^3	R^4	L*	% ee	(Conf.)	ref.
H	H	CH_2CO_2H	H	DIPAMP	38	(*R*)	138
		$CH_2CO_2^{\ominus}$			78	(R)	138
		CH_2CO_2Me	Me		88	(*R*)	138
			H		55	(*R*)	138
		CH_2CO_2H	Me		88	(*R*)	138
		$CH_2CONHBn$	H		90	(*R*)	138
		$(CH_2)_2CO_2^{\ominus}$			10	(*R*)	138
		$(CH_2)_2CO_2Me$	Me		3	(*R*)	138
CO_2H		Me	H	BPPM	33	(*S*)	79
				65	24	(*S*)	67
H	CO_2H	Me		BPPM	65	(*R*)	67
	Ph			DIPAMP	1	(*R*)	30
				DIOXOP	58	(*S*)	62
				$(R_P)-4$	61	(*R*)	134
				5	52	(*S*)	16
				31	15.4		57
				38	45	(*R*)	67
				57	14.3		57
				62	54	(*S*)	67
Me		H		$(R_P)-4$	70	(*S*)	134
H		Ph		*31*	12.2		57
Me	Me	$4-Cl-C_6H_4$	H	*7*	85	(*S*)	139
H	Ph	OAc	Et	DIPAMP	90	(*S*)	30
AcO		H	H	DIPAMP	<10		30

As a group of eminent practical importance not included in either place, substituted 2-acylaminocinnamic acids have to be mentioned. At Monsanto the hydrogenation of the acetoxy-4-methoxyphenyl derivative using the DIPAMP-rhodium complex as catalyst has been developed to an industrial process for the manufacture of (*S*)-3,4-dihydroxyphenylalanine [(*S*)-DOPA] [6, 141]. Enantioselectivity is slightly sensitive to substitution at the aromatic ring of the substrate. This is not due to steric effects but rather to subtle changes in the coordinating ability of the double bond.

Application of enantioselective hydrogenation to a wider range of acylenamines is barred by low yields in the preparation of precursors with an aliphatic C-3 substituent.

Correlation of substrate structure and enantioselectivity has been thoroughly investigated in the laboratories of Glaser, Brown, Halpern, and others. These

Table 2-5. Enantioselective Hydrogenation of Acylenamines with Chiral Rhodium-Phosphine Catalysts.

R^2–CH=C(R^1)–$NHCOR^3$

R^1	R^2	R^3	L*	% ee	(Conf.)	Ref.
CN	Ph	Ph	DIPAMP	89	(*S*)	30
Ph	H	Me	*32*	25	(*R*)	59
			33	5	(*S*)	59
			(−)-DIOP	45	(*R*)	140
	Me		*32*	35	(*R*)	59
			33	35	(*S*)	59
			(−)-DIOP	83	(*R*)	140
		Ph	(−)-DIOP	>73	(*R*)	140
		iPr	(−)-DIOP	85	(*R*)	140
	H	OEt	(−)-DIOP	14	(*S*)	140

studies have permitted the proposal of some cautions generalizations about the correlation of substrate structure and enantioselectivity.

(i) A polar group attached to the double bond promotes selectivity. The effect can be attributed primarily to electron attraction, but sometimes also to coordination with the metal center [130]. Accordingly, hydrocarbons have given poor enantioselectivities with all the rhodium-phosphine catalysts tested so far.

(ii) Carboxylic acids are generally preferred to esters, amides or nitriles, although occasionally esters may be equivalent to (*e. g.* with DIPAMP and *14*) or much better than the acids (*e. g.* with BPPFA [70]). On one occasion inversion of the configuration on changing to the ester was observed [44] (Fig. 2-20 (a)).

(iii) Usually *N*-acetyl- and *N*-benzoylenamines are employed as substrates, and generally ee differences between the two are marginal. There are, however, cases in which one of them (more often the benzoyl compound) performs much better (*cf. 14, 20, 21,* and *37* in Table 2-1). Results on a series of methyl 2-acylaminocinnamates (Fig. 2-20 (a)) obtained with the (+)-DIOP-rhodium complex [142] prompted Glaser and Geresh to come to the conclusion that selectivity drops with increasing bulk of the *N*-acyl group. Once again the validity of the correlation appears to be rather local, as shown by data obtained with another catalyst (*20*) both for the ester and the free acid. Sensitivity to the nature of the *N*-acyl group is not surprising, since it has been shown (*cf.* Fig. 2.2) that the amide carbonyl is coordinated to rhodium in the substrate-catalyst complex.

(iv) Experiments with a series of esters using DIOP [143] and *20* [44] (Fig. 2-20) as ligands demonstrated that in this field generalizations have to be accepted with much caution. With *20* as ligand, the methyl ester was a very poor substrate, while bulky esters were even better than the free acid. In turn, DIOP was rather indifferent to the nature of the alkyl group.

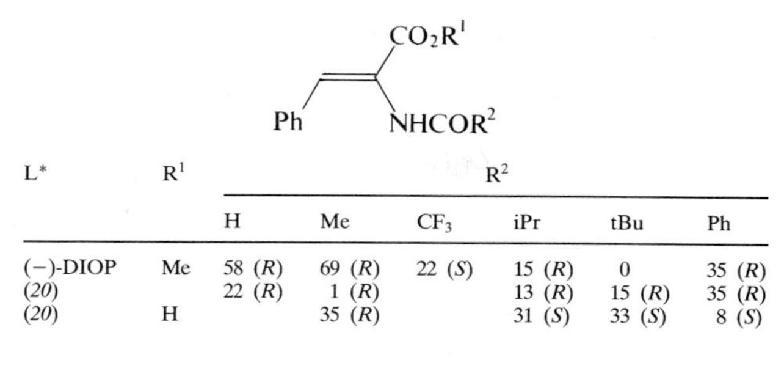

L*	R^1	R^2					
		H	Me	CF_3	iPr	tBu	Ph
(−)-DIOP	Me	58 (*R*)	69 (*R*)	22 (*S*)	15 (*R*)	0	35 (*R*)
(*20*)		22 (*R*)	1 (*R*)		13 (*R*)	15 (*R*)	35 (*R*)
(*20*)	H		35 (*R*)		31 (*S*)	33 (*S*)	8 (*S*)

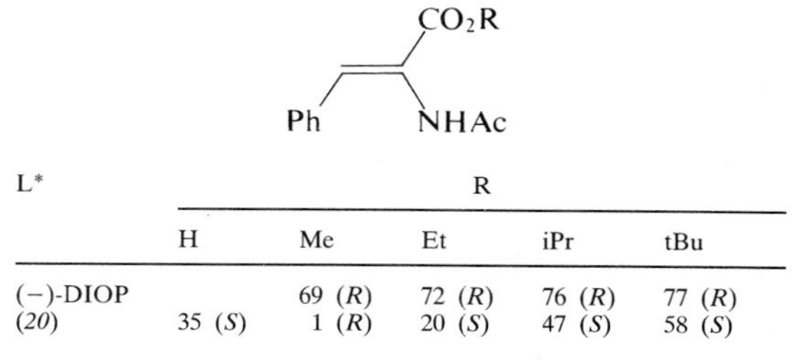

L*	R				
	H	Me	Et	iPr	tBu
(−)-DIOP		69 (*R*)	72 (*R*)	76 (*R*)	77 (*R*)
(*20*)	35 (*S*)	1 (*R*)	20 (*S*)	47 (*S*)	58 (*S*)

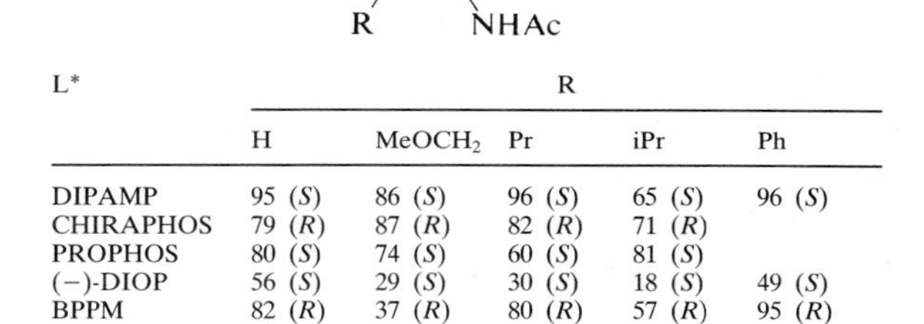

L*	R				
	H	$MeOCH_2$	Pr	iPr	Ph
DIPAMP	95 (*S*)	86 (*S*)	96 (*S*)	65 (*S*)	96 (*S*)
CHIRAPHOS	79 (*R*)	87 (*R*)	82 (*R*)	71 (*R*)	
PROPHOS	80 (*S*)	74 (*S*)	60 (*S*)	81 (*S*)	
(−)-DIOP	56 (*S*)	29 (*S*)	30 (*S*)	18 (*S*)	49 (*S*)
BPPM	82 (*R*)	37 (*R*)	80 (*R*)	57 (*R*)	95 (*R*)

Fig. 2-20. Enantioselective hydrogenation of olefins with chiral rhodium-phosphine catalysts (III). Data refer to %ee values, configuration of the product in excess in parentheses.

Although it is believed that the ester group is not bound to the metal in the substrate-catalyst complex, significant double induction was observed in the hydrogenation of (*Z*)-(−)-methyl-2-acetamidocinnamate: (+)-DIOP afforded 76.8% ee, while (−)-DIOP only 52.5% [144].

(v) Scott *et al.* subjected a series of methyl (*Z*)-3-substituted-2-acetamidoacrylates to hydrogenation with four efficient catalysts. Not logical pattern emerged from their results (Fig. 2-20 (c)) [145], except for a vague indication that ligands forming five-membered chelates are less susceptible to substrate variation than those giving seven-membered chelates.

(vi) Although of the stereoisomeric 2-acylamino-3-arylacrylic acids only the *Z* form is readily accessible and with the 3-alkyl analogues both diastereomers are difficult to prepare pure, in order to elucidate substrate-catalyst interactions it was of interest to compare the behavior of stereoisomeric (*E* and *Z*) substrates. It was observed with DIPAMP that in alcohols (*Z*)-2-acylaminocinnamic acids and esters were hydrogenated both much faster (by a factor of 16–100) and more selectively than the corresponding (*E*)-acids [30] (*e. g.* 94 *vs.* 47% ee for benzoylamino acid). One of the reasons why the *E* isomers are inferior to *Z* ones may be their different mode of complexation. Thus (*E*)-2-benzamidocinnamic acid was shown to bind with the carboxy CO to rhodium, while with the *Z* isomers the amide CO was coordinated [146]. Since it was recognized that rhodium-phosphine catalysts also promote $E \rightarrow Z$ isomerization of the substrate, the real significance of these data remained obscure until the isomerization process could be monitored. This was accomplished in an elegant way by Koenig and Knowles [147] who reduced *E* and *Z* substrates with dideuterium, thereby creating a second chiral center at C-3. Thus different diastereomers arose from the *E* and *Z* isomers, and these could be identified by NMR (Fig. 2-21).

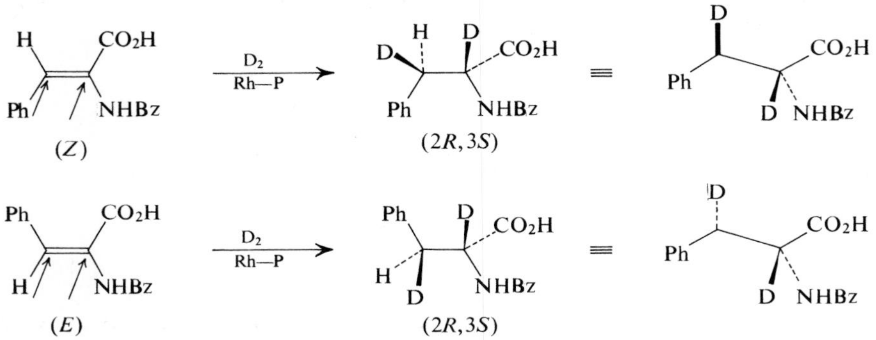

Fig. 2-21. Identification of substrate configuration (*E* or *Z*) by the addition of dideuterium.

These studies revealed the following interesting facts: (a) Stereoselectivity for (*E*)-2-benzamidocinnamic acid, when corrected for isomerization, was even less (30%). (b) No isomerization took place in benzene in the presence of any of the four catalysts examined. (c) In benzene, enantioselectivities for the *E* isomer were only marginally lower than for the *Z* isomer.

A similar study by Kagan *et al.* [148] with DIOP and 2-benzamidocinnamic acids in ethanol-benzene 3:1 also demonstrated $E \rightarrow Z$ isomerization. In this case the (*Z*)-acid gave the *S* product in 70% ee, while the corrected value for the (*E*)-acid was 20% ee for the *R* product.

$E-Z$ isomers of models shown in Fig. 2-20 (c), *i. e.* those with an aliphatic group at C-3, show no consistent behavior. Depending on the catalyst, either the *E* or the *Z* isomer gives higher ee values, but the prevailing configuration is the same for both [145].

High enantioselectivities achieved with simple amino acid precursors stimulated the adaptation of rhodium-chiral phosphine catalyzed hydrogenation to di- and tripeptide precursors. Most of these are in fact diastereoselective transformations and thus formally do not belong under the heading of the present chapter, but, since the same catalyst systems are used, it would be impractical to treat them separately.

The first example in Fig. 2-22 is a true enantioselective hydrogenation, whereas the second is in fact a diastereoselective one, since it is unlikely that the saturation of the two centers should be simultaneous. As compared with an acetyl group, the *N*-acetylglycine moiety impaired the selectivity of both DIOP and BPPM substantially [149]. DIOP as ligand proved to be rather unselective with *97* (10% de) [150].

The influence of the existing chiral center in compounds of type *98* has been thoroughly investigated, and the results are summarized in Table 2-6. It can be seen that asymmetric induction plays a secondary role; it modifies diastereoselectivity but does not control the prevailing configuration at the new chiral center. For the hydrogenation of dipeptide precursors, DIOXOP is the ligand of choice since it gives products with the natural configuration. Unfortunately, the most efficient ligands BPPM and *48* give the unnatural configuration. The snthesis of their antipodes is known [75] but extremely laborious.

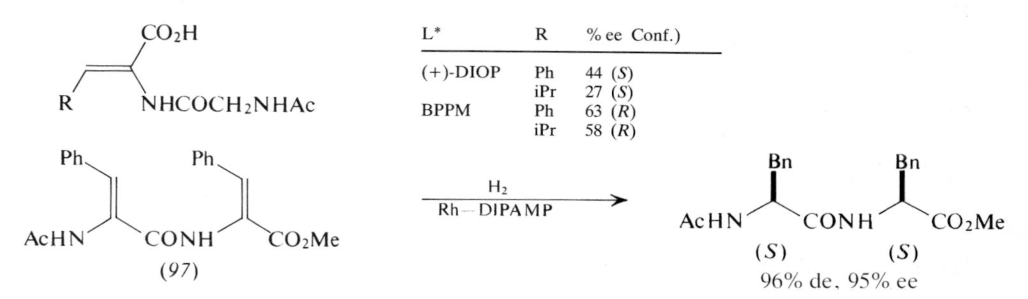

Fig. 2-22. Rhodium-phosphine catalyzed hydrogenation of peptide precursors.

Diastereoselective hydrogenation, of course, does not require that the ligand should be chiral. Results obtained with the achiral and acyclic analogues *99* and *100* of the eight-membered chelate-ring-forming ligand DIOXOP give an impression about the magnitude of asymmetric induction alone [152].

In a study of the hydrogenation of tripeptide precursors DIPAMP, *51, 54* and *65* proved to be the best ligands; selectivities in excess of 90% de in favor of the *S* isomer at the new center were achieved [155, 158].

Finally, some examples for diastereoselective hydrogenations with the achiral catalyst $[(Rh(NBD)(Ph_2P(CH_2)_4PPh_2)]^+BF_4^-$ should be mentioned: 3-methylenecyclohexanol [156], 3-methylcyclohex-2-enol and 4-methylcyclohex-3-enol [157] all gave more than 98% of the *trans* product.

In summary, the effects of substrate and catalyst structure, as well as of experimental parameters are rather complex and, as will be seen in the next section, only part of them can be rationalized by the mechanism of the reaction.

Table 2-6. Diastereoselective Hydrogenation of Dehydro-Dipeptides with Chiral Rhodium-Phosphine Catalysts.

Ph–CH=C(NHCOR^2)–CONH–$\overset{*}{C}$H(R)–CO_2R^3 (98)

Ph_2P–$(CH_2)_2$–X–$(CH_2)_2$–PPh_2 (99) X = CH_2; (100) X = O

R[1]	(Conf.)	R[2]	R[3]	Ligand	% ee	(Conf.)	Ref.
Ph	(*S*)	Me	Me	*49*	96	(*R*)	151
Me	(*S*)			*49*	98	(*R*)	151
	(*S*)		H	DIOXOP	72	(*S*)	152
	(*R*)			DIOXOP	60	(*S*)	152
iPr	(*S*)			DIOXOP	80	(*S*)	152
Bn	(*S*)			DIOXOP	86	(*S*)	152
	(*R*)			DIOXOP	20	(*S*)	152
	(*S*)			*99*	34	(*S*)	152
	(*S*)			*100*	20	(*R*)	152
	(*S*)			(+)-DIOP	90	(*S*)	153
	(*R*)			(+)-DIOP	60[a]	(*S*)	153
	(*S*)			BPPM	90	(*R*)	153
				DIPAMP	>90	(*S*)	153
Me	(*S*)			(+)-DIOP	82	(*S*)	154
	(*R*)			(+)-DIOP	76	(*S*)	154
iPr	(*S*)			(+)-DIOP	50	(*S*)	154
	(*R*)			(+)-DIOP	90	(*S*)	154
Bn	(*S*)	Ph	Me	BPPM	97.4	(*R*)	20
	(*S*)			*49*	98.4	(*R*)	20
Me	(*S*)	Me	H	*65*[b]	94	(*S*)	95
	(*R*)			65	86	(*S*)	95
Ph	(*S*)			65	98	(*S*)	95
			Me	65	0		

[a] Calculated from the result obtained with (−)-DIOP.
[b] The 4−$MeOC_6H_4$ analogue.

2.1.1.5 Mechanistic considerations

It has often been stated that, as regards efficiency, rhodium-chiral phosphine catalysts emulate enzymes. Their capricious response to variations of substrate structure and reaction conditions is also reminiscent of enzymes. No wonder that the elucidation of their mechanism is a formidable task, and after the publication of well over 30 research papers devoted partly or completely to this subject the problem has not yet been solved satisfactorily (for reviews see refs. 120, 132).

Mechanistic studies have been carried out with almost all of the more important catalysts, such as DIPAMP [30, 135, 138, 159], CHIRAPHOS [160–162], DIOP [135, 162, 163], DIOXOP [64, 173], and BPPM [164–166].

Before the origin of stereoselection could be unraveled, some fundamental knowledge about hydrogenation with rhodium-diphosphine ligands had to be accumulated. For this purpose Halpern and his coworkers studied the hydrogenation of methyl 2-acetamidocinnamate (MAC) catalyzed by the rhodium complex of an achiral ligand, 1,2-bis(diphenylphosphino)ethane (DIPHOS) [132, 161, 167, 168]. Their findings are summarized in Fig. 2-23. The ionic diene-adduct catalyst precursor (*CP*) enters the catalytic cycle by taking up two moles of H_2. On saturation the diene loses its affinity to the metal and is replaced by solvent molecules. The solvated catalyst is an unstable species whose structure was derived mainly from ^{31}P-NMR data. When isolated, it transforms to a binuclear complex containing no solvent [167]. On addition of the substrate a complex (*C–Su*) is formed at high rate in which the substrate is coordinated by its double bond and amide group. The structure of *C–Su* has been established both by NMR and X-ray crystallography [169], and this coordination pattern proved to be quite general for the 2-acylaminoacrylic acid and -cinnamic acid and ester complexes of other catalysts, too [55, 120, 159, 161, 162, 166, 170, 171], but is not necessarily valid for the corresponding anions [79, 172]. By the way, coordination of the carboxyl group would involve a four-membered chelate ring, and this is unfavorable.

At room temperature, oxidative addition of H_2 to *C–Su* giving the species *C–Su*–H_2 was the rate-determining step, and therefore there was no chance of detecting any of the downstream intermediates at this temperature. Due to its higher activation enthalpy, the product-forming step becomes rate-limiting at low temperature and, in fact, at −75°C the hydridoalkyl complex *C–Su*H–H accumulated and could be characterized by NMR [168].

It should be borne in mind that the scheme shown in Figure 2-23 leads exclusively to an *R* product, but, since this is only one of the two possible schemes related as mirror images, the actual product is racemic.

Formation of the hydrido-alkyl complex *C–Su*H–H is irreversible, as demonstrated by deuteration experiments with BPPM as catalyst [79].

After the above results the stage was set for studying the hydrogenation of the same substrate with a chiral rhodium-diphosphine catalyst. The situation is less complex for chiral ligands with C_2 symmetry, such as CHIRAPHOS or DIPAMP, and is shown in Fig. 2-24. Now, because they are no longer mirror images, both possible modes of coordination have to be constructed, giving two unequally populated sequences (*R* and *S*).

Experiments with rhodium-(*S,S*)-CHIRAPHOS, which hydrogenates ethyl (*Z*)-2-acetamidocinnamate (EAC) with over 95% ee to the (*R*)-amino acid, furnished an astonishing result, namely, that the predominant catalyst-substrate complex (>95%) was the one which was supposed to lead to the *S* product, *i. e.* $C^*{-}Su_S$, [159, 161]. Analogous results were obtained with the Rh-(DIPAMP)-MAC system,

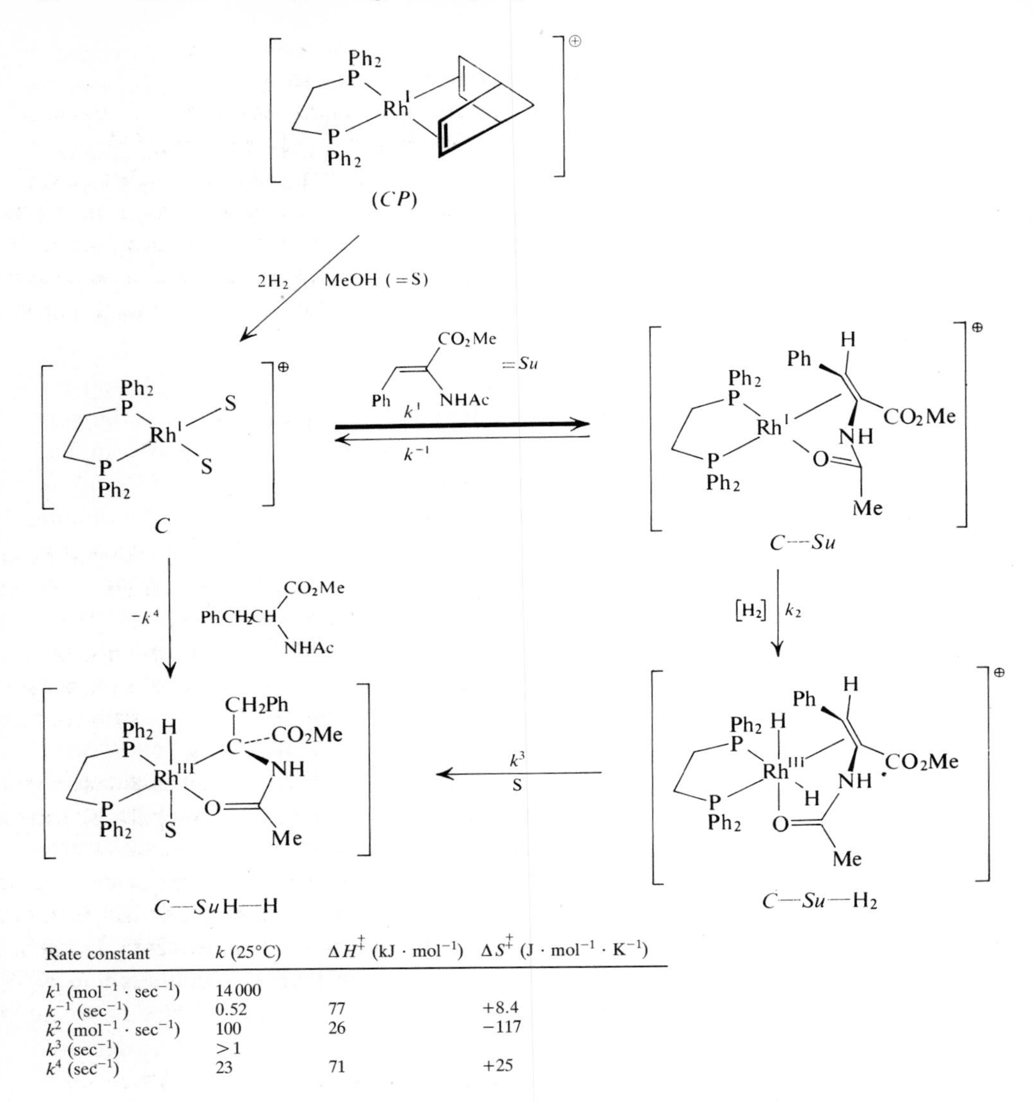

Rate constant	k (25°C)	$\Delta H^{\ddagger}$ (kJ · mol^{-1})	$\Delta S^{\ddagger}$ (J · mol^{-1} · K^{-1})
k^1 (mol^{-1} · sec^{-1})	14000		
k^{-1} (sec^{-1})	0.52	77	+8.4
k^2 (mol^{-1} · sec^{-1})	100	26	−117
k^3 (sec^{-1})	>1		
k^4 (sec^{-1})	23	71	+25

Fig. 2-23. Kinetics of hydrogenation with a rhodium-phosphine catalyst containing an achiral diphosphine.

where the minor diastereomer could be detected also [132, 162]. Two conclusions may be drawn from these findings, namely, that either the proposed mechanism is wrong, or that the thermodynamically favored substrate-catalyst complex (C^*-Su_S) is kinetically disfavored and the main pathway proceeds *via* a hidden intermediate *i. e.* C^*-Su_R. Halpern opted for the second alternative, which could be supported by the following arguments.

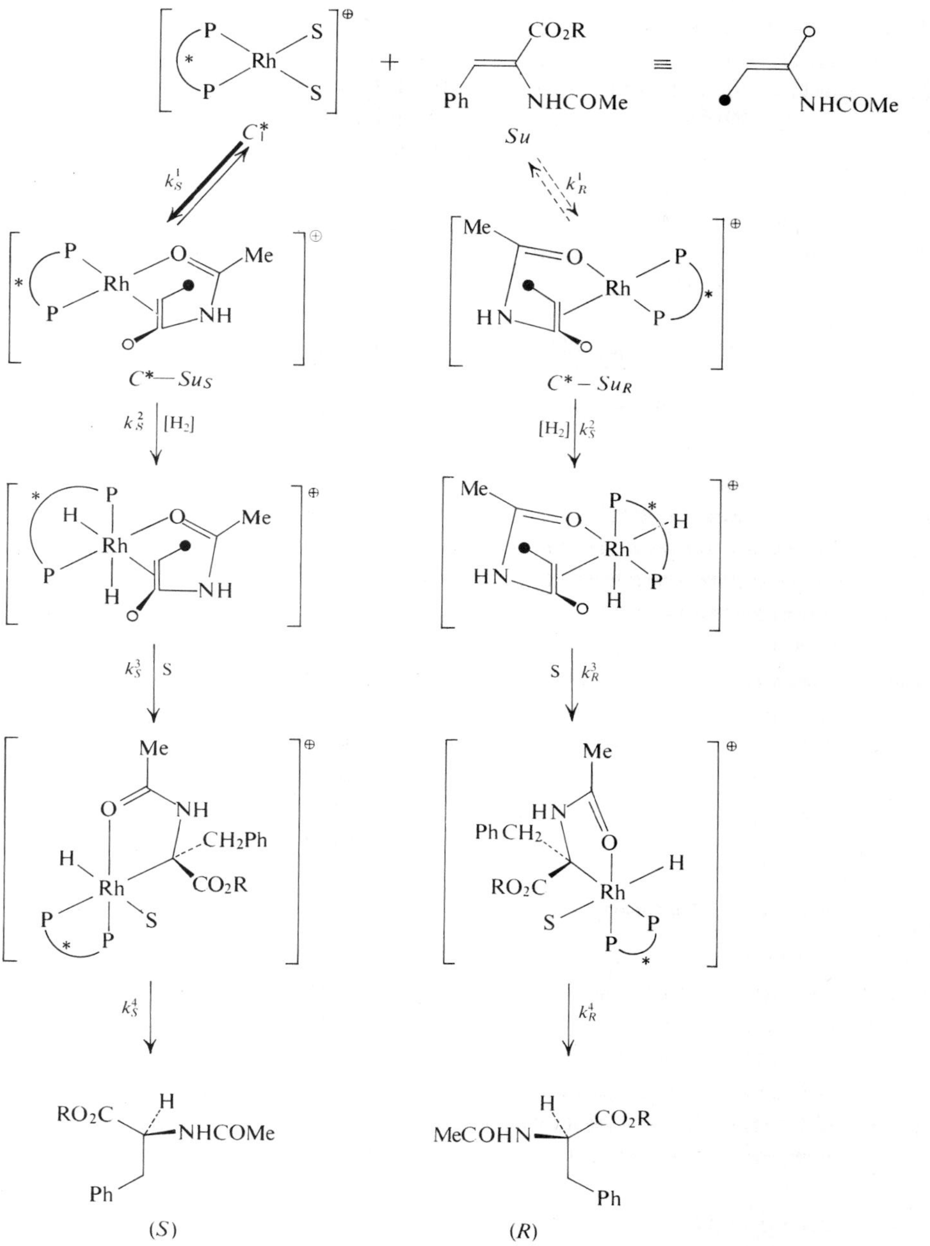

Fig. 2-24. Hydrogenation of an alkyl (*Z*)-2-acetamidocinnamate with a catalyst containing a chiral diphosphine of C_2 symmetry (P—*—P) (S = solvent).

It has been mentioned that selectivity decreased with increasing hydrogen pressure. Since, being a bimolecular process, oxidative addition of hydrogen depends on H_2 concentration and on increasing the pressure, a point may be reached where the rate at which the complex is formed becomes comparable to the rate of its oxidation. Then k_S^1/k_R^1 would gain control of the process, and this places the *S* pathway at an advantage.

Interconversion of diastereomeric catalyst-substrate complexes has a much higher activation enthalpy than hydrogen uptake (see above). This means that on lowering the temperature the rate of interconversion declines much more rapidly than that of hydrogen uptake. This fact again favors the minor pathway (*S*). An increase in selectivity on elevation of the temperature has been observed [79, 138]; on one occasion it increased from 0% at 0°C to 60% ee at 100°C [64]. The foregoing rationalization is a striking demonstration of the Curtin-Hammett principle (*cf.* Section 1.5), but unfortunately no explanation based on structural features can be offered as yet for the discrepancy between stability and reactivity of the catalyst-substrate complexes.

Halpern's scheme is in accordance with many of the phenomena encountered in rhodium-phosphine catalyzed hydrogenations, but does not apply to any substrate on any catalyst, and his views about the background of the pressure effect are not shared by certain authors. Based on a careful study of this effect involving several catalysts, substrates, and experiments when triethylamine was added to the system, Ojima and his coworkers concluded that two competing pathways shown below are operative [79].

olefin ⇌ Rh(olefin) ⇌(H_2) Rh(olefin)(H_2) → (Rh) product (A)

H_2 ⇌ Rh(H_2) ⇌(olefin) Rh(H_2)(olefin) → (Rh) product (B)

Another reason for a change in selectivity could have been that due to the acceleration of the $C^*{-}Su \rightarrow C^*{-}Su{-}H_2$ transformation, it is the next step which becomes rate-limiting, and, if this is reversible, equilibration with a diastereomer of $C^*{-}Su{-}H_2$ leading to an enantiomeric product may take place. This possibility was excluded by showing that using D_2 only a 2,3-dideuterio product was formed from 2-acetamidoacrylic acid, the substrate which showed the highest pressure-effect. It should be noted that the increase in selectivity on elevating the temperature may also be used as an argument in support of the dual mechanism [64].

As evidence for the emergence of pathway B at high pressure, it was mentioned that this was the established mechanism of olefin hydrogenation catalyzed by $(PhP)_3RhCl$. Although the crucial intermediate in the high pressure pathway ($C^*{-}H_2{-}Su$) could not be captured*, support for Ojima's hypothesis was invoked

* A DIOXOP–Rh–H_2 complex of type $C^*{-}H_2$ was detected by Brown *et al.* [173].

from the base-effect (*cf.* Section 2.1.1.3), a phenomenon which Halpern's theory failed to explain. It was argued that the anion generated by the addition of triethylamine has a much greater affinity to rhodium than the corresponding acid, and thus diversion to route B is discouraged. No experimental proof for stronger binding of the anions to rhodium has, however, been presented.

In order to determine in which step chiral recognition took place, hydrogenation of the dicarboxylic acids *101*–*103*, all giving 2-methylsuccinic acid (Fig. 2-25) with an ionic rhodium-BPPM catalyst, was examined. When, according to the mechanism in Figs. 2-23 and 2-24, hydrogen is first transferred to C-3, and stereoselectivity is controlled by the structure of *C**–*Su*H–H or in subsequent steps, then the outcome of the reaction should be indifferent to the configuration of the substrate. In other words *101, 102* and *103* should all give the same product composition. Since this is not the case, chiral recognition must take place in the step of olefin complexation. The three dicarboxylic acids also respond differently to triethylamine addition. While 0.1 equivalent of the base enhances selectivity with all three, 2.0 equivalents are beneficial only with itaconic acid (*101*). It was assumed that with *101* the CH_2CO_2-group was preferentially complexed even in the dianion, whereas with *102* and *103* competition between the two carboxylate groups for complexation impaired selectivity.

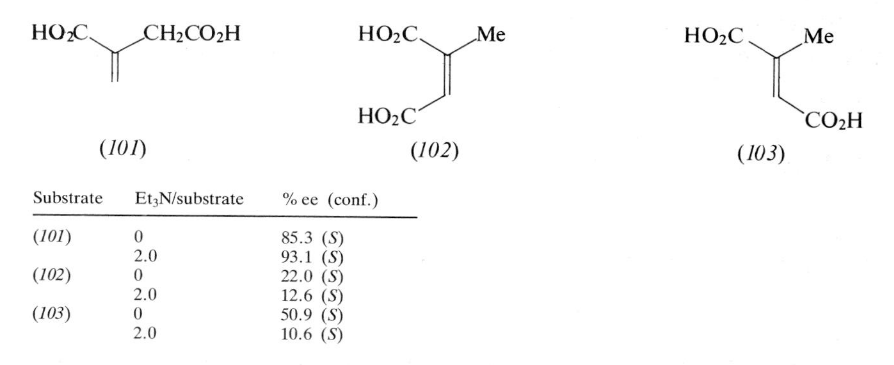

Substrate	Et_3N/substrate	% ee (conf.)
(*101*)	0	85.3 (*S*)
	2.0	93.1 (*S*)
(*102*)	0	22.0 (*S*)
	2.0	12.6 (*S*)
(*103*)	0	50.9 (*S*)
	2.0	10.6 (*S*)

Fig. 2-25. Effect of triethylamine on enantioselectivity in rhodium-BPPM catalyzed hydrogenation.

2.1.1.6 Hydrogenation of olefinic bonds with soluble catalysts containing metals other than rhodium

It should be pointed out right at the beginning that results with chiral complexes of other metals have not been as good by far as those achieved with rhodium catalysts.

Ruthenium*

With olefinic substrates, ruthenium-chiral phosphine complexes are generally inferior to their rhodium analogues. Using as catalyst precursors HRuCl[(+)-DIOP]$_2$, $RuCl_2$[(+)-DIOP]$_2$, and $RuCl_4$[(+)-DIOP]$_3$, selectivities up to 40% ee (*R*) were achieved with α,β-unsaturated carboxylic acids and 59% ee (*S*) with 2-acetamidoacrylic acid [174, 175]. The cluster complexes $H_4Ru_4(CO)_8$[(−)-DIOP]$_2$ and $Ru_6(CO)_{18}$[(−)-DIOP]$_3$ were also tested [176]. The latter gave slightly better results, up to 61% ee (*S*) with α,β-unsaturated carboxylic acids [176]. Selectivity of cluster catalysts depends on the Ru/DIOP ratio. Thus, from 2-acetamidocinnamic acid a 2:1 ratio gave rise to an excess of the *R* product (19% ee), while with a 1:1 ratio the *S* product (40% ee) was obtained [177]. Enantioselectivities produced with the phosphinite ligand *62* were in the same range [178].

A special feature of ruthenium is that it can form chiral catalysts with sulfoxides racemic at the sulfur center but which are linked to non-racemic chiral groups. Again, enantioselectivities are well under the limit of practical applicability (≤22% ee) [179].

Ruthenium phosphine complexes catalyze hydrogen transfer reactions between a secondary alcohol as donor, and a ketone or olefin as acceptor. When the phosphine ligand is chiral, hydrogen transfer may be enantioselective in several ways: (i) kinetic resolution of a chiral racemic donor alcohol; (ii) enantioselective reduction of a prochiral ketone or olefin; (iii) combination of processes (i) and (ii); (iv) use of a chiral donor (double induction). Transformations proceed at high temperatures, and selectivities have been very poor [180]. The best result obtained so far is shown below [181].

CO_2H / Me, Me (*104*) + $PhCH_2OH$ —[Ru_2Cl_4[(−)-DIOP]$_3$, 180 °C, 6h]→ Me / Et, CO_2H 16% ee

With 1,2-isopropylidene-α-D-glucofuranose as a chiral donor and Ru_2Cl_4[(−)-DIOP]$_3$ as the catalyst, 104 was reduced giving 22% ee of the *R* product [182]. With the achiral catalyst $RuCl_2(PPh_3)_3$ this dropped to 9% [182], but with 2,5,5-trimethylcyclohexenone the same system gave rise to the *S* product in 34% ee [289].

Cobalt

Cobalamin (*Co**), the Co(I)-containing porphirin complex contained in vitamin B_{12}, has been shown to be capable of mediating enantioselective hydrogenations [183, 184]. An example is as follows:

* For a review about asymmetric homogenous catalysis by ruthenium complexes *cf.* ref. [180].

Me
Ph(CH2)2 COMe
Co*, H2O--AcOH
0°C
[Ph(CH2)2 CH2COMe Me Co*]
Zn
Me
Ph(CH2)2 COMe 33% ee

Reductive cleavage of the alkyl-cobalt intermediate has been shown to proceed with 97% retention; thus stereoselection is controlled in the coordination step.

Addition of a chiral amine to a bis(dimethylglyoximato)Co(II) complex gives rise to a chiral complex capable of enantioselective catalysis [185]. Up to 34% ee could be obtained with such systems [186].

Titanium

Cyclopentadienyl complexes of titanium catalyze the hydrogenation of olefins. Bis[η^5-(−)-menthylcyclopentadienyl]-titanium dichloride is a unique catalyst in two ways: (i) It contains as a chiral ligand a hydrocarbon, and (ii) in contrast to other catalysts it only transfers hydrogen to hydrocarbons [15% ee (*S*) with 2-ethylstyrene] [187]. Interestingly, when one of the ligands is cyclopentadiene, the prevailing configuration inverts [10% ee (*R*)] and, finally, reduction of this complex with $LiAl(OtBu)_3H$ leads to the best catalyst in this series giving 23% ee (*S*) [188].

2.1.2 Enantioselective Catalytic Hydrogenation of Ketones

In this section reductions with hydrogen are described. Catalytic hydrosilylation, which leads ultimately to secondary alcohols as well, will be described in Section 2.2.

Results with ketones are not as spectacular as with amino acid precursors, and the method is not yet competitive with enantioselective reductions by chiral hydride reagents. The most important catalysts are again rhodium-chiral phosphine complexes, and an overview of the results obtained with a series of simple ketones is given in Table 2-7. The ligands have all been described before (*cf.* Figs. 2-1 to 2-3), except the chiral Schiff-base *105*.

As with the hydrogenation of olefins, the best results were obtained with substrates containing two groups capable of coordination with the catalyst. Thus, the method was most efficient when applied to the preparation of pantolactone

Table 2-7. Reduction of Prochiral Ketones (R^1COR^2) to Chiral Secondary Alcohols by Catalytic Hydrogenation with Chiral Rhodium-Phosphine Catalysts.

R^1	R^2	% ee	(Conf.)	Ligand	Ref.
Me	Et	12	(*R*)	CAMP/iPrCO_2H	189
	iPr	60	(*R*)	*45*/C_6H_6	19
	tBu	39	(*R*)	*45*/C_6H_6	19
	Hex	48	(*R*)	*45*/C_6H_6	19
	cHex	7.8	(*R*)	*83*	190
	1–Naph	84		(+)-DIOP + Et_3N	191
	2–Naph	69		(+)-DIOP + Et_3N	189
	CO_2Pr	76	(*R*)	BPPM/TMF	189
	Bn	20	(*R*)	CAMP/iPrCO_2H	126
	$CHPh_2$	32	(*R*)	CAMP/iPrCO_2H	126
Ph	Me	80	(*R*)	(+)-DIOP + Et_3N	126
	Et	20	(*R*)	*105* + KOH	192
	iPr	19	(*R*)	*105* + KOH	192
	tBu	9	(*R*)	*105* + KOH	192

(*106*). With a neutral complex of BPPM in benzene the (*R*)-lactone was obtained in 87% ee [125, 193], and in a patent [22] 98.9% ee was claimed when the ligand was the *N*-phenyl analogue *50*. It should be noted that the parent pyrrolidine (*107*) gave the (*S*)-lactone in 15% ee [124]. *106* is a key intermediate in the industrial synthesis of pantothenic acid.

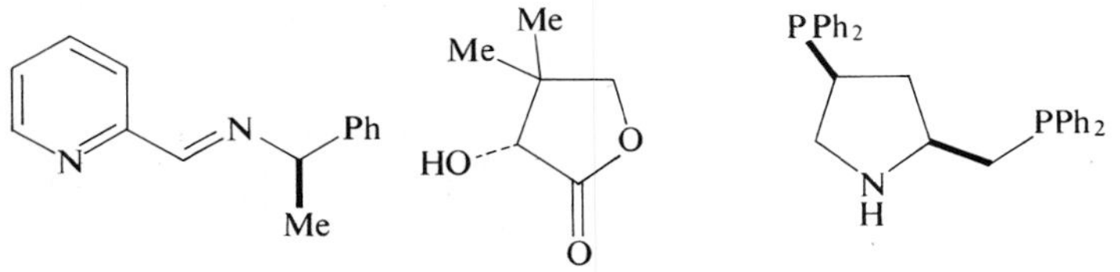

(*105*) (*R*)-(*106*) (*107*)

Selectivities in the hydrogenation of α-aminoketones with the ferrocenyl alcohol *40* as ligand were also high (*cf.* data in Fig. 2-26 given for the ionic catalyst) [194]. The products are important intermediates in the synthesis of adrenergic drugs.

Several other ligands were tested for ketone hydrogenations, but they all proved to be less selective than the ones listed in Table 2-7 [17, 18, 195–197].

As with enantioselective olefin hydrogenation, not much progress has been made with catalysts containing metals other than rhodium. Thus, with Ru_2Cl_2-[(–)-DIOP]$_3$ as catalyst, the (*S*)-carbinol was obtained from acetophenone in but 34% ee [198], and the iridium complex of *105* was even less selective than its rhodium analogue [199, 200].

One of the very rare examples of selection between enantiotopic groups in a non-enzymatic process has already been shown in Fig. 1-25 (c) [201].

R^1	R^2	R^3	% ee
OMe	OMe	H	92
H	H	H	60
H	OH	H	69
OH	OH	Me	95

Fig. 2-26. Enantioselective hydrogenation of α-aminoketones.

On addition of a chiral amine (B*) to the planar bis(dimethylglyoximato)cobalt(II) (*108*) a chiral catalyst is obtained (*cf.* Fig. 2-27). Several alkaloids were employed as cocatalyst, of which quinine was the most effective [202–205].

Although outside the scope of this section, it should be mentioned here that acetophenone benzylimine could be reduced to the corresponding (*S*)-amine with rhodium catalysts prepared with the diphosphines *26–28* (64–72% ee). Both DIOP and DIPAMP were quite unselective [206], as were ruthenium catalysts [207].

R	Ph	Me	OEt	OiPr
% ee	62	56	19	11

Fig. 2-27. Enantioselective ketone hydrogenation with a chiral cobalt catalyst.

2.2 Catalytic Hydrosilylation*

Hydrosilylation is the addition of a silicon hydride accross a C=C, C=O or C=N bond (Fig. 2-28). The *O*- and *N*-silylated products derived from ketones and imines respectively can be readily hydrolyzed, and therefore the final result is equivalent to hydrogenation. The reaction is catalyzed by noble metal complexes, for olefin hydrosilylation chloroplatinic acid is most commonly used [211, 212]. It was discovered by Ojima in 1972 that $Rh(Ph_3P)_3Cl$ was an efficient catalyst for ketone hydrosilylation [213, 214]. The use of chiral phosphines as ligands, resulting in enantioselective hydrosilylation was first reported in 1971 [215].

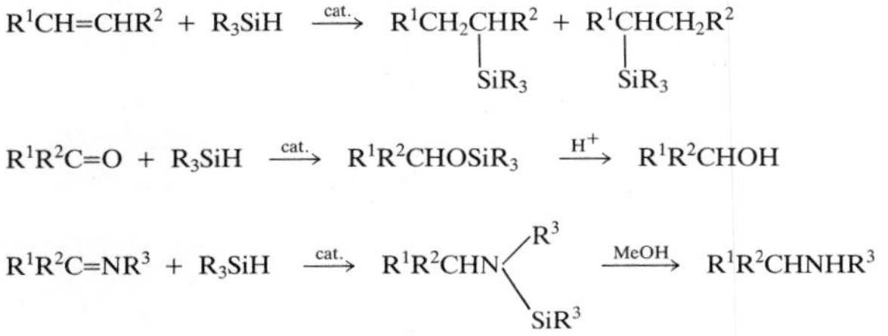

$$R^1CH{=}CHR^2 + R_3SiH \xrightarrow{cat.} R^1CH_2CH(SiR_3)R^2 + R^1CH(SiR_3)CH_2R^2$$

$$R^1R^2C{=}O + R_3SiH \xrightarrow{cat.} R^1R^2CHOSiR_3 \xrightarrow{H^+} R^1R^2CHOH$$

$$R^1R^2C{=}NR^3 + R_3SiH \xrightarrow{cat.} R^1R^2CHN(R^3)(SiR^3) \xrightarrow{MeOH} R^1R^2CHNHR^3$$

Fig. 2-28. Hydrosilylation of olefins, ketones and azomethines.

Owing to poor enantioselectivity (generally <10% ee), the hydrosilylation of olefins has not yet acquired much practical significance. The results shown in Fig. 2-29 are the best so far obtained [208, 216].

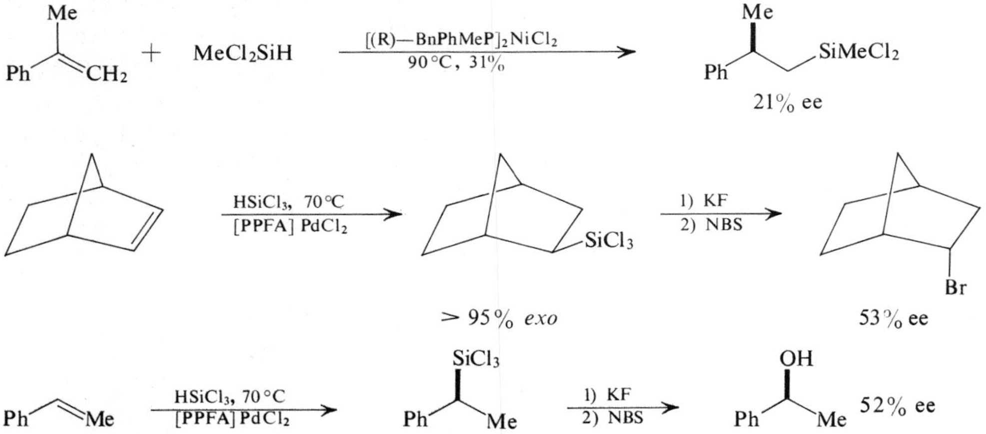

$$Ph(Me)C{=}CH_2 + MeCl_2SiH \xrightarrow[90\,°C,\ 31\%]{[(R){-}BnPhMeP]_2NiCl_2} PhCH(Me)CH_2SiMeCl_2 \quad 21\%\ ee$$

$$\text{norbornene} \xrightarrow[[PPFA]\,PdCl_2]{HSiCl_3,\ 70\,°C} \text{2-norbornyl-}SiCl_3\ (>95\%\ exo) \xrightarrow[2)\ NBS]{1)\ KF} \text{2-norbornyl-Br} \quad 53\%\ ee$$

$$PhCH{=}CHMe \xrightarrow[[PPFA]\,PdCl_2]{HSiCl_3,\ 70\,°C} PhCH(SiCl_3)Me \xrightarrow[2)\ NBS]{1)\ KF} PhCH(OH)Me \quad 52\%\ ee$$

Fig. 2-29. Enantioselective hydrosilylation of olefins.

* For reviews, see refs. [107, 208–210].

The most thoroughly studied aspect of stereoselective hydrosilylation is that of ketones catalyzed by rhodium-chiral phosphine complexes. Not counting the solvent, three components, *i. e.* the substrate, the silane and the catalyst participate in the reaction, and the structures of all three are important. Three vectors of this three dimensional matrix are shown in Fig. 2-30 to illustrate the complexity of the situation. (Data taken from ref. [208].)

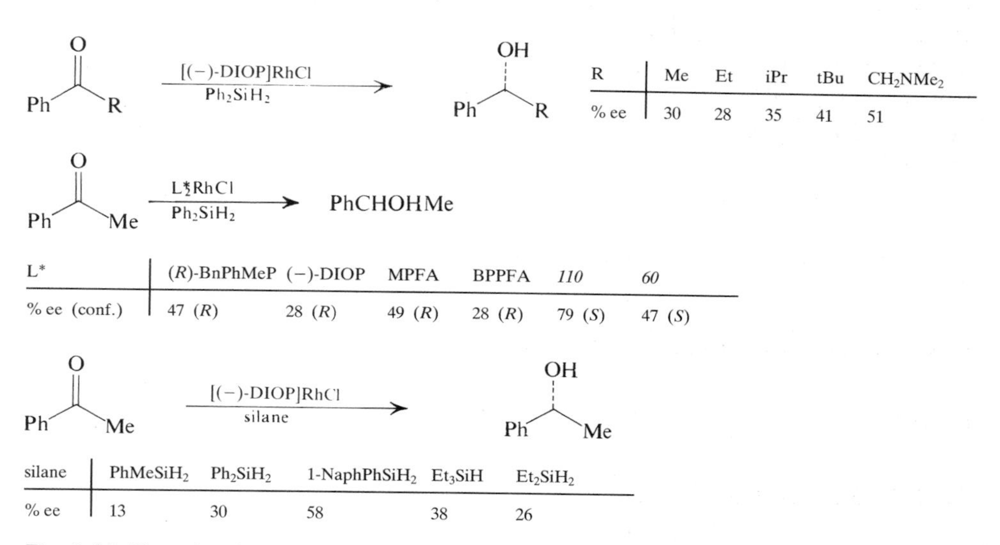

R	Me	Et	iPr	tBu	CH_2NMe_2
% ee	30	28	35	41	51

L*	(*R*)-BnPhMeP	(−)-DIOP	MPFA	BPPFA	*110*	*60*
% ee (conf.)	47 (*R*)	28 (*R*)	49 (*R*)	28 (*R*)	79 (*S*)	47 (*S*)

silane	$PhMeSiH_2$	Ph_2SiH_2	1-NaphPhSiH_2	Et_3SiH	Et_2SiH_2
% ee	13	30	58	38	26

Fig. 2-30. Enantioselective hydrosilylation of ketones (I).

While with the rhodium-phosphine catalyzed hydrogenation of amino acid precursors monophosphines were much inferior to diphosphines, this was not so with hydrosilylation. There are some ligands, like the series of chiral ferrocenyl-dimethylphosphines [217] [*e. g.* BPPFA and MPFA (*109*)], the chiral Schiff-bases (*110*–*112*) [218, 219, 220] and the thiazolidine (*113*) [219, 221] which were specially developed for hydrosilylation. Non-chelating ligands are assumed to be coordinated *trans* to each other, while chelating ligands must be coordinated in a *cis* manner; therefore the two types of ligands cannot be forced into a common mechanistic scheme.

Me NMe2 Fe PMe2

(R_C, S_{Fe}) – (*109*)
MPFA

N R Me Me Me

(*110*) R=H
(*111*) R=Me
(*112*) R=Ph

N S HN CO_2H

(*113*)

The outcome of hydrosilylation is much dependent on the structure of the silane. Methyldichlorosilane is a reagent of poor selectivity. It is, however, of mechanistic interest because in the presence of $[(R)\text{-BPPM}]_2PtCl_2$ it attacks 1-methylstyrene and acetophenone from opposite faces [222].

The monosilane $PhMe_2SiH$ is out of line inasmuch as in combination with $[(R)\text{-BMPP}]RhCl$ it gives a product with predominant configuration opposite to the one obtained with all the other silanes. No such effect was observed when $PhMe_2SiH$ was used with a (−)-DIOP based catalyst [223]. Several studies have been initiated to find the best hydride donor [208, 224, 225] and the order $Et_3SiH < PhMeSiH_2 < Et_2SiH_2 < Ph_2SiH_2 < \text{1-NaphPhSiH}_2$ obtained with acetophenone and [(−)-DIOP]RhCl may serve as a guideline, but is subject to changes with other substrate-catalyst combinations. Thus, *e.g.*, Ph_2SiH_2 is more selective than 1-Naph$PhSiH_2$ with PhCOiPr [224].

Predictions concerning the configuration of the major product based on models, with no or very little experimental background, have been proposed for hydrosilylations catalyzed by $[(R)\text{-BMPP}]_2RhCl$ and [(+)-DIOP]RhCl by Ojima, Yamamoto and Kumada [208]* and Glaser [226], respectively. The first suggestion operates with the relative size of the silane and the groups attached to the carbonyl group, while the second contents itself with comparing the size of the two groups attached to the carbonyl.

Aralkyl ketones are generally more selectively reduced than dialkyl ketones, but with a proper choice of ligand the difference is not as conspicuous as with other enantioselective carbonyl reductions (*cf.* Sections 2.1.2 and 3.1).

$MeCOCO_2Pr$ → (1-NaphPhSiH_2, [(+)-DIOP]RhCl) → $MeCH(OH)CO_2Pr$ 85% ee

$PhCOCO_2R$ → (1-NaphPhSiH_2, [(−)-DIOP]RhCl) → $PhCH(OH)CO_2R$

R	Et	cHex	(−)-Ment
% ee	39	47	77

$MeCOCH_2CO_2Et$ → (1-NaphPhSiH_2, [(−)-DIOP]RhCl) → $MeCH(OH)CH_2CO_2Et$ 26% ee

$MeCOCH_2CH_2CO_2iBu$ → (1-NaphPhSiH_2, [(−)-DIOP]RhCl) → $MeCH(OH)CH_2CH_2CO_2iBu$ 84% ee

Fig. 2-31. Enantioselective hydrosilylation of ketones (II).

* The authors arrived at their prediction by comparing, what they claimed to be, three different conformations of a single diastereomer of the substrate-silane-catalyst complex. In fact, only two of them are related as conformers, while the third, the very one leading to the alternative configuration, is a different diastereomer.

A highly interesting situation arises with α,β-unsaturated ketones. Monohydrosilanes give, *via* 1,4-addition, silylenolethers and, after hydrolysis, saturated ketones. With dihydrosilanes, normal hydrosilylation of the carbonyl function takes place to provide, after hydrolysis, chiral allylic alcohols [227–229]. Enantioselectivities are poor, especially with 1,4-addition.

Ketoesters are good substrates for hydrosilylation; 2- and 4-ketoesters can be reduced more selectively than 3-ketoesters [225, 230]. Enantioselectivity is dependent on the structure of the ketoacid, the alcohol moiety, the silane and the catalyst. Some of the best results are shown in Fig. 2-31, more data are listed in Table 2-7. It should be noted that double induction [R = (−)-Ment] effectively improves selectivity with the phenylglyoxylate but not with the pyruvate and levulinate.

As mentioned above, Schiff-bases of ketones undergo rhodium catalyzed hydrosilylation to yield, after methanolysis, secondary amines. Conditions are usually mild (0–100°C, depending on the silane) and yields are good (>85%) [231]. Using a chiral catalyst enantioselectivity was moderate with acetophenone-derived Schiff-bases and poor when both substituents of the ketone were aliphatic [140, 232].

$$R^1C(Me)=NR^2 \xrightarrow[\text{[(+)-DIOP]RhCl}]{PhSiH_2} R^1CH(Me)NHR^2$$

R^1	Ph	Ph	Bn
R^2	Bn	Ph	Bn
% ee	65	47	12

Hydrosilylation of oximes with a (−)-DIOP-rhodium catalyst gave <19% ee [233].

Enantioselectivities in hydrosilylation for a series of substrates obtained with the best combination of silane and catalyst, further, a few additional data of interest are compiled in Table 2-8.

The catalytic cycle involved in the hydrosilylation of ketones catalyzed by rhodium-phosphine complexes can be envisaged as shown in Fig. 2-32 [210].

The rhodium(I) catalyst is applied as a phosphine-olefin complex, or, when prepared *in situ*, it contains coordinated solvent molecules. In the first step the olefin or the solvent is displaced by the silane, which becomes coordinated to

$L_nRh^I(S)$ (*A*) $\xrightleftharpoons{+\ \gt SiH}$ $L_nRh^I(\gt Si)H$ (*B*)

B $\xrightarrow{+\ \gt C{=}O}$ $L_nRh^{III}(\gt Si)(\gt C{=}O)H$ (*C*)

C $\longrightarrow$ $L_nRh^{III}(\gt C{-}OSi\lt)H$ (*D*)

D $\xrightarrow{-\ \gt C(H)OSi\lt}$ *A*

Fig. 2-32. The catalytic cycle in hydrosilylation of ketones with rhodium-phosphine complexes.

Table 2-8. Enantioselective Hydrosilylation of Ketones (R^1COR^2) in the Presence of Chiral Rhodium Catalysts.

R^1	R^2	Silane	Ligand	% ee	(Conf.)[a]	Ref.
Me	Et	1−NaphPhSiH_2	(−)-DIOP	42	(*R*)	234
	Bu	Et_2SiH_2	(*R*)-BnPhMeP	30	(*R*)	223
	tBu	Ph_2SiH_2	*111*	61	(*S*)	220
	Bn	Ph_2SiH_2	*112*	56	(*S*)	220
tBu	CH_2OBz	1−NaphPhSiH_2	(−)-DIOP	50	(*S*)	287
Ph	Me	Ph_2SiH_2	(*R*)-BnPhMeP[b]	47	(*R*)	235
		$PhMe_2SiH$	(*R*)-BnPhMeP	44	(*S*)	223
		Ph_2SiH_2	*110*	79	(*S*)	220
		Ph_2SiH_2	*113*	97	(*R*)	221
		1−NaphPhSiH_2	(−)-DIOP	58	(*R*)	236
	Et	Ph_2SiH_2	(*R*)-BnPhMeP	42	(*R*)	223
		$PhMe_2SiH$	(*R*)-BnPhMeP	50	(*S*)	223
		1−NaphPhSiH_2	*60*	61	(*S*)	237
	Pr	1−NaphPhSiH_2	*60*	51	(*S*)	237
	iPr	$PhMe_2SiH$	(*R*)-BnPhMeP	56	(*S*)	223
	Bu	1−NaphPhSiH_2	*60*	39	(*S*)	237
	iBu	Et_2SiH_2	(−)-DIOP	37	(*R*)	208
	tBu	$EtMe_2SiH$	(*R*)-BnPhMeP	56	(*R*)	223
		$PhMe_2SiH$	(*R*)-BnPhMeP[b]	62	(*S*)	288
	CH_2Cl	1−NaphPhSiH_2	(−)-DIOP	63	(*S*)	287
	CH_2Br	1−NaphPhSiH_2	(−)-DIOP	365	(*S*)	287

[a] Configurations obtained with (+)-DIOP and (*S*)-BnPhMeP were inverted.
[b] Cationic catalyst.

rhodium in an oxidative addition establishing simultaneously a metal-hydrogen and metal-silicon bond. Thereafter, the ketone becomes coordinated and inserted into the metal-silicon bond. Finally, internal hydrogen transfer from metal to carbon and dissociation to a silyl ether concludes the cycle. This is a plausible scheme but, unfortunately, very little experimental work has been undertaken to explore its details. Kolb and Hetflejš studied the kinetics of the hydrosilylation of the non-enolizable ketone tBuCOPh with Ph_2SiH_2 in benzene catalyzed by [Rh(COD)(−)-DIOP]$^+ClO_4^-$ [238] and [Rh[(−)-DIOP]$_2$]$^+ClO_4^-$ [239]. The substrate was chosen so as to avoid complications associated with enolether formation. The authors observed an induction period in the hydrosilylation which disappeared if the ketone was added 30 min after the catalyst and the silane had been mixed. The induction period was associated with the oxidative addition of silane to rhodium which could be followed by UV spectroscopy. Initial rates showed a linear dependence on catalyst and silane concentration, but correlation with substrate concentration was non-linear. The kinetic behavior of the system suggested that the rate determining step was either the $B \rightarrow C$ or the $C \rightarrow D$ transformation, the latter being more likely. The absence of a significant deuterium isotope effect when

Ph_2SiD_2 was used excluded process $A \rightarrow B$ as the rate determining step. Temperature dependence of rate was also in agreement with the proposed scheme.

No information can be gleaned from the above experiments concerning the origin of stereoselection, which may be associated either with step $B \rightarrow C$ (steric approach control), as suggested by Glaser, with the insertion step ($C \rightarrow D$), as is probably the case with rhodium-chiral phosphine catalyzed hydrogenation (*cf.* Section 2.1.1.5), or finally with the hydrogen transfer step ($D \rightarrow A$), as proposed by Ojima (product development control). In want of relevant experimental contributions other than substrate structure–enantioselectivity correlations, it appears to be premature to discuss this problem any further.

Catalytic hydrosilylation of cyclic ketones can be remarkably diastereoselective, and the predominant mode of approach is influenced by the bulk of the silane. In Fig. 2-33 (a) the ratio of axial and equatorial (or *endo* and *exo*) attack is shown for hydrosilylations of camphor and menthone catalyzed by $(Ph_3P)_3RhCl$ [240].

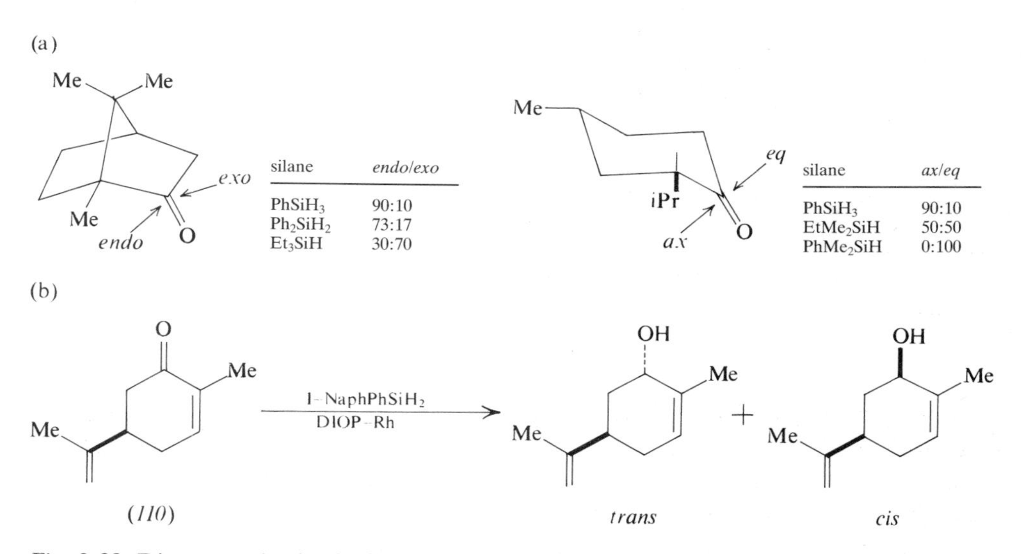

Fig. 2-33. Diastereoselective hydrosilylation of ketones.

Reduction of (*R*)-carvone (*120*) with 1-NaphPhSiH_2 in the presence of a DIOP-rhodium catalyst (Fig. 2-33 (b)) revealed that in this double-induction experiment it was the chirality of the catalyst and not that of the substrate which controlled stereoselectivity: with (+)-DIOP the *trans/cis* ratio was 77:23, while with (−)-DIOP it was 21:79 [229]. Double induction also enhanced stereoselectivity in the hydrosilylation of *N*-2-ketoacyl-2-aminoesters [241].

2.3 Heterogeneous Stereoselective Catalytic Hydrogenations

In this section hydrogenations over classical solid phase catalysts will be discussed; metal-complex catalysts immobilized by linking to a polymer matrix have already been mentioned in Section 2.1.1.1.

For efficient chiral catalysis a highly ordered and very specific structure of the catalyst is necessary, and it is very unlikely that traditional non-complexed metal catalysts could ever meet this requirement. The most such a catalyst can contribute to stereoselectivity is a *syn* transfer of dihydrogen to the substrate, and in fact this is performed with very high selectivity. All the rest has to come from the substrate side, and we shall see that while diastereoselective hydrogenations are common, examples for genuine enantioselective heterogeneous catalysis in hydrogenation can hardly be found.

2.3.1 Enantioselective Heterogeneous Catalytic Hydrogenations

The development of enantioselective heterogeneous catalysts was pioneered by Izumi and his school, who prepared a silk fibroin-supported palladium catalyst [290]. Up to 70% ee could be realized with cyclic amino acid precursors, but the method could not be developed to a practical procedure.

An enormous amount of work has been invested in studies with chirally modified catalysts by Izumi, Harada and their coworkers in Japan (38 papers up to 1984) and by Klabunovski and his group in the Soviet Union (more than 100 papers up to 1983) [252].

Both groups were primarily concerned with the hydrogenation of β-dicarbonyl compounds (almost exclusively of methyl or ethyl acetoacetate) on metal catalysts modified with (−)(2*R*,3*R*)-tartaric-acid. While the Japanese group worked with nickel only, the Soviet team also investigated a variety of bimetallic systems.

Harada *et al.* found that the enantioselectivity of tartaric acid-modified Raney nickel was greatly enhanced by the addition of salts [242], of which NaBr proved to be the best. In Fig. 2-34 some results obtained with this system are shown. Enantioselectivities with linear β-ketoesters are in the useful range [242, 243], those with a β-ketoalcohol [244, 245], a ketone [246] and β-ketosulfonic acids [247]

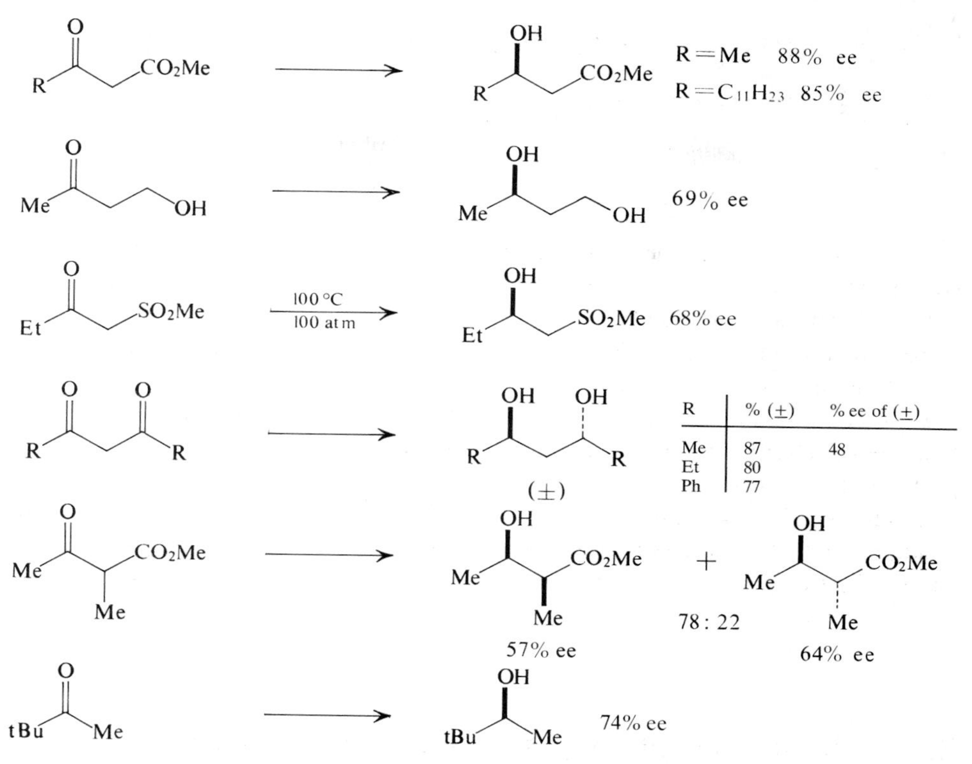

R	% (±)	% ee of (±)
Me	87	48
Et	80	
Ph	77	

Fig. 2-34. Stereoselective hydrogenations over Raney nickel modified with (2*R*,3*R*)-tartaric acid and NaBr.

are less good. With both 1,3-dicarbonyl compounds [248, 249] and 2-alkyl-3-ketoesters [250] diastereomers are formed. In the latter substrate the center of chirality is unstable, and the stoichiometry of the product reveals kinetic preference for the hydrogenation of the 2*S* substrate.*

It was supposed by Tai *et al.* [251] that on the surface of the catalyst a chelate complex, as shown below, is formed between the substrate and tartaric acid which is then preferentially attacked from the *Si* face. Thus, on closer inspection we are dealing with the diastereoselective hydrogenation of a chiral complex rather than with an enantioselective hydrogenation by a chiral catalyst.

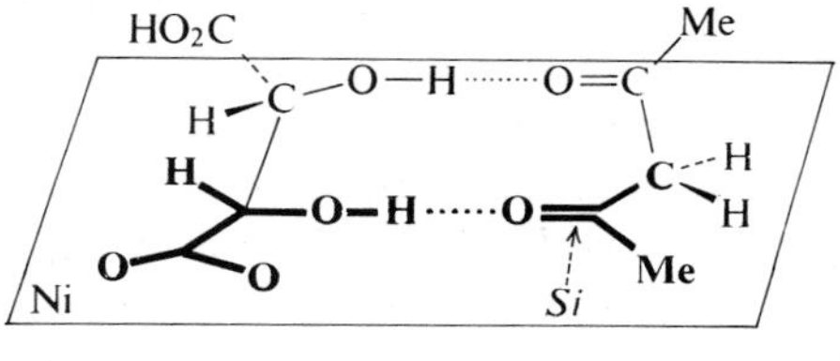

* The product mixture contains a total of 79% of 2*S* stereoisomers.

The favorite models of Klabunovski and his coworkers were also the acetoacetates, and their results were essentially similar [252], although they reported somewhat lower enantioselectivities than the Japanese authors. It appears that considerable know-how is needed for the preparation of an efficient catalyst, and therefore this method seems to be more suited for industrial than for laboratory application.

2.3.2 Diastereoselective Heterogeneous Catalytic Hydrogenations

Hydrogenation over classical metal catalysts of substrates having diastereotopic faces is a well established stereoselective procedure which involves the *syn* addition of dihydrogen preferentially to the less hindered face of the substrate. In this section we concentrate on studies which were aimed at the synthesis of optically active amino acid. In some of the examples (*e. g.* the hydrogenation of dehydropeptides) the inducing group became an integral part of the product, while in others it was removed after it had fulfilled its task (chiral auxiliaries). The basic requirement for a chiral auxiliary is that its inducing power should be high. It is also desirable that its chirality should not be destroyed in the course of the synthesis, *i. e.* it should be either recoverable or recyclable after simple transformations. In the following examples all three situations will be illustrated.

Hydrogenation of enamines

Cyclic enamines of type *111* (Fig. 2-35 (a)) can be readily prepared from azlactones. If R^1 is an aromatic group and the chiral auxiliary proline, hydrogenation proceeds with high enantioselectivity, while with R^1 being alkyl the optical purity of the product is low [253]. This is somewhat contradictory to later reports, which claimed high diastereoselectivity for 2-alkylidene diketopiperazines unsubstituted at N(1) [254–256].

The best result so far with cyclic enamines has been achieved with the oxazine *112*, which is readily accessible from the aminoalcohol and dimethyl acetylene dicarboxylate. In *112* one face of the ring is blocked by the axial phenyl group. Unfortunately, the chirality of the auxiliary aminoalcohol is destroyed on its removal (Fig. 2-35 (b)) [257, 258]. Technically important is the stereoselective hydrogenation of the biotin precursor *113* (Fig. 2-35 (c)) [259].

Hydrogenation of chiral but acyclic amino acid [260] and peptide precursors [261] and of the chiral esters of 3-methyl-3-phenylacrylic acid [262], further the aluminum-amalgam reduction of a chiral 3-alkylidene-2-oxazinone [263] were all of

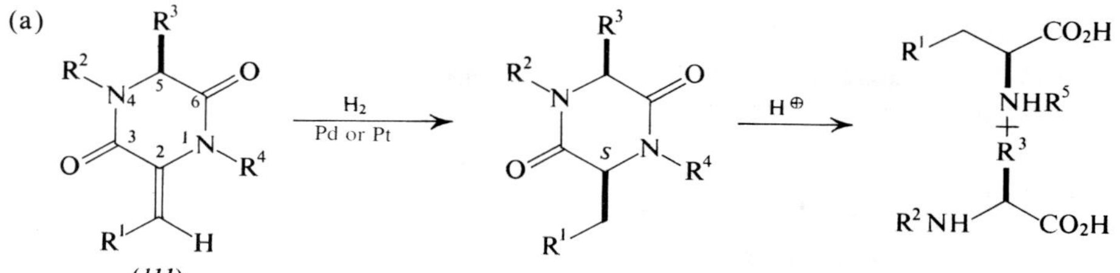

(a)

H_2, Pd or Pt; $H^{\oplus}$

(*111*)

R^1	R^2	R^3	R^4	% de	ref.
Ph	$-(CH_2)_3-$		Ac	90	253
H	H	Me	H	94.6	255
		iPr		98.4	255
		iBu		94.6	255
		Bn		94.6	255
	$-(CH_2)_3-$			85.0	255
Me	H	Me		99	256

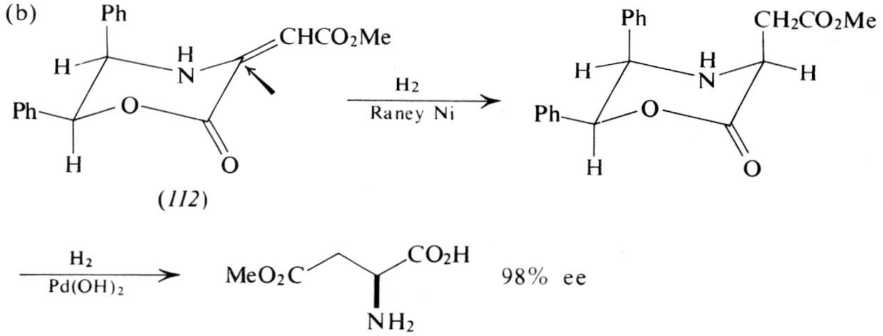

(b)

H_2, Raney Ni; H_2, $Pd(OH)_2$

(*112*)

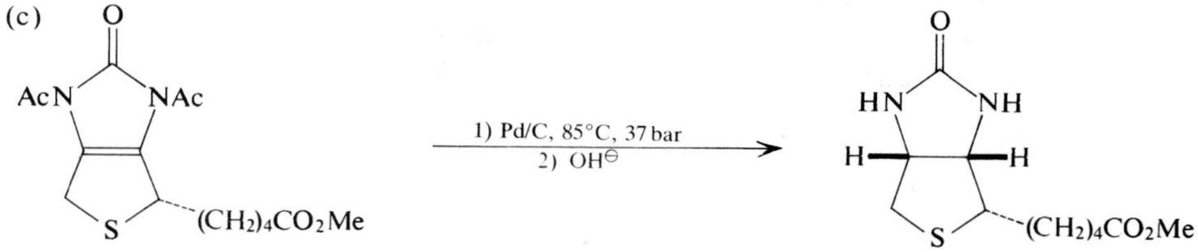

98% ee

(c)

1) Pd/C, 85°C, 37 bar
2) $OH^{\ominus}$

(*113*)

(–)-biotin, 90% de

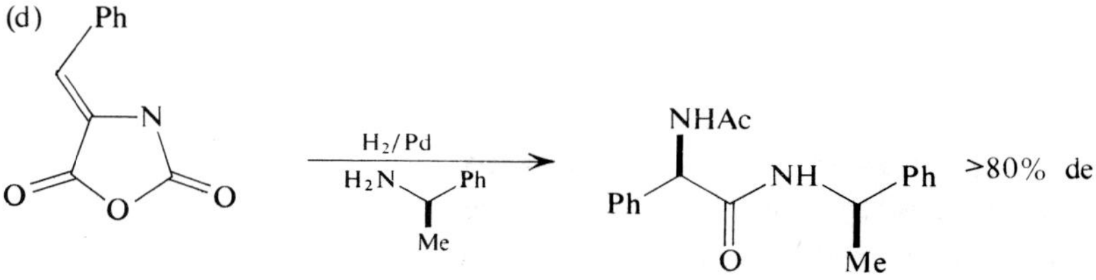

(d)

H_2/Pd

(*114*) (*115*)

>80% de

Fig. 2-35. Diastereoselective hydrogenation of enamines.

low selectivity. An exception seems to be the hydrogenation of the azlactone *114* in the presence of (*R*)-1-phenylethylamine (Fig. 2-35 (d)) which gives the amide *115* with acceptable diastereoselectivity.

Hydrogenation of ketones

Surprisingly few studies deal with the catalytic reduction of prochiral ketones. A series of chiral pyruvamides were investigated, of which the (*S*)-1-(1-naphthyl)-ethylamide could be hydrogenated with the highest selectivity giving, at −30°C, the (*S,S*)-lactamide in 96% de [264].

Hydrogenation of C=N bonds

Pyridoxal phosphate-mediated transamination involving 2-ketoacids and 2-amino acids is a fundamental reaction of amino acid metabolism [265]. Many attempts have been made to transplant the process into the laboratory, but only very few of the methods elaborated are real parallels to the pyridoxal-catalyzed transformation. One of them involves the planar chiral pyridoxamine analogue *116* in which the chiral protein environment is replaced by a non-invertable hydrocarbon chain [266]. Uptake of deuterium from the solvent indicated close analogy with the biosynthetic process.

CH_2NH_2 HO S N S Me Me + $RCOCO_2H$ $\xrightarrow[MeOD]{Zn^{2\oplus}}$ D NH_2 R CO_2H + R*—CHO

(*116*)=R*–CH_2NH_2

R	iPr	iBu	Bn
% ee	80	94	40

A similar enantioselective transamination was realized with a copper complex containing as ligands a chiral amine and a Schiff-base formed from pyridoxamine and 2-ketoacids, but in this case ee was ≤45% [267].

Transfer of chirality from a chiral amine to a ketone to form a new chiral amine has been thoroughly investigated.* The general scheme of this procedure is shown in Fig. 2-36 (a). The chiral amine is most often 2-phenylethylamine, in which the nitrogen atom is linked to a benzylic position and therefore the inducing group can be removed by catalytic hydrogenation. This requires more stringent conditions than the saturation of the C=N bond and destroys the original chiral center.

* The method is often called asymmetric transamination, but the analogy to the biochemical process is only formal.

When applied to α-arylketones the method is suitable for the preparation of psychotomimetic amines. Diastereoselectivities are usually moderate when simple chiral amines are used, but can be upgraded by recrystallization of the diastereomeric intermediates [268, 269].

(a)

$R^1COR^2 + H_2N{-}R^* \longrightarrow$ imine $\xrightarrow{H_2/cat^1}$ $R^1CH(NHR^*)R^2$ $\xrightarrow{H_2/cat^2}$ $R^1CH(NH_2)R^2 + R^*{-}H$

(b)

(±) 2-R-cyclohexanone + Ph(Me)CH–NH₂ ⟶ imine $\xrightarrow{H_2 / \text{Raney Ni}}$ secondary amine $\xrightarrow{H_2 / \text{Pd-C}}$ (+) 2-R-cyclohexylamine

R	Me	Et	Ph	Bn
% ee	96	92	94	98

(c)

Me(Bn)C=N–CH(R¹)CO₂R² $\xrightarrow{H_2 / \text{Pd–C}}$ Me(Bn)CH–NH–CH(R¹)CO₂R $\xrightarrow{1)\ tBuOCl;\ 2)\ B^{\ominus}}$ Me(Bn)CH–N=C(R¹)CO₂R² $\xrightarrow{H^{\oplus}}$ (S) Me(Bn)CH–NH₂

R¹	Me	Me	iPr	iPr
R²	Et	tBu	Et	tBu
% ee	66	85	50	87

(d)

R(CO₂H)C=N–CH(Me)Ph (117) ⟶ ⟶ R–CH(NH₂)CO₂H

R	Me	Et	Ph	Bn	CH₂CO₂H
% ee	67	39	30	14	58

Fig. 2-36. Transfer of chirality between amines and ketones.

Much improvement can be brought about by using as chiral amines 2-amino acid esters with bulky groups (Fig. 2-36 (c)) [270]. In this case the inducing group was removed *via N*-chlorination, elimination and hydrolysis. With acetophenone results were similar.

Transformation of racemix 2-alkylcyclohexanones to the less stable *cis*-amines with high enantioselectivity has been reported (Fig. 2-36 (b)) [271, 272]. This result can only be explained if one assumes equilibration of the enantiomeric ketones or of the diastereomeric azomethines.

Harada and his school investigated the transamination of 2-ketoacids very thoroughly. Their findings can be summarized as follows: (i) In a series of Schiff-bases formed with (*S*)-2-phenylethylamine selectivity depends highly on the nature of the 2-ketoacid [273, 274], but there is no clear-cut correlation with the bulk of the R group. (ii) As amine component, 2-phenylethylamine is preferable to 2-phenylpropylamine or 2-(1-naphthyl)-ethylamine [274] except for $R=CH_2CO_2H$, when the latter is more advantageous. (iii) The use of 2-ketoesters instead of the acids does not bring about distinct advantages, even when the alcohol is chiral [275]. (iv) Surprisingly, in the single case when it has been examined [276], the *E*/*Z* ratio of the substrate had no influence on the stereoselectivity of hydrogenation. (v) Stereoselectivity is very sensitive to changes in temperature. Thus, *117* (R=Me), as the ethyl ester afforded (*S*)-alanine in 60% ee at −20°C, while at 45°C (*R*)-alanine in 43% ee was formed [277]. A similar phenomenon was observed when the chiral amine was phenylglycine [278, 279]. (vi) As solvent, benzene, ethyl acetate, tetrahydrofuran or dioxane give better results than methanol or water [275, 276, 280, 281].

Asymmetric transamination resembles rhodium-phosphine catalyzed enantioselective hydrogenation in so far as conditions have to be optimized separately for each substrate. Note, for example, that the best selectivity recorded so far (83% ee) has been attained in the hydrogenation of the Schiff-base prepared from benzyl pyruvate and 1-(1-naphthyl)-ethylamine in hexane at room temperature [274], *i. e.* neither the substrate nor the conditions corresponded to those found to be optimal in systematic experiments.

Selected sets of experimental results can be interpreted in terms of a preferred conformation of the substrate at the surface of the catalyst. Since neither can all facts be accommodated with a single hypothesis nor has there been any verification of the assumed predominant conformation presented, such transition state models will not be discussed here.

A variation of the method giving less than 64% ee of the end product is that first an amide is formed with a 2-amino acid ester, which is then converted to the Schiff-base of an achiral amine and hydrogenated [282].

With none of the preceding methods can the chiral auxiliary be recycled, since its center of chirality is destroyed during hydrogenolytic cleavage. A recyclable chiral amine available in both enantiomeric forms is *118*, obtained by an 8 step synthesis and resolution [283]. Some of its applications are shown in Fig. 2-37. Analogues of

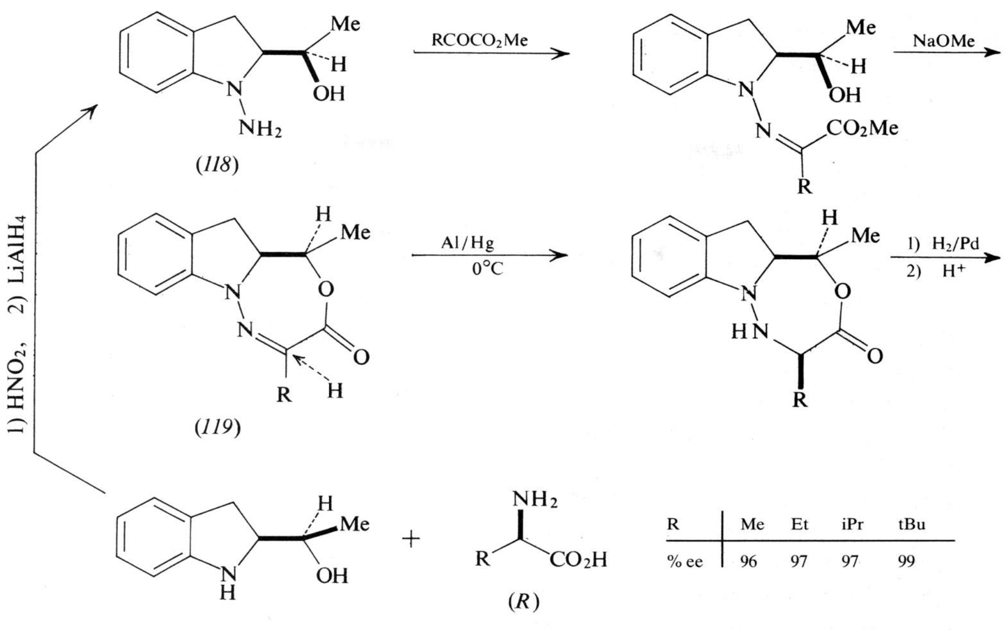

R	Me	Et	iPr	tBu
% ee	96	97	97	99

Fig. 2-37. Asymmetric induction in reductive amination by a recyclable amine.

118 are also valuable chiral auxiliaries [284]. A crucial feature of this scheme is that reduction involves a cyclic substrate (*119*) of which, according to X-ray analysis, only the face opposite to the methyl group is accessible.

Hydrazines derived from the alkaloid anabasine [285] or from various terpenic amines [286] which did not form a cyclic Schiff-base with the 2-ketoacid provided only modest stereoselectivity (ee $\leqslant 47\%$).

3 Stereoselective Non-Catalytic Reductions

The subject of this chapter is almost exclusively the non-catalytic reduction of carbonyl compounds to alcohols. Conversion of ketones to secondary alcohols, as well as of azomethines to amines by catalytic hydrogenation, and the hydrosilylation of ketones have already been discussed in Chapter 2.

3.1 Enantioselective Reductions

Two main approaches to the enantioselective reduction of C=X double bonds (mainly of C=C bonds) are known: (i) the use of aluminum or boron hydrides having chiral ligands, and (ii) transfer of chirality by exchange of a hydride ion attached to a chiral center. The latter requires activation of the C–H bond usually by a metal atom (*e. g.* Al or Mg) in β-position or by an alkoxide group (*e. g.* $OAl_{1/3}$) in α-position. The first method is much more efficient and also more generally applicable.

3.1.1 Chiral Lithium Aluminum Hydrides*

3.1.1.1 Reduction of Carbonyl Groups

The idea that lithium aluminum hydride (LAH) can be converted to a chiral reagent by exchanging hydrogens for chiral alcohols can be credited to Bothner-By (1951) [2]. He reacted LAH with (+)-camphor and with the resulting reagent reduced prochiral aliphatic ketones. Although it was shown later that the optical

---

* For a recent review see ref. [1].

activity of the product was due to contamination by the reagent [3, 4], the idea was born, and chiral lithium aluminum alkoxides have been developed to one of the most efficient enantioselective reducing agents. The factors controlling selectivity are still not well understood; the main reason for this is the complexity of the LAH-alcohol system. Reagents are usually prepared by mixing LAH with the appropriate alcohol in a solvent and are used without isolation.

The equation for the formation of the reagent *i. e.*:

$$\mathrm{LiAlH_4} + n\mathrm{R^*{-}OH} \rightarrow \mathrm{LiAl(OR^*)}_n\mathrm{H}_{4-n} + 0.5n\mathrm{H_2}$$

is only formal since several species with different numbers of alkoxide groups may be formed consecutively, and the situation is further complicated by disproportionation producing achiral LAH, *e. g.*:

$$2\,\mathrm{LiAlH_3OR^*} \rightleftarrows \mathrm{LiAlH_2(OR^*)_2} + \mathrm{LiAlH_4}$$

Also, the alkoxides often precipitate from solution and may undergo ageing [*e. g.*, 5]. Thus, different results can be obtained not only with different molar ratios of LAH and alcohol but also with a freshly prepared reagent and with one left standing. Therefore, careful optimization of conditions is necessary in order to achieve high selectivities. The experimental protocol followed by the majority of authors is (i) testing a series of reagents with a single substrate, (ii) optimization of the conditions with the best reagent(s), (iii) testing the best reagent(s) under optimum conditions on a range of substrates.

Bothner-By's choice of a chiral auxiliary, a monohydric alcohol with no other functionality, was unfortunate, and ever since attempts to develop, efficient reagents with such alcohols have been unsuccessful [4, 6–12]. An exception seems to be the reduction of some β-aminoketones with (−)-menthol-LAH (3:1)* [13].

Ph–CO–CH$_2$CH$_2$–NMe$_2$ → [(−)-menthol + LAH (3:1), Et$_2$O, 0 °C] → Ph–CH(OH)–CH$_2$CH$_2$–NMe$_2$ (77% ee)

Selectivity was somewhat less (66 and 59%) with the piperidino and morpholino analogues or when the temperature was raised to 35°C. Selectivity was sharply reduced with a molar ratio of 2:1 and was practically nil with a ratio of 1:1. Relatively high selectivities may be explained by the participation of the amino group, as suggested by later results of Yamaguchi and Kabuto [11], who conducted one of the few substrate oriented studies. They intentionally selected (−)-menthol because this was a chiral ligand with no other functional group than the complex-forming one and recorded ee values of the alcohols *1* (all homochiral except for Y = OMe, $n = 1$).

* Expressed here and later as molar ratios.

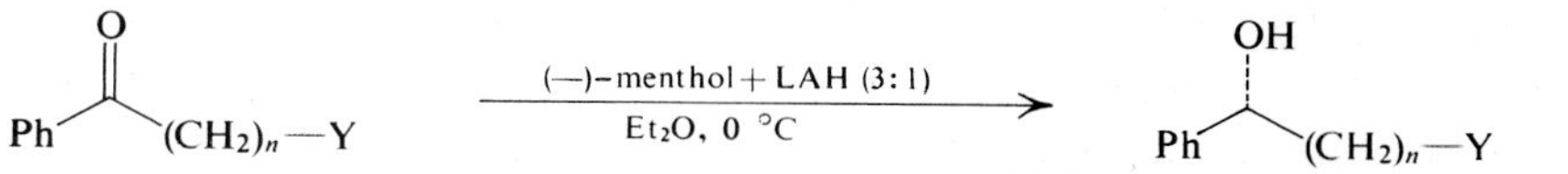

$$\text{Ph-CO-(CH}_2)_n\text{-Y} \xrightarrow[\text{Et}_2\text{O, 0 °C}]{(-)\text{-menthol} + \text{LAH (3:1)}} \text{Ph-CH(OH)-(CH}_2)_n\text{-Y}$$

(*1*) (n=1–4, Y=NMe_2, OMe, SMe, Et)

Best selectivities were found for Y = OMe and $n = 2$ and 3 (35 and 38% ee resp.), less for NMe_2 (25%, $n = 2$) and poor for SMe (12%, $n = 2$) and Et (19%, $n = 2$). Coordination with Li^+ to form a cyclic intermediate (six-membered for $n = 2$) may explain the observed effect. Indeed, addition of 2 moles of *N,N,N′,N′*-tetramethylethylenediamine, which competed for Li^+ coordination, lowered ee for *1*($n = 2$, Y = piperidino) from 23 to 12%. It should be noted that the ee recorded for *1* ($n = 2$, Y = NMe_2) (25%) in this study was much lower than that claimed by Andrisano *et al.* [13] (77%).

The application of bifunctional compounds as chiral ligands was pioneered by Landor and his coworkers in the sixties. Not only was a highly enantioselective reagent discovered, but the ideas on which it was based could later be successfully adapted to the preparation of other reagents.

In order to block the reappearence of LAH by disproportionation, instead of a monohydric alcohol, diols and triols derived from carbohydrates were used as chiral ligands [4, 14]. Among them, 3-*O*-benzyl-1,2-*O*-cyclohexylidene-α-D-glucofuranose proved to be the most enantioselective. The parent triol, the 3-*O*-methyl and 3-*O*-ethyl analogues, as well as three other similar carbohydrate diols and a triol gave products with less than 14% ee. With the 3-*O*-benzyl derivative, propiophenone was reduced to (*S*)-1-phenylpropan-1-ol in 57% ee. The presence of the 3-*O*-benzyl group seems to be essential for acceptable selectivity, and, in fact, the 1:1 complex (*2*) may take up a conformation in which one of the transferable hydrogens (H^1) is shielded by the phenyl group and therefore the other

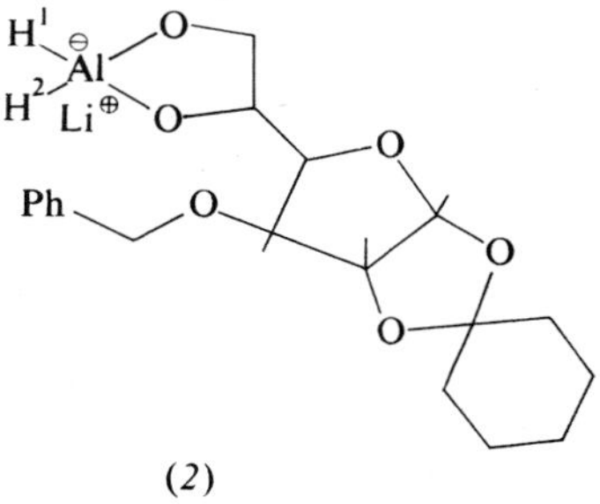

(*2*) 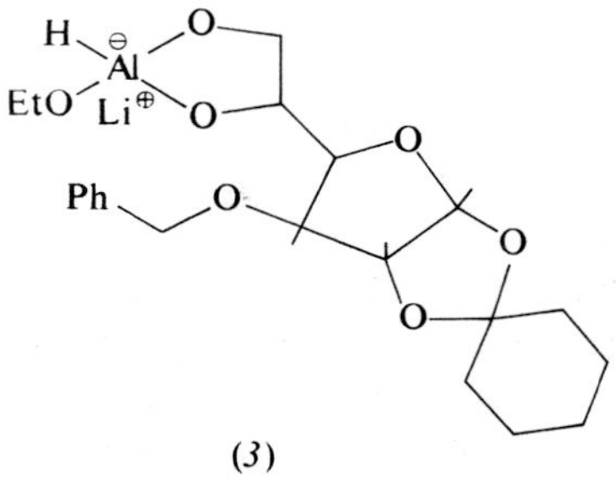 (*3*)

one is preferentially transferred to the substrate, for which a preferred orientation should be assumed as well. An alternative explanation is that interaction of the substrate with the benzyl group assists the approach of the former in a specific orientation. Although it was not rigorously proved which of the hydrogens was more active, the hypothesis was fruitful and suggested that replacement of the more exposed hydrogen by an alkoxy group would give a reagent which was less active but showed increased and inverted selectivity. In fact when ethanol (1.3–2.5 mol) was added to *2* a monohydride with the tentative structure *3* was formed, and a dramatic increase in selectivity with inverted configurational preference was experienced [15].

A further increase in the bulkiness of the complex proved to have an adverse effect [16]. Enantioselectivities of up to 71% were realized, and therefore the method is only of historic interest.

Some monohydroxy sugar derivatives gave only very low enantioselectivities [17]. The selectivity of *3* could not be reached with complexes prepared from other carbohydrate-based [17, 18] or terpenic diols [19–22].

The front runner among LAH-diol complexes, and one of the most selective chiral reagents at all, is based on 2,2′-dihydroxy-1,1′-binaphthyl and was discovered by Noyori *et al.* [23, 24]. Addition of an alcohol to provide a third ligand, preferably methoxy or ethoxy, was indispensable; the reagent with binaphthyl as the only ligand was non-selective. The reagent prepared from the (*S*)-diphenol can be formulated as *4*. Results were optimized using acetophenone as substrate. High selectivity required low temperatures. Replacement of ethanol by 2,6-di-tert-butylphenol not only reduced enantioselectivity to 44% but inverted the configuration of the major product to *S*. Even more mysterious was the effect of 3,3,3-trifluoroethanol, which also gave the (*S*)-alcohol in 42% ee [25].

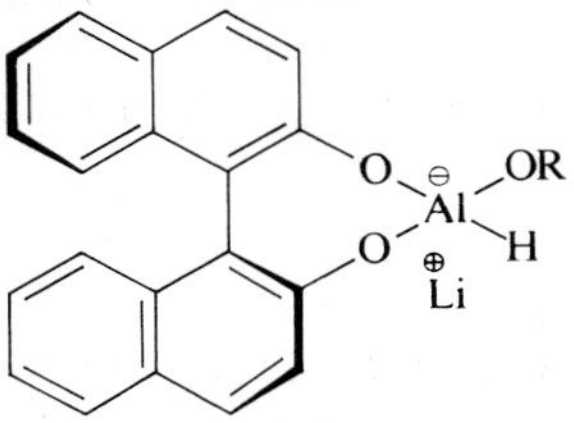

(*4*) (*S*)-BINAL–H (R=Me, Et)

BINAL-H reduced with high selectivity not only aralkyl ketones [23] but also alkynyl alkyl ketones [26, 27], deuteriated aldehydes [23, 28] and vinyl alkyl ketones [24], some of which were useful as prostaglangin synthons [29] and prostaglangins proper [24]. Note that the microbiological reduction of bromovinyl pentyl ketone gave the corresponding (*S*)-alcohol in but 80% ee and 10% yield [29]. Representative examples are shown in Table 3-1. Simple dialkyl ketones gave low enantiomeric excesses [23]. It is convenient that both enantiomers of 2,2′-dihydroxy-1,1′-binaphthyl are available. The absence of an ageing effect and the linearity of the ln*R*/*S* *vs.* 1/*T* plot suggested that a single reducing species was involved. This and the C_2 symmetry of the diol may be the clues to the exceptional selectivity of BINAL-H. Similar reagents with 2,2′-dihydroxy-6,6′-dimethylbiphenyl [30] and a phenanthrene analogue [31] were prepared, but proved to be only marginally better than BINAL-H.

Table 3-1. Reduction of Carbonyl Compounds (R^1COR^2) with (*S*)-BINAL-H (*4*) to (*S*)-Alcohols[a,b].

R^1	R^2	% ee	Ref.
Ph	D	82	23
(*E*)-$Me_2C{=}C(Me){-}CH{=}CH$		84	27
(*Z*)-$Me_2C{=}C(Me){-}CH{=}CH$		72	27
Ph	Me	95	25
	Et	98	25
	Pr	100	25
	iPr	71	25
	Bu	100	23
	tBu	44	25
n-$C_{18}H_{17}$	$(CH_2)_2CO_2Me$	84	26
$HC{\equiv}C$	Pent	84	27
	Oct	90	27
	iPr	57	26
$BuC{\equiv}C$	Me	84	27
	Pent	90	27
$MeO_2CC{\equiv}C$	Oct	87	26
(*E*)-BuCH=CH	Me	79	24
	Pent	91	24
(*E*)-cPentCH=CH		92	24
(*E*)-CH=CH		96	29

[a] Experiments were performed in THF at −100°C. As additive, EtOH was used in ref. [25] and MeOH in ref. [27].

[b] In some experiments (*R*)-BINAL-H was used and thus the (*R*)-alcohols were obtained.

Fundamental work in the field of enantioselective reductions was carried out by Červinka and his school, who first employed aminoalcohols, notably alkaloids, as chiral ligands [32]. Although enantioselectivities were modest (up to 39% ee with PhCOtBu and 48% with PhCOMe), interesting correlations between the configurations of the aminoalcohol and product were discovered (for details see ref. [33]).

Efficient aminoalcohol-type chiral auxiliaries are all tertiary amines and replace one hydrogen of LAH per mole. The role of the nitrogen atom in enhancing selectivity is not yet clear, it may lend a more rigid structure to the complex or to the transition state, or may inhibit disproportionation of the complex to regenerate LAH.

The (−)-quinine-$LiAlD_4$ complex proved to be relatively selective in reducing aromatic aldehydes (ee's of the *S* alcohol up to 35%) [34].

Landor's idea of adding a second but achiral ligand also proved to be fruitful in the hands of Vigneron and coworkers [35, 36]. While reaction of LAH with 1, 2 and 3 moles of (−)-*N*-methylephedrine (*5*) gave reagents with increasing but low selectivities (40% ee of the *R* alcohol at best with acetophenone), the addition of 2 moles of an aliphatic alcohol (LAH : (−)-*5* : ROH 1:1:2) increased enantiomeric purity to 56% (PrOH). Phenols were even better additives and, out of 17 different phenols tested, orcinol (3,5-dimethylphenol) proved to be the best. Interestingly, however, unlike with *3*, addition of the third component did not bring about a change of the prevailing configuration, except for 2,2′-dihydroxy-1,1′-biphenyl and -binaphthyl. The reaction has an optimum temperature at around −15°C, ee values drop sharply at both higher and lower temperatures. This points to the existence of several reducing species of different selectivity at equilibrium, the concentration of the most selective being at its maximum at the optimum temperature. Characteristic examples for reductions with LAH-(−)-*N*-methylephedrine-

Table 3-2. Reduction of Ketones (R^1COR^2) with Reagents Prepared from LAH and Chiral Aminoalcohols. (A: LAH + (−)-*N*-methyl-ephedrine (*5*) + 3,5-dimethylphenol (1:1:2), Et_2O, −15°C; B: LAH + (−)-*N*-methylephedrine (*5*) + *N*-ethylaniline (1:1:2), Et_2O, −78°C; C: LAH + Darvon-alcohol (*6*), (1:2:3) −78°C.)

R^1	R^2	A		B		C	
		% ee	(Conf.)	% ee	(Conf.)	% ee	(Conf.)
Ph	Me	84	(*R*)	88[a]	(*S*)	75[b]	(*R*)
4-MePh		63	(*R*)				
Ph	CF_3	30	(*R*)				
	Et	85	(*R*)	90	(*S*)		
	Pr	89	(*R*)			62[c]	(*R*)
	iPr	17	(*R*)	78	(*S*)	28[c]	(*R*)
	Bu	78	(*R*)	78	(*S*)		
	iBu	84	(*R*)				
	tBu	31	(*S*)				
	cHex	11	(*R*)			35[c]	(*R*)
1-Tetralone		57[d]	(*R*)	51	(*S*)		
iPr	Me	41[e]	(*S*)				
tBu		20[e]	(*S*)				
Bn		45[e]	(*S*)	41	(*S*)		
cHex				35	(*S*)		
HC≡C		79	(*R*)				
	Et	87	(*R*)				
	iPr	86	(*R*)				
	Bu	85	(*R*)				
	iBu	88	(*R*)				
	tBu	90	(*R*)				
	Pent	84	(*R*)			72[f]	(*R*)
MeC≡C	iBu	82	(*R*)			91[g]	(*R*)
$Me_3SiC≡C$	Pent					66[f]	(*R*)
Pent−C≡C						82[f]	(*R*)

[a] At −100°C. [b] At −65°C with fresh reagent, with aged reagent at r. t. 75% ee but *S*! [c] At 0°C with fresh reagent. [d] 63% ee at 0°C. [e] At 0°C. [f] At −78°C. [g] from ref. [45].

3,5-dimethylphenol 1:1:2 are displayed in Table 3-2. High selectivity was not only achieved with aralkyl ketones [35, 36] but also with α,β-acetylenic alkyl ketones [37], and moderate selectivity with 2-alkyl-1,3,4-cyclopentatriones [38, 39], intermediates in the synthesis of allethrolones. As usual, the reagent failed with dialkyl ketones (ee 14–19% with unbranched methylketones, somewhat higher with others, see Table 3-2) [36].

Enantioselectivity of the LAH-(−)-*N*-methylephedrine complex was also enhanced by the addition of *N*-alkylanilines, most effectively by *N*-ethylaniline (up to 90% ee (*S*) with aralkyl ketones) [40]. A variation of the method, in which the modifier is 2-ethylaminopyridine, is of greater interest, since this enables the reduction of cyclic α,β-unsaturated ketones to the (*R*)-alcohol with 73–98% enantioselectivity (98% for cyclohex-2-enone) [41].

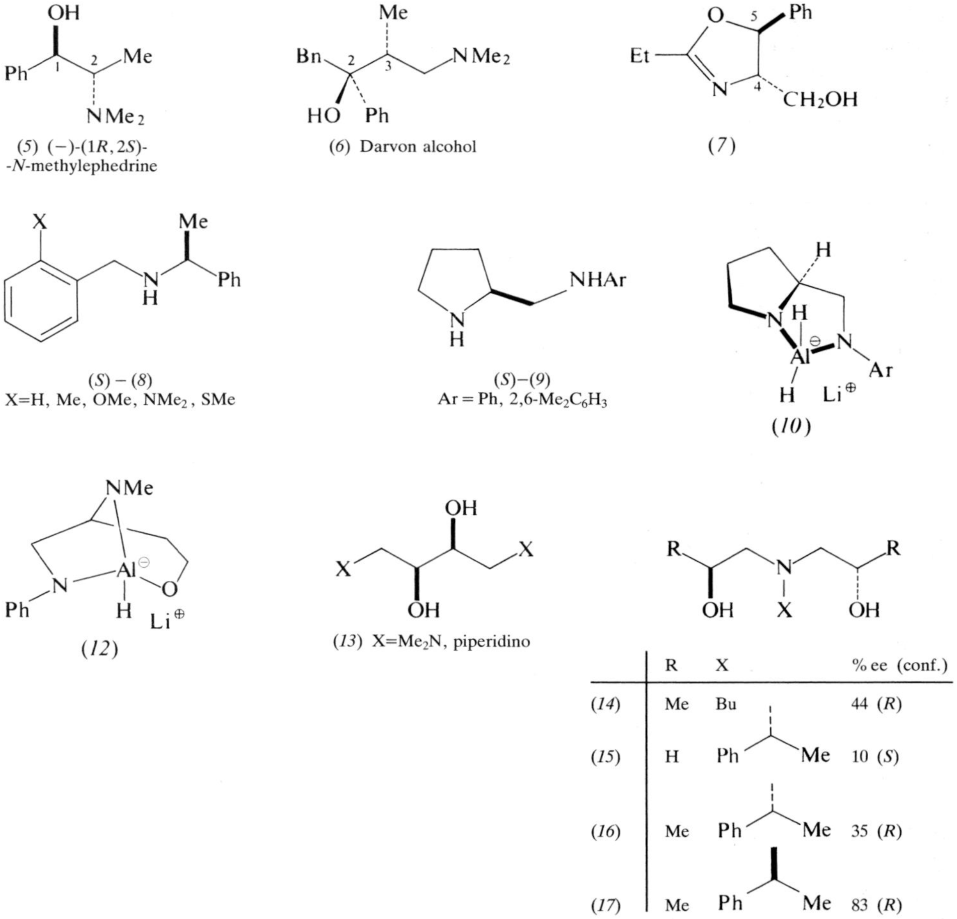

(*5*) (−)-(1*R*,2*S*)-*N*-methylephedrine

(*6*) Darvon alcohol

(*7*)

(*S*)–(*8*)
X=H, Me, OMe, NMe_2, SMe

(*S*)–(*9*)
Ar = Ph, 2,6-$Me_2C_6H_3$

(*10*)

(*12*)

(*13*) X=Me_2N, piperidino

	R	X	% ee (conf.)
(*14*)	Me	Bu	44 (*R*)
(*15*)	H	Ph Me	10 (*S*)
(*16*)	Me	Ph Me	35 (*R*)
(*17*)	Me	Ph Me	83 (*R*)

Fig. 3-1. Chiral amines and aminoalcohols used for the preparation of chiral lithium aluminium hydride complexes.

The first aminoalcohol which was found to form a highly selective reagent with LAH without the aid of a third component was Darvon-alcohol (*6*) [42] and was used for the reduction of aralkyl ketones by Yamaguchi, Mosher and Pohland [5, 43] (*cf.* Table 3-3). Its real potential, namely, the enantioselective reduction of α,β-acetylenic ketones was discovered by Brinkmeyer and Kapoor [44] and was exploited for the synthesis of tocopherol [45]. The triple bond seems to play the same role as aryl groups, since the configuration of the major enantiomer is the same in both cases and selectivity drops sharply when the triple bond is replaced by a double bond or an aliphatic group. The reagent performs well, even with highly complex polyenic substrates [44, 45]. Ten analogues of Darvon alcohol were prepared and tested later but none of them came even close to *6* [45].

LAH-Darvon alcohol displays all the caprices characteristic for modified LAH reagents. On addition of the aminoalcohol to a solution of LAH in ether, a precipitate was formed and when this was used immediately at −65°C (*R*)-alcohols were obtained in excess. On standing, the precipitate dissolved and at room temperature this reagent gave a product with opposite rotation with some of the substrates, (*e.g.* PhCOPr, PhCOiPr and tBuCOMe) but not with others ($PhCOCF_3$ and PhCOtBu) [5].

In comparison to stereoselective C−C bond-forming reactions (*cf.* Section 6.2.1), the role of chiral oxazolines as auxiliaries is less important. Even their best representative (*7*) gave not more than 65% ee with aralkyl ketones [46].

Reaction of LAH with chiral secondary amines gives chirally modified reagents similar to those obtained with alcohols. Yamaguchi and his coworkers prepared a series of reagents from LAH and 1−3 moles of the amines *8* [47]. As may be anticipated, amines with no second binding sites (X = H, Me) provided very poor reagents, while those with polar X groups (OMe, SMe and NMe_2) and, surpris-

Table 3-3. Reduction of Ketones (R^1COR^2) with Reagents Prepared from LAH and Chiral Secondary Amines. (A: LAH + (*S*)-*N*-(2-dimethylaminophenylmethyl)-1-phenylethylamine (1:3), Et_2O, 0°C; B: LAH + (*S*)-2-anilinomethylpyrrolidine (1:1.16), Et_2O −100°C; C: LAH + (*S*)-2-(2,6-dimethylphenyl)methylpyrrolidine (1:1), Et_2O −100°C; D: LAH + (*S*)-anilino-3-methyl-1-butanol (1:1), THF −100°C.)

R^1	R^2	A		B		C		D	
		% ee	(Conf.)	% ee	(Conf.)	% ee	(Conf.)	% ee	(Conf.)
Ph	Me	43	(*R*)	92	(*S*)	95	(*S*)	51	(*S*)
	Et	52	(*R*)	85	(*S*)	96	(*S*)	68	(*S*)
	iPr			50	(*S*)	78	(*S*)	77	(*S*)
	tBu	47	(*R*)					86	(*S*)
	cHex	14	(*R*)						
1-Tetralone				50	(*S*)	86	(*S*)	88	(*S*)
$PhCH_2$	Me			31	(*S*)	11	(*S*)		
Hex				13	(*S*)	26	(*S*)	33	(*S*)

ingly, the mesitylmethyl derivatives were much better but still unsatisfactory (with X = NMe_2, LAH : *8* 1:3 and propiophenone 52% ee (*R*) at 0°C) (for other substrates see Table 3-3). As with other LAH complexes, anomalies were abundant with this system also. In the reduction of acetophenone, for example, some reagents performed better at −78°C while others, *e.g.* X = NMe_2, at 0°C. The prevailing configuration was temperature dependent for X = H and SMe and changed with LAH-amine ratio for X = NMe_2 and OMe. At 0°C all the reagents gave an excess of (*R*)-carbinol except X = OMe. Finally, selectivities at 0°C were higher for a 1:3 than for a 1:2 LAH-amine ratio in the case of X = H, NMe_2 and the mesitylmethyl compound, and lower for X = Me, OMe and SMe. This again suggests that we are not dealing with a single reagent but with a complex equilibrium of several species.

A reagent of unknown structure prepared by Hawkins and Sharpless by mixing bis[(*S*)-2-phenylethylamino]sulfone, *N*-methylbutylamine and LAH was found to be moderately selective [48].

A series of 12 chiral auxiliaries with two secondary amine functions (*9*) were prepared from L-proline by Mukaiyama *et al.*, of which the ones with Ar = Ph [49, 50] and Ar = 2,6-dimethylphenyl [51] were the best (for details *cf.* Table 3-3). Only one hydrogen is available for hydride transfer in the 1:1 complex, the other, presumably the one in *endo* position of the rigid bicyclic structure (*10*), is efficiently shielded both by the aryl group and the pyrrolidine ring. Li^+ may also be an essential part of the complex, since addition of agents complexing with Li^+ (1,2-dimethoxyethane or TMEDA) lowered selectivity.

A tridentate ligand (*11*) was prepared from aspartic acid [52]. Its remarkable feature was that it reduced the black sheep substrate PhCOtBu with not only the highest selectivity reported (86% ee) but more selectively than it did other ketones (*cf.* Table 3-3). The reagent is supposed to have the rigid structure *12*. Cyclohex-2-enone was reduced by *12* with complete enantioselectivity (*S*) [53].

A further increase in the number of functional groups seems to do more harm than good. Thus Seebach and his coworkers prepared a series of ten 1,4-bis-dialkylamino-2,3-butanediols from (*R,R*)-tartaric acid and reacted them with LAH [54, 55]. Among them, only two (*13*, X = Me_2N and piperidino) were useful. They were rather unselective with the usual substrates (ee for acetophenone 42 and 30% resp.), but surprisingly selective with mesityl methyl ketone, giving 75 and 85% ee of the *S* alcohol.

Morrison *et al.* [56] were interested in the role and interaction of chiral centers in a series of diethanolamine type ligands (*14–17*). Results obtained on reducing acetophenone with LAH modified with these aminodiols are shown in Fig. 3-1. With ligands *14* and *15* the effects of the carbinol centers and of the center adjacent to nitrogen alone could be explored, while with *16* and *17* counteracting and cooperative effects were at work. The effects are of course not additive, as shown by the high selectivity achieved with ligand *17*.

3.1.1.2 Reduction of C=N and C≡C bonds

Although the first successful experiment for enantioselective reduction with a chiral LAH complex involved the reduction of immonium salts with LAH-(−)-menthol reagent by Červinka in 1961 [57], not very much attention was paid to reducing C=N bonds with chiral LAH complexes. In fact, early results had been rather disappointing, reduction of immonium salts with LAH-(−)-menthol complex gave ≤6% ee [57], that of imines >10%. Even *3* failed to produce a useful degree of enantioselectivity in the reduction of azomethines [58].

The same reagent and a close analogue [59] were, however, moderately selective with ketoximes, which all gave (*S*)-amines in 9.5−56% ee and >60% yield [60]. Essentially the same results were obtained with *O*-methyl- and *O*-tetrahydropyranyl oximes. It was remarkable that with aliphatic ketoximes selectivity was higher than with the aralkyl analogues. Reduction with the ethanol-modified reagent gave the (*R*)-amines.

3.1.2 Chirally Modified Borohydrides*

Much less attention and success is associated with sodium borohydride-based enantioselective reductions than with those involving LAH.

Only a few chiral ligands have been reported as yet which lend acceptable enantioselectivity to $NaBH_4$. 1,2:5,6-Diisopropylidene-α-D-glucofuranose (*18*) was selected out of six monosacharide derivatives by Hirao *et al.* [61]. Even this was

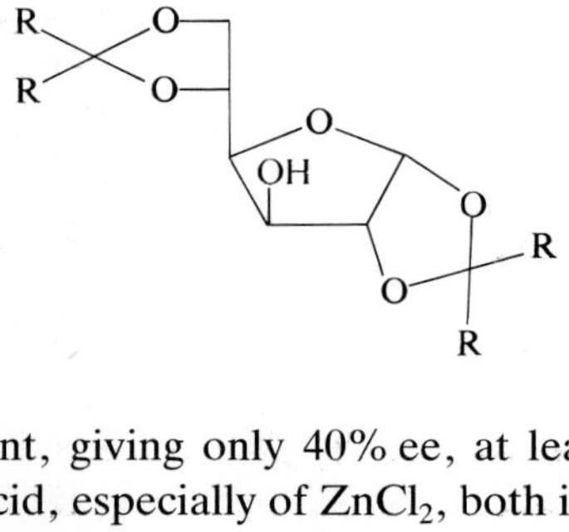

(*18*) R=Me
(*19*) R= $-(CH_2)_5-$

inefficient, giving only 40% ee, at least with aralkyl ketones [61]. Addition of a Lewis acid, especially of $ZnCl_2$, both improved selectivity and inverted the prevailing configuration from *R* to *S* [62, 63]. Improvements of similar magnitude but without a change of configuration could be achieved when, as a third component, a carboxylic acid (preferably isobutyric acid) was added [64]. Temperature, solvent and molar ratio of acid were found to influence yields and ee in opposite directions,

* For a review see ref. [97].

Table 3-4. Reduction of Ketones (R^1COR^2) with Chirally Modified Sodium Borohydrides. (A: $NaBH_4$ + 1,2:5,6-di-*O*-isopropylidene-α-D-glucofuranose (*18*) + $ZnCl_2$ (1:2:0.33), THF, 30°C; B: $NaBH_4$ + (*18*) + $iPrCO_2H$ (1:2:1.2), THF, 0°C; C: $NaBH_4$ + bovine serum albumin 1.7 mM/H_2O.)

R^1	R^2	A		B		C	
		% ee	(Conf.)	% ee	(Conf.)	% ee	(Conf.)
Ph	Me	50	(*S*)	78	(*R*)	45	(*R*)
	Et	68	(*S*)	85	(*R*)	78	(*R*)
	Pr	58	(*S*)	74	(*R*)	27	(*R*)
	iPr	28	(*S*)	57	(*R*)	66	(*R*)
	Bu					14	(*S*)
	tBu					22	(*R*)
2-Naph	Me	20	(*S*)			66	(*R*)
iPr		37	(*S*)				

and a compromise was found on using reagent composed of $NaBH_4$, *18* and $iPrCO_2H$ (1:2:1.2) at 0°C in THF [65] (for details *cf.* Table 3-4).

Morrison, Grandbois and Howard investigated the reduced aralkyl ketones with a reagent obtained from $NaBH_4$, *19*, and 3-methylpentanoic acid (1:2:1) in THF [66], but the results were somewhat inferior to those obtained with *18*.

Recently Soai *et al.* prepared fairly selective reducing agents by reacting (*R*)-*N*-benzoylcysteine and -cystine [68] with $LiBH_4$ and tert-butanol, which gave up to 90% ee of (*R*)-aralkyl carbinols. Sodium triacyloxyborohydrides prepared from *N*-benzoyl- or *N*-benzyloxycarbonylproline reduced dihydroisoquinolines with up to 86% ee [69].

The solubility of $NaBH_4$ in water and its stability to alkali make it an obvious candidate for phase transfer catalysis experiments, and several attempts at enantioselective reductions with $NaBH_4$ in the presence of chiral phase transfer catalysts have been reported.*

Being a process operating with catalytic ammounts of relatively cheap and recoverable chiral agent, phase transfer catalysis is economically very attractive. Unfortunately, so far results have not been very encouraging. This is not quite unexpected, since the involvement of the chiral cation in the hydride transfer step may not be very intimate.

Quaternary ammonium salts prepared from alkaloids such as (−)-*N*-alkyl-*N*-methylephedrinium bromide [70–72], (−)-*N*-benzylquininium chloride [73] and phosphonium salts derived from tartaric acid [74] have been employed as catalysts. Since a claim [70] for 39% ee in the reduction of acetophenone in the presence of (−)-*N*-dodecyl-*N*-methylephedrinium bromide was refuted [73] (only 4.5% was found), the highest selectivity (35% ee of (*R*)-carbinol) has been found in the two-

* The active species in these reductions is BH_4^- associated with a chiral cation and this can therefore be regarded as a reaction involving a chiral reagent rather then a true catalytic process.

phase reduction of phenyl t-butyl ketone in the presence of (−)-*N*-benzylquininium chloride [73].

The water solubility of $NaBH_4$ permits the use of proteins as chiral aids in reductions. While results were modest with lecithin [75] (no ee values reported), selectivities achieved in the presence of bovine serum albumin were remarkable [76] (*cf.* Table 3-5). Ee values increased with protein concentration and dropped sharply on denaturation.

3.1.3 Chiral Boranes

Reaction of borane with α-pinene gives diisopinocampheylborane (IPC_2BH), a highly enantioselective hydroborating agent (*cf.* Section 8.3). Reduction of ketones with both (+)- and (−)-IPC_2BH gave carbinols with relatively low ee, and there was a controversy as to the configuration of the predominant isomer [77−79]. The reaction was later reinvestigated using (−)-IPC_2BH [(−)-*20*] of 100% optical purity [79]. Even then selectivities were low.

(−)-IPC—BBN $\xrightarrow[\text{THF, −78°C}]{\text{tBuLi}}$ [ ]$^{\ominus}$ $Li^{\oplus}$ +

(−)-(*20*) (*24*) (*21*)

(−)-Monoisopinocampheylborane [(−)-$IPCBH_2$], a reagent which is more difficult to prepare than (−)-*20*, gives comparable ee values and the same predominant configuration, except for acetophenone [81, 82].

A chiral boron hydride with tetracoordinate boron (*21*)* was prepared from B-(3α-pinanyl)-9-borabicyclo[3.3.1]nonane (*24*) (itself a reducing agent) by reacting it with tert-butyl lithium. *21* is a very active reducing agent with moderate enantioselectivity consistently giving the (*R*)-carbinols in excess (up to 37% ee) [83].

A similar reagent prepared from the borane *25* (see later) showed unexceptional selectivity in the reduction of aralkyl ketones but was relatively effecient with dialkyl ketones (70−79% ee with unbranched ketones). The prevailing configuration (*S*) was opposite to that obtained with the parent borane [84].

---

* The parachute-like symbol represents the carbon skeleton of 9-borabicyclononane (9-BBN).

Boranes, as Lewis acids, readily form complexes with amines and, when the amine is chiral, the complex is a chiral reducing agent. Despite continuous efforts, no highly enantioselective amine-borane reagent could be developed. The first attempts by Fiand and Kagan with ephedrine derivatives gave barely perceptible enantioselection (ee ≤5%) [85]. Complexes of 1-phenylethylamines [86, 87] or of α-aminoesters, even when combined with activation of the carbonyl group (by BF_3) [88, 89], gave but up to 23% ee. The borane complex of the *N*-benzoyl-(*S*)-proline sodium salt was reported to give 50% ee of the (*S*)-carbinol on reduction of propiophenone [90].

When aminoalcohols were used as complexants, one of the hydrogen atoms in borane was exchanged for an alkoxy group and reagents with much improved selectivity were formed [91, 92, 93], the most efficient of which (*22*) gave >94% ee of (*R*)-carbinols on reduction of aralkyl ketones (Fig. 3-2) [94].

(*22*) R^1COR^2 —(22), THF, 30°C→ $R^1CH(OH)R^2$ (*23*)

R^1	R^2	% ee
Me	iPr	60
	Bu	55
	iBu	61
	tBu	78
	Ph	94

R^1	R^2	% ee
Me	2-Naph	52
Et	iPr	94
	Ph	60
Bu		>99

Fig. 3-2. Enantioselective reduction of ketones with a borane-aminoalcohol complex.

Although formally it is outside the scope of this section, the aminoalane *23* obtained by reacting LAH with the corresponding amine hydrochloride [95] is a borane analogue. It probably exists as the dimer and in ether at −70°C it reduces acetophenone and phenyl benzyl ketone to the (*S*)-carbinols with remarkable selectivity (84.5 and 61% ee resp.).

3.1.4 Enantioselective Reductions with Hydride Transfer from Carbon

Carbon atoms may transfer a hydride ion to a carbonyl group when activated by a metal atom two bonds away which also provides a pair of electrons to form a double bond. This general pattern is illustrated by one of the first processes of this kind discovered, the Meerwein-Pondorf-Verley reduction. The six-membered cyclic transition state for the mechanism of this and similar reactions is a very obvious, and in most cases adequate, representation, though exceptions are known.

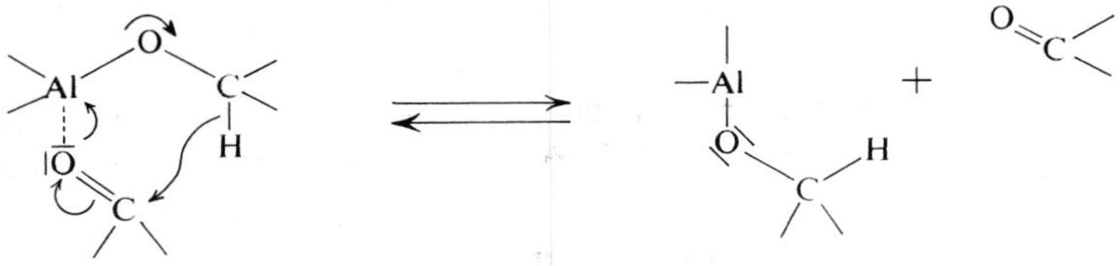

The activating atom can also be attached to carbon, and enantioselective reductions by transfer of hydrogen from a β-carbon in trialkyl boranes as well as in alkyl beryllium, aluminum and magnesium compounds are known.

3.1.4.1 Chiral trialkyl boranes*

Usually trialkylboranes show reducing properties only under vigorous conditions [96]. Midland and his coworkers have developed a reagent, B-(3α-pinanyl)-9-borabicyclo[3.3.1]nonane (*24*, Alpine-borane®**), readily availably from 1,5-cyclooctadiene, borane and α-pinene, which reduces carbonyl groups to alcohols

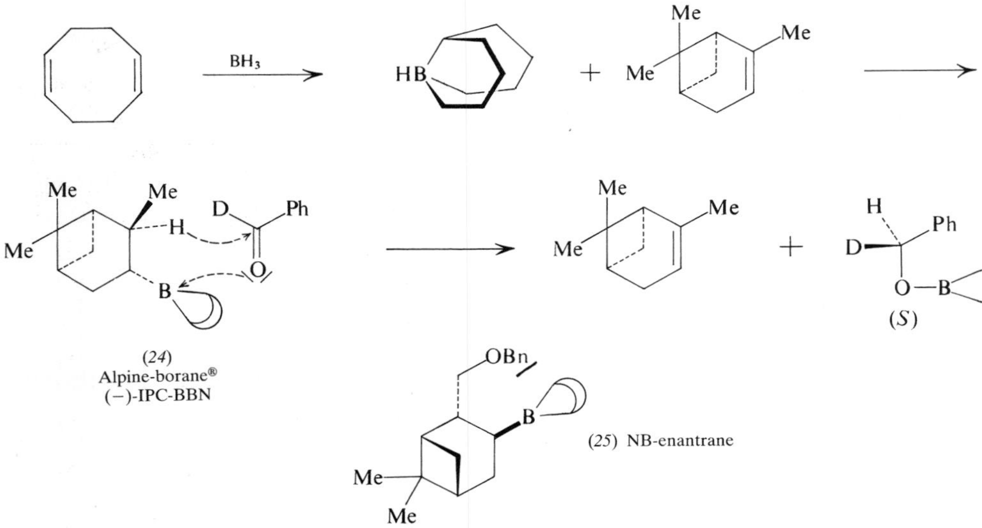

(*24*) Alpine-borane® (−)-IPC-BBN

(*25*) NB-enantrane

Fig. 3-3. Enantioselective reductions with chiral trialkylboranes.

* For recent reviews see refs. [81] and [97].
** Trade name of Aldrich Chemical Co.

under very mild conditions and generally with high stereoselectivity (Fig. 3-3 [98–102].

For example, at room temperature *24* reduced deuteriobenzaldehyde to (*S*)-benzyl-α-d-alcohol of 100% ee [99]. Reduction of the substrate is coupled with the elimination of α-pinene, which may then be recycled.

As terpene partners, (−)-β-pinene, [103] (−)-camphene and (+)-3-carene were also tested, but all produced less selective reagents [99]. Only the borane derived from *O*-benzylnopol (*25*, NB-enantrane) was of comparable selectivity, but it was less reactive [104, 105].

Alpine-borane® (*24*) is outstandingly selective in the reduction of aliphatic, allylic and aromatic aldehydes [99] and of α,β-acetylenic ketones [100–102]. With other ketones reduction is too sluggish and accompanied by the dissociation of the reagent to α-pinene and borabicyclo[3.3.1]nonane. Since the latter is an achiral reducing agent, enantioselectivity suffers [106]. This can be remedied by performing the reaction at high pressure [107] or by applying the reagent neat which permits the reduction of any kind of ketone, albeit with varying enantioselectivity [108]. The neat reagent reduces α-ketones to α-hydroxyesters with almost complete enantioselectivity [109]. The most important results obtained with Alpine-borane® (*24*) are summarized in Table 3-5. Note that the reduction products of α,β-acetylenic ketones are all homochiral.

Table 3-5. Reduction of Carbonyl Compounds (R^1COR^2) with (+)-B-(3α-Pinanyl)-9-borabicyclo[3.3.1]borane (*24*) at Room Temperature in THF or Neat (*).

R^1	R^2	% ee[a]	(Conf.)[b]	R^1	R^2	% ee[a]	(Conf.)[b]
Pr	D	100	(*S*)	(*E*)-PhCH=CH	Me	97*	(*S*)
Pent		89	(*S*)	1-Cyclohexenyl		64*	(*S*)
$tBuCH_2$		98	(*S*)	Ph		83*	(*S*)
(*E*) PhCH=CH		84	(*S*)			92[c]	(*S*)
Ph		98	(*S*)	Et		43*	(*S*)
$4-NO_2-C_6H_4$		100	(*S*)	iPr		62*	(*S*)
$4-MeO-C_6H_4$		82	(*S*)			83[c]	
Ph	C≡CBu	89	(*S*)	tBu		1*	(*S*)
Me	C≡CPh	100*	(*R*)	Me	CO_2Me	86*	(*S*)
iPr	C≡CH	99*	(*R*)		CO_2tBu	100*	(*S*)
tBu		99	(*S*)	Et		100*	(*S*)
Me	$C≡CCO_2Et$	77	(*R*)	iPr		100*	(*S*)
Pent		92	(*R*)	Ph		100*	(*R*)
Ph		100	(*S*)				

[a] Corrected for enantiomerically pure reagent. [b] All products derived from α,β-acetylenic ketones are homochiral. [c] At 600 atm.

3.1.4.2 Chiral metal alkyls

The standard reaction of metal alkyls with carbonyl compounds is the addition of the alkyl group to form carbinols. It was, however, recognized very early that Grignard reagents with a hydrogen available at the β-carbon, especially those with a bulky alkyl group, may transfer this hydrogen to the substrate with the concomitant formation of an olefin. Trialkyl aluminum, dialkyl zinc and dialkyl berrylium compounds are also amenable to reduction of carbonyl compounds. When the alkyl groups are chiral, the reduction may be enantioselective. Selectivity is generally poor and therefore, although of much theoretical interest, such reagents hardly qualify for practical application.

It is convenient to envisage the mechanism of hydrogen transfer as involving a six-membered transition state in which the electron deficient metal atom is coordinated to the carbonyl oxygen. Polarization of the C=O bond by coordination facilitates the transfer of a hydride ion to the carbon atom. This model permits the prediction of the prevailing enantiomer by evaluating steric hindrance in the two possible mutual orientations of substrate and reagent. For a reagent with a chiral center in β-position, an example (a) [110] and its generalization (b) are shown in Fig. 3-4. This model is a gross simplification of the real situation, since it does not accrount for (i) coordination with the solvent (crucial with Grignard reagents), (ii) for the role of the other alkyl group(s) (in dialkyl and trialkyl metals), and (iii) for the possibly oligomeric structure of the reagent. In addition, Grignard reagents may disproportionate to dialkyl magnesium and magnesium halide.

Enantioselective reactions with Grignard reagents have been extensively reviewed by Morrison and Mosher who covered the literature up to 1968 [33]. The problems in this field are rather similar to those encountered with other metal alkyls and will now be discussed briefly.

The potential of Grignard reagents for enantioselective reduction was recognized as early as 1946 by Vavon and coworkers who used a mixture of bornyl- and isobornylmagnesium chlorides, the latter being the active agent [111]. Systematic studies were later conducted mainly by Mosher and his school [33].

The model in Fig. 3-4 predicts the configuration of the product in excess as soon as one can make up ones mind which one of the ligands is "larger". Thus for the reduction of phenyl alkyl ketones with (*S*)-2-methylbutylmagnesium chloride Ph > tBu [112], while the opposite should be presumed when the reagent is (*S*)-2-phenylpropylmagnesium chloride (22% ee of the (*R*)-carbinol) [113]. But before anyone should contemplate an explanation for this result, he should be reminded that 2-phenyl-3-methylbutylmagnesium chloride reduces t-butyl phenyl ketone with the highest enantioselectivity observed as yet for a Grignard reagent (91%) to the (*S*)-carbinol [113]! The simple model in Fig. 3-4 also requires that CF_3 should be taken as being larger than phenyl, an assumption with clearly no physical content.

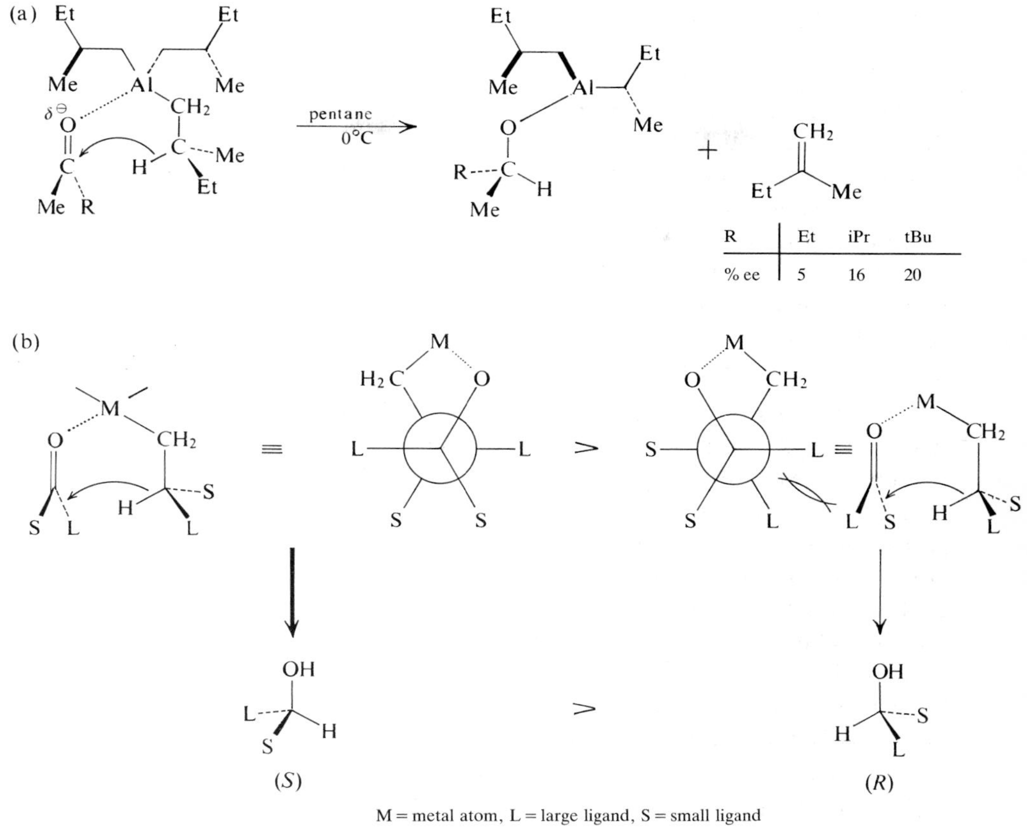

R	Et	iPr	tBu
% ee	5	16	20

M = metal atom, L = large ligand, S = small ligand

Fig. 3-4. Enantioselective reduction of ketones with alkylmetals.

The importance of electronic effects in enantioselective reductions with Grignard reagents was demonstrated by Capillon and Guetté, who investigated the reduction of 4-substituted benzophenones [114] and aralkyl ketones [115] with a series of 2-aryl-butylmagnesium chlorides (*26*) (Fig. 3-5). No unambiguous interpretation of these results is as yet possible.

A comparison of a series of Grignard reagents of the type Me-$CHRCH_2MgCl$ and their homologues, *i. e.* Me-$CHR(CH_2)_2MgCl$, in the reduction of aralkyl ketones [116] also sheds doubt on the validity of the model in Fig. 3-5. While selectivity moderately improved when the size of the R group increased, no such regularity was observed on the substrate side. Reagents with a chiral center in γ-position [(*S*)-$MeCHRCH_2CH_2MgCl$, R=Et, iPr, tBu] were generally of low selectivity.

Since a generally valid interpretation of enantioselective reductions by chiral Grignard reagents is not yet available, we have contented ourselves with compiling some of the representative data in Table 3-6.

Table 3-6. Enantioselective Reduction of Carbonyl Compounds R^1COR^2 by Chiral Alkyl Magnesium Chlorides (R*MgCl). (Data from refs. [111-113] and [116–118].)

R^1	R^2	R* (S)-EtCH(Me)CH_2 % ee	(Conf.)	Isobornyl % ee	(Conf.)	(S)-iPrCH(Me)CH_2 % ee	(Conf.)	(S)-tBuCH(Me)CH_2 % ee	(Conf.)	(S)-PhCH(Me)CH_2 % ee	(Conf.)	(S)-PhCH(Et)CH_2 % ee	(Conf.)	(S)-PhCH(iPr)CH_2 % ee	(Conf.)
Ph	D	17	(*S*)									67	(*S*)		
	Me	4	(*S*)	36	(*R*)					38	(*S*)	47	(*S*)		
	Et	6	(*S*)	19	(*R*)	14	(*S*)	25	(*S*)	38	(*S*)	52	(*S*)	66	(*S*)
	iPr	24	(*S*)	55	(*R*)	67	(*S*)	65	(*S*)	59	(*S*)	82	(*S*)	80	(*S*)
	tBu	16	(*S*)	52	(*R*)	2	(*R*)	36	(*S*)	22	(*R*)	16	(*S*)	91	(*S*)
	CF_3^a	22	(*S*)							47	(*S*)	38	(*S*)	22	(*S*)
tBu	D	12	(*S*)									29	(*S*)		
	Me	13	(*S*)												
	Et	11	(*S*)												
	iPr	0													
	CF_3^a	0								74	(*S*)	63	(*S*)		
Me	CF_3^a	3	(*S*)							67	(*S*)	64	(*S*)		

[a] Note the change of priority of the ligands.

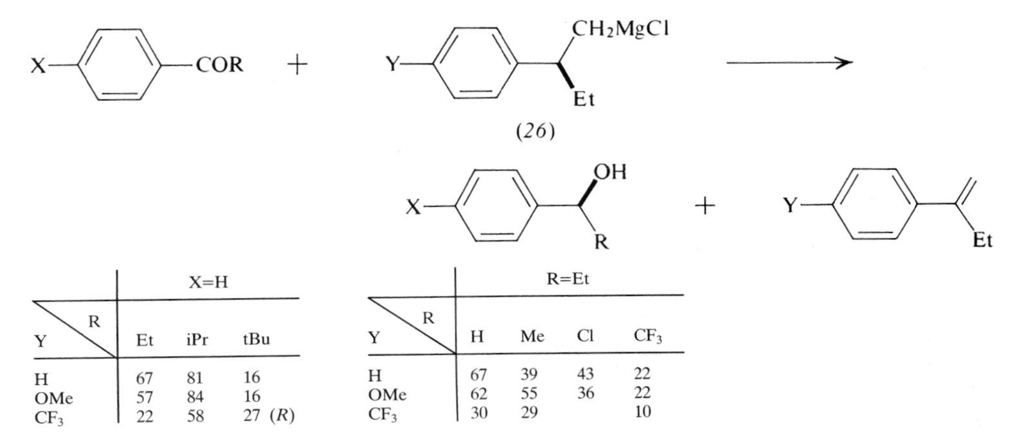

Y \ R	X=H Et	iPr	tBu
H	67	81	16
OMe	57	84	16
CF_3	22	58	27 (*R*)

Y \ R	R=Et H	Me	Cl	CF_3
H	67	39	43	22
OMe	62	55	36	22
CF_3	30	29		10

Fig. 3-5. Reduction of aromatic ketones with chiral Grignard reagents.

Data on reductions with metal alkyls carrying two chiral ligands are scarce. Di-(*R*)-2-methylbutyl beryllium was reported to reduce aralkyl ketones with up to 46% ee [119]; the analogous zinc reagent showed only marginal enantioselectivity (15%) [120].

Giacomelli, Lardicci, Menicagli and others [110, 119, 121–126] as well as Kretchmer [49] thoroughly investigated enantioselective carbonyl reductions with chiral trialkyl aluminum compounds, first of all with tris-(*S*)-2-methylbutyl-aluminum. Enantioselectivities achieved with these reagents did not exceed 50%; the mechanism proposed and the difficulties to accommodate some of the experimental results with it were all very similar to those already indicated for reductions with Grignard reagents. Selected results taken from refs. [116, 122, 123, and 128] are shown in Fig. 3-6. An interesting feature is the inversion of the prevailing configuration (again not without an exception) on changing the R group in the reagent from Et to iPr or tBu [128]. Application of the diethyl etherate of the reagent does not offer distinct advantages [123].

R^1COR^2 + $(R\text{–}CH(Me)\text{–}CH_2)_3Al$ $\xrightarrow{\text{pentane}}$ R^1CHOHR^2 + $R\text{–}C(Me){=}CH_2$

R=Et, R^1=Ph, −60°C

R^2	% ee (conf.)
Me	8 (*S*)
Et	15 (*S*)
iPr	44 (*S*)
tBu	30 (*S*)
CF_3	12 (*S*)

R^1=Ph, 0°C

R	R^2	% ee (conf.)
iPr	Et	22 (*R*)
	iPr	56 (*R*)
	tBu	26 (*R*)
tBu	Et	23 (*R*)
	iPr	20 (*R*)
	tBu	13 (*S*)

Fig. 3-6. Enantioselective reduction of ketones with chiral aluminium alkyls.

A chiral dialkyl aluminum dichloride (*27*) was prepared from β-pinene in two laboratories concurrently [103, 129] and used for the reduction of the usual set of aralkyl ketones. Enantioselectivity was modest, but unlike with hydride reagents, it improved on increasing the size of the alkyl group [129].

3.1.4.3 Chiral metal alkoxides

Metal alkoxides with a hydrogen atom available in α-position, *i. e.* two bonds away from the metal atom, are capable of transferring a hydride ion to a cabonyl group. For practical purposes secondary alcoholates of aluminum and magnesium have proved to be of value.

Reduction with achiral aluminum alcoholates (the Meerwein-Ponndorf-Verley reduction) has long been exploited as a diastereoselective method. Aluminum alkoxides formed with chiral secondary alcohols are very interesting for mechanistic studies but not for practical purposes, since their enantioselectivity is generally very low (<22%) [130].

Chiral alkoxyaluminum dichlorides, obtained by mixing 1 mole of $LiAlH_4$ with 3 moles of $AlCl_3$ and thereafter adding 4 moles of chiral alcohol [131], are much more efficient reagents, but are still not competitive with LAH-complexes except for special cases [132]. Thus (−)-menthylaluminum dichloride (*21*) reduces phenyl trifluoromethyl ketone to the (*S*)-carbinol in 77% ee [133], and (−)-bornylaluminum dichloride was used for the enantioselective reduction of aryl aminoalkyl ketones (60–92% ee) [134].

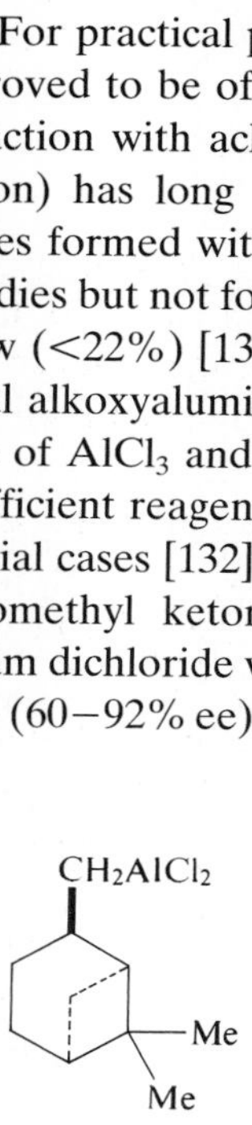

CH_2AlCl_2 ... Me, Me (*27*) $OAlCl_2$ ... iPr, Me (*28*)   Me, Me, Me, OMgBr, D (*29*)

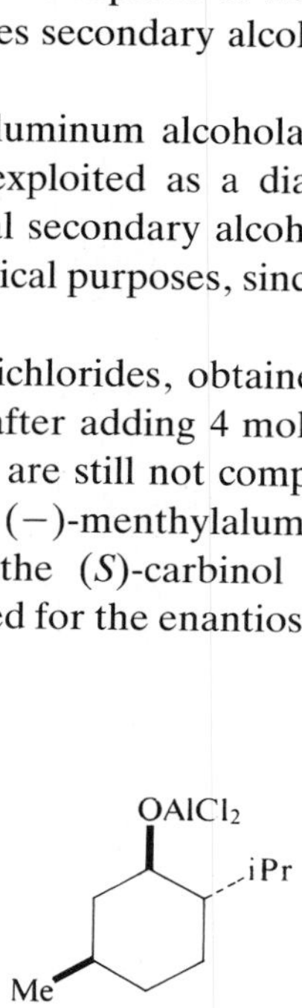

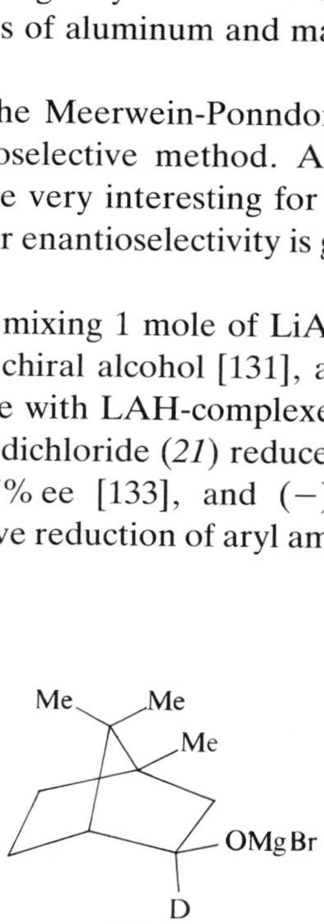

Chiral halomagnesium alkoxides are rather similar to the alkoxyaluminum dichlorides, both in their scope and efficiency. A reagent derived from a primary alcohol, namely (*S*)-2-methylbutoxymagnesium bromide, was unsatisfactory [135], but the (−)-isobornyl analogue proved to be useful for the reduction of aldehydes. For example, the deuteriated reagent (*29*) reduced acetaldehyde to (*S*)-1-deuterioethanol of 44% ee [136], and the unlabelled reagent reduced deuteriobenzaldehyde to (*R*)-benzyl-α-deuterioalcohol of 65% ee [131]. This remarkable deuterium isotope effect supports the cyclic transition state model shown in Fig. 3-4.

3.1.4.4 Chiral 1,4-dihydropyridines

Dihydropyridines, primarily nicotinamide-adenine-dinucleotide phosphate (NADPH), play a central role in biochemical redox processes. The reaction, which is shown schematically below, has been the subject of highly interesting model studies, and although the results are not yet of immediate practical interest, they deserve to be surveyed briefly.

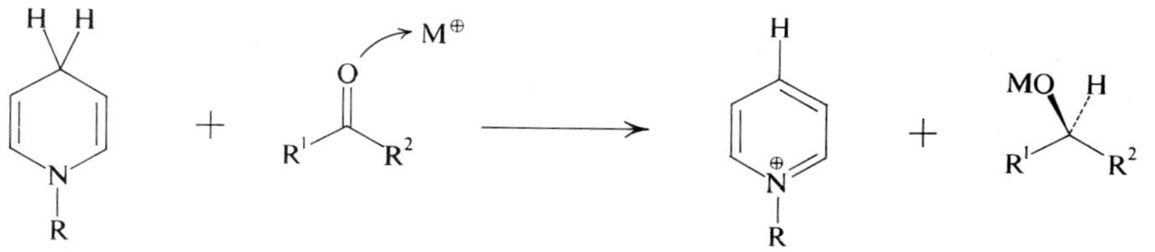

Since dihydropyridines themselves are not potent hydride donors, activation of the acceptor, usually by a divalent cation such as Mg^{2+} or Zn^{2+}, is necessary. Enantioselectivity in the hydride transfer step requires, of course, that some chiral functionality should be attached to the dihydropyridine nucleus. With NADPH this is a dinucleotide moiety linked to the nitrogen.

The standard experiment for the testing of NADPH models is the reduction of ethyl benzoylformate to ethyl phenyllactate in the presence of $Mg(ClO_4)_2$ in acetonitrile. In Fig. 3-7 the structure of some model compounds, in Table 3-7 enantioselectivities achieved in standard experiments are shown.

Not unexpectedly, highest selectivity was observed with *30*, in which the hydride-donating carbon was also a chiral center. The configuration at the side chain was irrelevant to both configuration and ee. Thus (*R,R*)- and (*S,R*)-*30* both gave the (*S*)-alcohol in nearly the same ee (97.6 and 96% resp.) [137], while the analogue (*31*) with no chiral center in the dihydropyridine ring was quite unselective (ee 16%). Without added Mg^{2+} selectivity dropped to 52%, which suggested that Mg^{2+} not only activated the acceptor group but contributed to the rigidity of the activated complex.

Enantioselectivities comparable to that achieved with *30* could only be realized with two compounds, *32* and *33*, belonging to the bridged series (*32*–*37*). In both, the dihydropyridine rings are linked by a sequence of six carbon atoms. In this case again selectivity increases on the addition of Mg^{2+}, until one equivalent of Mg^{2+} is provided [138]. Another series in which optimum selectivity could be achieved by a systematic modification of molecular geometry was that of crown ethers, *e.g.* *38* and *39*.

Chiral dihydropyridines were also used for the reduction of several other substrates. Among them conversion of $tBuCOCO_2Me$ with total selectivity to the (*R*)-alcohol and of $PhCOCF_3$ with 70% ee to the (*R*)-alcohol by the 2-methyl analogue of *30* [137] should be mentioned.

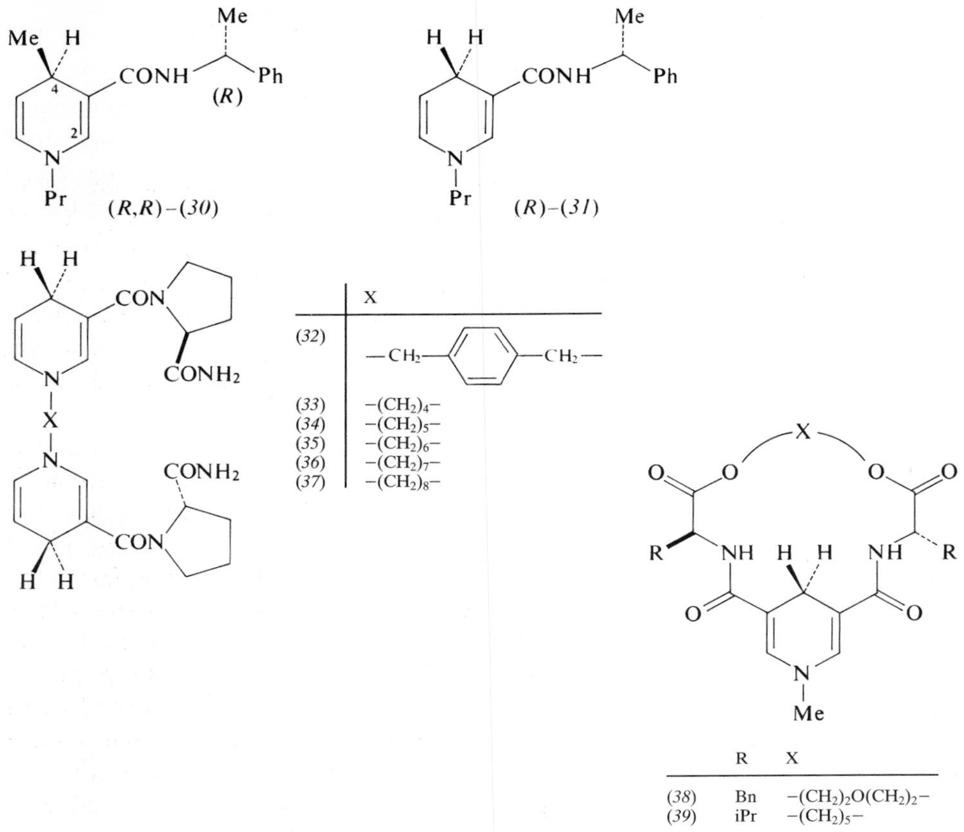

(R,R)-(30) (R)-(31)

	X
(32)	$-CH_2-C_6H_4-CH_2-$
(33)	$-(CH_2)_4-$
(34)	$-(CH_2)_5-$
(35)	$-(CH_2)_6-$
(36)	$-(CH_2)_7-$
(37)	$-(CH_2)_8-$

	R	X
(38)	Bn	$-(CH_2)_2O(CH_2)_2-$
(39)	iPr	$-(CH_2)_5-$

Fig. 3-7. Dihydropyridines (NADH models) used for the enantioselective reduction of ketoesters.

Table 3-7. Reduction of Ethyl Benzoylformate to Ethyl Phenyllactate with Chiral Dihydropyridines in Acetonitrile in the Presence of $Mg(ClO_4)_2$ at r. t.

Compound	% ee	(Conf.)	Ref.	Compound	% ee	(Conf.)	Ref.
(*R, R*)-*30*[a]	97.5	(*R*)	137	*35*	95.6	(*R*)	138
(*S, R*)-*30*[a]	96.5	(*S*)	137	*36*	59	(*R*)	138
31	16	(*R*)	137	*37*	81	(*R*)	138
32[b]	93.5	(*R*)	138	*38*	87	(*S*)	139
33	40	(*R*)	138	*39*	90	(*S*)	139
34	43	(*R*)	138				

[a] The methyl ester was reduced.
[b] Yield 80%, at 66% conversion ee 98.1%.

3.1.5 Correlation of Substrate Constitution and Enantioselectivity

One of the aims of studies concerned with enantioselective reductions has always been the elucidation of the mechanism by which enantiotopic faces of a prochiral carbonyl group are distinguished. An obvious way to gather some information about the nature of the transition state is to determine selectivites with a range of substrates under identical conditions. In our case this approach exposed certain tendencies but failed to provide a real insight into the process. In Table 3-8 enantioselectivities obtained with the most important chiral reducing agents and a selected set of carbonyl compounds were compiled.

The substrates given in Table 3-8 have, as the only group capable of coordination, the carbonyl function, and the substituents attached to it are either apolar or of medium polarity. Therefore it is not unreasonable to anticipate that the structure of the transition states should be controlled by steric effects. Thus one of the enantiotopic faces of the carbonyl compound should be the more preferred the greater the difference between the effective size of the substituents. Thus stereoselectivities should increase in the order Et < iPr < tBu in the methyl ketone series, and decrease in the same order in the phenyl ketone series. This expectation was not borne out by experiment. In the dialkyl ketone series the general trend tends to be Et < iPr > tBu and in the aromatic series Me ≅ Et > iPr, but in both exceptions are common. Since tBu is certainly more bulky than phenyl, with PhCOtBu there should be a change in the prevailing configuration relative to PhCOMe, but this occurs only rarely. Perhaps the irregular behavior of tert-butyl ketones can be traced back to the fact that this group cannot turn out in a way that a C–H bond should point towards the inside of the transition state complex (as is possible, *e. g.*, with the isopropyl group). The behavior of trifluoromethyl phenyl ketone is also capricious. The product of the reduction is sometimes homochiral, in other cases, however, heterochiral to that obtained from acetophenone.

The rod-like ethynyl group is one of the sterically least demanding groups, nevertheless enantioselectivities with α,β-acetylenic ketones are generally high and the products are (with one exception) all homochiral when the same reagent is used. A common trait of the phenyl and ethynyl groups is high π-electron density, which may be the clue to the strong stereodirecting effect characteristic for both.

In order to be able to resolve the anomalies associated with enantioselective hydride transfer, a model based on some experimental evidence other than product distribution is called for. The retrospective construction of transition state models "so as to fit results" is especially futile in this field, where very often the exact structure of the reagent is uncertain.

Table 3-8. Enantiomeric Excess (% ee) and Configuration of the Major Enantiomer in Reduction of Carbonyl Compounds with Chiral Hydride-Donating Reagents.

Reducing agent	MeCOEt	MeCOiPr	MeCOtBu	PhCDO[a]	PhCOMe	PhCOEt	PhCOiPr	PhCOtBu	$PhCOCF_3$[b]	$R^1C{\equiv}CCOR^2$[c]
LAH + (*R*)-2,2-dihydroxy-1,1′-binaphtol + EtOH [23]				82 (*R*)	95 (*R*)	98 (*R*)	71 (*R*)			57–90 (*S*)
LAH + 3-*O*-benzyl-1,2-cyclohexylidene-α-D-glucopyranose + EtOH [15]			18 (*R*)		71 (*R*)	46 (*R*)				
LAH + (2*S*,3*S*)-1,4-bis-(dimethylamino)butane-2,3-diol [54, 55]	24 (*S*)		12 (*S*)	8 (*S*)	42 (*S*)	44 (*S*)	27 (*S*)	21 (*S*)		
LAH + quinine [34, 140]	0	6 (*S*)	11 (*S*)	35 (*R*)	48 (*R*)			2 (*S*)		
LAH + (2*S*,3*R*)-4-dimethylamino-1,2-diphenyl-3-methyl-2-butanol (Darvon alcohol) fresh reagent [5, 43, 44]			28 (*R*)		75 (*R*)		30 (*R*)	36 (*R*)		
aged reagent [5]			21 (*S*)		75 (*S*)		20 (*S*)	28 (*R*)		66–91 (*R*)
LAH + (1*R*,2*S*)-*N*-methylephedrine + 3,5-dimethylphenol [35–37]	14 (*S*)	41 (*S*)	21 (*S*)		84 (*R*)	85 (*R*)	17 (*R*)	63 (*S*)	30 (*R*)	79–90 (*R*)
LAH + (1*R*,2*S*)-*N*-methylephedrine + *N*-ethylaniline [40]					84 (*S*)	90 (*S*)	78 (*S*)			
LAH + (*S*)-anilino-3-methyl-1-butanol [52]					51 (*S*)	68 (*S*)	77 (*S*)	86 (*S*)		
LAH + (*S*)-*N*-methyl-1-phenylethylamine [141]			23 (*S*)		85 (*S*)					

Table 3-8. Continued.

Reducing agent	MeCOEt	MeCOiPr	MeCOtBu	PhCDO[a]	PhCOMe	PhCOEt	PhCOiPr	PhCOtBu	$PhCOCF_3$[b]	$R^1C{\equiv}CCOR^2$[c]
LAH + (*S*)-*N*-(2-dimethylaminobenzyl)-1-phenylethylamine [47]					43 (*R*)	52 (*R*)		47 (*R*)		
LAH + (*S*)-2-anilinomethylpyrrolidine [49]					92 (*S*)	85 (*S*)	50 (*S*)			
LAH + (4*S*,5*S*)-2-ethyl-4-hydroxymethyl-5-phenyl-2-oxazoline [46]					65 (*R*)	62 (*R*)	43 (*R*)			
$LiBH_4$ + *N*-benzoyl-(*R*)-cysteine[4] + tBuOH [67]					87 (*R*)	88 (*R*)	57 (*R*)			
LAH + (*S*)-2-(2,6-dimethylphenylamino)methylpyrrolidine [51]					95 (*S*)	96 (*S*)	89 (*S*)			
(−)-Monoisopinocampheylborane [81]	22 (*S*)	46 (*S*)	21 (*S*)		15 (*S*)					
(−)-Diisopinocampheylborane [79]	13 (*S*)	37 (*S*)	20 (*S*)	30 (*R*)	9 (*R*)					
Lithium-B-(3α-pinanyl)-9-borabicyclo[3.3.1]nonylhydride [83]	29 (*S*)		3 (*R*)		17 (*R*)	13 (*R*)				
$NaBH_4$ + 1,2:5,6-diisopropylidene-α-D-furanose										
(a) + $ZnCl_2$ [62, 63]		37 (*S*)			50 (*S*)	68 (*S*)	28 (*S*)			
(b) + $iPrCO_2H$ [64]					78 (*R*)	85 (*R*)	57 (*R*)			
(c) + serum albumin [76]					45 (*R*)	78 (*R*)	66 (*R*)	22 (*R*)		

Table 3-8. Continued.

Reducing agent	MeCOEt	MeCOiPr	MeCOtBu	PhCDO[a]	PhCOMe	PhCOEt	PhCOiPr	PhCOtBu	$PhCOCF_3$[b]	$R^1C{\equiv}CCOR^2$[c]
(+)-B-(3α-Pinanyl)-9-borabicyclo[3.3.1]borane (Alpine-borane) [99, 101, 108]	43 (*S*)	62 (*S*)	0.7 (*S*)	98 (*S*)	85 (*S*)					77–100 (*R*)
NB-Enantride [84]	76 (*S*)	2 (*S*)			70 (*S*)				50 (*R*)	10 (*S*)–30 (*R*)
(*S*)-2-Methylbutylmagnesium chloride [116]			13 (*S*)	17 (*S*)	4 (*S*)	6 (*S*)	24 (*S*)	16 (*S*)	22 (*S*)	
(*S*)-2-Phenylbutylmagnesium chloride [113]					38 (*S*)	38 (*S*)	59 (*S*)	22 (*S*)	47 (*S*)	
Tris-[(*S*)-2-methylbutyl]aluminum [122, 123]					8 (*S*)	15 (*S*)	44 (*S*)	30 (*S*)	12 (*S*)	
Tris[(*S*)-2,3,3-trimethylbutyl]aluminum [123]						23 (*R*)	20 (*R*)	13 (*S*)		
(−)-α-Pinanyl-aluminum dichloride [129]			10.5 (*R*)		32 (*R*)	56 (*R*)	63 (*R*)	82.6 (*R*)		64–85 (*R*)

[a] Results obtained with deuterated reagents were transformed as if obtained with PhCDO and the protio reagent.

[b] Note the change in priority: $CF_3 > Ph$.

[c] R^1 and R^2 alkyl groups, configuration given for priority of C≡C over R^1.

3.2 Diastereoselective Reduction of Carbonyl Groups

Diastereoselective reduction processes are highly interesting both for theory and praxis. The practically important cases involve almost exclusively the reduction of C=N and C=O bonds. The diastereoselective reduction of C=N bonds is most often carried out by catalytic hydrogenation and was discussed in Section 2.3.2. Thus, here we only deal with the reduction of the carbonyl group and discuss even this subject much more briefly than its importance wold justify. Not only is our space limited, but also the most important aspects of this subject have already been assimilated by the chemical community and relatively few important novel contributions have been made recently to this field.

Chemical reduction of a carbonyl group can be discussed in terms of a nucleophilic addition of hydride ion to an electrophilic carbon center and is therefore closely related to carbon-carbon bond-forming reactions involving nucleophilic addition. In fact the theoretical background for the two processes is very similar and in this way it was unavoidable to include some examples of C–C bond forming reactions in the present chapter.

3.2.1 Stereochemistry of Diastereoselective Ketone Reductions

Although the necessary requirement for a diastereoselective reduction of a carbonyl group is that its faces should be diastereotopic (*cf.* Section 1.3), a significant degree of selectivity is generally only achieved when the carbonyl group is a member of a ring or when there is a chiral center adjacent to the carbonyl group (1,2-asymmetric induction). There is a sharp contrast in the behavior of the two systems, and much effort has been invested to offer an interpretation which would apply to both.

A series of typical examples for asymmetric 1,2-induction in acyclic carbonyl compounds is shown in Fig. 3-8 [142].*

Example (a) [143] illustrates the effect of the constitution of the inducing group on selectivity, in example (b) [144] the achiral ligand of the carbonyl group is

---

* Here and throughout this book, when discussing diastereoselection involving a chiral compound, only one enantiomer is shown, but arguments apply equally both to its antipode and the racemic mixture.

(a) MeO_2C–CH(R)–C(=O)–Ph $\xrightarrow[MeOH]{KBH_4}$ *syn* + *anti*

R	Me	iPr	tBu
% *syn*	27	66	95

(b) Ph–CH(Me)–C(=O)–R $\xrightarrow[35\,°C]{LiAlH_4}$ *anti* + *syn*

R	Me	Et	iPr	tBu
% *anti*	74	76	85	98

(c) Ph–CH(Me)–CHO $\xrightarrow[-50\,°C]{RLi,\ Et_2O}$ *syn* + *anti*

R	Me	iPr	tBu	Ph
% *syn*	76	87	86	84

Fig. 3-8. Asymmetric 1,2-induction in nucleophilic additions to acyclic carbonyl compounds.

varied, and finally in example (c) [145] the same substrate is reacted with a series of different nucleophiles.

In Fig. 3-9 examples for the addition of nucleophiles to cyclic ketones are shown [146, 147]. It can be seen that even with quite bulky hydride reagents the more stable equatorial alcohols are formed by axial attack from t-butylcyclohexanone while with metal alkyls equatorial attack is preferred. With 2,5,5-trimethylcyclohexanone, however, equatorial attack prevails with any reagent.

Inspection of Figs. 3-8 and 3-9 should convince anybody that the interpretation of stereoselection, even in such simple systems, is difficult, and the situation becomes even more complex when the whole range of reported examples, the effects of solvent, temperature, molar ratio *etc.* are also drawn into the discussion.

From the fifties on, some of the best organic chemists have been challenged to give rationalizations both for the direction and the degree of stereoselections observed. The first successful attempt to predict the direction of stereoselection in acyclic systems was Cram's rule published in a paper by Cram and Elhafez in 1952 [149]. A similar interpretation of 1,4-asymmetric induction in nucleophilic addi-

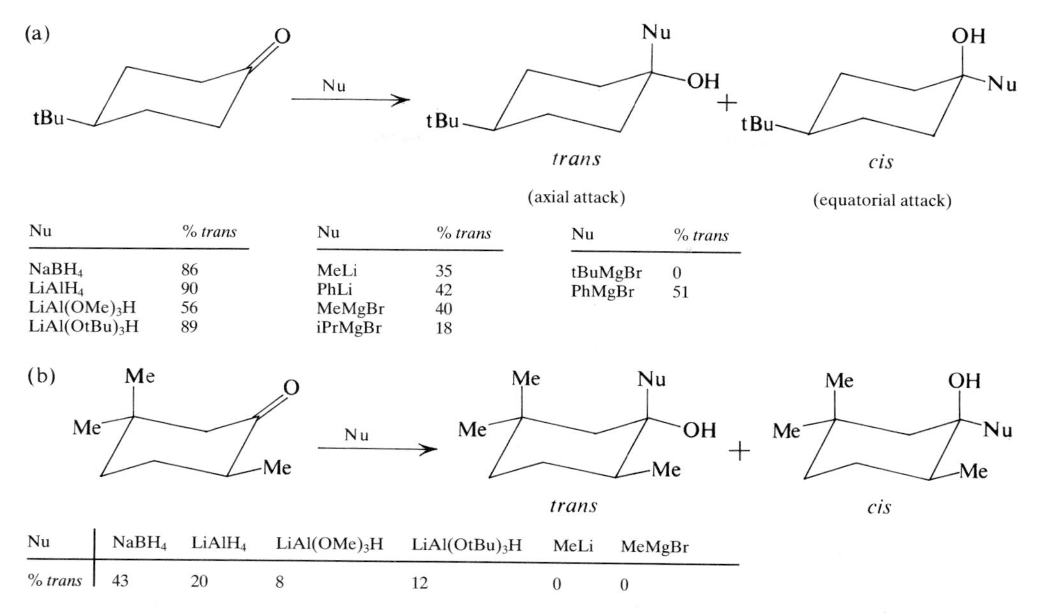

Nu	% *trans*	Nu	% *trans*	Nu	% *trans*
$NaBH_4$	86	MeLi	35	tBuMgBr	0
$LiAlH_4$	90	PhLi	42	PhMgBr	51
$LiAl(OMe)_3H$	56	MeMgBr	40		
$LiAl(OtBu)_3H$	89	iPrMgBr	18		

Nu	$NaBH_4$	$LiAlH_4$	$LiAl(OMe)_3H$	$LiAl(OtBu)_3H$	MeLi	MeMgBr
% *trans*	43	20	8	12	0	0

Fig. 3-9. Diastereoselective addition of nucleophiles to cyclic ketones.

tions to the α-ketoesters of chiral alcohols was independently provided by Prelog in 1953 [150].

Cram's rule is summarized in Fig. 3-10. Substituents attached to the chiral center were classified as being small (*S*), medium (M) and large (L), and it was assumed that, due to complexation with the reagent, the carbonyl group became a bulky substituent and therefore in the most stable conformation the C=O bond bisected the S–C–M angle. Preferred attack from the side of the small (S) substituent follows from this model and was actually observed in a large number of additions [142]. Cram's rule is an example for a not uncommon situation in which a correct conclusion is reached on incorrect premises. First, as was pointed out later by Cornforth [151] and Karabatsos [152], the conformation in which R is eclipsed with the L substituent is not the one preferred either in the ground state or the transition state. Second, the free energy difference between ground states is small compared with the free energy of activation. Therefore, the Curtin-Hammett principle (*cf.* Section 1.5) is applicable, and ground state conformations have little relevance to selectivity. Finally, substituents influence the direction of attack not only by their size, but also by electronic interactions with the incoming nucleophile. Actually the preferred conformation for rotation around an $sp^3 - sp^2$ bond (*e. g.* in propene or acetaldehyde) is the one in which one of the ligands at the sp^3 center is eclipsed with the double bond [153].

Corollaries of Cram's concept are that: (i) stereoselectivity should increase when the difference in size between M and S substituents becomes larger and (ii) when

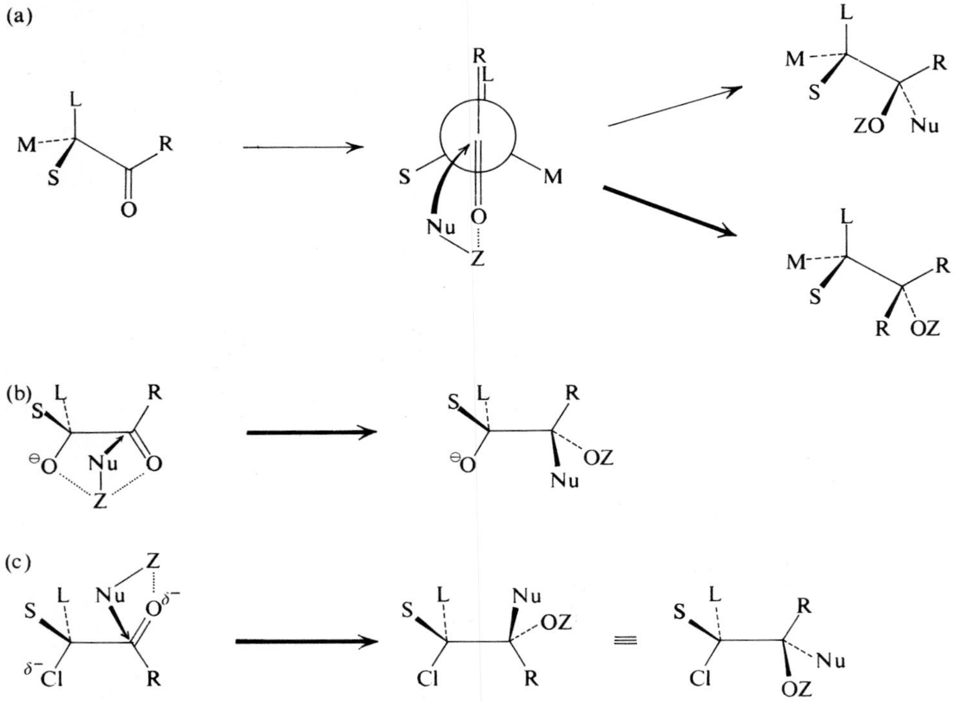

Fig. 3-10. Cram's rule. (a) The basic model. (b) The cyclic model (chelation control). (c) The dipolar model (Cornforth' rule).

more bulky reagents are applied; further, (iii) selectivity should diminish with increasing size of R since this would force ligand L out of its eclipsed conformation. There are not enough data available in the literature either to confirm or refute the first prediction. A confirmation of prediction (ii) is, *e. g.*, the addition of methyl-, ethyl- and phenylmagnesium bromide to 2-phenylacetaldehyde giving the expected diastereomers in 33, 50, and 60% de resp. [149]. Prediction (iii) is clearly incorrect, as shown by example (b) in Fig. 3-8.

Cram's rule, as first formulated [149], is only valid when there is no chelating group in the substrate other than the carbonyl group, and none of the groups attached to the chiral center is highly polarizable. For cases not fulfilling these conditions, the "cyclic model" [154] and the "dipolar model", shown in Figs. 3-10 (b) and (c), respectively, were suggested, the latter by Cornforth in 1959 [151]. Some reactions which obey these models are shown in Figs. 3-11 (a) and (b). The validity of the cyclic and dipolar models (b) and (c) is even more subject to reaction conditions than that of the acyclic model (a). This is illustrated by example (a) in Fig. 3-11. Not only is the degree of induction dependent on the nature of the nucleophile but the predominant configuration may be inverted too. While the limitations of Cram's rule are obvious, its outstanding importance for the development of the theoretical background of stereoselectivity cannot be questioned.

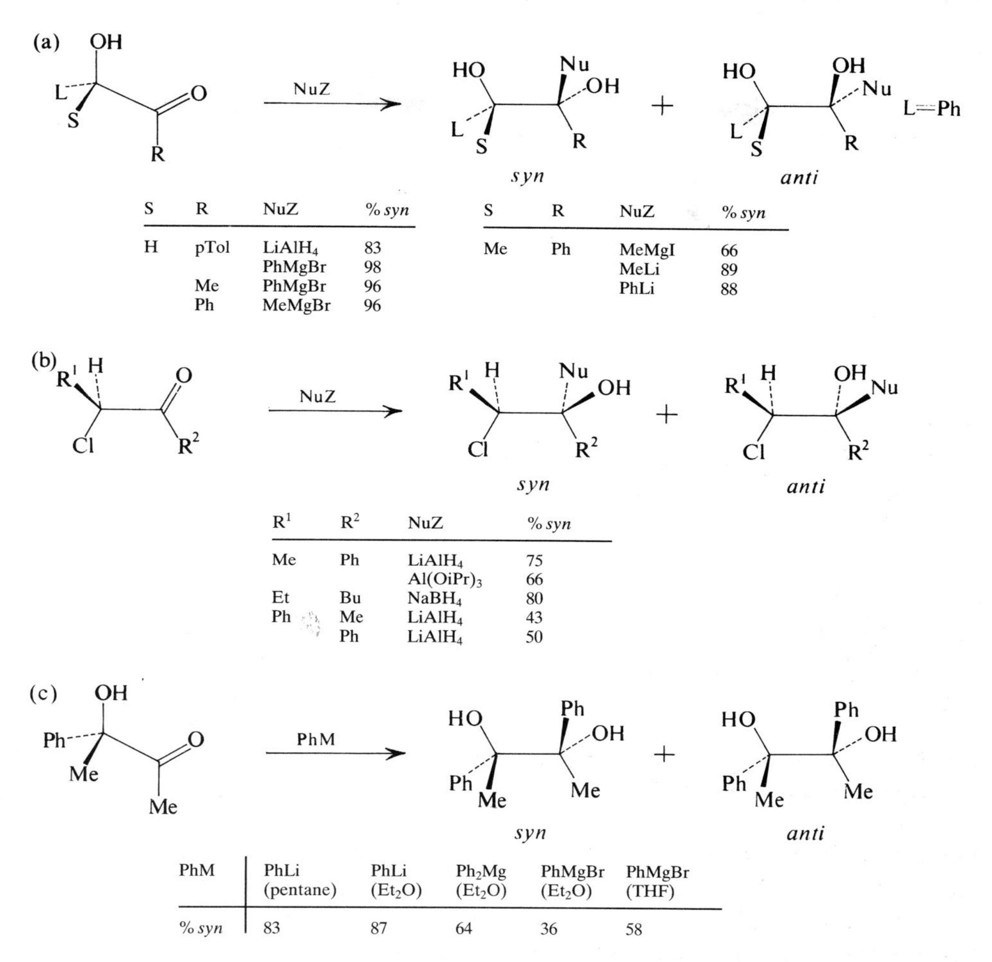

S	R	NuZ	% *syn*
H	pTol	$LiAlH_4$	83
		PhMgBr	98
	Me	PhMgBr	96
	Ph	MeMgBr	96

S	R	NuZ	% *syn*
Me	Ph	MeMgI	66
		MeLi	89
		PhLi	88

R^1	R^2	NuZ	% *syn*
Me	Ph	$LiAlH_4$	75
		$Al(OiPr)_3$	66
Et	Bu	$NaBH_4$	80
Ph	Me	$LiAlH_4$	43
	Ph	$LiAlH_4$	50

PhM	PhLi (pentane)	PhLi (Et_2O)	Ph_2Mg (Et_2O)	PhMgBr (Et_2O)	PhMgBr (THF)
% *syn*	83	87	64	36	58

Fig. 3-11. Dependence of diastereoselectivity on reaction conditions and the nature of the reagent.

In 1967 Karabatsos suggested an alternative approach to the problem [152], still based on ground state conformations, but taking into account newly acquired information about the conformations of carbonyl compounds [153]. He postulated that attack is most likely over the small group and this can occur in two conformations. In the major form the carbonyl group is eclipsed with the medium size group, while in the minor one with the large group (Fig. 3-12(a)). It was also supposed (provided that R was not too bulky) that complexation took place *syn* to R, as shown in the figure, and therefore steric interaction with the eclipsing group was small. An analysis of the interactions in conformations *A, B,* and *C* between groups S, M, L, and R, as well as with the nucleophile, revealed that from combinations giving rise to the product predicted by Cram's rule (*X*) rotamer *A* and from those leading to its diastereomer (*Y*) rotamer *C* were the most stable.

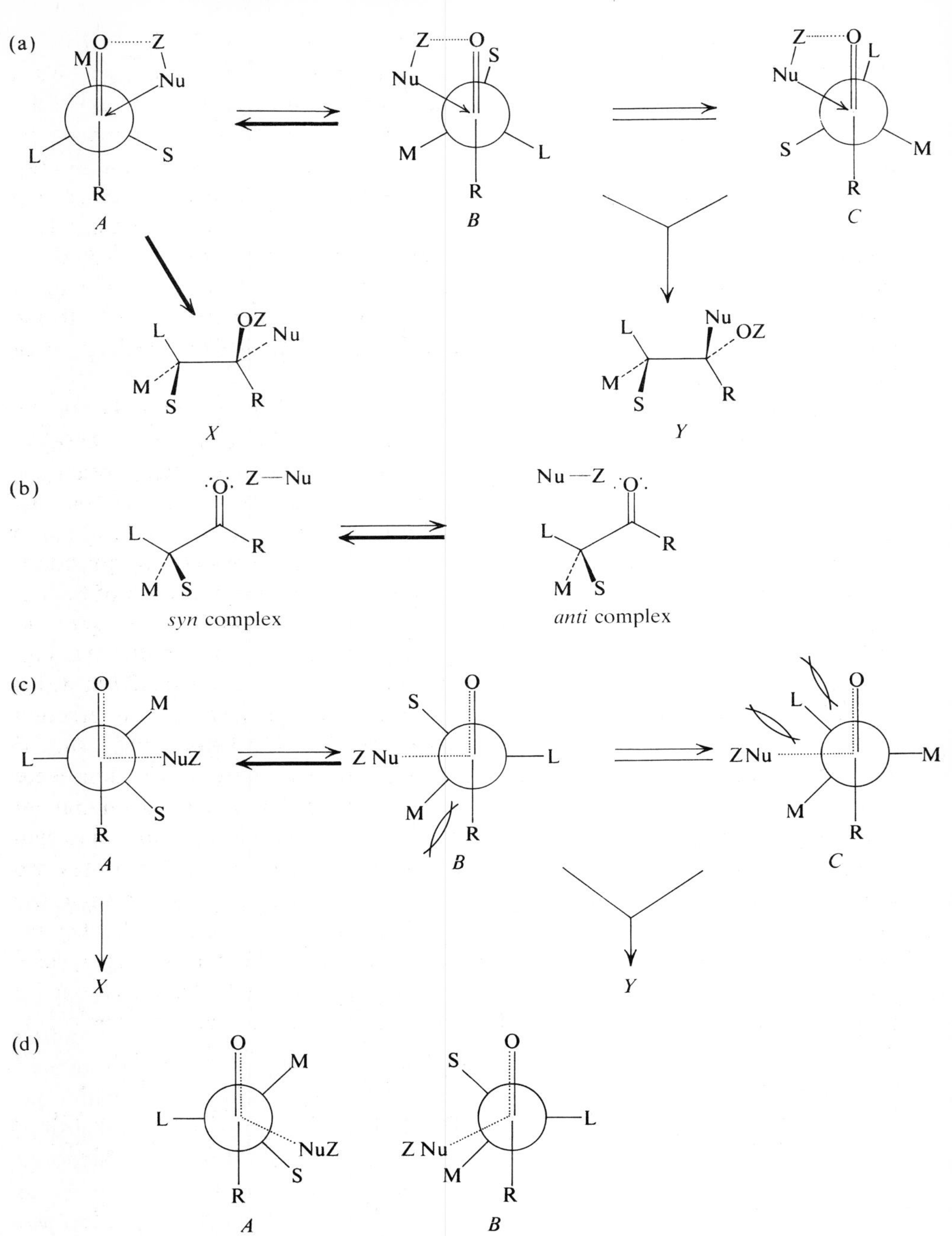

Fig. 3-12. Advanced stereochemical models for the addition of nucleophiles to acyclic carbonyl compounds.

Karabatsos calculated that with R = Me rotamer *B* was about 3.3 $kJ \cdot mol^{-1}$ less stable than *A* and therefore only a minor component of the equilibrium. Thus product ratios can be correlated with the relative stabilities of *A* and *C*. Approximate relative stabilities were calculated and were found to correctly reflect the trends of stereoselectivities [145, 152], but because they were based on a ground-state model, they were unable to account for the effect of the reagent. When group R is bulky, *anti* complexation may become important, by which rotamer *C* is further destabilized (Fig. 3-12 (b)). This may account for enhanced selectivity when the size of R increases (*cf.* Fig. 3-8 (a)), although this effect was not reflected in Karabatsos's calculations. In fact, considering only the interaction of R with the directing groups, destabilization of rotamer *A* and diminishing selectivity with increasing size of R could be anticipated.

Felkin and his coworkers were the first who not only directed attention to the transition state but also attempted to rationalize both acyclic and cyclic stereoselection by a common concept [144, 155]. They, too, adopted the hypothesis that, owing to the pronounced exothermic nature and high rate of the reaction, the transition state was essentially substrate-like rather than product-like, but also considered torsional strain generated by interaction between existing full bonds and the partial bond Nu · · · C · · · O. The requirement that the approach of the nucleophile should be both perpendicular to the R–CO–C_α plane *and* staggered to the directing group, can only be fulfilled by conformations of the type shown in Fig. 3-12 (c). Here, out of the six possible rotamers those three are depicted in which the bond in formation is close to the small group. The smallest steric interaction occurs in rotamer *A*, which leads to the product predicted by Cram's rule, while *B* and *C* give rise to its epimer. To these considerations two more assumptions were added, namely, (i) that unlike with the Cram and Karabatsos models, the dominant interaction was that between the entering group and the achiral group R attached to the carbonyl, and (ii) that due to polar effects those transition states are stabilized in which separation between NuZ and the electronegative group at the directing center was the largest (*i. e.* the L group or the one of the highest electronegativity). The latter is a necessary amendment to the dipolar model, since it was shown that in the ground state the carbonyl group and the halogen atom are eclipsed rather than antiperiplanar [156]. The validity of this statement, however, depends on the relative size of the polar group and the M group, as was demonstrated by Canceill and Jacques [143] for a series of ketones shown below. With R=Me and X=CO_2Me or CN the major product arose from an attack antiperiplanar to X, while with R=iPr or tBu, X took up the role of the M group.

O R X H R $\xrightarrow[\text{MeOH}]{\text{KBH}_4}$ OH R Ph X *syn*

X=	CO_2Me	$CONH_2$	CN
R		% *syn*	
Me	73	43	55
iPr	34	5	20
tBu	5	0	16

Felkin's model correctly predicts the trend observed when increasing the size of R and the lack of selectivity experienced with methyl 2-butyl ketone (L=Et, M=Me, S=H) for which rotamers *A* and *C* are almost of the same stability.

Ab initio calculations by Anh and Eisenstein for the approach of $H^{\ominus}$ to 2-chloro- and 2-methylpropanal [157, 158] established that transition state *A* according to Felkin represented an energy minimum, while those proposed by Cram, Cornforth and Karabatsos were all of higher energy (at least by 11 $kJ \cdot mol^{-1}$), *i. e.* they account for less than 1% of the total yield. Anh's calculations showed that in the favored transition state the incoming nucleophile is antiperiplanar to one of the groups attached to the vicinal atom. The magnitude of this "antiperiplanar effect" follows the order $Cl > CH_3 > H$.

A weak point in Felkin's argumentation was that the significant preference for rotamer *A* over *B* lacked adequate experimental support. An explanation was provided later by the hypothesis of "non-perpendicular attack" based on X-ray evidence [159] and *ab initio* calculations [160], from which it was concluded that the nucleophile approached the carbonyl at an angle of about 107° rather than at 90°. This gives a clear preference for rotamer *A* over *C* (*cf.* Fig. 3-12 (d)).

The behavior of cyclic ketones in stereoselective nucleophilic additions cannot be rationalized by any of the models proposed by Cram, Cornforth or Karabatsos [161]. Although the general observation that reduction with $NaBH_4$ or $LiAlH_4$ gives predominantly the equatorial alcohol with unhindered, and the axial alcohol with hindered cyclohexanones had been disclosed by Barton in 1953 [162], a global rationalization of this phenomenon has not been forwarded and it is rather unlikely that it will ever be possible. An excellent analysis of the situation was given by Wigfield [163] who reviewed not less than seven different treatments of the problem, to which he added that of his own.

Dauben, Tonken and Noyce argued that with unhindered ketones the transition state was product-like and the direction of addition was controlled by product stability ("product development control"), while with hindered ketones axial approach of the reagent was hampered ("steric approach control") [164] (for cyclohexanone *cf.* Fig. 3-13 *A* and *B*). This rationalization was, however, not adequately supported by facts. The most pertinent criticism out of many was that kinetic preference often surpassed thermodynamic preference for an equatorial alcohol in the product [165].

On the other hand, steric approach control was adequately documented by experiments and therefore became widely accepted. Thus cyclohexanones with an axial substituent at C-3 are preferentially attacked equatorially, and this selectivity increases with bulky reagents (*cf.* Fig. 3-9 (b)).

Preference for axial attack in unhindered ketones was attributed by Richer [166] to repulsion by axial hydrogens at C-2 and C-6 (Fig. 3-13, *C*). If we take into account that the attack is non-perpendicular, this interaction is even more significant than was assumed by Richer. A similar conclusion can be reached by applying the Felkin model to cyclohexanones [155]. The reagent approaching the CO group

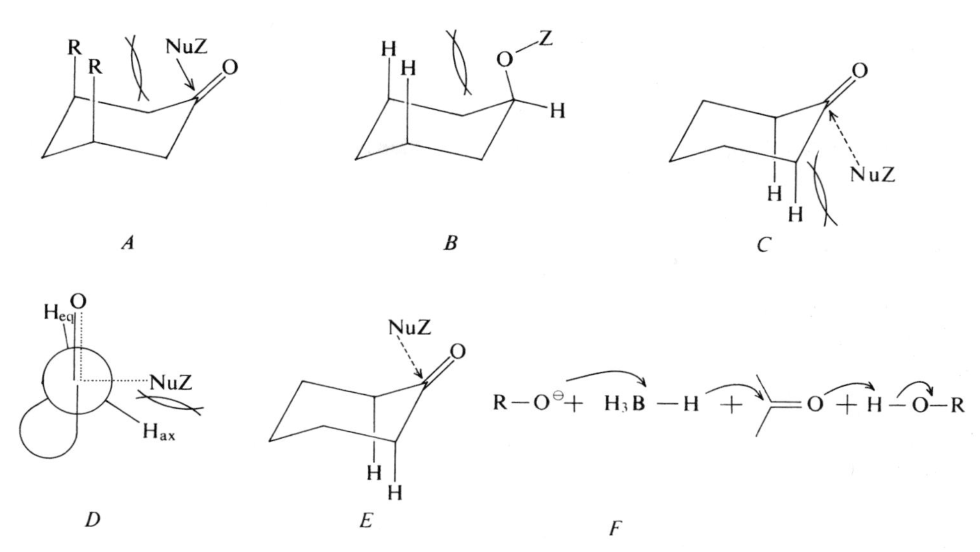

Fig. 3-13. Models for the interpretation of stereoselectivity in nucleophilic additions to cyclic ketones (demonstrated for cyclohexanone). *A*: Interactions in steric approach control. *B*: Interactions in product development control. *C*: Steric repulsion by vicinal axial hydrogens. *D*: The torsional strain effect. *E*: The antiperiplanar effect. *F*: Mechanism of borohydride reductions.

at 90° becomes almost eclipsed with axial C–H bonds at C-2 and C-6 (shown for C–H_{ax} in Fig. 3-13, *D*).

These arguments were inverted by Anh *et al.* [167] who argued that due to the above mentioned axial C–H bonds equatorial attack was not inhibited, but, due to the antiperiplanarity effect, axial attack was promoted. Flattening of the ring enhances this effect, and in fact axial attack is slower in some rigid polycyclic ketones in which ring flattening is difficult. It should be noted that in a non-perpendicular mode of attack the antiperiplanarity effect is much diminished.

In an important theoretical treatment of the problem Wipke and Gund developed empirical functions to quantify steric congestion (*i. e.* ground state steric hindrance) and torsional effects at the reaction center. Results strongly supported that the effect of axial α-hydrogens was torsional in nature. Calculations for 52 cyclohexanones, cyclopentanones, steroidal and polycyclic ketones were carried out, and correlation with experimental values was good [168]. A calculation based on the linear combination of steric strain and product stability was recently published by Rei [169].

It was pointed out by Wigfield in his critical review [148] that none of the above rationalizations seriously took the mechanism of the reaction into account. In a series of studies concentrated mainly on sodium borohydride reductions Wigfield established that it was futile to conceive any single rationalization even for hydride reductions, since the mechanism of the reaction was changing with the reagent.

More specifically, borohydride reductions in alcohols proceeded by an acyclic mechanism, probably also involving the solvent (Fig. 3-13, *F*) *via* a product-like transition state. The mechanism of reduction by $LiAlH_4$ was not discussed but it could be established that a substrate-like transition state was involved and, finally, the transition state with $LiAl(OtBu)_3H$ was midway between the two types. Lack of product development control in reductions by $LiAlH_4$ was demonstrated for bicyclic ketones by Ashby and Noding [170]. Kinetic isotope effects with $LiAl(OtBu)_3H$ showed no variation with ketone structure and indicated that the character of the transition state did not change with the substrate.

The ratio of rates for equatorial attack in 3,3,5-trimethyl- and 3,3,5,5-tetramethylcyclohexanone should reflect the contribution of product development control to the transition state, since in the latter the effect should be enhanced by increased 1,3-diaxial interaction. In fact no such effect was observed in $LiAlH_4$ reduction, while the ratio with $LiAl(OtBu)_3H$ and $NaBH_4$ was 1.8 and 5.8 respectively.

It must be emphasized that interactions in a product-like transition state cannot be equated with those in the product, since in the former the reagent is still present. Thus in borohydride reductions of unhindered cyclohexanones interaction with hydrogens at C-2 and C-6, while in hindered cyclohexanones interaction with axial group(s) at C-3 and C-5 are dominant (*cf.* Fig. 3-14).

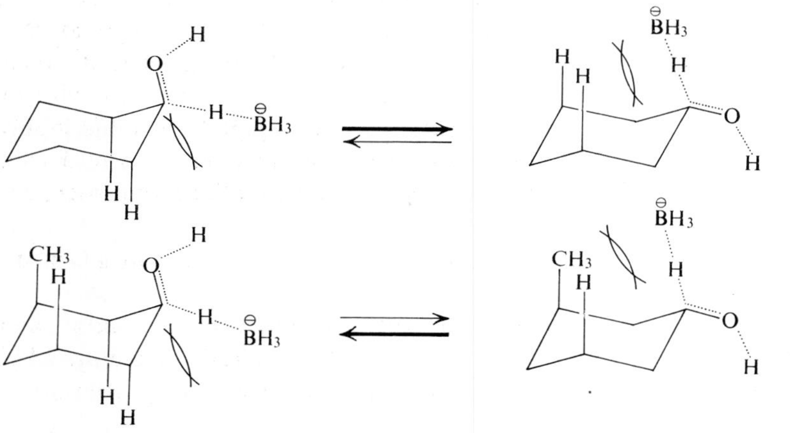

Fig. 3-14. Borohydride reduction of unhindered and hindered cyclohexanones.

In $LiAlH_4$ reductions there is an intrinsic preference for axial attack due to the antiperiplanarity effect. This may be compensated for or overruled by steric interaction in the case of hindered ketones.

3.2.2 Practical Aspects of Diastereoselective Ketone Reductions

1,2-Induction in acyclic systems

Although studies on diastereoselective nucleophilic addition to ketones were of outstanding importance for the development of theories about asymmetric induction, their contribution to the solution of practical problems is less impressive.

While methods of excellent stereoselectivity have been developed in the past decade for the nucleophilic addition of carbon to carbonyl in acyclic systems (*cf.* Chapter 5), stereoselectivity in the reduction of acyclic carbonyl compounds is seldom at a high level. In Table 3-9 the most important data available in this field have been compiled. Interpretation of the results in terms of the concepts discussed in the preceding section is not always straightforward. The role of reagent size is rather elusive. Thus $K(sBu)_3BH$ is more selective than $LiAlH_4$, but $Na(MeO)_3BH$ and the highly hindered 9-BBNH are less so. The outcome of the reduction of 3-ketoesters is highly dependent on the counter ion: KBH_4 gave a moderate excess of the *anti* product, while this was the only product with $Zn(BH_4)_2$.

Since Zn^{2+} is a much better complexant than alkali metals, it is not surprising that $Zn(BH_4)_2$ is superior to other complex hydride donors [202] in the reduction of α-hydroxy- [187], β-hydroxy- [195], α-carboxamido- [199], and α-methylthio- and phenylthioketones [196]. Unexpectedly, with the latter $Zn(BH_4)_2$ gave the *anti*-alcohol, while $Li(sBu)_3BH$ afforded the *syn* product. The stereoselectivity of $Zn(BH_4)_2$ is only paralleled by L-Selectride®, *i. e.* $Li(sBu)_3BH$, which is highly selective not only in the presence of a polar neighboring group, but also with vinylketones [198]. This property has been exploited in the synthesis of prostaglandins.

Recently, efficient stereodirection in the borohydride reduction by an α-sulfonium group was demonstrated [197].

Combination of enolate acylation in the presence of a chiral auxiliary group and subsequent diastereoselective reduction permitted the synthesis of 2-alkyl-3-hydroxycarboxylic acids in high enantiomeric and diastereomeric purity [203].

CH_2OCH_2OMe R^1 N O CH_2OCH_2OMe

1) BuLi, THF, −78 °C
2) R^2COCl, 3) $Zn(BH_4)_2$

CH_2OCH_2OMe R^1 R^2 3 2 N OH O CH_2OCH_2OMe

% *syn* 98%
% ee >94%

R^1=Me, Et, Bn, R^2=Et, iPr, tBu, Ph

With α-aminoketones selectivity is much influenced by substitution at the nitrogen. Owing to the importance of arylethanolamines to the pharmaceutical

Table 3-9. Diastereoselective Reduction of Acylic Ketones with a Chiral Center Adjacent to the Carbonyl Group[a].

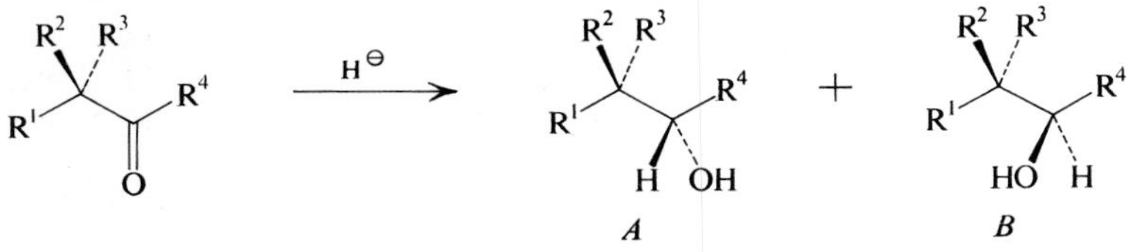

R^1	R^2	R^3	R^4	Reagent	%*A*	Ref.
Apolar neighboring groups[b]						
Ph	Me	H	Me	$LiAlH_4/Et_2O$, 35°C	74	144
				$LiAlH_4/Et_2O$, −70°C	85	171
				$Al(OiPr)_3$/iPrOH	61	171
				$Li(sBu)_3BH$	95	172
				$[iPrC(Me)_2]_2BH$	20	172
			Et	$LiAlH_4/Et_2O$, 35°C	76	144
			iPr	$LiAlH_4/Et_2O$, 35°C	85	144
			tBu	$LiAlH_4/Et_2O$, 35°C	98	144
				$LiAlH_4/Et_2O$, −70°C	~100	144
			Ph	$LiAlH_4$	80	149
	iPr		iPr	$LiAlH_4$	96	174
Ph	Et	Me	Ph	$LiAlH_4$	69	174
			Me	$LiAlH_4$	75	175
			Et	$LiAlH_4$	75	175
Et	Me		Me	$Al(OiPr)_3$	56.5	171
				$LiAlH_4$, −70°C	51	171
				$LiAlH_4$, 35°C	48.5	171
cHex				$NaBH_4$	62.5	177
				$LiAlH_4$	61.5	144
				$Al(OiPr)_3$	34.5	177
cHex	Me	H	Et	$LiAlH_4$	66.5	144
			iPr	$LiAlH_4$	80	144
			tBu	$LiAlH_4$	61.5	144
cHex			Me	$LiAlH_4$/THF	65	178
				$LiBH_4$/THF	69	178
				$Li(sBu)_3BH$/THF	78	178
Ph				$AlHCl_2/Et_2O$	86	178
				$LiAlH_4$/THF	82	178
				$LiBH_4$/THF	80	178
				$Li(sBu)_3BH$/THF	94	178
				$K(sBu)_3BH$/THF	85.4	200
				$Al(OiPr)_3$/iPrOH[c]	61.5	200
				$Na(MeO)_3BH/Et_2O$	67.1	200
				9-BBNH/THF	53.7	200
				$NaBH_4/Et_2O$ + HO^-/aq	67.3	200
				$Na(MeOCH_2CH_2O)_2AlH_2$/ /C_6H_6	69.5	200
				$NaB(CN)H_3$/MeOH + H^+	81.1	200

Table 3-9. Continued

R^1	R^2	R^3	R^4	Reagent	%*A*	Ref.
$H_2C{=}CH$			$Ph(CH_2)_2$	$Li(sBu)_3BH$	96	198
(*Z*)-BuCH=CH				$Li(sBu)_3BH$	99	198
(*E*)-BuCH=CH				$Li(sBu)_3BH$	99	198
Polar neighboring group(s)[c]						
Cl	Et	H	Bu	$NaBH_4$	80	151
	Ph		Me	$LiAlH_4$	43	179
	Me		Ph	$LiAlH_4$	75	179
				$Al(OiPr)_3$	66	179
OH	Ph	H	pTol	$LiAlH_4$	84	180
OMe	Me		Ph	$NaBH_4$	73	181
	CH_2OH			$NaBH_4$	58	181
OH	Me		H	$LiAlD_4$, −42°C	60[d]	185
			Pent	$LiAlH_4$	64	186
				$ZnBH_4$	77	187
			iPr	$(iBu)_3Al$	61	201
				$Zn(BH_4)_2$	85	187
			tBu	$LiAlH_4$	87	187
			Ph	$LiAlH_4$	87	187
				$Zn(BH_4)_2$	98	201
	Et		Bu	$ZnBH_4$	89	187
	Pr		Pr	$ZnBH_4$	>99	187
	Bu		Et	$ZnBH_4$	87	187
	Pent		Me	$LiAlH_4$	70	186
				$Zn(BH_4)_2$	87	187
	iPr			$(iBu)_3Al$	83	186
	tBu			$Zn(BH_4)_2$	85	187
				$(iBu)_3Al$	78	201
				$LiAlH_4$	75	186
	Ph			$LiAlH_4$	80	201
		Me		$(iBu)_3Al$	82	186
				$Zn(BH_4)_2$	90	187
	iPr			$Al(iBu)_3$	85	186
	tBu			$Al(iBu)_3$	91	186
CO_2Me	Me	H	Ph	$LiAlH_4$	90	143
				KBH_4	27	183
CO_2Me	Me	H	Ph	$Zn(BH_4)_2$	97	195
			(*E*)-CH=CHMe	$Zn(BH_4)_2$	96	195
			$Ph(CH_2)_2$	$Zn(BH_4)_2$	56	184
CONHPh			Me	$Zn(BH_4)_2$	98	199
SMe	Et		Ph	$Zn(BH_4)_2$	6	196
				$Li(sBu)_3BH$	>99	196
	iPr			$Zn(BH_4)_2$	4	196
				$Li(sBu)_3BH$	96	196
	CH_2OH			$Zn(BH_4)_2$	93	196
	CH_2OAc			$Zn(BH_4)_2$	27	196
				$Li(sBu)_3BH$	96	196

Table 3-9. Continued

R^1	R^2	R^3	R^4	Reagent	%*A*	Ref.
SPh	Bu		Me	$Zn(BH_4)_2$	82	196
	iPr			$Li(sBu)_3BH$	98	196
				$Zn(BH_4)_2$	61	196
				$Li(sBu)_3BH$	99	196
$S^{\oplus}Me_2$	Et		Ph	$NaBH_4$	1	197
$S^{\oplus}MePh$	Bu		$Ph(CH_2)_2$	$NaBH_4$	20	197
	iPr			$NaBH_4$	1	197
Nitrogen as neighboring group[c]						
NH_2	CH_2NH_2	H	Ph	$NaBH_4$	36	181
	CH_2OH			$NaBH_4$	43	181
NHBn	Me			$NaBH_4$	12	181
NMe_2				$NaBH_4$	54	181
NMeBn				$NaBH_4$	75	181
NHAc				$NaBH_4$	23	181
NHBz				$NaBH_4$	23	181
NHAc			$4\text{-}NO_2\text{–}C_6H_4$	$Al(OiPr)_3$	28	188
N-Phthaloyl				$Al(OiPr)_3$	67	188
NHAc	CH_2OH		Ph	$Al(OiPr)_3$	47	189
NH_2	Me	H		$NaBH_4$/EtOH	5	190
	CH_2NH_2			$NaBH_4$/EtOH	50	190
	CH_2OH			$NaBH_4$/EtOH	19	190
	CH_2OEt			$NaBH_4$/EtOH	8	191
NHMe	Me			$NaBH_4$/EtOH	7	191
NHBn					0	191
NMetBu				$NaBH_4$/EtOH	97	191
NBn_2				$LiAlH_4/Et_2O$	100	191
$NiPr_2$	Ph			$LiAlH_4/Et_2O$	100	191
NMeBn	Me			$NaBH_4$/EtOH	88	192
Piperidino				$NaBH_4$/MeOH	94	193
CO_2Me	NBn_2		Me	$NaBH_4$/EtOH	93	194
			Ph	$NaBH_4$/EtOH	98	194
			Hex	$NaBH_4$/EtOH	96	194

[a] For reviews, see refs. [192] and [204].

[b] If $R^1 = L$, $R^2 = M$ and $R^3 = H$ *A* is the *anti* and *B* the *syn* isomer. *A* is preferred according to Cram's rule.

[c] If R^1 is a polar group (OH, CO_2R, SR, NR^2 *etc.*) and $R^3 = H$, *A* is the *syn* and *B* the *anti* isomer. Under chelation control *A* is formed preferentially.

[d] Replace H by D in formulas *A* and *B*.

[e] Under kinetic conditions.

industry, research on this subject has ben intensive [191, 204]. Lack of an adequate rationalization did not prevent utilization of the effect of *N*-substitution. Generally, product *B* (*cf.* Table 3-9) (being the *anti* isomer with amines) is the prevailing product with primary and secondary amines, while with tertiary amines *A* is in

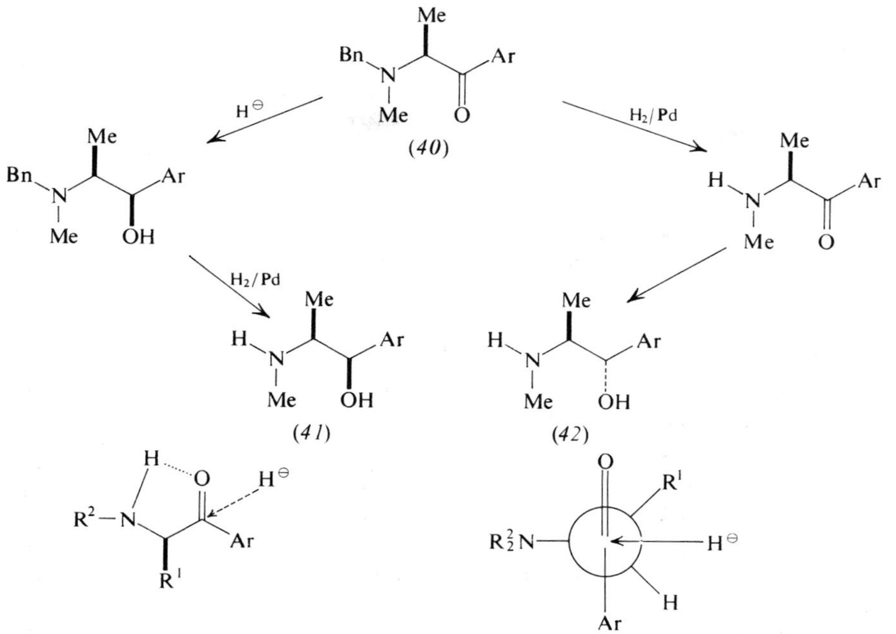

Fig. 3-15. Reduction of α-aminoketones.

excess. Thus reduction of the aminoketone *40* gives the *syn* isomers (debenzylated to *41*) in excess, while the parent amine gives mainly the *anti* product (*42*) (Fig. 3-15). As an explanation it was proposed that hydrogen bonding stabilizes conformation *43* in primary and secondary amines, while the transition state for tertiary amines has to be envisaged as *44*. Interestingly, even tertiary amines gave the *anti* alcohols as the major product when the hydrochlorides were reduced in methanol at less than 25°C [196].

1,3- and 1,4-Induction in acyclic systems

1,3-Induction, *i. e.* the stereodirecting effect of a chiral center separated by two bonds from the carbonyl group, plays an important role in C–C bond-forming reactions (*cf.* Section 5.1.2), but in hydride transfer reactions the effect is generally too weak for practical application. For substrates without a polar directing group slight to moderate excess (up to 56% de) of the 1,3-*anti* product was found in reductions with $LiAlH_4$ [205].

A chiral sulfur atom is a moderately effective directing group, which can then be removed by Raney nickel desulfurization to give an optically active alcohol. For this purpose both the sulfoxides and the sulfoximides are useful.

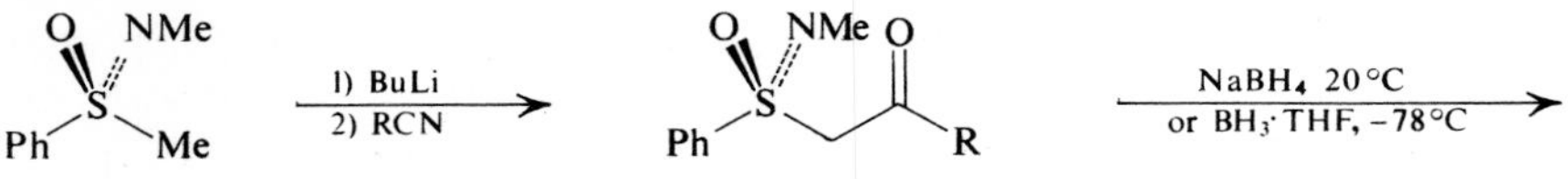

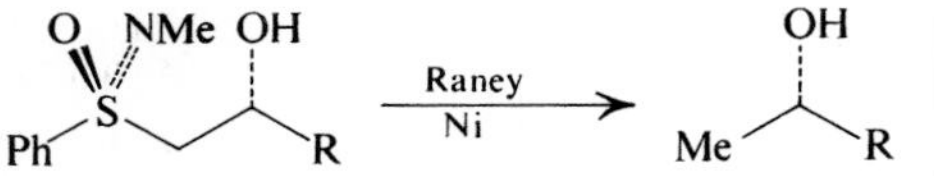

R	Me	Et	iPr	iBu	tBu	Hex	Ph
% ee	0	8	20	50	68	43	69

Fig. 3-16. Asymmetric induction by a chiral sulfur center in carbonyl reductions.

For the reduction of sulfoximides Annunziata *et al.* [206] used $NaBH_4$ at room temperature (R = Me, Et, iPr, tBu, Ph), while, out ten different hydrides, Johnson and Stark [207] selected $BH_3 \cdot THF$ at $-78°C$ as the optimal reagent (R = iBu, Hex, tBu, Ph). For comparable cases (tBu, Ph) the latter method is the more selective one (Fig. 3-16).

Results with a sulfoxide as the directing group and $LiAlH_4$ as the reducing agent are rather similar [208].

1,4-Induction played a very important role in the development of our understanding of stereoselective reactions. At the turn of the century it was discovered by McKenzie [209] that Grignard addition to esters of 2-keto acids with chiral alcohols gave, after hydrolysis, optically active 2-hydroxy-2-alkyl acids. In 1953 Prelog reviewed the accumulated data and formulated a prediction known as Prelog's rule, by which the prevailing configuration of the product could be deduced from the constitution and configuration of the chiral alcohol [150]. As in Cram's rule, that of Prelog also operates by drawing the molecule in a fixed

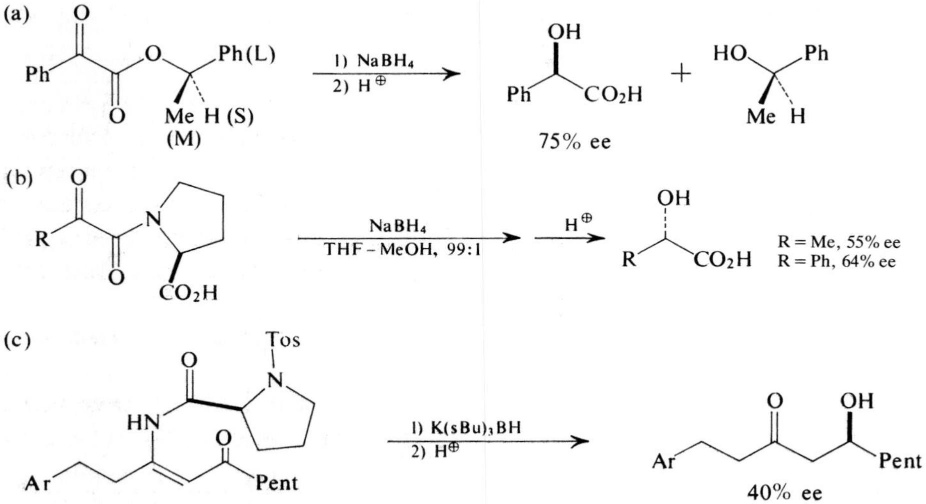

Fig. 3-17. Asymmetric 1,4- and 1,6-induction in carbonyl reductions.

conformation and classifying substituents at the inducing center as small, medium and large. The rule is illustrated in Fig. 3-17 (a). Since in reductions selectivities are generally less than 20%, the method is unsuited for practical purposes and will not be discussed further. Recently its adaptation to chiral amides resulted in somewhat better selectivities (*cf.* Fig. 3-17 (b)) [210]. A rare case of 1,6-induction is shown in Fig. 3-17 (c) [211].

Reduction of cyclic ketones

With cyclic ketones it is impractical to discuss 1,2- 1,3- and 1,4-induction separately, so these cases will be discussed together.

With cyclohexanones, including their complex derivatives such as steroids and terpenes, there are useful practical methods available to produce both the equatorial and the axial alcohols. It can be seen from the foregoing discussion that a different approach has to be applied with unhindered and hindered cyclohexanones.

When the objective is to prepare an equatorial alcohol from an unhindered ketone $LiAlH_4$ or $NaBH_4$ is the reagent of choice, while very high selectivity for the axial alcohols can be achieved with very bulky reducing agents, *e. g.* with $Li(sBu)_3BH$ (L-Selectride) [212] (*cf.* Fig. 3-9 (a)). Hindered ketones give axial alcohols preferentially with any hydride donor, and this effect can be enhanced by applying a bulky hydride donor (*cf.* Fig. 3-9 (b)). Finally, equatorial alcohols cannot be obtained from hindered cyclohexanones by a kinetically controlled process, but by equilibration, *e. g.* over Raney nickel [213], it is possible to convert an axial alcohol to the more stable equatorial alcohol.

An obvious limitation to stereoselectivity is when the substrate is conformationally not homogeneous. Lowering of temperature may then be an effective measure. It was shown [218] that at −78°C a methyl group is almost as good a conformation biasing group as tert-butyl. Selectivities achieved by bulky hydrides at low temper-

Table 3-10. Reduction of Alkylcyclohexanones with Bulky Hydride Donors.

Substituent(s) (Temp. °C)		% Isomer from equatorial attack			
		$Li(sBu)_3BH$ [214]	$LiMes_2BH_2$ [a] [215]	$Li(Mecp)_3BH$ [b] [216]	$Li(Siam)BH$ [c] [216, 217]
4-tBu	(0)	93	94		
	(−78)	96.5		>99	>99
4-Me	(0)	80.5	94		
	(−78)	90		98	98
3-Me	(0)	85	99		
	(−78)	94.5		99	>99
2-Me	(0)	99.3	99	99	>99
3,3,5-Me_3	(0)	>99			>99

[a] Mes = 2,4,6-trimethylphenyl. [b] Mecp = *trans*-2-methylcyclopentyl. [c] Siam = 3-methyl-2-butyl.

ature are impressive, as demonstrated by data in Table 3-10. Surprisingly, $Li(iBu)_2tBuH$, which should not be less bulky than $Li(Siam)_3H$, is a less selective reagent [219, 220].

A series of reducing agents obtained by mixing LiH with magnesium alkoxides, prepared from hindered phenols or alcohols such as 2,6-di-t-butylphenol and 2,2,6,6-tetramethylcyclohexanol, also exhibited almost total selectivity for axial attack with 2-methyl- and 3,3,5-trimethylcyclohexanone and over 80% preference with 4-t-butylcyclohexanone [221, 222]. $LiAl(OMe)_3H$ and even $LiAl(OtBu)_3H$ are in turn only moderately stereoselective hydride donors, presumably because of the intervening oxygen atom [67].

Potassium 9-(2,3-dimethyl-2-butoxy)-9-boratobicyclo[3.3.1]nonane, prepared by Brown *et al.* [80], was a highly selective reagent, but less so than *e. g.* $Li(Siam)_3BH$.

Cyclopentanones are conformationally mobile molecules and thus it is more difficult to reduce them stereoselectively than cyclohexanones. Usually only one of the epimeric alcohols can be obtained in great excess. Some results are shown in Fig. 3-18 (a). Product distributions for the reduction with $LiAlH_4$ of a series of 2-alkylcyclopentanones was succesfully predicted using the Ruch-Ugi approach [127] and in the case of 2-methylcyclopentanone by Wipke's calculations [168].

Camphor can be regarded as a typical sterically hindered bicyclic ketone and has been used as a test substrate for the evaluation of hydride donors. *Endo* attack is preferred with all of them. Typical results are shown in Fig. 3-18 (b).

(a) $H^{\ominus}$ → *trans* + *cis*

reagent	% *trans*	ref.
$LiAlH_4$	84	176
$Li(OtBu)_3AlH$	72	182
$Li(sBu)_3BH$	98	214
$Li(Mes)_2BH$[a]	98	215

(b) $H^{\ominus}$ → *exo* + *endo*

reagent	% *exo*	ref.
$LiAlH_4$	91	176
$Li(OtBu)_3AlH$	93	182
$Li(sBu)_3BH$	98	219
$Li(Mes)_2BH$	>99	215

[a] Mes = 2,4,6-trimethylphenyl

Fig. 3-18. Stereoselective reduction of 2-methylcyclopentanone and camphor.

4 Stereoselective Oxidations

For structural and mechanistic reasons the range of stereoselective oxidations is rather narrow. Oxidation of a hydroxy group to a carbonyl group is a stereodestructive transformation, while oxidation of hydrocarbons is a radical process of high activation energy with little chance to be conducted stereoselectively.

Oxidation of sulfides to sulfoxides and of amines to amine oxides may create a chiral center, but this aspect has been studied very little. The stereoselective oxidation scene is dominated by studies on epoxidation. This is not only a mild reaction which can be carried out in organic solvents at low temperature, but, in addition, epoxides can be easily transformed to many other functional groups.

4.1 Enantioselective Oxidations

4.1.1 Epoxidation with Chiral Oxidants

Henbest was the first to use a chiral peracid, percamphoric acid, for enantioselective epoxidation [1]. Enantioselectivity was disappointingly low (4% with styrene), but it was shown later by Pirkle and Rinaldi [2] that the reagent used by Henbest, and later by others, was a mixture of constitutional isomers. Purified peroxycamphoric acid (*1*) gave 1.4–2.7 times higher ee values with several substrates, but this was still very low (<15%). Optically active peracids, most often *1*, have been employed for the enantioselective oxidation of olefins to epoxides [3], of prochiral sulfides to sulfoxides [4], and of azomethines to oxaziridines [5, 6], but poor selectivities (<66% ee) prevented practical application.

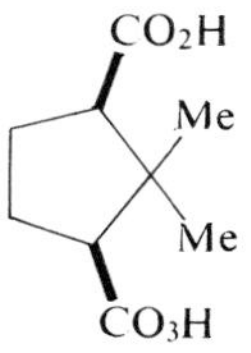

(*1*)

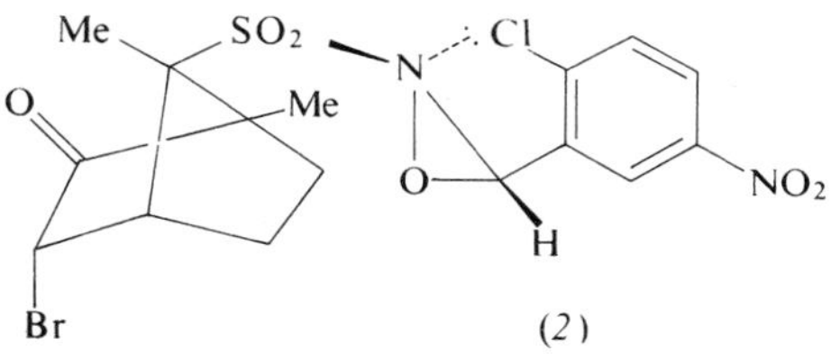

(*2*)

N-sulfonyl oxaziridines are epoxidizing agents, and with the chiral oxaziridine *2* developed by Boschelli [7] it was possible to obtain products with up to 40% ee [8].

4.1.2 Oxidations in the Presence of Chiral Catalysts

Epoxidation in the presence of a chiral phase transfer catalysts was actively investigated by Wynberg and his coworkers. Their studies revealed interesting correlations between catalyst and product configuration [9], but failed to turn out a practically useful method. An example is shown in Fig. 4-1 (a) [10].

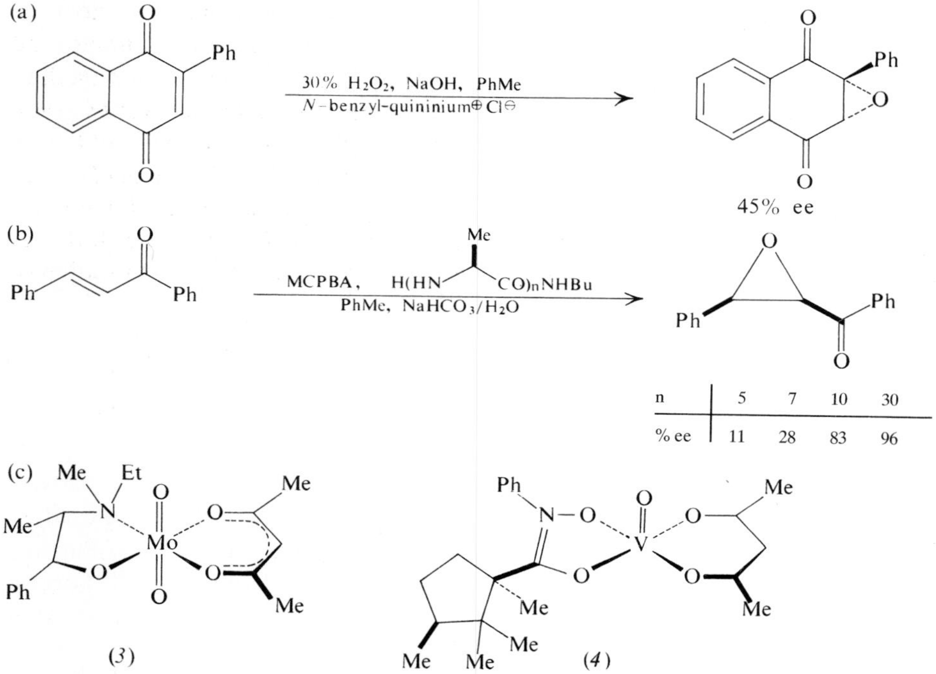

(a) 30% H_2O_2, NaOH, PhMe; *N*-benzyl-quininium$^{\oplus}$ $Cl^{\ominus}$ → 45% ee

(b) MCPBA, H(HN–CH(Me)–CO)$_n$NHBu; PhMe, $NaHCO_3/H_2O$

n	5	7	10	30
% ee	11	28	83	96

(c) (*3*) (*4*)

Fig. 4-1. Enantioselective epoxidations.

Surprisingly high enantioselectivities were observed in the epoxidation of chalcones in the presence of poly-(*S*)-alanine *N*-butylamide using a triphasic system. Optical purity depended both on the molecular weight of the peptide and on substrate structure and was in the range of 50–86% with n ≅ 10, and 96% with unsubstituted chalcone and n ≅ 30 (Fig. 4-1 (b)) [11]. Similar polypeptides containing other amino acids were less effective [12].

In 1976 a method for the vanadyl-acetylacetonate catalyzed epoxidation of allylic alcohols with tert-butyl hydroperoxide was published by Terashini and his coworkers [13]. Shortly thereafter Yamada *et al.* [14] and Sharpless *et al.* [15, 16] concurrently disclosed that similar complexes of molybdenum and vanadium respectively containing chiral ligands (*3* and *4*) catalyzed enantioselective epoxidation of allylic alcohols. Up to 33% enantioselectivity was attained in the presence of *3* using cumene hydroperoxide and up to 50% with *4* using tert-butyl hydroperoxide. These processes deserve mention because they were among the first examples of efficient asymmetric catalysis.

Kagan *et al.* later used a similar, but non-catalytic system, the molybdenum (VI) diperoxo-(*S*)-*N*,*N*-dimethyllactamide complex, for the epoxidation of simple olefins with up to 34% ee [17].

All these results were soon overshadowed by the discovery (Katsuki and Sharpless 1980) that in the presence of titanium (IV) isopropylate and (+)- or (−)-diethyl tartrate allylic alcohols could be epoxidized with very high enentioselectivity [18, 19]. A selection of epoxidations carried out with this simple and relatively inexpensive method is shown in Fig. 4-2. It is apparent that the method not only gives high yields and excellent enantioselectivities, but is also very flexible as regards the structure of the allylic alcohol. In the original work five out of the six possible substitution patterns of mono and disubstituted allylic alcohols were successfully tested [18]. The procedure proved to be extremely useful for the synthesis of insect pheromones and other natural products [20–26]. Since tartrates are available as both enantiomers, there is no restriction as to the configuration of

(a) $R^2R^1C{=}C(R^3)CH_2OH$ —[$tBuO_2H$, $Ti(OiPr)_4$, CH_2Cl_2, −20 °C; (+)-diethyl tartrate, 45–87%]→ epoxy alcohol

R^1	R^2	R^3	% ee
H	Et	Me	>95
	Me	H	>95*
$C_{10}H_{21}$	H		>98
H	$C_{10}H_{21}$		91
	Ph	Ph	99

R^1	R^2	R^3	% ee
H	H	cHex	95
	$MeO_2C(CH_2)_2$	H	>95*
Me	$Me_2C{=}C(CH_2)_2$		95
$Me_2C{=}C(CH_2)_2$	Me		94

* With diisopropyl tartrate.

(b) $R^2R^1C{=}CH{-}CH_2CH_2OH$ —[$tBuO_2H$, $Ti(OiPr)_4$, CH_2Cl_2; (+)-diethyl tartrate, −20°C]→ epoxy alcohol

HO, CONHBn / HO, CONHBn (*5*)

R^1	H	Et	H	Me
R^2	H	H	Et	Me
% ee	55	50	41	27

Fig. 4-2. Enantioselective epoxidation of allylic alcohols with the Sharpless method.

the product. Anhydrous conditions are essentiell [27], and sometimes diisopropyl tartrate gives better results than the ethyl ester. The same reagents combined with molybdenum (VI) gave poor results [28].

Since participation of the hydroxy group of the substrate is evidently essential for selectivity, it is not surprising that homoallylic alcohols are oxidized much less selectively than allylic alcohols and give the opposite prevailing configuration (Fig. 4-2 (b)) [29].

Sharpless and his coworkers studied the mechanism of the oxidation and established the following characteristics of the reaction [30]:

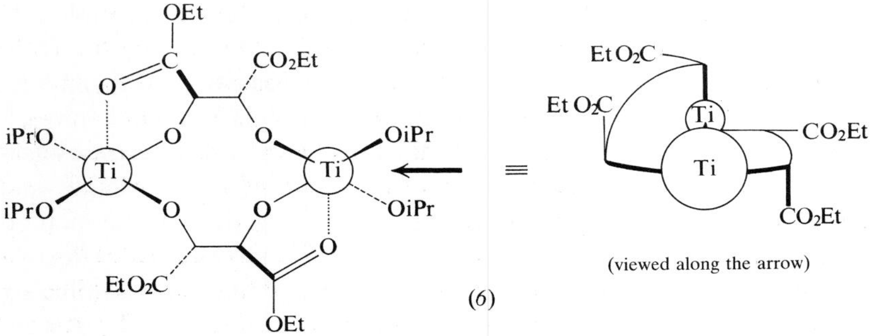

OEt
C
O
CO2Et
iPrO
O
O
OiPr
Ti
Ti
iPrO
O
O
OiPr
O
EtO2C
OEt
(6)

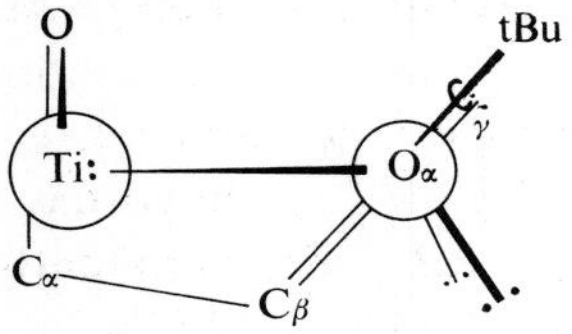

EtO2C
EtO2C
Ti
CO2Et
Ti
CO2Et
(viewed along the arrow)

O
tBu
Ti:
Oα
Cγ
Cα
Cβ
(7)

Fig. 4-3. The stereochemical model of the Sharpless epoxidation.

(i) Enantioselectivity is at its optimum at a 1:1 molar ratio of $Ti(OiPr)_4$ and tartrate ester.

(ii) Mixing of the two reagents results in a rapid release of 2 moles of alcohol whereby a dimeric species, a ten-membered cyclic ester of the tentative structure *6* (Fig. 4-3), is formed. *6* is a fluxional structure with rapid exchange of the carbonyl groups coordinated to titanium. C_2 symmetry about titanium is apparent from the figure.

(iii) As shown by kinetic experiments*, on addition of the substrate and the oxidant (both are alcohols) the remaining two isopropyloxy ligands are displaced by allyloxy and tert-butylhydroperoxy groups. Epoxidation takes place, as the rate

---

$$* \text{ rate} = \frac{k\,[\text{tBuO}_2\text{H}][\text{Ti(OiPr)}_2(\text{tartrate})][\text{allylic alcohol}]}{[\text{inhibitor alcohol}]^2}$$

determining step, in the complex by displacement of the product and the spent oxidant by new substrate and hydroperoxide molecules. The reactants are thus assembled on the metal in a chiral environment prior to reaction.

The importance of the 1 : 1 (actually 2 : 2) complex *6* was underlined by experiments with the amide *5* and 2,3-diphenylallyl alcohol. On changing the Ti/*5* ratio from 1 : 1.2 to 2 : 1 the configuration of the major product became inverted [31]. This was explained by assuming a non-cyclic complex in the case of the higher ratio. Though further argumentation by Sharpless has not been supported by direct experimental evidence, his ideas should be discussed briefly. The loaded titanium complex and transfer of oxygen is subject to the following contraints: (i) The substrate must occupy one of the two open quadrants available around each titanium atom. (ii) The hydroperoxide is not only fixed to titanium by the O_β-Ti bond but also by coordination to one of the enantiotopic lone pairs at O_α. In this way chirality of the catalyst is extended onto the oxidant. (iii) Transfer of oxygen requires perpendicular attack of O_β at the π bond, while the π^* orbital must overlap with one of the lone pairs at O_β. A model constructed along these lines (*7*) is in accordance with the observed enantioselectivity.

Interestingly, chiral catalysis in the oxidation of sulfides to sulfoxides has not been reported, but several chiral media have been tested for this purpose [32–34]. Among them, the bovine serum albumin-H_2O_2 system was the most effective and gave, albeit in poor yields, up to 90% ee with aryl-alkyl sulfides [35]. Unexpectedly the prevailing configuration was opposite for phenyl-alkyl and *p*-tolyl-alkyl sulfides [36].

In 1978 Feringa and Wynberg reported that oxidation of 2-naphthol with $Cu(NO_2)_2$ in the presence of (*S*)-2-phenylethylamine gave 2,2′-dihydroxy-1,1′-binaphthyl in but 2.8% ee in 63% yield [37], while in 1983 under apparently the same conditions Brussee and Janssen achieved 95% ee in 85% yield [38].

Osmium tetroxide is well-known as a reagent for the *syn* hydroxylation of olefins. Hentges and Sharpless developed an enantioselective version of the method by adding, as base, dihydroquinine acetate instead of pyridine [39]. Except for stilbene, enantioselectivities were low.

Ph–CH=CH–Ph → 1) OsO_4, MePh, −78°C, dihydroquinine acetate; 2) LAH → Ph–CH(OH)–CH(OH)–Ph

4.2 Diastereoselective Oxidations

4.2.1 Diastereoselective Epoxidation

Fundamental studies in this field were initiated in the late fifties by Henbest and his coworkers, who investigated both simple cycloolefins and cyclic allylic alcohols [40, 41]. They established that with simple cyclic olefins or allylic ethers epoxidation by peracids was influenced primarily by steric hindrance and modified by strong solvent effects. With the free allylic alcohols, in turn, results could be interpreted by assuming stereodirection by hydrogen bonding between the peracid and the hydroxy group. Later studies were practically restricted to the epoxidation of allylic alcoholic.

Two techniques of diastereoselective epoxidation will be discussed here, *viz*. that using peracids and that using hydroperoxides in the presence of transition metal catalysts.

With more advanced techniques of analysis at hand, Whitham *et al.* [42, 43] and Itoh *et al.* [14, 44] established that cyclic allylic alcohols gave predominantly the *cis*-epoxides with the $VO(acac)_2{-}H_2O_2$ (or $-tBuO_2H$) system, while with peracids cyclohex-2-enol and cyclohept-2-enol gave the *cis*-, the eight- and nine-membered analogues mainly the *trans*-epoxides (Fig. 4-4 (a) [42, 43].

When the hydroxy group is blocked by acetylation stereodirection by means of hydrogen bonding with the peracid, or ester formation with the metal does not operate anymore and the *trans*-epoxide is formed in slight excess. If a bulky group interferes with *cis* epoxidation, as in *cis*-5-tert-butylcyclohex-2-enol, using the $VO(acac)_2$-$tBuO_2H$ system, instead of a *trans*-epoxide 5-tert-butylcyclohex-2-enone was formed [42]. Potassium peroxomonosulfate in the presence of a crown ether gave, from cycloalken-2-ols in a two-phase system, mainly *trans*-epoxides [45].

Pierre, Chautemps and others studied the epoxidation of acyclic allylic alcohols with 4-nitro-perbenzoic acid and found that stereoselectivity depended on the substitution pattern of the double bond (Fig. 4-4 (b)) [46, 47]. Results were interpreted by considering the relative stability of those ground state conformations in which one of the bonds attached to the α-carbon was eclipsed with the double bond.

Sharpless, Nozaki and others reported on the $VO(acac)_2$ and $Mo(CO)_6$ catalyzed epoxidation of the same compounds by $tBuO_2H$ [48, 49]. Catalysis by Mo^{6+} proved to be less selective than that by V^{5+}. With certain substrates stereochemical preference is opposite and selectivity higher than with peracids, whereas with

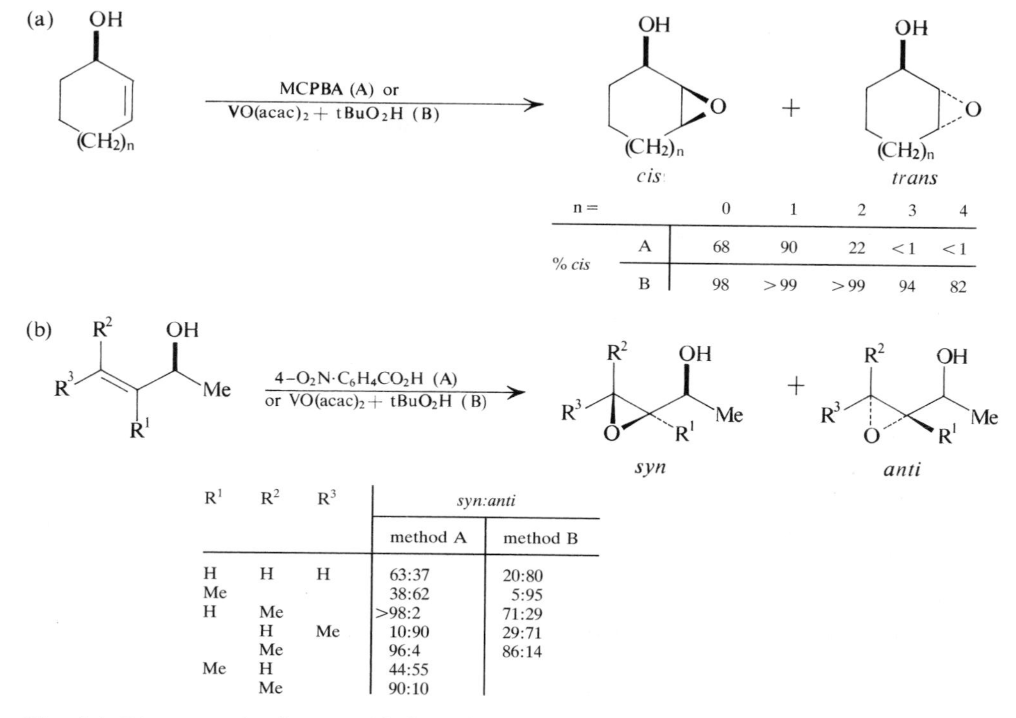

	n =	0	1	2	3	4
% *cis*	A	68	90	22	<1	<1
	B	98	>99	>99	94	82

R^1	R^2	R^3	*syn:anti*	
			method A	method B
H	H	H	63:37	20:80
Me			38:62	5:95
H	Me		>98:2	71:29
	H	Me	10:90	29:71
	Me		96:4	86:14
Me	H		44:55	
	Me		90:10	

Fig. 4-4. Diastereoselective epoxidation of allylic alcohols.

others metal catalysis also gives the *syn*-epoxide in excess but in lower selectivity (Fig. 4-4 (b)). Similar results were recorded by Mihelich [50]. Takai *et al.* described the epoxidation of allylic alcohols in the $tBuO_2H$-$(tBuO)_3Al$ system, but, except for special cases, selectivity was lower than in V^{5+} catalysis [51]. An interpretation of the stereochemistry of allylic epoxidation by transition state models based merely on product analysis was proposed by Narula [52, 53].

The Sharpless method of enantioselective epoxidation was also applied to substrates with diastereotopic faces. As expected, the reaction is highly sensitive to the constitution of the substrate. While high *syn* selectivity was observed by Kishi *et al.* with β-substituted allylic alcohols [54] *anti*-epoxides were obtained from α-methyl allylic alcohols by Isobe and others [55]. In the above substrates the oxygenated carbon was not a chiral center.

Exceptionally effective kinetic resolution ($k_S/k_R = 16-104$) of racemic allylic alcohols of type *8* were realized by Sharpless and his coworkers, permitting the convenient preparation of allylic alcohols in high optical purity (>96% ee at 50% conversion). As in other double-induction experiments, diastereoselectivity is highly dependent on the configuration of the substrate (or *vice versa* of the reagent) (Fig. 4-5) [56]. In the absence of diisopropyltartrate diastereoselectivity dropped sharply with β-unsubstituted or (*E*)-allylic alcohols but was high (>95% *syn*) with (*Z*)-allylic alcohols.

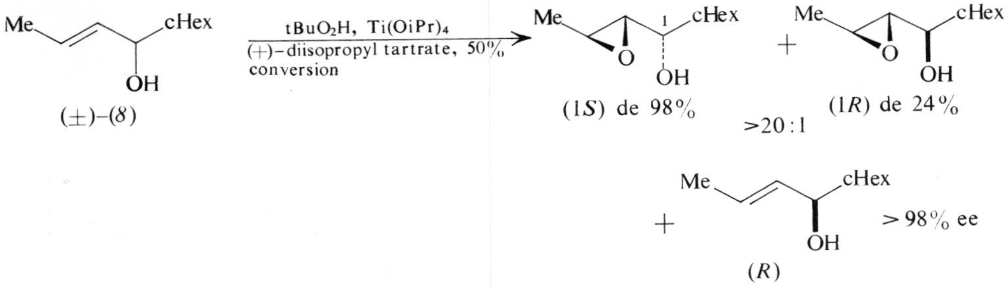
Me
cHex
OH
(±)-(8)
tBuO$_2$H, Ti(OiPr)$_4$
(+)-diisopropyl tartrate, 50% conversion
Me
1
cHex
O
OH
(1S) de 98%
+
>20:1
Me
cHex
O
OH
(1R) de 24%
Me
cHex
+
OH
(R)
> 98% ee

Fig. 4-5. Kinetic resolution by allylic epoxidation.

Epoxidation of homoallylic alcohols acquired great importance in the synthesis of macrocyclic antibiotics, first of all in the hands of Kishi and coworkers. As a prelude, Sharpless and Michaelson found almost complete *cis* selectivity in Mo^{6+} catalyzed epoxidation [57]. Later Kishi applied both the tBuO$_2$H-VO(acac)$_2$ system and *m*-chloroperbenzoic acid (MCPBA) to various chiral homoallylic alcohols and obtained the *anti*-epoxides (or their transformation products) in excess [58–60]. Some examples are shown in Fig. 4-6.

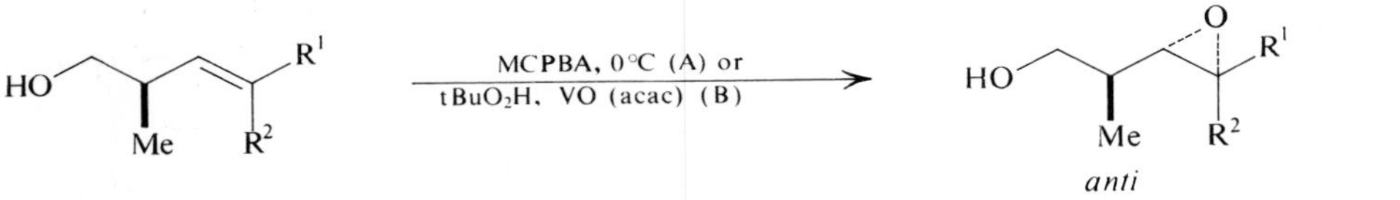
HO
R^1
Me
R^2
MCPBA, 0°C (A) or
tBuO$_2$H, VO (acac) (B)
HO
O
R^1
Me
R^2
anti

R^1	R^2	% *anti* method A	% *anti* method B
Me	CH_2OH	>96	>96
CH_2OH	Me	80	83
CH_2OBn		83	>96

Fig. 4-6. Diastereoselective epoxidation of homoallylic alcohols.

Despite repeated efforts no efficient methods for the enantio- and diastereoselective preparation of oxaziridines has become available.

4.2.2 Various Stereoselective Oxidations

Dialkyl sulfides having a chiral center in the α-position are preferentially transformed to *syn*-sulfoxides by peracids. As an explanation the synclinal disposition of the large substituent was assumed, for which some NMR and X-ray evidence could be furnished (Fig. 4-7) [61]. Periodate oxidation of 1,3-dithianes proceeds with strong equatorial preference [62].

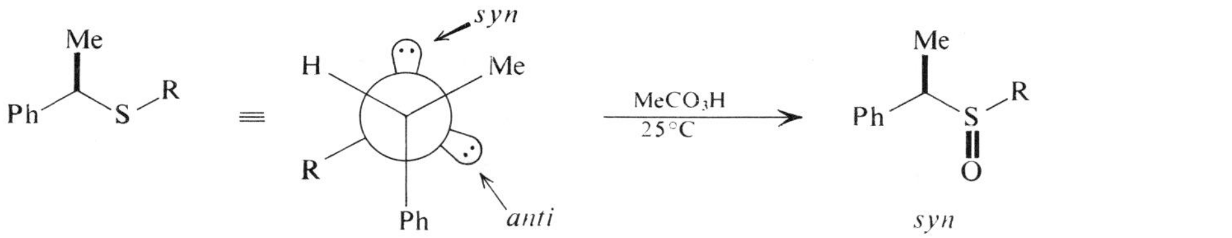

R	Me	Et	iPr	tBu	Bn	Ph
% *syn*	75	76	78	98	66	71

Fig. 4-7. Diastereoselective oxidation of sulfides.

Kishi and his coworkers explored the osmium tetroxide oxidation of chiral allylic alcohols. Besides confirming the well-known *syn* stereochemistry of hydroxylation, they established *anti* preference, as related to the inducing center and the adjacent hydroxyl. 2,3-*Anti* selectivity was higher with polar directing groups and (*Z*)-allylic alcohols (Fig. 4-8) [63].

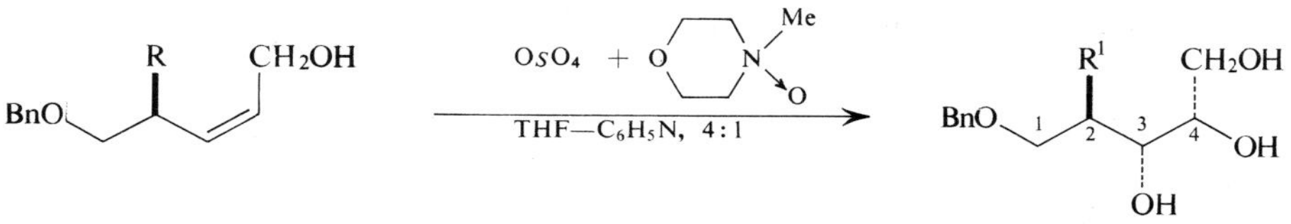

R = OBn, 89% 2,3-*anti*
R = Me, 50% 2,3-*anti*

Fig. 4-8. Stereochemistry of the osmium tetroxide oxidation of allylic alcohols.

Oxidative coupling of the pure *S* enantiomer of a chiral phenol (*9*) was claimed by Wynberg to be totally selective for the *l,l* diastereomer (Fig. 4-9) [64]. Diastereoselectivity in respect of coupling products containing identical chiral centers (*i. e. u, l/l, l*) dropped to 89% when the racemic substrate was oxidized. This was attributed to an "auto-solvent" effect, *i. e.* to the interaction of antipodal substrates in solution.

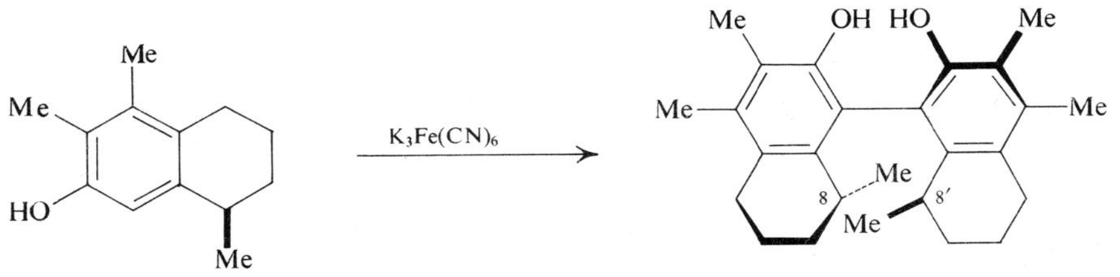

(*S*)-(*9*) (*S*)-(8*S*,8′*S*) (*l,l*) 100% de

$(\pm)-9 \xrightarrow{K_3Fe(CN)_6} l,l\ (66\%) + u,l\ (8\%) + l,u\ (26\%)$

Fig. 4-9. An example for diastereoselective oxidative phenol coupling.

5 Stereoselective Carbon-Carbon Bond Forming Reactions by Nucleophilic Addition to Carbonyl Groups

While diastereoselective C–C bond formation in cyclizations and in the transformation of cyclic systems is a traditional area of stereoselective synthesis, some fifteen years ago no efficient methods for the enantioselective formation of such bonds or procedures for diastereoselective C–C bond formation in acyclic systems were available. Stimulated by the rapid advancement of stereoselective synthesis in general, further by an endeavor to elaborate methods for the synthesis of macrocyclic antibiotics and similar natural products containing a multitude of chiral centers in a non-cyclic environment, a whole arsenal of highly selective enantio- and diastereoselective methods has been developed. The abundance of information available poses serious problems of systematization. Our own approach to classification was based on types of reactions rather than on target molecules. Stereoselective pericyclic reactions are dealt with in Chapter 7, while the rest of the material is distributed between Chapters 5 und 6. Chapter 5 is essentially devoted to nucleophilic additions to C=O and C=N groups, whereas Chapter 6 is a collection of diverse methods ranging from Michael addition to Grignard cross-coupling and other catalytic enantioselective reactions.

5.1 Addition of Simple Nucleophiles to Carbonyl Compounds

5.1.1 Enantioselective Additions

One of the fundamental methods to establish a C–C linkage is the addition of simple carbon nucleophiles, such as Grignard reagents, lithium alkyls, or a cyanide ion to a carbonyl group. While diastereoselective addition involving first of all chiral carbonyl compounds but often also chiral nucleophiles has been the subject of extensive studies and led to the discovery of the rules of Cram and Prelog (*cf.* Sections 3.2.1 and 3.2.2), enantioselective addition to prochiral carbonyl compounds remained, until recently, a rather neglected field. Very few examples for such reactions have been reported and in most of them enantioselectivity was impractically low.

Enantioselective additions can be realized in two ways: (i) an achiral nucleophile is used in the presence of a chiral additive, such as a chiral solvent, complexant or catalyst, or (ii) by transfer of an achiral alkyl group from a chiral reagent.

Additions in the presence of a chiral complexant

It is well known that Grignard reagents readily form complexes with ethers, while alkyl lithiums coordinate to tertiary amines. Therefore, in the presence of chiral complexants such alkyl metals may add enantioselectively to prochiral carbonyl compounds. Such an effect, albeit very small, was first demonstrated with (+)-2,3-dimethoxybutane by Cohen and Wright in 1953 [1]. Better selectivity [70% ee for the (R)-carbinol] was achieved in the reaction of cyclohexylphenyl ketone with MeMgBr in the presence of 1,2:5,6-di-*O*-isopropylidene-α-D-glucofuranose [2], while other combinations of ketones and Grignard reagents, as well as other sugar derivatives, gave poor results.

No real success in enantioselective additions of Grignard reagents has been recorded as yet, and in a study on the addition of diethylzinc to benzaldehyde catalyzed by chiral cobalt(II) and palladium complexes or aminoalcohols [3] 58% ee was scored as the maximum [4].

Addition of lithium alkyls in the presence of chiral ditertiary diamines, in turn, could be developed, at least for certain models, into a reasonably enantioselective procedure. Some of the ligands used are shown in Fig. 5-1.

Compounds *1-3* were selected by Seebach *et al.* from 21 tartaric acid derivatives [5–8]. Mukaiyama and his coworkers first discovered that *5* was an excellent chiral

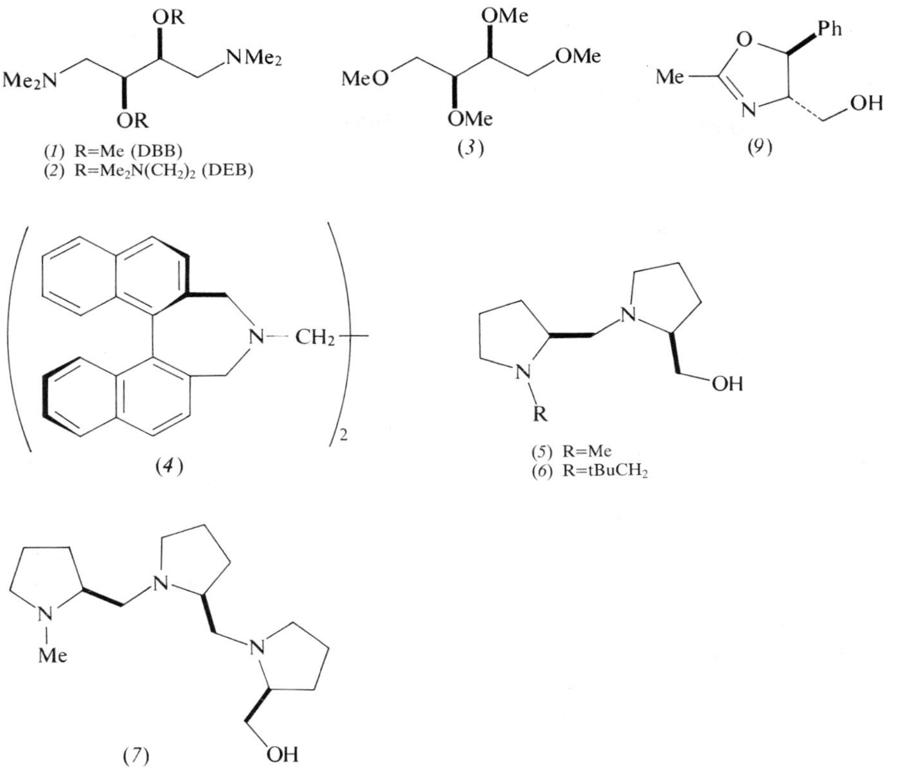

(*1*) R=Me (DBB)
(*2*) R=$Me_2N(CH_2)_2$ (DEB)

(*3*)

(*9*)

(*4*)

(*5*) R=Me
(*6*) R=$tBuCH_2$

(*7*)

Fig. 5-1. Chiral additives used in the enantioselective addition of alkylmetals to carbonyl compounds.

aid [9] and thereafter prepared 8 less successful analogues [10]. Selected results for the enantioselective addition of alkyllithium compounds to aldehydes in the presence of the amines shown in Fig. 5-1 have been compiled in Table 5-1.

Addition of a carbanion prepared from allyl p-tolylsulfone to acetone in the presence of *5* proceeded with 80% ee (*S*) [14].

Complexes of the chiral oxazolin *9* with methyl- and phenylmagnesium bromide were reacted with ketones to give a series of homochiral ketones in 9–25% ee [129].

However, the scope of the reaction could be successfully extended to lithium acetylides; the products were converted into optically active butenolides (Fig. 5–2) [15, 16].

Chiral reagents donating alkyl groups

By reacting a metal tetraalkyl aluminate with a chiral alcohol, a chiral reagent capable of enantioselective alkyl transfer to carbonyl groups is formed:

$$MAlR_4 + R^*{-}OH \rightarrow MAlR_3OR^* + RH$$

The reagent obtained from (−)-*N*-methylephedrine and $NaAlBu_4$ gave from acetophenone the (*S*)-carbinol in but 44% ee; (+)-Darvon alcohol, which was an excellent chiral aid in LAH reductions, did not produce better results [17].

$Me_3Si{-}C{\equiv}C{-}Li$ —(RCHO, (*5*); Me_2O, −123°C)→ R–CH(OH)–C≡C–$SiMe_3$

R	Et	iBu	Pent	Oct
% ee	68	65	76	80

—(R = Oct, 1) $OH^{\ominus}$ 2) BuLi, 3) CO_2)→ Oct–CH(OH)–C≡C–CO_2H —(1) H_2 2) $H^{\oplus}$)→ Oct-substituted butenolide (=O)

Fig. 5-2. Enantioselective addition of a lithium acetylide to aldehydes.

Table 5-1. Enantioselective Addition of Alkylmetals to Aldehydes in the Presence of Chiral Ligands (L*).

$$R^1{-}CHO + R^2{-}Li \xrightarrow{L^*} R^1{-}CH(OH){-}R^2$$

R^1	R^2	L*	% ee	(Conf.)	Solvent	Temp. °C	Ref.
Ph	Me	*5*	40	(*R*)	$(MeO)_2CH_2$	−123	11
	Et	*5*	54	(*S*)	$(MeO)_2CH_2$	−123	11
		4	67	(*R*)	Et_2O	−120	8
	Pr	*5*	60	(*S*)	$(MeO)_2CH_2$	−123	11
	iPr	*1*	11	(*S*)	pentane	−120	5
	Bu	*1*	40	(*S*)	pentane	−150	6
		2	15	(*R*)	pentane	− 78	7
		3	30	(*R*)	pentane	− 78	6
		4	89	(*R*)	Et_2O	−120	8
		5	95	(*S*)	$(MeO)_2CH_2$	−123	11
		6	54	(*R*)	$(MeO)_2CH_2$	−123	11
		7	68	(*R*)	$(MeO)_2CH_2$	−123	11
		8	36	(*R*)	$(MeO)_2CH_2$	− 85	12
	$PhSCH_2$	*5*	60		$(MeO)_2CH_2$	−123	10
Me	Bu	*2*	46	(*R*)	pentane	− 78	7
Et		*2*	35	(*R*)	pentane	− 78	7
iPr		*2*	53	(*S*)[a]	pentane	− 78	7
tBu		*3*	23	(*R*)	pentane	− 78	13
cHex		*2*	48	(*S*)	pentane	− 78	7
CH_2=CH		*2*	24	(*S*)	pentane	− 78	7

[a] Note change of priority as compared with preceding entry!

Chiral trialkoxy-alkyltitanium reagents were prepared using a series of chiral alcohols by Seebach and his coworkers [18, 19], of which the one modelled after the highly selective LAH-complex *4* (p. 108), namely *10,* was the most efficient (Fig. 5-3).

(*10*)

R	H	4-Me	2-NO_2
% ee	59	88	76

Fig. 5-3. Transfer of an alkyl group to a carbonyl group from a chiral reagent.

Finally, chiral catalysis in cyanohydrin synthesis should be mentioned. Enantioselective addition of hydrogen cyanide to benzaldehyde in the presence of quinine, discovered in 1912 by Breding and Fiske [20], was probably one of the first enantioselective carbon-carbon bond forming reactions ever reported. Despite extensive investigations, among others by Prelog and Wilhelm [21], enantioselectivities did not reach practical levels. Recently Oku and Inoue found that enantioselectivity increased when, instead of simple amino acids or linear dipeptides, the reaction was catalyzed by *cyclo*-(*S*)-phenylalanyl-(*S*)-histidine [22, 23]. The real degree of selectivity could not be established due to racemization in the medium, but 90% ee (*R*) was reported at 40% conversion.

5.1.2 Diastereoselective Additions

In this section addition of simple nucleophiles, mainly of organometallic compounds and carbanions, to carbonyl groups with diastereoselective faces, *i. e.* to chiral carbonyl compounds and substituted cyclic ketones, will be discussed. First, asymmetric induction by an adjacent chiral center in acyclic substrates (1,2-induction) will be described, followed by induction by more remote centers in acyclic systems (first of all 1,3-induction) and additions to cyclic ketones. Finally, the addition of chiral nucleophiles to prochiral carbonyl compounds will be covered.

1,2-Asymmetric induction

Along with reductions by hydride donating reagents, a study of the stereochemistry of the addition of carbon nucleophiles to carbonyl groups with an

Table 5-2. Diastereoselective Addition of Simple Organometallic Reagents to Acyclic Carbonyl Compounds with a Chiral Center Adjacent to the Carbonyl Group.

R^1	R^2	R^3	R^4	R^5M	% A^a	Ref.
Ph	Me	H	H	MeMgBr	70	24
				EtMgBr	75	26
				PhMgBr	80	26
				$MeTi(OiPr)_3$	88	32
	Et			MeMgI	71	26
				EtMgI	75	30
	iPr			iPrMgBr	69	27
	Et	Me	H	PhMgBr	66	28
			Me	EtLi	71	29
			Et	MeLi	89	30
Et	Me	H	H	MeMgBr	60	24
			Me	iBuMgBr	64	24
cHex				MeMgI	65	31
Cl	Bu		H	EtMgBr	70	35
	Et			BuMgBr	80	35
Ph	OH	Me	Ph	MeLi	89	33
	OMe			MeLi	66	34
	OH		Me	PhLi	91	33
				PhMgCl	76	33
	OMe			PhLi	90	34
				PhMgBr	95	34
cHex	OH	Me	cHex	MeLi	89	36
Ph		H	Tol	PhMgBr	98	33
			Me	PhMgBr	96	33
			Ph	MeMgI	96	33
		Me	Ph	MeLi	89	33
				MeMgI	66	33
			Me	PhLi	91	33
Hept	OBn	H	Me	BuMgBr	100	42
	OMEM		Bu	MeMgCl	0	42
$BnOCH_2OCH_2$	Me		H	Me_2CuLi	97	43
				MeLi	41	43
				Bu_2CuLi	94	43
				$CH_2{=}CHCuBu_3P$	89	43
$BnOCH_2O$				MeMgBr	91	43
Me	NH_2		Ph	TolMgBr	99	37
Ph			$4\text{-}MeOC_6H_4$	PhMgBr	50	37
Et	N (ring)		H	EtMgBr	100	38
	N O (ring)				66	38

Table 5-2. Continued

R^1	R^2	R^3	R^4	R^5M	% *A*[a]	Ref.
	NEt_2				48	38
	$NiPr_2$				0	38
Ph	NMe_2			PhMgBr	100	38
					55	38
Me	NMe_2		Me	EtMgBr	100	39
	NMePh				0	40
		H	Me	PhMgBr	93	41
				PhMgBr + $MgBr_2$	10	41
			Ph	MeMgBr	88	41

[a] Normalized to % A + % B = 100.

adjacent chiral center played a central role in the development of theories concerning asymmetric induction.* The essence of these theories has already been discussed in Section 3.2 and need not be restated here. The most important results obtained under optimized conditions have been compiled in Table 5-2. To this the following comments may be added:

(i) For compounds with carbonyl as the only polar group, Cram's rule applies. Stereoselectivities are, by present standards, moderate. Asymmetric induction in 2-chloroaldehydes can be interpreted by the antiperiplanar effect (*cf.* Section 3.2.1).

(ii) A practically useful degree of stereoselectivity can be realized under chelation control (*cf.* Section 3.2.1). An oxygen atom (in alcohol or ether function) in α-, possibly also in β-position, or an amino group in α-position can serve as an anchor for the metal atom of the reagent. Since the solvent may compete with the reagent for chelation and the coordinating ability of various organometallic compounds is also different, such reactions have to be carefully optimized for solvent, reagent, temperature, *etc.* [33, 44]. Thus, addition of butyllithium to an α-alkoxyketone gave 41−75% of the *syn*-diol (type *A*) (lowest de in THF), while that of butylmagnesium bromide afforded 90−100% (highest de in THF!) [42]. As was shown, among others by Still *et al.*, the nature of the protecting group was also important, and the 2-methoxyethoxy-methyl (MEM) group was particularly useful owing to its "crown-ether" effect. The authors realized very high stereoselectivities with models having a chiral center in α- and an ether function in α- or in β-position [42, 43].

Comparison of models *11*−*13* revealed that the more remote chiral center played a secondary role in stereodirection (Fig. 5-4 (a)). Interestingly, no induction was

* For a detailed discussion of the literature up to 1968 *cf.* ref. [24], pp. 84−132. For a recent review see ref. [25].

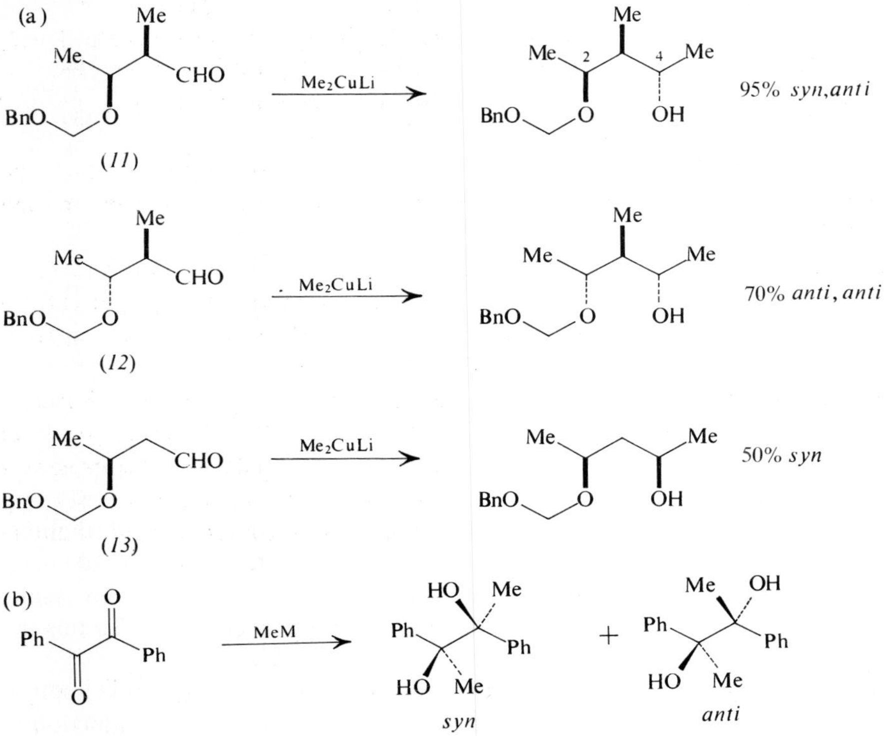

M	Li	MgBr	Zn(OiPr)$_3$	Ti(OiPr)$_3$	Ti(OEt)$_3$	TiMe$_3$	TiMe(OiPr)$_2$
% *syn*	15	28	19	>98	60	98	93

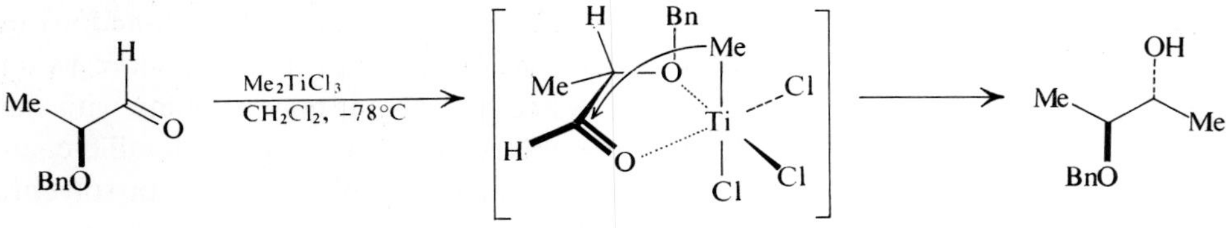

Fig. 5-4. Asymmetric 1,2-induction in the addition of metal alkyls to aldehydes.

exerted by the β-center alone, but its effect on induction by the α-center was significant [42].

Addition of Grignard reagents to α-aminoketones is much influenced by the substitution of the amino group [45]. Primary amines and dialkylamino compounds in general give an excess of the diastereometer expected under chelation control. An exception was 2-diisopropylaminopropanal, which gave exclusively product *B*. The behavior of phenylmethylaminoketones was rather unpredictable, and stereoselectivity could be inverted by the addition of $MgBr_2$.

The coordinating ability and bulkiness of Ti(IV) compounds can be modified by varying the ligands. This was exploited by Reetz and his coworkers who studied the addition of various metal alkyls to benzil (Fig. 5-4 (b)) [46]. Clearly no chelation control was operative in the addition of methyl-alkoxytitanium reagents. Note, however, that with phenylacetaldehyde as substrate $MeTi(OiPr)_3$ only gave 88% of the *syn* product [32]. It is also remarkable that the zirconium analogue led to the *anti*-diol in moderate excess [46].

On the other hand, with $MeTiCl_3$, in which coordination is promoted by the electrophilic chlorine atoms, the chelation-controlled product was formed in high excess [47].

Finally, examples for 1,2-asymmetric induction by removable chiral auxiliary groups should be discussed (Fig. 5-5 (a) and (b)). The ketone *14* was prepared from relatively inexpensive mannitol; Grignard addition proceeded under chelation control [48].

Eliel and his coworkers utilized the chiral dithianes *15* [49], *16* [50], *17* and *18* [51] as auxiliaries (*cf.* Section 6.2.2). *16* was prepared from (+)-champhor, *17* from (+)-pulegone. It was supposed that Mg, as a hard acid, would complex preferentially with a hard base (oxygen) rather than with a soft base like sulfur. The starting ketones were prepared by addition of the lithiated oxathianes to benzaldehyde

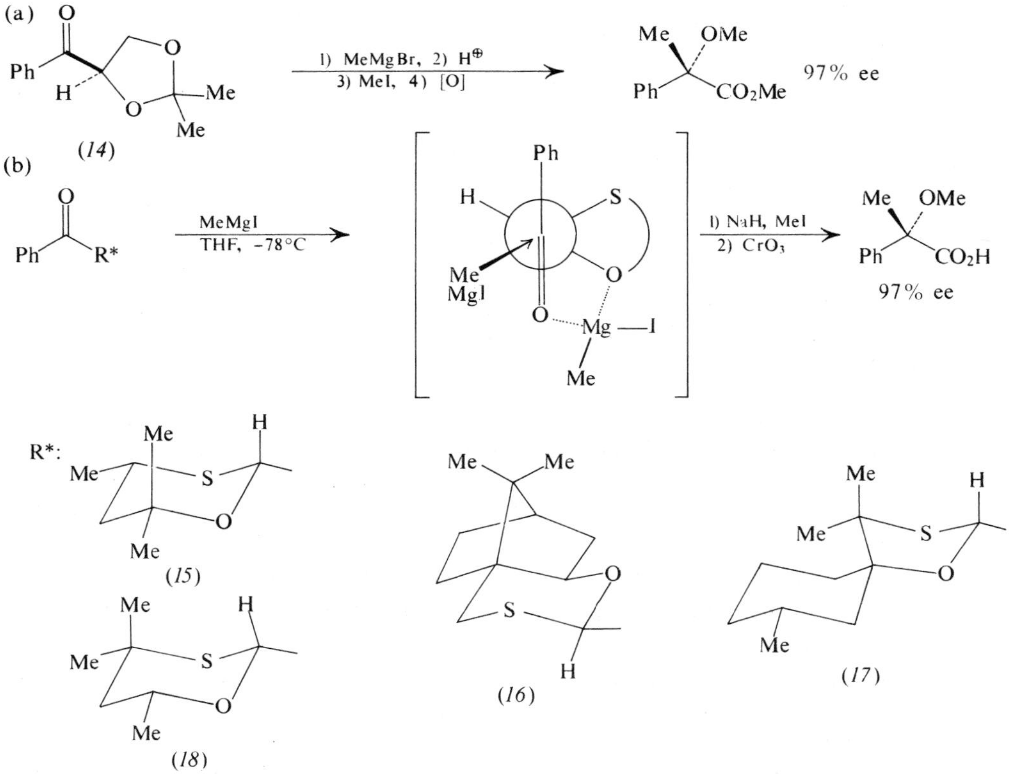

(a) (14) 1) MeMgBr, 2) $H^{\oplus}$ 3) MeI, 4) [O] → 97% ee

(b) MeMgI THF, −78°C → [chelate] → 1) NaH, MeI 2) CrO_3 → 97% ee

R*: (15) (16) (17) (18)

Fig. 5-5. Addition of alkylmetals to ketones attached to a removable chiral auxiliary group.

Table 5-3. Addition of Metalorganic Reagents to Chiral Dithianes.

R*—H ⟶ R*COR¹ ⟶ 1) R²M 2) [O] ⟶ R¹C(R²)(OH)CO₂H (R¹>R²)

R¹	R²M	R*	% ee[a]	R¹	R²M	R*	% ee[a]
Ph	MeMgI	*15*	98	Me	Bu_2Mg	*18*	74
		17	96		(pentenyl)MgBr		82
		18	96		≡-MgBr		88
	EtMgI	*15*	98		PhMgBr		80
		17	94	(allyl)	MeMgI		90
		18	98	iPr	MeMgI	*15*	52
	iPrMgI	*15*	86		PhMgBr	*18*	98
		18	98	tBu	MeMgI		92
Me	EtMgI	*18*	94				
	iPrMgI		34				

[a] Prevailing configuration as shown above.

followed by oxidation of the mixture of epimeric alcohols. Addition gave the equatorial products exclusively, but it was unselective regarding the enantiotopic faces of benzaldehyde.

The procedure has been recently successfully extended to other ketones and nucleophiles, and methods which enabled the isolation of the hydroxyacids and conversion to glycols [53] have been developed [51, 52] (*cf.* Table 5-3). The high selectivity obtained with the combination $R^1 = Me$, $R^2 = Et$ suggests that it is not the diference in the steric requirements of groups R^1 and R^2 which is essential for stereoselectivity.

There is hardly a field of enantioselective synthesis where proline derivatives do not play an important role. Indeed, Mukaiyama and his coworkers found that addition of Grignard reagents to aminals prepared from the anilide *19* [derived from (*S*)-proline] and 2-ketoaldehydes was highly diastereoselective. After hydrolysis, the chiral auxiliary could be recovered (Fig. 5-6 (a)). In the case of R = Ph the aminal could be prepared by direct condensation [54], otherwise it was obtained from the ester *20* by Grignard addition [55]. The aminals obtained by the two methods consist of a single diastereomer. Both enantiomers of the end product can be prepared by reversing the order of the reagents [56]. When the ester *20* was reduced to an aldehyde chiral 1,2-diols could be prepared with equally high enantioselectivity (Fig. 5-6 (b)) [143].

1,3-Induction

Until recently it appeared that in acyclic compounds a chiral center in β-position relative to a carbonyl group could not effectively control the stereochemistry of addition to the latter. Some characteristic data are presented in Table 5-4. The

(a)

R^1	R^2	% ee
Me	Ph	99
	Et	78
		93
Et	Ph	100
	Me	78

R^1	R^2	% ee
iPr	Ph	>94
Ph	Me	95
	Et	94
	iPr	>95
		96

(b)

R	Et	iPr	Bu	Pent
% ee	96	97	94	83

Fig. 5-6. Addition of Grignard reagents to chiral acylated aminals.

trend of diastereoselection for simple ketones can be predicted by a modified acyclic Cram model, while that for hydroxyketones is in accordance with the chelate model. No straightforward rationalization can be recommended yet for stereoselection with β-aminoketones [45].

Recently, however, very good selectivities have been achieved under chelate-control by Reetz and Jung employing $MeTiCl_3$ or a combination of Bu_2Zn and $TiCl_4$ [68] (see Table 5-4).

In summary, addition of simple nucleophiles to acyclic ketones under 1,3-asymmetric induction is generally not a promising method for preparative utilization.

Asymmetric induction over three or more bonds

The first example of asymmetric 1,4-induction, the addition of a Grignard reagent to a chiral 2-ketoester, was described by McKenzie in 1904 [61]. These and later results by both McKenzie and other authors, as well as his own data on this reaction, were systematized in 1953 by Prelog and formulated in a rule, which has become known as Prelog's rule [62]* (*cf.* Section 3.2.1). For a long time, due to the low level of its stereoselectivity, the method was used for the determination of the

* For developments in this field up to 1968 see ref. [24], pp. 50–83.

Table 5-4. Diastereoselective Addition of Simple Metalorganic Reagents to Acyclic Carbonyl Compounds with a Chiral Center in β-Position to the Corbonyl Group.

R^1–CH(R^2)–CH$_2$–CO–R^3 $\xrightarrow{R^4M}$ (A) + (B)

R^1	R^2	R^3	R^4M	% A [a]	Temp. °C	Ref.
Ph	Me	Me	PhMgBr	35	r. t.	57
				83	−110	57
		Ph	MeMgBr	42	r. t.	57
Ph	OH	Me	PhMgBr	12	−75	58
		Ph	MeMgBr	44	−78	58
	OMe	Me	PhLi	57	r. t.	58
		Ph	MeMgBr	26	−78	58
Me	NHMe	Ph	MeMgBr	22	r. t.	59
			MeLi	61	r. t.	59
Ph			MeLi	87	r. t.	59
Me	NMe_2		MeLi	69	r. t.	59
Me	OBn	H	$MeTiCl_3$	10	−78	60
			$ZnBu_2$ + $TiCl_4$	9	−78	60

[a] Normalized to % A + % B = 100.

absolute configuration of the chiral alcohol rather than for synthetic purposes. Enantioselectivities were in fact rarely higher than 50% ee [24], except for the phenylglyoxylates of (*R*)-2-hydroxy-l,l′-binaphthyl [63] and a 2-hydroxybiphenyl (*21*) [64] (Fig. 5-7(a)), for which 85 and 93% ee were reported in the case of methylmagnesium iodide addition. Excellent selectivities were observed recently by Whitesell *et al.* using the more readily available chiral auxiliary (−)-8-phenylmenthol [65]. Additions to the corresponding pyruvate and phenylglyoxylate also proceeded with 90% de [66].

Meyers and Slade investigated 1,4-induction in a series of chiral 2-oxazolidinyl ketones, which were readily available by oxidizing the anions of the corresponding alkyl compounds [67]. Selectivities with oxazolines prepared from (*R*)-phenylalanine, (*R*)-phenylglycine and (*S*)-mandelic acid were inferior to that for the example shown in Fig. 5-7(b).

Complexation by titanium proved to be useful in 1,4-induction, too, as shown in Fig. 5-7(c) [47].

Addition of Grignard reagents to chiral acyclic ketoesters with the inducing center more than three bonds away proceeds with very poor stereoselectivity.

Additions to cyclic ketones

Over the decades following the pioneering work of Barton on the problem which is now called stereoselectivity in cyclic systems, an impressive amount of material

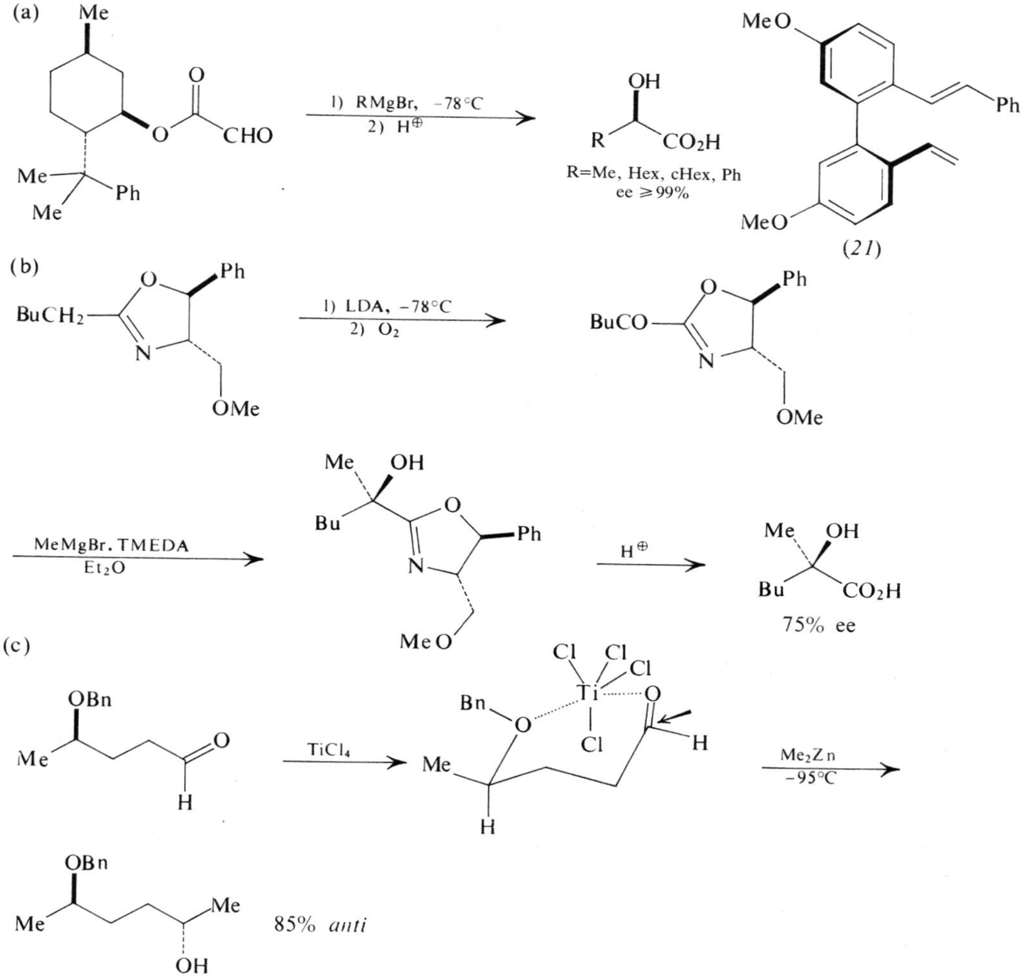

(a) 1) RMgBr, −78 °C 2) $H^{\oplus}$ → R=Me, Hex, cHex, Ph ee ≥99% (21)

(b) 1) LDA, −78 °C 2) O_2 → MeMgBr.TMEDA Et_2O → $H^{\oplus}$ → 75% ee

(c) $TiCl_4$ → Me_2Zn −95 °C → 85% *anti*

Fig. 5-7. Asymmetric induction over three or more bonds in the addition of alkylmetals to carbonyl compounds.

has accumulated on the addition of organometallic compounds to cyclic ketones.* The theoretical background of organoalkyl addition is very much the same as that of hydride reduction (*cf.* Section 3.2.1), and here we only wish to point out some distinguishing features.

(A) Whereas in the hydride reduction of unhindered cyclohexanones axial attack is generally preferred, in organometal addition equatorial attack is predominant both with hindered and unhindered ketones. The only exception is the

---

* Results on ketones with no other functional groups have been summarized up to 1974 in ref. [69], while cyclic aminoketones have been reviewed up to 1981 in ref. [45].

addition of acetylide anion to unhindered ketones in ionizing solvents in which attack is predominantly axial. Axial attack was also highly preferred when LiC≡CH was added to a conformationally stable 2,3-cyclohexanone [68].

(B) Addition to ketones with an axial substituent at C-3 is totally selective for equatorial attack, even with the acetylide anion, while with hydride reduction selectivity for the same approach is high but not total.

Some typical results for organometal additions to cyclohexanones are summarized in Table 5-5.

For aliphatic Grignard reagents and unhindered cyclohexanones preference for equatorial attack (*A*) increased with the bulk of the alkyl group (Fig. 5-8 (a)) [70], while PhMgBr only exhibited stereoselectivity with hindered ketones [69]. The factors governing selectivity are not always clear. Thus addition of MeLi to tert-butylcyclohexanone at −78°C proceeded with 68% equatorial attack while a 3:2 mixture of Me_2CuLi and MeLi showed 88% equatorial preference according to one report [71] and 94% according to another [72]. The non-chelating titanium reagent $MeTi(OiPr)_3$ was tested as a methyl source but no improvement over existing methods could be recorded [32, 73, 74].

The situation with the bicyclic ketones shown in Fig. 5-8 (b) ist straightforward: organometallic reagents add with 100% or near to 100% diastereoselectivity from the direction indicated by the arrows [69].

Generalizations are less easy to make in the cyclopentanone series. Steric approach control seems to be the rule for the addition of alkyllithiums and Grignard reagents, except acetylenic ones, when the behavior of 2-methylcyclopentanone is considered (Fig. 5-8 (c)). One would expect the improvement of selectivity when the substrate is 2-tert-butylcyclopentanone, however, the percentage of the *A*-type product does not exceed 66%.

Table 5-5. Addition of Organometallic Reagents to Substituted Cyclohexanones [69].

Reagent	%Equatorial attack[a] Substituent(s)				
	4-tBu	4-Me	3-Me	2-Me	3,3,5-Me_3
MeLi/Et_2O	65			84[c]	100
PhLi	58	53	44	88	
HC≡CLi/THF + NH_3	12[b]	53	18	45	
MeMgBr/Et_2O	53			84	100
EtMgBr/Et_2O	71		68	95	100
PhMgBr/Et_2O	49	54	59	91	100
HC≡CMgBr				45	100

[a] Normalized to % eq + % ax = 100.
[b] For HC≡CNa.
[c] From ref. [71].

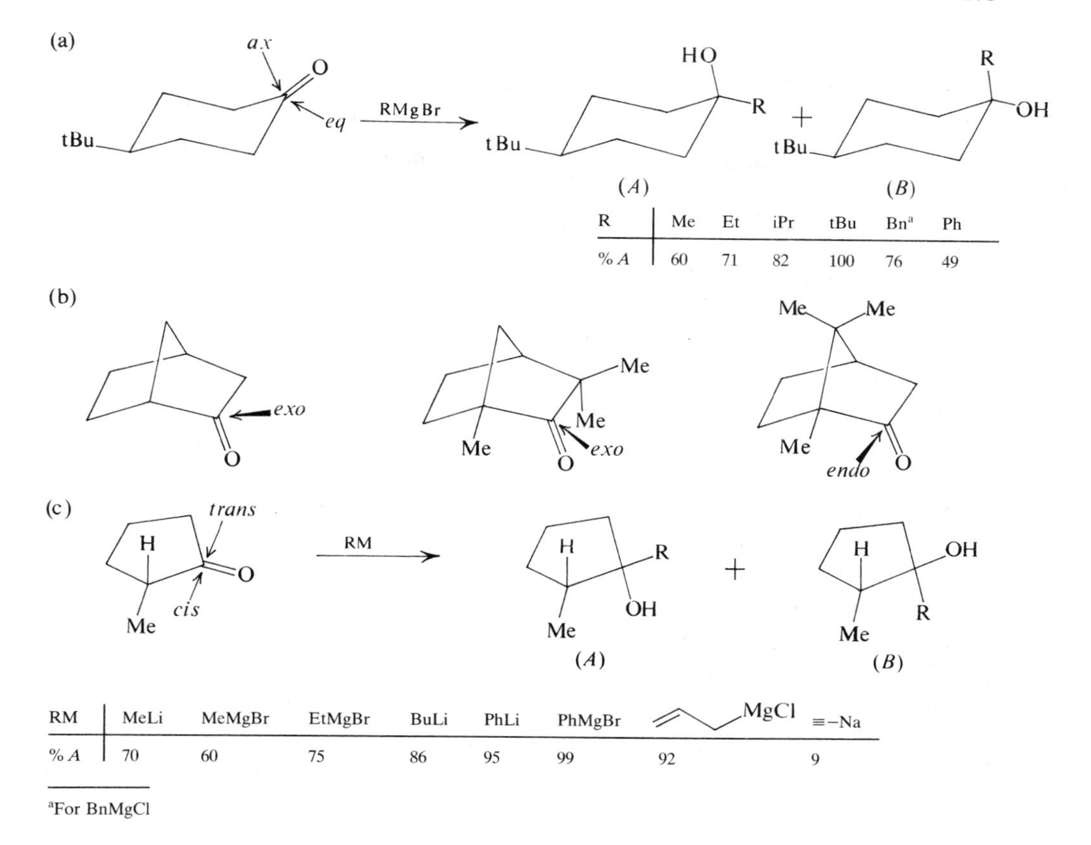

R	Me	Et	iPr	tBu	Bn[a]	Ph
% *A*	60	71	82	100	76	49

RM	MeLi	MeMgBr	EtMgBr	BuLi	PhLi	PhMgBr	allyl-MgCl	≡–Na
% *A*	70	60	75	86	95	99	92	9

[a] For BnMgCl

Fig. 5-8. Diastereoselective addition of alkylmetals to cyclic ketones.

Torsional strain takes control over the approach of the nucleophile in the case of ethynyl reagents which add preferentially in the *cis* mode. These are linear molecules with but a single other atom bonded to the entering carbon atom. To 2-methoxycyclopentanone in liquid ammonia, however, HC≡CLi (in fact $HC{\equiv}C^-$) adds predominantly in the *trans* mode due to repulsion of the negative charge and the oxygen atom. In THF the reagent is not ionized and again *cis* attack predominates [69].

Since with organoaluminum compounds, which have a hydrogen atom in β-position, alkyl addition to ketones is coupled with reduction by hydrogen transfer [75], this method lacks generality and has no special advantages even when reduction is not possible (*e. g.* Me_3Al, Ph_3Al).

Addition of chiral nucleophiles to prochiral carbonyl compounds

It has been demonstrated before (p. 24) that the permanent association of a chiral reagent with a substrate with enantiotopic groups or faces results in the formation of a pair of diastereomers. The selective formation of one of the diastereomers may be an end in itself, but often the original chiral center can be removed later to give a pair of enantiomers with one of them in excess. This removal may results in the destruction of the original chiral center, but ideally it should be conserved and the chiral auxiliary recycled. For the exploration of selectivity experiments with a racemic chiral aid yield valuable information, since the distribution of the diastereomers, formed in the first step, puts a limit to the enantiomeric purity of the end product.

Chiral sulfoxides should be well suited for the transfer of chirality because (i) they can be relatively easily prepared in an optically active form [76], (ii) hydrogens in the α-position are acidic and therefore α-metalated sulfoxides are readily formed, and (iii) the sulfur center can be smoothly removed by Raney nickel. Addition of lithiated methyl-*p*-tolylsulfoxide to simple ketones was, however, rather unselective (de 0–40%) [77]. Recently diastereoselectivity was much

(a) 1) $ZnCl_2$, 2) PhCHO; Ra—Ni; 80% de

(b) 1) tBuMgBr, THF, −78 °C; 2) R^1COR^2, 3) Al/Hg

R^1	H	H	Me	Me	CF_3
R^2	Ph	Hept	Ph	cHex	Ph
% ee	91	86	68	95	20

(c) 1) BuLi, 0°C; 2) RCHO, −100 °C; 51–73%; $H^{\oplus}$

R	Et	iPr	Bu	Me ⁄⁄ (propenyl)
% ee	88	90	87	20

Fig. 5-9. Addition of chiral nucleophiles to carbonyl compounds with enantiotopic faces.

improved by converting the lithio-compound to a zinc alkyl (Fig. 5-9 (a)) [78]. Solladié *et al.* found that α-sulfinylesters metalated with tBuMgBr added with acceptable selectivity to both ketones and aldehydes (Fig. 5-9 (b)) [79, 80].

(S)-2-Anilinomethyl-pyrrolidine, a versatile chiral auxiliary, was used by Asami and Mukaiyama to prepare chiral phthalides in high optical purity (Fig. 5-9 (c)) [81]. The use of the (4*S*,5*S*)-4-methoxymethyl-5-phenyl-2-oxazolinyl group instead of the above aminal group gave poor diastereoselectivities, although the products could be separated by crystallization [82].

(a)

R	Ph	Pr	iPr
% *syn*	95	95	87

(b)

R	Et	iPr	tBu
% *syn*	26	64	100

Fig. 5-10. Diastereoselective addition of racemic metalorganic compounds to aldehydes with enantiotopic faces.

In the following examples chiral but racemic metalorganic compounds were added to aldehydes (Fig. 5-10 (a) and (b)). The method of Nakamura and Kuwajima is a typical kinetically controlled process, if the reaction mixture is not quenched immediately diastereoselectivity rapidly vanishes [83]. A method for the preparation of *syn* vicinal diols by McGarvey and Kimura was also based on organotin compounds but was less selective [84].

5.2 Additions Involving Allylmetal and Allylboron Compounds*

5.2.1 General Aspects of Allylmetal Addition

The majority of stereoselective methods developed in the last decade for application to acyclic systems was based on the following scheme:

(A)

One realization of scheme (A) is the aldol reaction (B) (Section 5.3) and another, to be discussed here, is the nucleophilic addition of an allyl group to a carbonyl

OX (B)

X (C)

group with concomitant allylic rearrangement (C), which is the carbo analogue of scheme (B). A typical reaction of this kind is the following [87]:

Et H Me (S) + Me B O O Me Me Me Ph (−)-(Z)

OB* Et 5 4 3 Me *anti, syn* + 92 : 8 OB* Et Me Me *anti, anti*

---

* For recent reviews see refs. [85] and [86].

The reaction proceeds with complete 3,4-*anti* diastereoselectivity as a consequence of its cyclic mechanism (see below) and with 92% 4,5-*syn* diastereoselectivity due to cooperative asymmetric induction, on the one hand by the chiral center of the aldehyde (1,2-induction) and on the other by the chiral boronate ester moiety of the allylic partner (1,6-induction). By ozonolysis, epoxidation and a score of other reactions the vinyl group can be transformed to a wide variety of useful derivatives.

Before embarking on a detailed discussion of the allyl addition to carbonyl compounds, let us first discuss (i) the symmetry and stereochemistry of the reaction and (ii) the role of the leaving group.

Regarding the symmetry and constitution of the partners several different situations can be envisaged in which stereoselectivity is possible.

The following structural features of the reactants have to be considered: (i) The allyl partner can either be substituted at C-1 or not. Substitution at C-2 does not contribute to the stereochemical complexity of the reaction. (ii) The allyl component can be achiral or chiral. (Either the leaving group or C-3 may be chiral.) (iii) The faces of the carbonyl partner can be enantiotopic or diastereotopic.

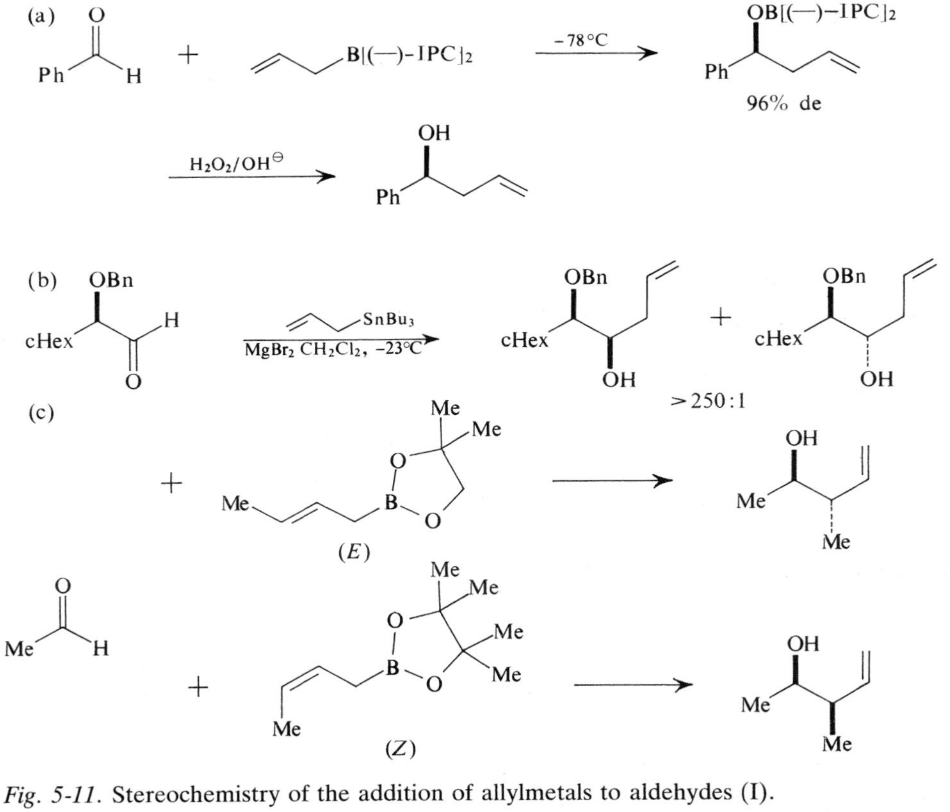

(a) Ph–CHO + allyl–B[(−)-IPC]$_2$ —(−78 °C)→ Ph–CH(OB[(−)-IPC]$_2$)–CH$_2$–CH=CH$_2$ (96% de) —($H_2O_2/OH^{\ominus}$)→ Ph–CH(OH)–CH$_2$–CH=CH$_2$

(b) cHex–CH(OBn)–CHO + allyl–SnBu$_3$ —($MgBr_2$ CH_2Cl_2, −23°C)→ cHex–CH(OBn)–CH(OH)–CH$_2$–CH=CH$_2$ + cHex–CH(OBn)–CH(OH)–CH$_2$–CH=CH$_2$ (>250:1)

(c) Me–CHO + (*E*)-Me–CH=CH–CH$_2$–B(pinacol-type boronate, Me, Me) → Me–CH(OH)–CH(Me)–CH=CH$_2$

Me–CHO + (*Z*)-Me–CH=CH–CH$_2$–B(O–CMe$_2$–CMe$_2$–O) → Me–CH(OH)–CH(Me)–CH=CH$_2$

Fig. 5-11. Stereochemistry of the addition of allylmetals to aldehydes (I).

When a simple allyl compound is reacted with a carbonyl compound having enantiotopic faces, the reaction can only be enantioselective when the leaving group is chiral (*E. g.* Fig. 5-11 (a)) [88]. If the carbonyl partner is a compound with diastereotopic faces, such as a chiral aldehyde, diastereomers are obtained as a result of asymmetric induction, regardless of whether the allyl compound is chiral or achiral (Fig. 5-11 (b)) [89].

With allyl compounds substituted at both ends an additional chiral center is created at the point of attachment and even with achiral carbonyl partners a mixture of diastereomers is obtained. When none of the partners is chiral, the products are racemic and, provided that geometrical isomers of the allyl partner are stable under the conditions of the reaction, the relative configuration of the major diastereomer is determined by the configuration of the allyl partner (*E* or *Z*) (Fig. 5-11 (c)) [90].

With chiral 1-substituted allyl compounds non-racemic diastereomers arise, and the enantiomeric purity of the individual diastereomers may differ. The chirality of the leaving group is transmitted less efficiently than that of C-3 (*cf.* Figs. 5-12 (a) and (b)) [91, 92]. Note that in example (b) chirality at the silylated center is completely transferred to the C-Me center, whereas diastereoselectivity is high but not complete.

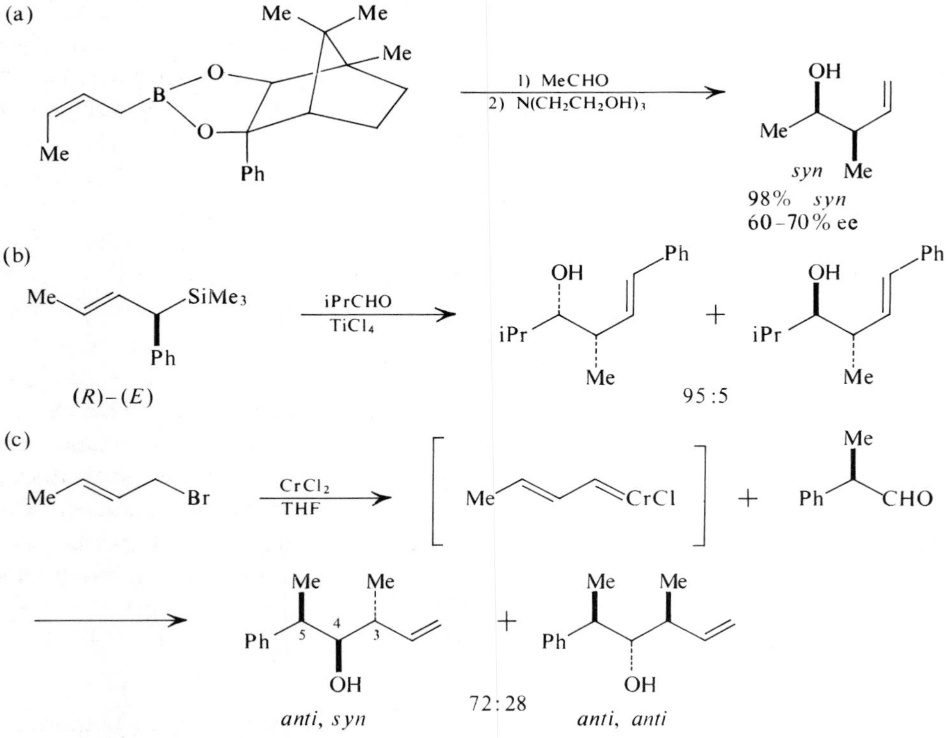

Fig. 5-12. Stereochemistry of the addition of allylmetals to aldehydes (II).

When a chiral carbonyl compound is attacked by a 1-substituted allyl reagent the relative configuration at the ends of the new bond (at C-3 and C-4) is primarily a consequence of the mechanism (mostly cyclic) of the addition, while the relative configuration of C-4 and the chiral center of the carbonyl partner is controlled by the asymmetric induction of the latter. The two effects interact, of course. In our example (Fig. 5-12 (c)) [93] 2,3-*anti* selectivity was complete, while 3,4-*syn* selectivity was modest.

The most complex combination is the reaction of a chiral 1-substituted allyl compound and a chiral carbonyl compound. This has already been exemplified in Section 5.2.1 and obeys the rules of double induction, *i. e.* diastereoselectivities are different for the *ul* (*R,S* or *S,R*) and *lk* (*R,R* or *S,S*) combinations of the partners.

For an understanding of stereoselection in allyl additions the mechanism of the process has to be analyzed [86]. It has been established that under kinetic conditions the reaction proceeds with allylic rearrangement. When addition is uncatalyzed and the metal atom in the leaving group is able to coordinate with the non-bonding pair of the carbonyl group, a six-membered cyclic transition state is highly probable. In this case the major product can most often be deduced by evaluating steric interactions in such a transition state. Although a boat-type transition state cannot be dismissed as totally improbable, for our purposes it is sufficient to compare the chair-type transition states shown for a (*Z*)-allyl compound in Fig. 5-13 (a).

If $R^2 = H$, transition state (C_1) predominates, in two forms. These are enantiomeric when the partners are both achiral and diastereomeric when either of them is chiral. In the latter case stereoselectivity can only be partially deduced from the present model.

If R^2 is other than hydrogen *gauche* interaction between R^1 and R^2 in (C_1) should be weighed against 1,3-diaxial interaction between R^1 and L in (C_2). Similar considerations apply to (*E*)-allyl compounds. A consequence of this model is that high diastereoselectivity can only be expected with aldehydes but not with ketones for which a 1,3-diaxial interaction would be present in (C_1), too.* Also, as was recognized by Evans *et al.* [156], this interaction should be more severe when the metal-oxygen distance in (C_1) is shorter, and this distance can be inferred from ground state bond lengths. Thus a decrease of diastereoselectivity in the series boron, silicon, titanium, tin, and zirconium can be anticipated.

These considerations may not be applicable to Lewis-acid catalyzed additions in which the catalyst is competing with the metal in the leaving group for coordination with the carbonyl oxygen. A linear mechanism, as shown in Fig. 5-13 (b), may be preferred in such cases, and then the *syn* diastereomer should be predominant regardless of the configuration of the allyl compound.

---

* Fortunately reactions do not read the literature and sometimes diastereoselectivity may be quite high with ketones too [94].

Experiments by Denmark and Weber [95, 96] on a model which could not take up the arangement required for the linear mechanism demonstrated that the latter was not exclusive for Lewis-acid catalyzed additions (Fig. 5-13 (d)).

Although a cyclic transition state is also improbable with ate complexes, reaction of such a compound (*22*) even with acetophenone gave *anti* products in high excess (Fig. 5-13 (c)) [97].

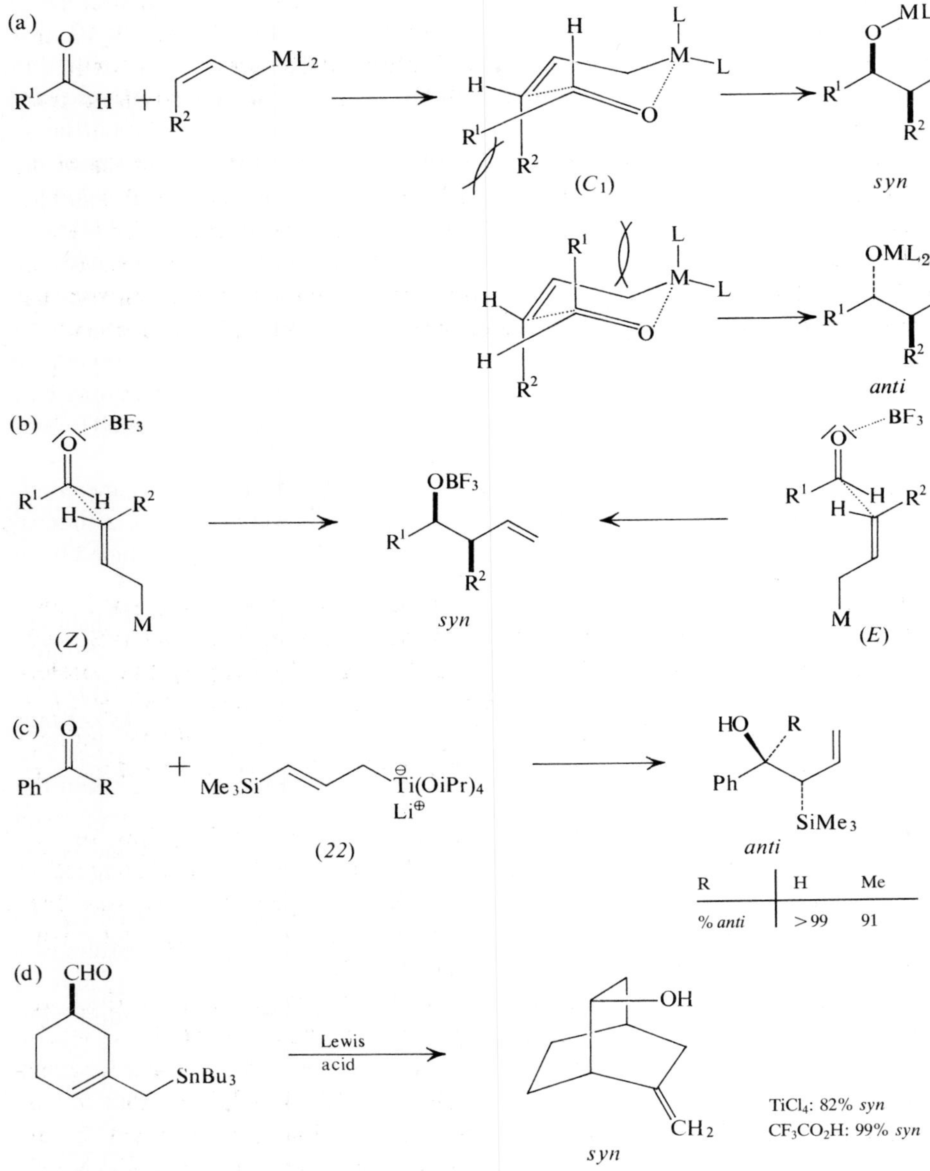

R	H	Me
% *anti*	>99	91

$TiCl_4$: 82% *syn*
CF_3CO_2H: 99% *syn*

Fig. 5-13. The mechanism of allylmetal addition to aldehydes.

When working with 1-substituted allyl compounds it is important that one should be able to prepare the geometrical isomers pure and that they should be configurationally stable. Allylmetals can be either *monohapto* (η^1) or *trihapto* (η^3) bonded, *i. e.* the metal-carbon bond is either localized at one end or delocalized over all three carbon atoms. Both types may be configurationally stable, stability being much dependent on the character of the metal and its substituents. Allyllithium, -magnesium, -zinc and -cadmium compounds undergo fast *E*/*Z* isomerization *via* a *trihapto* intermediate and there is not much one can do about it. Dialkyl-allyl boron compounds also undergo fast isomerization, but the introduction of one amine or two oxygen substituents at boron completely inhibits *E*/*Z* isomerization at room temperature.

As is apparent from the foregoing discussion, allylmetal addition to carbonyl compounds is a highly versatile method which can be tuned to the actual task by a proper choice of the metal, its ligands and the reaction conditions. There is a bewildering amount of information available about this topic and our own way of ordering it is division according to the leaving group. This is not ideal since it cuts across classification according to the stereochemical outcome of the reaction.

5.2.2 Addition of Allylboron Compounds

Relatively few applications of dialkyl-allyl boranes have been reported. Brown and Jadhav described the enantioselective synthesis of homoallylic alcohols with the aid of bis[(+)-diisopinocampheyl]-allylborane (Fig. 5-14 (a)) [88]. Stereoselectivities with the 2-methallyl analogue were in the same range [98]. B-Allyl-diisocaranylborane behaved similarly [99]. Addition of B-allyl-9-borabicyclo[3.3.1]nonane to azomethines of 2-phenylacetaldehyde gave ≥96% of the syn-amines [100].

Since simple dialkyl-allylboranes undergo fast *E*/*Z* isomerization, they are not very promising reagents for diastereoselective additions. Nevertheless the reaction can be selective when one of the geometrical isomers is highly preferred (*e. g.* Fig. 5-14 (b)) [101]. When the ligands are bulky enough, excellent diastereoselectivity can be attained even with ketones (Fig. 5-14 (c)) [102].

If the allylic carbon is substituted with two different groups, both of which are potential leaving groups, the more reactive one will be eliminated, while the remaining one can be subjected to further transformations. An example for such a reaction was reported by Yamamoto *et al.* (Fig. 5-14 (d)) [103]. Note that due to disubstitution of the allylic carbon *E*/*Z* isomerism is possible in the product. In this particular case, due to repulsion between the bulky borabicyclononyl and trialkylmetal groups, the latter was axially disposed in the transition state and therefore

(a) $[(+)\text{-IPC}]_2BH$ —1) MeOH, 2) allyl-MgBr→ $[(+)\text{-IPC}]_2B$-allyl —1) RCHO, −78°C, 2) $H_2O_2 + OH^-$→ R-CH(OH)-CH₂-CH=CH₂

R	Me	Et	iPr	tBu	Ph
% ee	93	86	90	83	96

(b) 1-Me-benzimidazol-2-yl-O-CH₂-CH=CH-Me —1) BuLi, THF, 2) Bu_3B, −100°C→ [Bu_2B-CH(Bu)-CH=CH-Me] —PhCOR→ Ph(R)C(OH)-CH(OH)-CH=CH-Bu

syn, *E*/*Z*>9:1

R	H	Me
% *syn*	>95	>95

(c) RO_2C-C(=O)-Me + Me-CH=CH-CH₂-B → RO_2C-C(OH)(Me)-CH(Me)-CH=CH₂

R	Me	Ph	2,6-tBu₂-4-MeC₆H₂
% *anti*	73	90	100

(d) Me-CH=CH-CH($SnBu_3$)-B —PhCHO, pyridine, −78°C→ [transition state: Me, H, R, H, B, O, Py, $SnBu_3$] → Ph-CH(OH)-CH(Me)-CH=CH-$SnBu_3$ —1) BuLi, 2) $(CH_2O)_n$→ Ph-CH(OH)-CH(Me)-CH=CH-CH_2OH 90% (*Z*)-*anti*

Fig. 5-14. Addition of allylboron compounds to carbonyl compounds (I).

the (*Z*)-*anti* diastereomer was formed in great excess. Addition of the α-trimethylsilyl derivative to methyl pyruvate was completely stereoselective in the same sense [102].

If the α-group was trimethylsilyl and instead of pyridine butyllithium was used for borate formation, addition became completely stereoselective but yields dropped to 50–60% [103].

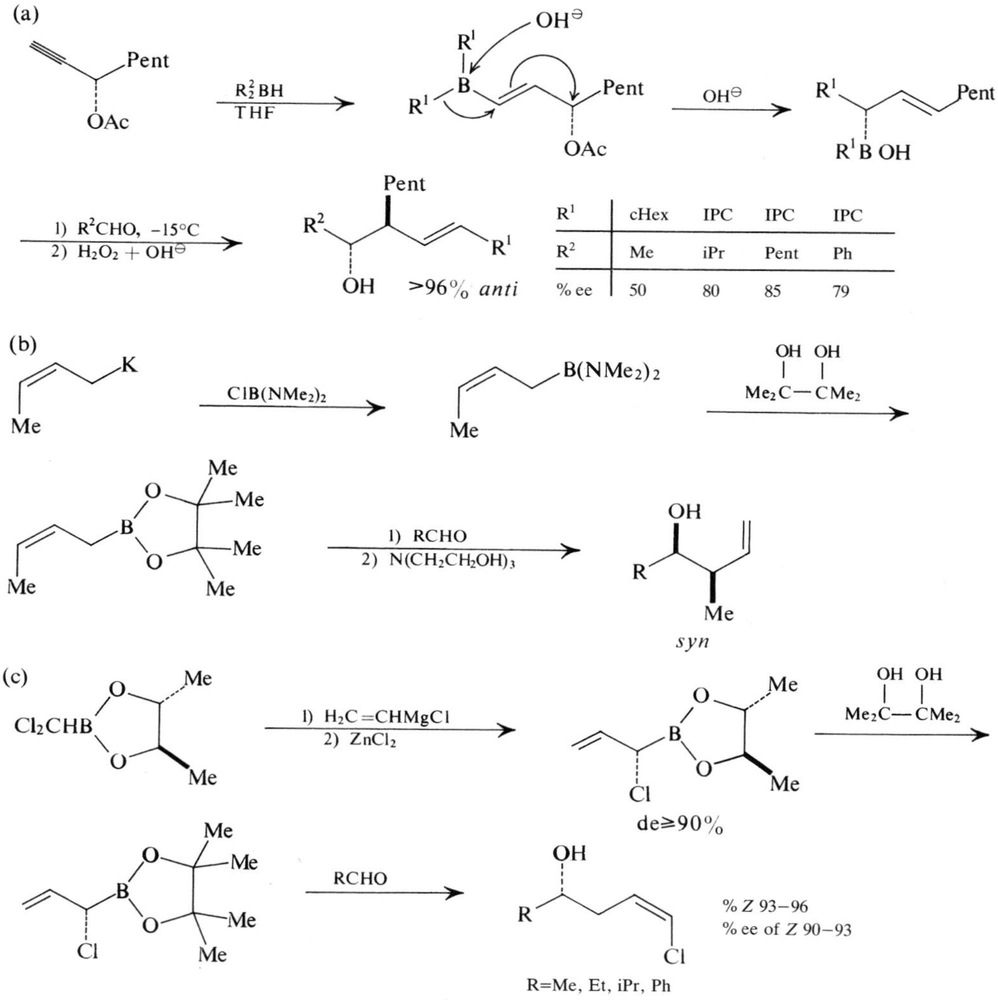

R^1	cHex	IPC	IPC	IPC
R^2	Me	iPr	Pent	Ph
% ee	50	80	85	79

Fig. 5-15. Addition of allylboron compounds to carbonyl compounds (II).

Reaction of chiral monooxygenated boranes with aldehydes was reported by Midland and Preston (Fig. 5-15 (a)) [104].

The addition of allyl boronates to aldehydes has been very thoroughly studied by Hoffmann and his group [86] and, since a wide selection of chiral and achiral diols is available, by a proper choice of the diol high dia- or enantioselectivity can be achieved. Preparation of the boronates is illustrated in Figs. 5-15 (b) and (c). Using pinacol as the diol component, addition to aldehydes was completely diastereoselective: the (*Z*)-crotylboronate gave the *syn*-, and the *E* isomer the *anti*-homoallylic alcohol. Enantiomeric purity was limited by the diastereomeric purity of the boronate (93–95%) [90, 105]. The method was recently extended to γ-alkylthio-allylboronates [279].

α-Substituted crotylboronates yield *E*/*Z* isomers of a homoallylic alcohol. Combination of a chiral, but less bulky, diol used for chirality transfer and pinacol for stereodirection in the addition step permitted the enantio- and diastereoselective synthesis of such alcohols (Fig. 5-15 (c)) [106]. Cooccurence of the indicated configurations is a proof for the cyclic mechanism of the reaction.

A search for chiral diols which lend both high enantio- and diastereoselectivity to the addition [107, 108] led to the introduction of a diol shown in Fig. 5-16 (a) prepared from (+)-champhor. Enantioselectivity was explored with allylboronates unsubstituted at C-1, and it was found to be less with the bulky or unsaturated aldehydes [109]. With the 2-methylallyl analogues enantioselectivity was not sensitive to the size of R (R=Me, 74% and R=tBu 70% ee) [107]. Reaction of the crotyl analogue with benzaldehyde was highly diastereoselective (98% *syn*) but not sufficiently enantioselective (60% ee) [91]. In the addition of the same to (*S*)-2-methylpropanol complete 2,3-*anti* and high 3,4-*syn* selectivity was observed (Fig. 5-

(a)

1) RCHO, −40°C
2) $N(CH_2CH_2OH)_3$

R	Me	Et	iPr	tBu	Ph	$H_2C{=}CH$
% ee	86	77	70	45	36	50

(b)

(*S*) + (*E*) → 92:8

(c)

(*R*) + (*Z*) → >98% *syn,anti*

(d)

R	Me	Et	iPr	Ph
% *anti*	65	70	78	79

Fig. 5-16. Addition of allylboron compounds to carbonyl compounds (III).

16 (b)), while the $R+E$, $S+Z$, and $R+Z$ combinations all gave poor 3,4-diastereoselectivities [110]. Interestingly, with (*R*)-2,3-*O*-isopropylidene-glyceraldehyde the $R+Z$ combination gave the highest diastereoselectivity (Fig. 5-16 (c)) [111]. The addition of an allenic boronate to aldehydes reported by Favre and Gaudemar is *anti* selective and probably involves a cyclic transition state, too (Fig. 5-16 (d)) [112].

5.2.3 Addition of Allyltitanium Compounds

Titanium can form *monohapto* tetracoordinate allyltitanium compounds, pentacoordinate ate complexes, further *trihapto* allyltitanium complexes and thus offers ample room for variation [113].

The *monohapto* amine complex *23* was found to add to a chiral aldehyde with high *syn* selectivity (Fig. 5-17 (a)) [74].

Several alkoxy and aryloxy titanium compounds were tested for their diastereoselectivity by Widler and Seebach, of which the triphenoxy compound was the best. Addition was *anti* selective and was studied in three structural "dimensions" (Fig. 5-17 (b–d)) [94, 114]. Addition to ketones was also selective and could be interpreted, similarly to ketone reductions, by considering the difference in the size of the substituents. The addition of crotyl-bis(cyclopentadienyl)-titanium halides (*24*) can be conducted either uncatalyzed or under boron trifluoride catalysis. In

(a) Ph–CH(Me)–CHO + CH2=CH–CH2–$Ti(NEt_3)_3$ (*23*) —pentane→ Ph–CH(Me)–CH(OH)–CH2–CH=CH2, 89% *syn*

(b) $ClTi(OPh)_3$ + Me–CH=CH–CH2–MgX → Me–CH=CH–CH2–$Ti(OPh)_3$ —RCHO, THF, −100°C→ R–CH(OH)–CH(Me)–CH=CH2 (*anti*)

R	iPr	Et_2CH	Ph	$4\text{-}MeOC_6H_4$	$4\text{-}NO_2C_6H_4$
% *anti*	98	99	92	99	90

(c) R–CH=CH–CH2–$Ti(OPh)_3$ —PhCHO, −100°C→ Ph–CH(OH)–CH(R)–CH=CH2 (*anti*)

R	Me	iPr	Bu
% *anti*	92	>98	97

Fig. 5-17. Addition of allyltitanium compounds to carbonyl compounds (I).

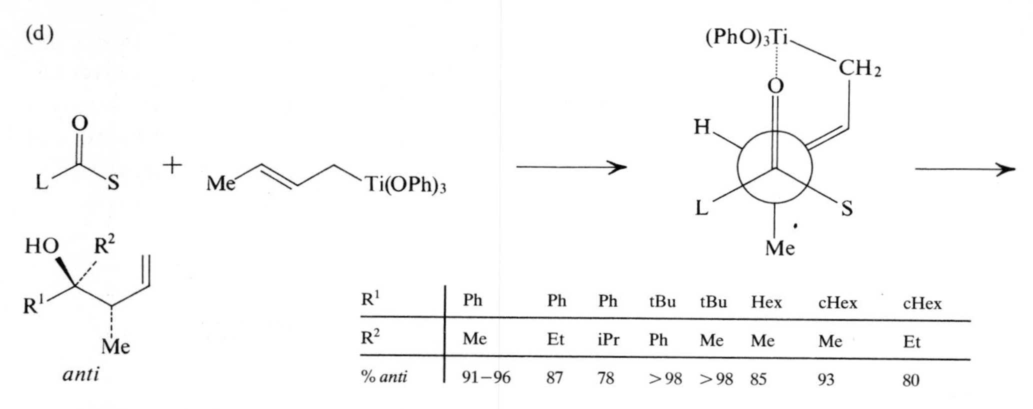

R^1	Ph	Ph	Ph	tBu	tBu	Hex	cHex	cHex
R^2	Me	Et	iPr	Ph	Me	Me	Me	Et
% *anti*	91–96	87	78	>98	>98	85	93	80

Fig. 5-17. Addition of allyltitanium compounds to carbonyl compounds (II).

the uncatalyzed reaction stereoselectivity depends on the nature of the halogen and is supposed to proceed *via* a cyclic intermediate. With X=Br the (*E*)-crotyl compound gave an *anti* product exclusively [115]. In the catalyzed reaction *syn* preference was observed (Fig. 5-18 (a)) independently of the *E*/*Z* ratio, an indication for a linear mechanism (*cf.* Fig. 5-13 (b)) [116].

Addition of the *ate* complex *22* has already been mentioned (p. 180).

Trihapto titanium complexes can be prepared *via* allyl anions, and the addition of 1,3-disubstituted representatives was studied by Sato *et al.* *Anti*-preference and an interesting dependence of double bond configuration on substitution at C-3 were observed (Fig. 5-18 (b)) [117].

Hoppe and Bröneke demonstrated that, in contrast to the α-lithiated compounds, α-titanated carbamates added stereoselectively both to pentanal and acetophenone (Fig. 5-18 (c)) [118].

A reaction related to allylmetal addition is depicted in Fig. 5-18 (d) [119]. Several metals were tested as ligand, but titanium provided the best selectivity. If azomethines were used as partners, the corresponding *anti*-amines were obtained, in most cases with total selectivity [120].

5.2.4 Addition of Allylsilanes

The addition of allylsilanes to carbonyl compounds requires activation by a Lewis acid, $TiCl_4$ being the most often used catalyst.

Enantioselective addition of a chiral allylsilane to a chiral aldehydes was recently described by Hayashi *et al.* (Fig. 5-19 (a))* [121]. Addition of the crotyl analogue

* Ee values corrected for the optical purity of the silane (95%).

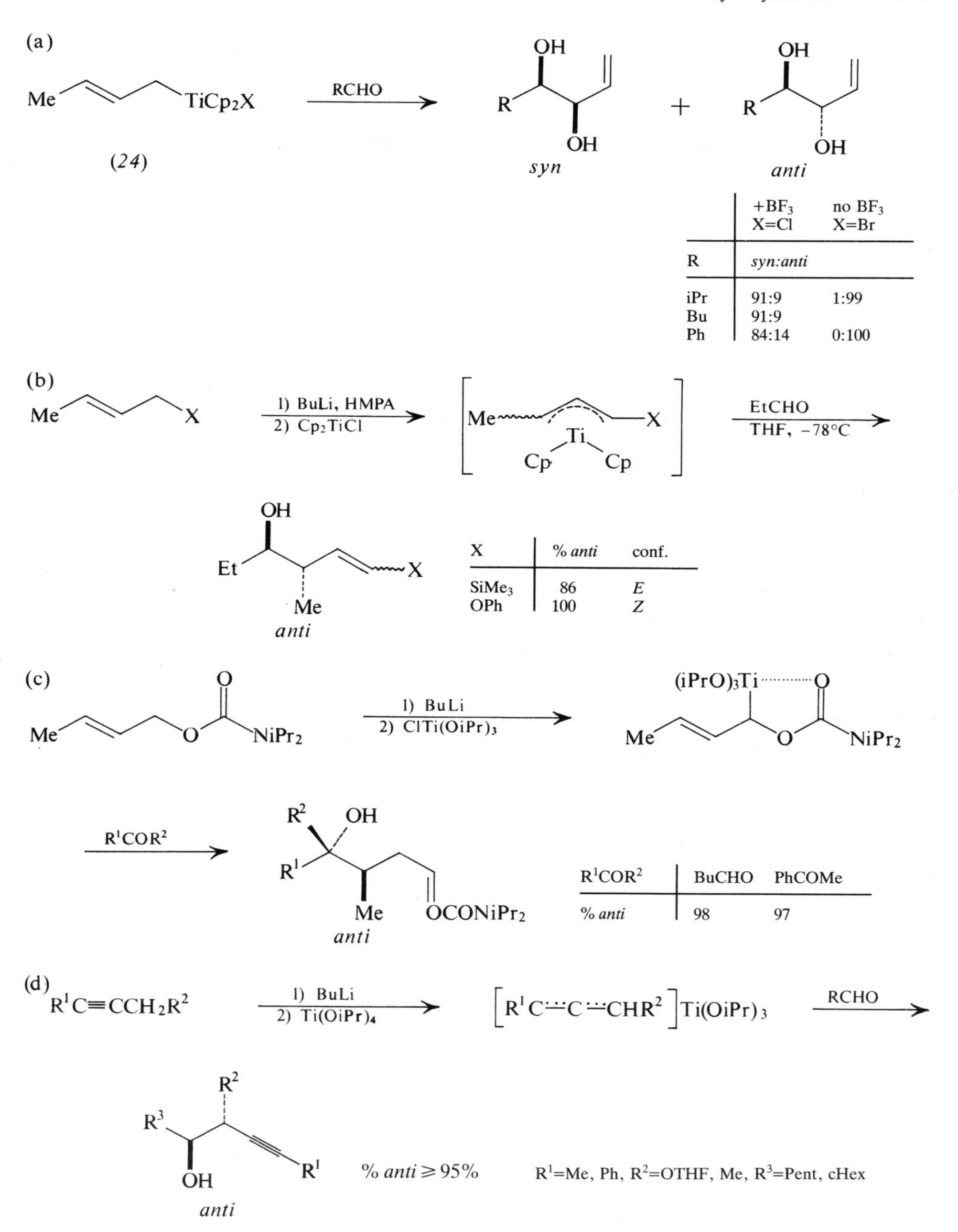

Fig. 5-18. Addition of allyltitanium compounds to carbonyl compounds (II).

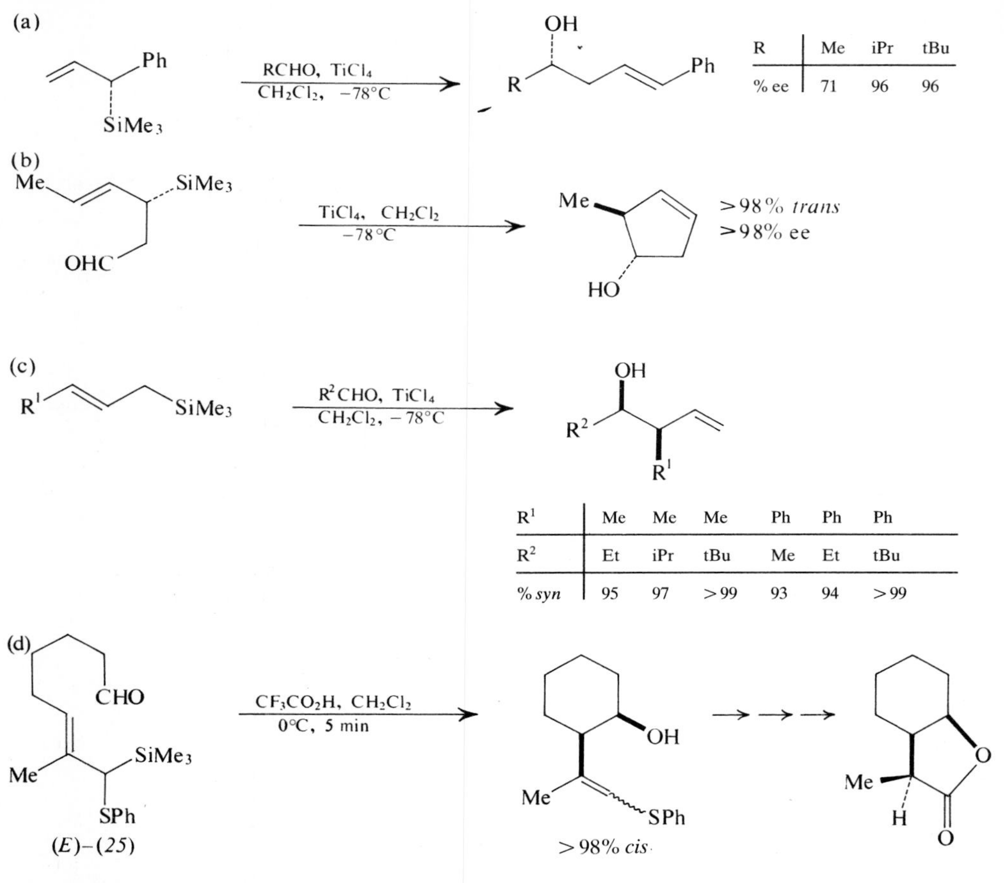

Fig. 5-19. Addition of allylsilanes to carbonyl compounds (I).

has already been mentioned (Fig. 5-12 (b)). Kishi and his coworkers demonstrated that in an intramolecular allyl addition chirality at the silylated carbon was completely transferred to the product (Fig. 5-19 (b)) [122].

Addition of both (*E*)- and (*Z*)-crotylsilanes to achiral aldehydes proceeds with *syn* preference, but selectivity is higher with the *E* isomers (Fig. 5-19 (c)) [123].

Sometimes protic acid catalysis should be preferred to that by Lewis acids, as shown by Itoh *et al.* (Fig. 5-19 (d)). The *Z* isomer of *25* gave the *trans*-lactone in >93% [124].

Addition of allylsilanes to chiral carbonyl compounds, both under 1,2- and 1,3-induction, was highly diastereoselective, while even this approach failed to produce useful results with chiral α-ketoesters [125]. Examples for 1,2-induction are depicted in Fig. 5-20 (a) [47, 126, 127] and for 1,3-induction in Fig. 5-20 (b) [60, 128].

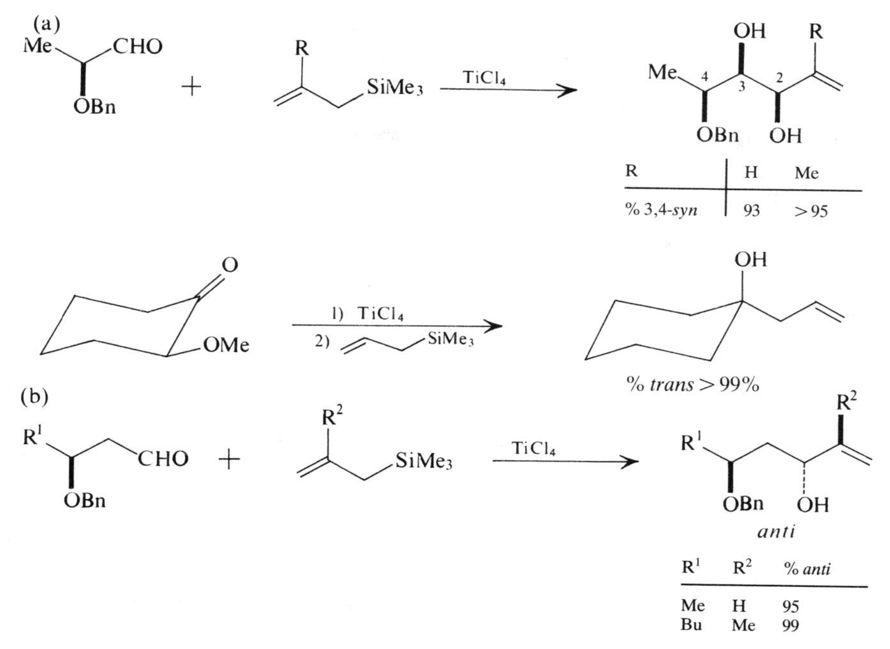

Fig. 5-20. Addition of allylsilanes to carbonyl compounds (II).

5.2.5 Addition of Allylstannanes

Allylstannanes can be induced to react with aldehydes in the absence of a Lewis acid at high temperature or by taking a very active aldehyde like chloral [130], but it is more convenient to activate the aldehyde, preferably with boron trifluoride. In both cases the *syn* products prevail [86].

A chiral stannane was prepared with the aid of chiral Grignard reagents, but enantioselectivities achieved with this reagent were generally less than 60% [131].

Allyl-tributylstannane adds to α-hydroxyaldehyde derivatives with excellent stereoselectivity, which can be inverted by changing the blocking group (Fig. 5-21 (a)) [89].

Achiral crotylstannanes add to achiral aldehydes with high *syn* preference, irrespective of the configuration around the double bond (Fig. 5-21 (b)) [130, 132–134]. Fig. 5-21 also shows the preparation of the stannane and one possible transformation of the product. Formation of the *syn* adduct became practically exclusive and the reaction was applicable to the geranly derivative when $SnBu_3$ was replaced by $SnPh_3$ [135].

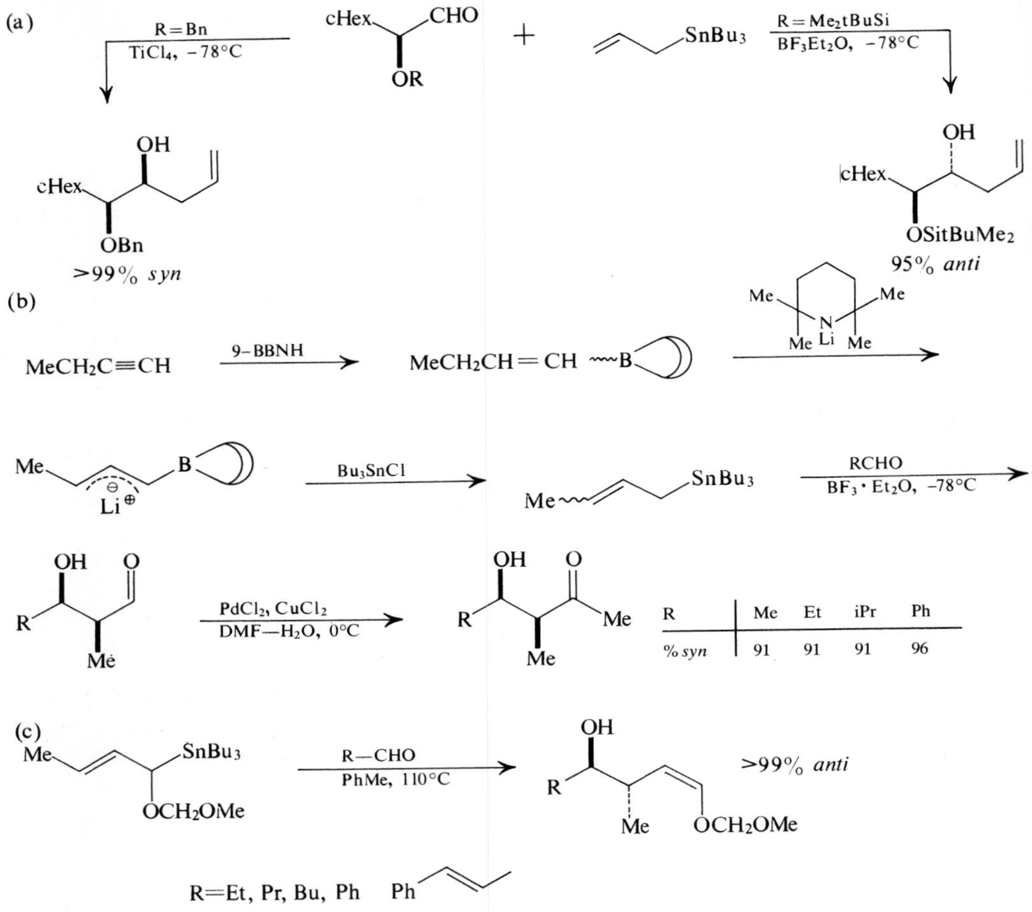

(a)

(b)

R	Me	Et	iPr	Ph
% *syn*	91	91	91	96

(c)

R=Et, Pr, Bu, Ph Ph

Fig. 5-21. Addition of allylstannanes to carbonyl compounds (I).

Stereoconvergence of the reaction for (*E*)- and (*Z*)-allylstannanes is attributed to the linear mechanism presented in Fig. 5-13 (b). In an uncatalyzed addition of an α-substituted stannane only the product anticipated by assuming a cyclic mechanism could be detected (Fig. 5-21 (c)) [136].

Yamamoto and his coworkers found that in the addition of a crotylstannane to glyoxylate esters diastereoselectivity increased with the size of the alcohol (90% *syn* with iPr), and when the alcohol was 8-phenylmenthol enantioselectivity was also acceptable (Fig. 5-22 (a)) [137].

In the addition of allyl-tributylstannane to α-oxygenated aldehydes Keck *et al.* observed that product configuration depended on the nature of the Lewis acid added (Fig. 5-22 (b)) [138]. Similar phenomena were observed in the addition of allyl-tributylstannane to α-alkoxymethyl aldehydes [139] and of crotylstannanes to α-alkoxyaldehydes [140].

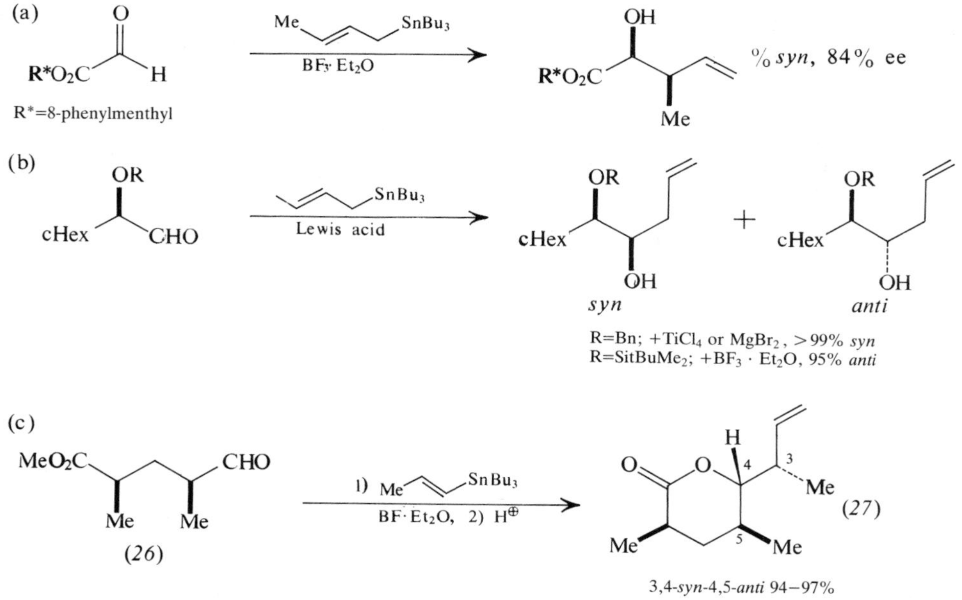

R*=8-phenylmenthyl

R=Bn; +$TiCl_4$ or $MgBr_2$, >99% *syn*
R=SitBuMe$_2$; +$BF_3 \cdot Et_2O$, 95% *anti*

3,4-*syn*-4,5-*anti* 94–97%

Fig. 5-22. Addition of allylstannanes to carbonyl compounds (II).

Efficient asymmetric induction was observed in the addition to *26* (Fig. 5-22 (c)). The main product (*27*)* was formed under chelation control; in addition 3–4% of the 4,5-*syn*-lactone and 2% of 3,4-*anti*-lactone were detected [142].

5.2.6 Addition of Other Allylmetals

It seems that no useful stereoselectivities can be realized with allylmetals in which the metal is lithium, halomagnesium or cadmium [85].

Yamamoto and Maruyama have shown, however, that, after introducing a stabilizing functional group, satisfactory stereoselectivity is possible even with lithio compounds (Fig. 5-23 (a)) [141]. Unfortunately about 10% of an α-adduct were also formed and in the case of the *E* stereoisomer diastereoselectivity in both respects (*E*/*Z*, *syn-anti*) was much lower.

According to Mulzer *et al.* [145] diallylzinc adds to isopropylidene-glyceraldehyde with 90% *anti* selectivity. Zweifel and Hahn found high *anti* selectivity for the addition of diallenylzinc to achiral aldehydes (Fig. 5-23 (b)) [146].

* *29* is a precursor of the famous Prelog-Djerassi lactone for which about a dozen syntheses have been published [141].

(a)

1) LDA
2) PhCHO

85% *syn*
100% *E*

(*E*)-*syn*

(b) $R^1CH{=}C{=}CH_2$

1) tBuLi, THF, −90°C
2) $ZnCl_2$, 3) R^2CHO

% *anti* 92–99%

R^1=Pr, tBu, Pent, Hex, cHex
R^2=H, Et, Pr, iPr, tBu

(c)

1) tBuLi,
2) $ClAl(iBu)_2$, −78°C

RCHO

(*E*)-*anti*

R	Me	iPr	tBu	Ph
% *anti*	91	94	97	83
% *E*	86	95	93	89

(d)

1) $CrCl_2$, THF
2) RCHO, −20°C

R	Pr	iPr	tBu	Pent	Ph
% *anti*	93	95	35	97	100

Fig. 5-23. Addition of allylmetals (M = Li, Zn, Al, Cr) to aldehydes.

Dialkyl-crotylaluminum compounds were studied by Hoppe and his coworkers [147, 148]. Both *anti* and *E* selectivities were high, especially with bulky aldehydes (Fig. 5-23 (c)).

Reaction of both (*E*)- and (*Z*)-crotylbromide with chromium(II) chloride gives a (not isolated) chromium-allyl compound which adds, as reported almost simultaneously by Buse and Heathcock [93] and by Okuda *et al.* [149], with high to total *anti* selectivity (except for pivalaldehyde) (Fig. 5-23 (d)). Asymmetric induction in the addition of this reagent is modest with simple α-chiral aldehydes [93], but with aldehydes containing several chiral centers 4,5-*anti* selectivity can amount to 95% [150].

It appears as if few metals have escaped from being tested as a ligand for allyl addition. Thus for example the addition of crotyl-biscyclopentadienyl zirconium chloride to aldehydes was also studied, but selectivities (*anti*) were lower than with the titanium analogue [60, 151].

5.3 Stereoselective Aldol Reactions*

An aldol reaction, also called aldol condensation, is, in the broadest sense, characterized by scheme (D) and involves the establishment of a C–C bond between two carbonyl compounds to form a β-hydroxycarbonyl compound, *i. e.* an aldol. The similarity to the addition of allylmetal compounds has already been pointed out and is not merely formal, but concerns the mechanism, stereochemistry and also the nature of M.

O H BM −H⊕ OM O OM O

(*D*)

The aldol reaction, discovered in 1838, is almost as old as organic chemistry itself, but its potential for stereoselective synthesis was only recognized in 1968 by the Dubois'. The idea was picked up by Heathcock, Masamune, Evans and many others and developed into a powerful method for the stereoselective construction of open chain compounds containing several adjacent chiral centers. The fact that this reaction remained dormant for so long should not be blamed on lack of imagination among chemists of yesterday but rather on the fact that the techniques (high field NMR, HPLC, flash chromatography, *etc.*) by which the complex mixtures of the heat sensitive products could be analyzed and separated were simply not available before. Also, earlier chemists thought it unrealistic to tackle such highly complex synthetic targets as, for example, macrocyclic antibiotics, which much stimulated the development of the aldol methodology.

* For reviews see refs. [44, 152–157].

The number of papers published on this topic is staggering, and at the present rate of publication any review becomes antiquated by the time it leaves the press.

As with allyl addition, several ordering principles can be invoked by which some shape can be given to the mass of facts. Such criteria may be (i) the symmetry of the partners (both achiral, one of them or both chiral), (ii) type of end product, (iii) the nature of metal component, and (iv) type of compound providing the enolate (ketone, ester, acid, *etc.*). We decided that a mixed approach would be least confusing, *i. e.* first we discuss how the stereochemistry at the ends of the newly formed bond can be controlled and then, in less detail, asymmetric induction by chiral centers in the carbonyl and the enolate partners.

5.3.1 Stereochemistry and Mechanism of the Aldol Reaction

Basic stereochemistry

The theoretical possibilities for stereoselection in the aldol reaction can be analyzed along the same lines as those of allyl addition and will be treated here only briefly.

(A) Addition of enolates, unsubstituted at the β-carbon, to achiral carbonyl compounds with enantiotopic faces gives no stereoselection unless the enolate is chiral. When chirality is imparted to the enolate by a chiral auxiliary, which is cleaved off after addition, enantiomers are obtained (Fig. 5-24 (a)) [158]. Otherwise the products are diastereomers (Fig. 5-24 (b)) [159]. Stereoselectivity in this category is usually modest or poor.

(B) Addition of enolates substituted at the β-carbon to carbonyl compounds with prochiral faces leads necessarily to diastereomers with configurations *syn* or *anti* at the ends of the new bond (Fig. 5-24 (c)) [160].

A chiral center in any of the partners, even when these are optically pure, doubles the number of possibilities, since this center is disposed either *syn* or *anti* to the nearest new center (Fig. 5-24 (d)) [161]. Note that in our example addition is completely *syn* selective concerning the centers at the new bond (no 2,3-*anti* products are detected), whereas asymmetric induction by the α-center of the aldehyde is incomplete.

(C) Reaction between achiral partners in the presence of a chiral additive may give rise to an optically active aldol. The Reformatsky reaction in the presence of (−)-spartein is a classical example [162]. So far chiral solvents have failed to induce appreciable enantioselectivity [5], while some diamines derived from proline

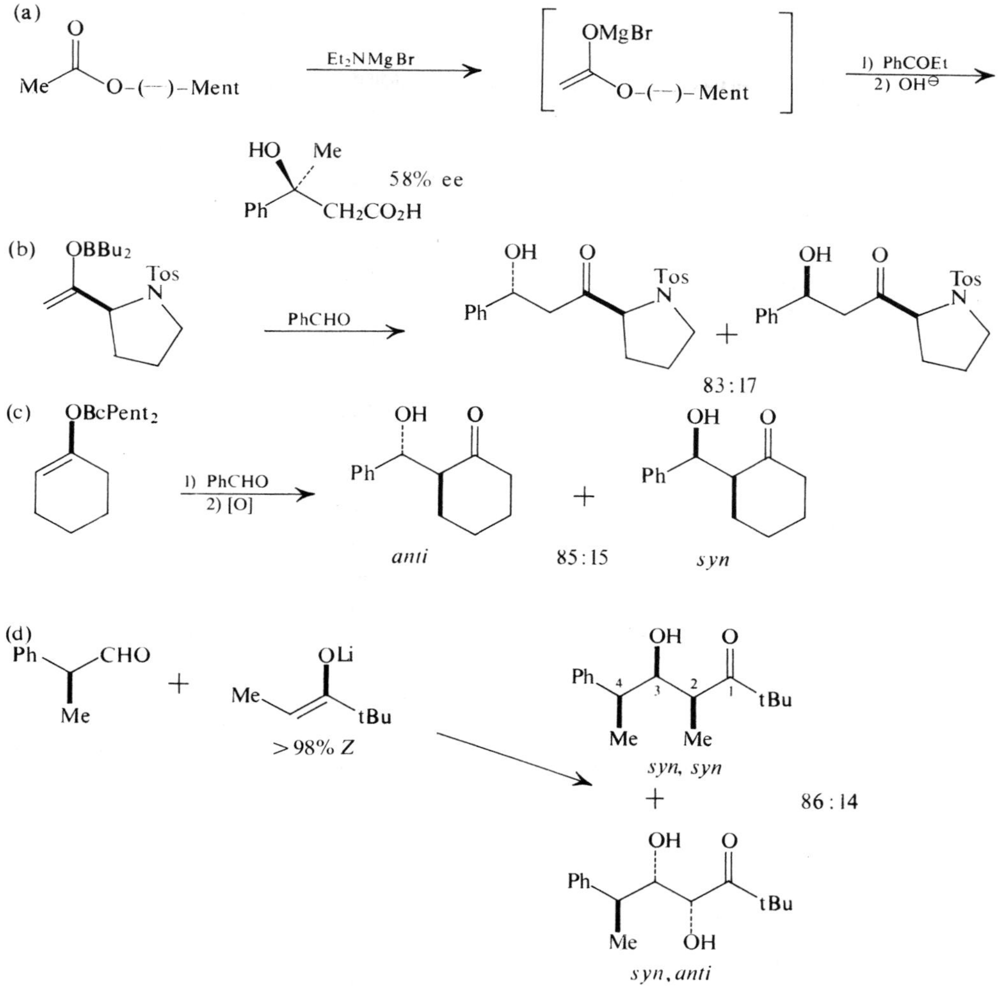

Fig. 5-24. The stereochemistry of the aldol reaction (I).

studied by Mukaiyama were highly effective with tin(II) enolates (Fig. 5-25(a) and (b)) [163]. Mulzer *et al.* found that a carboxylic acid dianion generated with a chiral lithium base added to benzaldehyde with appreciable enantioselectivity (Fig. 5-25 (c)) [164]*.

(D) Because simple 1,2-asymmetric induction generally failed to bring about high diastereoselectivity in the relationship of newly formed and existing centers of chirality, double induction in the aldol reaction has been thoroughly studied. It was assumed that a combination of chiral partners which produced the desired relative

* In want of more practically important examples this case will not be discussed any further.

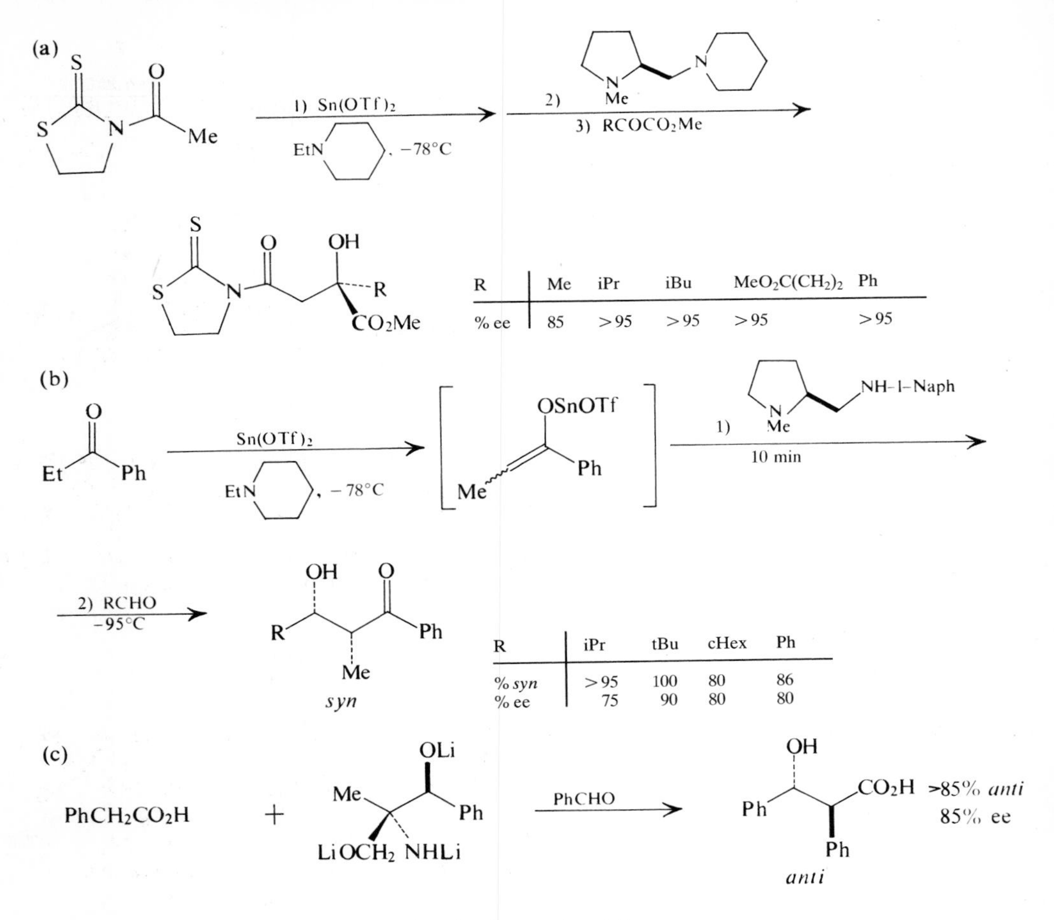

Fig. 5-25. The stereochemistry of the aldol reaction (II).

configurations in their reactions with achiral partners individually would provide high diastereoselectivities by synergism. Examples in Fig. 5-26 illustrate the confirmation of this hypothesis by experiment [165].*

Note that if the (*S*)-aldehyde was combined with the (*R*)-enolate diastereoselectivity was, as expected, poor and the *anti,syn,syn* product was formed in excess. Occasionally very high selectivities could be realized with complicated substrates [152, 165, 166]. Since the main field of application of double induction is the synthesis of complex natural products, very few simple examples can be quoted, and therefore no separate section will be devoted to this topic.**

* Multiplication of the individual stereoselectivities [154], whether the result agrees with experimental findings or not, is a gross simplification.

** For more details, see also refs. [167–172].

(a)

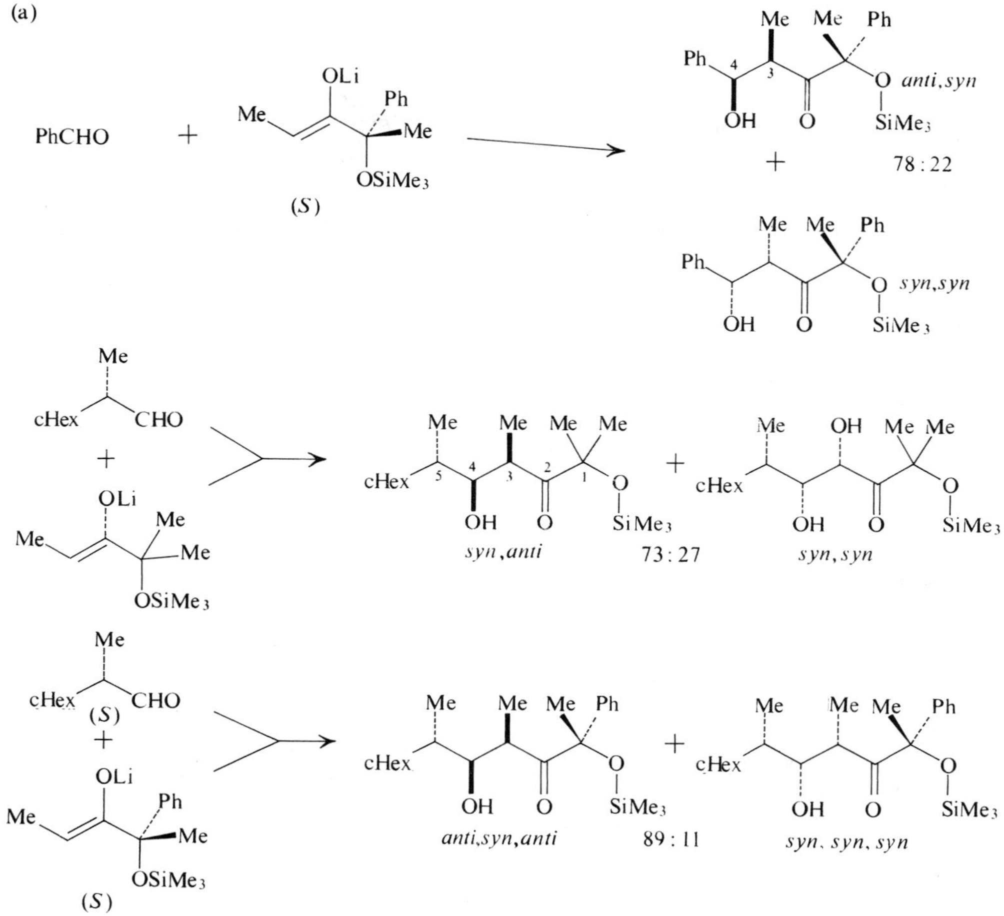

PhCHO + Me–CH=C(OLi)–C(Ph)(Me)(OSiMe3) (S) → anti,syn + syn,syn 78 : 22

cHex–CH(Me)–CHO + Me–CH=C(OLi)–C(Me)2(OSiMe3) → syn,anti + syn,syn 73 : 27

cHex–CH(Me)–CHO (S) + Me–CH=C(OLi)–C(Ph)(Me)(OSiMe3) (S) → anti,syn,anti + syn, syn, syn 89 : 11

Fig. 5-26. Double induction in the aldol reaction.

Transition state models

In contrast to allyl addition which is irreversible and thus under kinetic control only, the aldol reaction is reversible and may be controlled, according to conditions and the nature of partners, either kinetically or thermodynamically. The stereochemistry of addition under kinetic control can be successfully rationalized by considering the geometry of the transition state. In the majority of cases a cyclic transition state, as first proposed by Zimmerman and Traxler [173] is probable. In such a transition state the metal atom of the enolate is coordinated to the carbonyl oxygen of the partner. The associated product configuration is governed by three factors: (i) ring geometry (chair or boat), (ii) enolate configuration (Z or E), and (iii) mode of approach (*lk* or *ul*). A change in any of these factors results in a change of product configuration (*syn* or *anti*).

On circumstantial evidence the boat form is usually disregarded. The enolate configuration is mostly stable enough to be analyzed by spectroscopic methods. In Fig. 5-27 (a) cyclic chair-like transition states are depicted for the (*Z*)-enolates. For the (*E*)-enolates similar transition states can be drawn up and the result is complementary, C_1 leads to a *syn*- and C_2 to an *anti*-aldol. When R^2 is small $R^1 \leftrightarrow R^3$ 1,3-diaxial interactions overrule $R^2 \leftrightarrow R^3$ gauche interactions, C_1 is the favored transition state and therefore (*Z*)-enolates give *syn*-aldols and (*E*)-enolates *anti*-aldols as the major product. Also, stereoselectivity should increase as the size of R^3 increases. In transition state C_2 R^1 and R^3 get closer to each other when the M–O bond length is smaller and M···OC coordination is stronger. Indeed, boron enolates are more selective than lithium enolates.

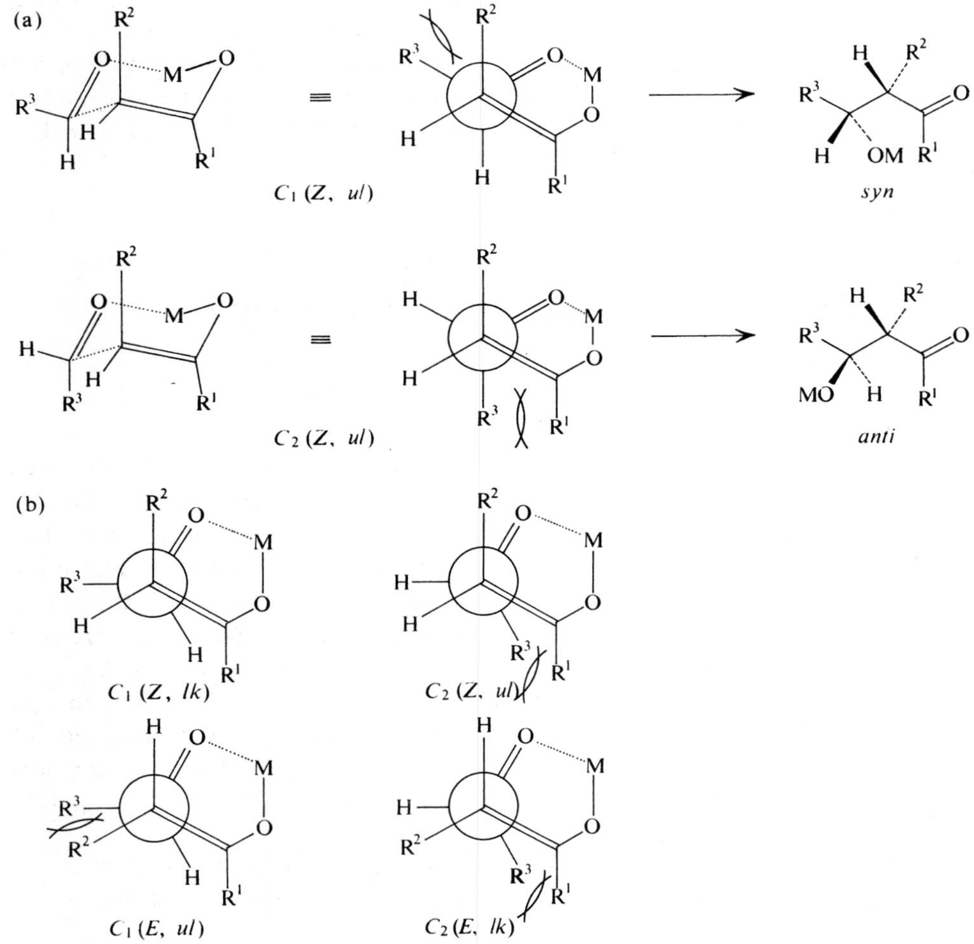

(a)

C_1 (*Z*, *ul*) *syn*

C_2 (*Z*, *ul*) *anti*

(b)

C_1 (*Z*, *lk*) C_2 (*Z*, *ul*)

C_1 (*E*, *ul*) C_2 (*E*, *lk*)

Fig. 5-27. Cyclic chair-like transition states for the addition of (*Z*)-enolates to aldehydes.

Table 5-6 provides ample evidence for the correctness of these predictions, but the diagrams in Fig. 5-27 (a) do not account for the fact that when R^1 is not large, *syn* selectivity with (*Z*)-enolates is usually much higher than *anti* selectivity with the (*E*)-enolates. This can be explained by assuming that the torsional angle of C=C and C=O is larger than the ideal 60°, which is a reasonable proposition for metals with a long M–O bond (Li, Mg, Zn). In such a "skewed" transition state (Fig. 5-27 (b)), while $R^1 \leftrightarrow R^3$ repulsion becomes more severe for both (*Z*)- and (*E*)-enolates, the advantage of C_1 (*E, lk*) over C_2 (*Z, ul*) is diminished due to increased $R^2 \leftrightarrow R^3$ gauche interaction in the latter. These effects are amplified by nonperpendicular attack (at 110°, *cf.* Section 3.2.1) and explain why selectivity deteriorates when R^2 is large. According to Evans *et al.* [156], a contribution by boat-type transition states cannot be excluded.

For enolates giving the same predominant product irrespective of their configuration, or for those requiring Lewis-acid catalysis, an open-chain transition state, similar to that for allyl addition (*cf.* Fig. 5-13 (b)), has to be postulated.

An interpretation of ester and acid enolate addition by orbital interactions was proposed by Mulzer *et al.* [222]. Molecular mechanics calculations on enolate additions to chiral aldehydes correctly predicted the trend of the reaction [174].

5.3.2 Generation of Enolates and their Addition to Achiral Aldehydes

In the knowledge of the mechanistic framework of enolate addition the problems to be solved for the stereoselective creation of the new bond are (i) the stereoselective generation of (*Z*)- and (*E*)-enolates and (ii) the conservation of enolate configuration in the course of addition.* Enolates can be generated from a variety of substrates, such as aldehydes, ketones, esters, carboxylic acids, thioesters, amides and thioamides using several metals, such as Li, Mg, Zn, Al, Ti, Zr, Si or B. The character of the oxygen-metal bond ranges from purely ionic (*e. g.* Li, K) to covalent (*e. g.* boronates). The most important facts about enolate addition to a selected set of achiral aldehydes are summarized in Table 5-6. The correlation of stereoselectivity with substrate structure, counter ion and experimental parameters constitutes a multidimensional matrix from which we have chosen a few vectors for more detailed discussion.

1. If under kinetic control a (*Z*)-enolate gives the *syn*-aldol in excess, then the (*E*)-enolate leads to an excess of the *anti* product. When this rule is not

* This does not apply to reactions in which aldol configuration is independent of enolate configuration.

Table 5-6. Addition of Chiral Enolates to Achiral Aldehydes. Data show the percentage of the major diastereomer under optimum conditions. Normalized to % *syn* + % *anti* = 100 (*s* = *syn*, *a* = *anti*).

$$R^3{-}CHO \; + \; R^2{-}CH{=}C(OM)R^1 \longrightarrow R^3CH(OH){-}CH(R^2){-}C(=O)R^1 \; (syn) \; + \; R^3CH(OH){-}CH(R^2){-}C(=O)R^1 \; (anti)$$

R^1	R^2	M	Enolate configu-ration (%)[a]	R^3: Me	Et	iPr	tBu	Ph	Comment	Ref.
(a) Ketone enolates										
$-(CH_2)_3-$		Li	*E*			>95 (*a*)				175
Et	Me		*Z*				88 (*s*)			176
			E				52 (*a*)			176
H	Me		*Z*					50		161
H	Me		*E*					65 (*s*)		161
Et			*Z* (66)					77 (*s*)		161
iPr			*Z* (>98)					90 (*s*)		161
			E					55 (*a*)		161
tBu			*Z* (>98)					>98 (*s*)		161
	Et		*Z*					>98 (*s*)	in THF at −78°C	214
								85 (*a*)	in pentane at 25°C	214
$Me_2COSiMe_3$	Me		*Z*			>98 (*s*)		>98 (*s*)		215, 216
Ph			*Z* (>98)				88 (*s*)			161
tBu	Me	MgBr	*Z*			100 (*s*)	100 (*s*)			177
	Et		*Z*			100 (*s*)				177
	iPr		*Z*			71 (*a*)				177
	tBu		*Z*			100 (*a*)				177
	Me		*Z*	100 (*s*)						177
	Et		*Z*	100 (*s*)						177
	iPr		*Z*	100 (*s*)						177

Table 5-6. Continued

R^1	R^2	M	Enolate configu-ration (%)[a]	R^3 Me	Et	iPr	tBu	Ph	Comment	Ref.
Me	tBu	$AlMe_2$	*Z*	100 (*a*)						178
			E	100 (*s*)						178
			b				100 (*a*)			178
$-(CH_2)_4-$		ZnCl	*E*				80 (*a*)			179
Me	tBu		c	100 (*s*)	88 (*s*)		50			180
Et	Me	$ZrCp_2Cl$	*Z* (92)				67 (*s*)			181
			E (86)				88 (*s*)			
Ph								90 (*s*)		181
$-(CH_2)_4-$			*E*					72 (*s*)		181
$-(CH_2)_3-$			*E*					64 (*s*)		181
		$Ti(NEt_2)_3$	*E*					85 (*s*)		182
$-(CH_2)_4-$			*E*					97 (*s*)		182
$-(CH_2)_5-$			*E*					90 (*s*)		182
Et	Me	$Ti(OiPr)_3$	b				81 (*s*)	89 (*s*)		182
Et	Me	$TiCl_3$	*Z*			75 (*s*)		81 (*s*)		183
	2,4,6-$Me_3C_6H_2$		*Z* (86)					63 (*s*)		183
$-(CH_2)_4-$			*E*					89 (*s*)		183
Ph	Me	BEt_2	*Z*	100 (*s*)						184
Me	Bu	BBu_2	*Z*					>95 (*s*)		185
			E					75 (*a*)		185
Bn			*Z*					>95 (*s*)		185
			E					80 (*a*)		185
Ph			*Z*		>95 (*s*)	>95 (*s*)	>95 (*s*)			185
			E					75 (*a*)		185
Et	Me	BBu_2	*Z* (>97)					>97 (*s*)		160, 186
iBu			*Z* (>99)					>97 (*s*)		160
iPr			*E* (55)					56 (*a*)		160
tBu			*Z*					>97 (*s*)		160

Table 5-6. Continued

R^1	R^2	M	Enolate configuration (%)[a]	R^3: Me	Et	iPr	tBu	Ph	Comment	Ref.
cHex		9-BBN	*Z*					>97 (*s*)		160
Ph		BBu_2	*Z*					>97 (*s*)		160
Me_3Si			c					81 (*a*)		160
Et			*Z* (96)					95 (*s*)		188
Bn			*Z* (95)					95 (*s*)		188
$CHEt_2$			*Z* (99)					>99 (*s*)		188
cHex			*Z* (98)					>99 (*s*)		
$-(CH_2)_4-$		BBu_2	*E*					87 (*s*)		160
		BcPent.Thex[d]	*E*					>96 (*s*)		160
Me	Me	$-B(O_2C_6H_4)$	*Z*		95 (*s*)	90 (*s*)		95 (*s*)	in petroleum ether at 25°C	217
Et		BClPh	*Z*			>99 (*s*)		>99 (*s*)	in CH_2Cl_2 at −28°C	96
Ph						>99 (*s*)		>95 (*s*)	in CH_2Cl_2 at −28°C	96
iPr						>99 (*s*)		>94 (*s*)	in CH_2Cl_2 at −28°C	96
Me		$-B(OCH_2CH_2O)$	*Z*					>99 (*s*)		219
Et		$-B(OCH_2CH_2O)$						>99 (*s*)		219
iPr		$-B(OCH_2CH_2O)$						99 (*s*)		219
PhS		$-B(OCH_2CH_2O)$						95 (*s*)		219
$-(CH_2)_4-$		$SiMe_3$	*E*			100 (*s*)[b]		75 (*a*)[f]		189, 190
$-(CH_2)_3-$			*E*			100 (*s*)[e]		50 (*a*)[f]		189, 190
tBu			*Z*					>98 (*a*)	under $BnMe_3N^{\oplus}F^{\ominus}$ catalysis	192
Bu	Pr		*Z*					88 (*s*)	under $BF_3 \cdot Et_2O$ catalysis	190
Et	Me		*Z*					86 (*s*)	under $BF_3 \cdot Et_2O$ catalysis	190

Table 5-6. Continued

R^1	R^2	M	Enolate configuration (%)[a]	R^3					Comment	Ref.
				Me	Et	iPr	tBu	Ph		
Ph			b					95 (*s*)	under $BF_3 \cdot Et_2O$ catalysis	190
$-(CH_2)_4-$		$SnPh_3$	*E*					71 (*s*)		193
		$SnBu_3$	*E*					80 (*a*)	at −78°C	194
			E					77 (*s*)	under $BnMe_3N^{\oplus}F^{\ominus}$ catalysis	194
		$SnEt_3$	*E*					92 (*s*)	at −78°C	194
			E					70 (*s*)	under $BnMe_3N^{\oplus}F^{\ominus}$ catalysis	194
		$SnMe_3$	*E*					92 (*a*)	at −78°C	194
$-(CH_2)_3-$		$SnPh_3$	*Z*					85 (*s*)	at −50°C	193
Et	Me		*Z* (92)[b]					82 (*s*)		193
Ph		$SnBu_3$	c			75 (*a*)		88 (*a*)		195
		$SnEt_3$	c			93 (*a*)		90 (*a*)		195
$-(CH_2)_3-$		$SnBu_3$	c					100 (*a*)	at −50°C	196
		SnBr	c			92 (*s*)		93 (*s*)		197
		SnOTf	c			91 (*s*)		>95 (*s*)		198
Et		SnBr	c			91 (*s*)		91 (*s*)		197
Me	Bn		c					>90 (*s*)		197
iPr	Me	SnOTf	c					91 (*s*)		198
$-(CH_2)_4-$			c					>95 (*s*)		198
(b) Ester and thioester enolates										
OMe	Me	Li	*E*	57 (*a*)		55 (*a*)		55 (*a*)		199
OCH_2OMe			*E*	67 (*a*)		90 (*a*)				199
OMEM			*E*			91 (*a*)		75 (*a*)		199
	iPr		*E*					50		161
OtBu	Me		*E*					51 (*s*)		161
ODMP[g]			*E*			>98 (*a*)	>98 (*a*)	88 (*a*)		200, 201

Table 5-6. Continued

R^1	R^2	M	Enolate configuration (%)[a]	R^3: Me	Et	iPr	tBu	Ph	Comment	Ref.
O-DMP[g]			*E*		86 (*a*)	>98 (*a*)				200
O-DMP[g]	Et		*E*			>98 (*a*)				200
O-BHT[h]	Me		*E*			>98 (*a*)		>98 (*a*)		200, 201
O-BHT[h]			*E*		>98 (*a*)	>98 (*a*)		>94 (*a*)		200
OtBu	Me_2N		*E*	50 (*s*)				75 (*s*)		206
OEt	Me	$OSiMe_3$	*Z*			52 (*a*)		67 (*a*)	under $TiCl_4$ catalysis	226
			E (85)			92 (*a*)		74 (*a*)	under $TiCl_4$ catalysis	226
	Ph	SnBr	c					59 (*s*)		203
	Me		c					55 (*s*)		203
	Et		c					57 (*s*)		203
OMe	Me	ZnBr	c					63 (*s*)		204, 205
	Et		c					54 (*s*)		204, 205
	iPr		c					53 (*a*)		204, 205
	cHex		c					50		204, 205
	tBu		c					69 (*a*)		204, 205
$OSiMe_3$	$(Me_3Si)_2N$	Li	*E*	100 (*a*)				100 (*a*)		207
OMe	Me	$ZrCp_2Cl$	*E* (95)					87 (*s*)		208
SPh		9-BBN	*Z* (>95)		>97 (*a*)			97 (*a*)		187
tBuS		BBu_2	>95 (*E*)			91 (*a*)		90 (*a*)		160
		BcPent				>95 (*a*)		95 (*a*)		160, 211
PhS		9-BBN	*E*[j]		>97 (*e*)	97 (*e*)		97 (*s*)		185
(c) Carboxylic acid dianions[i]										
OLi		Li					50	55 (*a*)		209, 161
	Et							55 (*a*)		209
	iPr							58 (*a*)		209
	tBu				52 (*a*)	90 (*a*)	>98 (*a*)	88 (*a*)		209

Table 5-6. Continued

R^1	R^2	M	Enolate configuration (%)[a]	R^3: Me	Et	iPr	tBu	Ph	Comment	Ref.
	1-adamantyl						>98 (*a*)			222
	Ph			64 (*a*)	64 (*a*)	97 (*a*)	>98 (*a*)	92 (*a*)		209
$OBCp_2$	Me	BCp_2						80 (*a*)		160
	OBn							98 (*a*)		160
(d) Amide enolates										
$NiPr_2$	Me	Li	*Z* (81)					63 (*s*)		161
piperidyl			*Z* (>95)					60 (*s*)		208
$NiPr_2$		$ZrCp_2Cl$	*Z* (81)					98 (*s*)		208
						>95 (*s*)		>95 (*s*)		208
(e) Thioamides and thioesters										
NMe_2	Me	Li	>97 (*Z*)		90 (*s*)			83 (*s*)		210
		MgBr	>97 (*Z*)			95 (*s*)		97 (*s*)		210
	Ph		>97 (*Z*)			72 (*s*)			at −78°C	210
						97 (*a*)			at r. t.	210
−N(Me)−$(CH_2)_2$−		Li	*E*			62 (*a*)				210
NMe_2	Me	$SiMe_3$	*Z*					95 (*s*)	under $Bu_4^{\oplus}F^{\ominus}$ catalysis	212
	iPr		*Z*					80 (*s*)	at −80°C	212
	Me		*Z*					85 (*a*)	under $TiCl_4$ + $Ti(OiPr)_4$ catalysis at −80°C	212
	iPr		*Z*					45 (*a*)	under $TiCl_4$ + $Ti(OiPr)_4$ catalysis at −80°C	212
PhNLi	Me	Li	*Z*	69 (*a*)		88 (*a*)		68 (*a*)		213
	iPr		*Z*			93 (*a*)		69 (*a*)		213
MeNLi	Ph	MgBr	*Z*			98 (*a*)		94 (*a*)		213
$MeOCH_2CH_2NLi$			*Z*					97 (*a*)		213

[a] Only quoted with mixtures. [b] Stereoselectivity independent of *E/Z* ratio. [c] Unknown. [d] 1,1,2-trimethyl-l-propyl. [e] Under $BF_3 \cdot Et_2O$ catalysis. [f] Under $TiCl_4$ catalysis. [g] DMP = 2,6-dimethylphenyl. [h] BHT = 2,6-ditert-butyl-4-methylphenyl. [i] No E/Z isometrism possible. [j] Note priority of S over O.

observed and/or product composition is independent of enolate configuration, either thermodynamic control or a non-cyclic mechanism can be suspected. Thermodynamic control is probable when an *anti* adduct is preferentially formed from a (*Z*)-enolate or from both, while a non-cyclic mechanism is likely when an (*E*)-enolate or both give a *syn* adduct. The Reformatsky reaction is usually under thermodynamic control. For example, the *syn*/*anti* ratio in the reaction of the (*Z*)-enolate (M = ZnBr) of propiophenone with benzaldehyde is a function of time (at −10°C 70 : 30 after 5 sec and 25 : 75 after 16 min) [161]. Lithium aldolates equilibrate even more rapidly. Kobayashi *et al.* studied the addition of the tributylstannyl enolates to benzaldehyde, which was *anti* selective at −50°C and rather non-selective at room temperature. This was shown not to be a result of equilibration, which was slow at room temperature, but due to a change of mechanism [196]. Titanium [182] and zirconium [181] enolates are typical if the case when the *syn* product predominantes irrespective of enolate configuration.

2. Stereoselectivity is generally higher with the (*Z*)-enolates than with the (*E*)-enolates (Fig. 5-28 (a)), and there are fewer methods available for the preparation of *anti*-aldols than for *syn*-aldols.
3. Stereoselectivity much depends on the nature of M. The preparation of lithium enolates is the easiest, but it is often necessary to resort to other metals. Although pure *Z* or *E* lithium enolates can be prepared from the corresponding trimethylsilyl enolates, when prepared directly lithium enolates tend to be formed as a *Z*-*E* mixture. The selection of the best M is not easy; no general guidelines can be given in this respect, since yields, ease of preparation, separability of products and last but not least the way in which the aldol can be transformed to the target compound all have to be pondered.
4. Correlation of enolate constitution and stereoselectivity is also rather complex. In accordance with the cyclic model, selectivity improves, both for (*Z*)- and (*E*)-enolates, with increasing size of R^1 (Fig. 5-28 (b)).

Few data are available for the evaluation of the role of R^2. When the aldehyde is bulky, on increasing the size of R^2 product composition shifts in favor of the *anti*-aldol (Fig. 5-28 (c)).

Finally, no generally valid correlation between the constitution of an achiral aldehyde and stereoselectivity has emerged yet.

A so far unique example involving a tetrasubstituted enolate reported by Tamuru *et al.* [220] is shown below. *Syn* selectivity was complete with R^1 = Et, iPr, Ph and R^2 = Me, iPr but nil with R^2 = Ph.

S Me NMe2 —R^1MgBr→ [SMgBr Me NMe2 R^1] —R^2CHO→ OH S R^2 NMe2 Me R^1

syn

After having surveyed the aldol reaction in general, we now turn to the generation of enolates. Lithium enolates are prepared by reacting a ketone with a lithium base, most often lithium diisopropylamide (LDA). Occasionally lithiated cyclohexylisopropylamine, hexamethyldisilylamine (LHMDS) and 2,2,6,6-tetramethylpiperidine (LTMP) are used. The enolates are formed as *Z*-*E* mixtures. LHDMS promotes the formation of (*Z*)-enolates, while LTMP that of the (*E*)-enolates [161, 215]. Pure (*E*)- and (*Z*)-enolates can be prepared by treating (*E*)- and (*Z*)-trimethylsilyl enolates with methyl lithium [176]. This, however, is only practical when (*E*)- or (*Z*)-silylenolates themselves can be obtained in sufficient purity [221].

Although esters give the (*E*)-enolates with LDA, *syn-anti* selectivity is poor except when the alcohol is a bulky phenol.

N,N-Dialkylamides and thioamides give mainly the (*Z*)-enolates, but *syn* preference is only at an acceptable level with thioamides.

Dianions of carboxylic acids are generated by treating the acids with LDA. The major product is the *anti*-aldol, but selectivity is only high when both R^2 and the

(a) R^2~CH=C(OM)R^1 —PhCHO→ *syn* + *anti*

R^1	R^2	M	% *syn* from *Z*	% *anti* from *E*
H	Me	Li	50	45
iPr			90	55
Me	Bu	BBu_2	>95	75
Bn			>95	80
Ph			>95	75

(b) (*Z*) Me–CH=C(OLi)R^1 —PhCHO→ *syn*

R^1	H	Et	iPr	tBu	Ph
% *syn*	50	90	90	100	87

(c) R^2–CH=C(OM)R^1 —R^3CHO→ *syn* + *anti*

R^2	R^1=tBu, M=MgBr: Me	Et	iPr	tBu
% *syn*	100	100	29	0

R^2	R^1=OMe, M=ZnBr: Me	Et	iPr	tBu	cHex
% *syn*	63	54	47	31	31

Fig. 5-28. Control of stereoselectivity in the aldol addition by the configuration and metal component of the enolate.

aldehyde are bulky [209, 222], or when the cation is potassium complexed with a crown ether.

Alkylamides and thioamides form dianions on treatment with butyllithium [213, 223]. A *Z* configuration has been assumed for the thioamide dianions, and modest to good *anti* preference is observed [223].

Bromomagnesium enolates are probably the intermediates when α-bromoketones are treated with magnesium and give excellent *syn* selectivity when the substrate is a tert-butyl ketone [177]. Bromomagnesium enolates are also accessible by treating the substrate with Et_2NMgBr [158, 224] or a Grignard reagent, *e. g.* tBuMgBr. Stereoselectivity may depend on the nature of the Grignard reagent [220] and is often better than with the lithium enolates [173, 182, 210].

Bromozinc enolates are believed to be the intermediates in the Reformatsky reaction and are prepared by reacting α-bromoesters with zinc. House *et al.* transformed lithium enolates by the addition of zinc chloride to zinc enolates, which were sometimes more selective than the parent compounds [179, 180], while in other cases only the yields increased at the expense of stereoselectivity.

The (*Z*)- and (*E*)-dimethylaluminum enolates of methyl neopentyl ketone prepared by Jeffery *et al.* [178] gave the pure *anti*- and *syn*-aldols respectively with acetaldehyde, but with benzaldehyde both stereoisomers led to the pure *anti* adduct.

Boron enolates have been intensively studied by Masamune and his group. Such compounds can be prepared from diazoketones with trialkylboranes as the *E* isomer and isomerized by lithium phenolate to the (*Z*)-enolate (Fig. 5-29 (a) and (b)) [185]. More conventiently, according to Evans *et al.* [160, 225] the ketones may be reacted with dialkylboron trifluorosulfonates (triflates) in the presence of a hindered tert-amine to give, in high purity, the (*Z*)-enolates from ketones and (*E*)-enolates from thioesters.* The triflates can be replaced, without loss of selectivity, by $PhBCl_2$ [218]. The transmission of enolate configuration to aldolate configuration depends somewhat on the constitution of the partner but is in general very efficient (% *syn* >95). Boronates formed from phenylthiopropionate with the bulky 9-borabicyclononyl (9-BBN) group have a *Z* configuration and give pure *syn*-aldols [187].

Hoffmann and Ditrich prepared pure (*E*)-and (*Z*)-enolborates [217], while Gennari *et al.* obtained mixtures of them [219]. In an uncatalyzed addition both gave 81–99% of the *syn*-aldols.

Enolsilanes can be prepared by trapping lithium enolates with trimethylsilyl chloride [*e. g.* 212, 221, 226] or by treating a ketone with the ester *28* (Fig. 5-29 (c, d)) [221]. Enolsilanes only react in the presence of a catalyst. In example (c) the pure (*Z*)-enolsilane shows poor *anti* selectivity under $TiCl_4$ catalysis, whereas the *E* isomer (85% pure) exhibits medium to high preference for the *anti*-aldol. In the

---

* In the latter, since S > O, the disposition of oxygen and R^2 is *trans* in the classical sense and gives, as expected, the *anti*-aldols in excess.

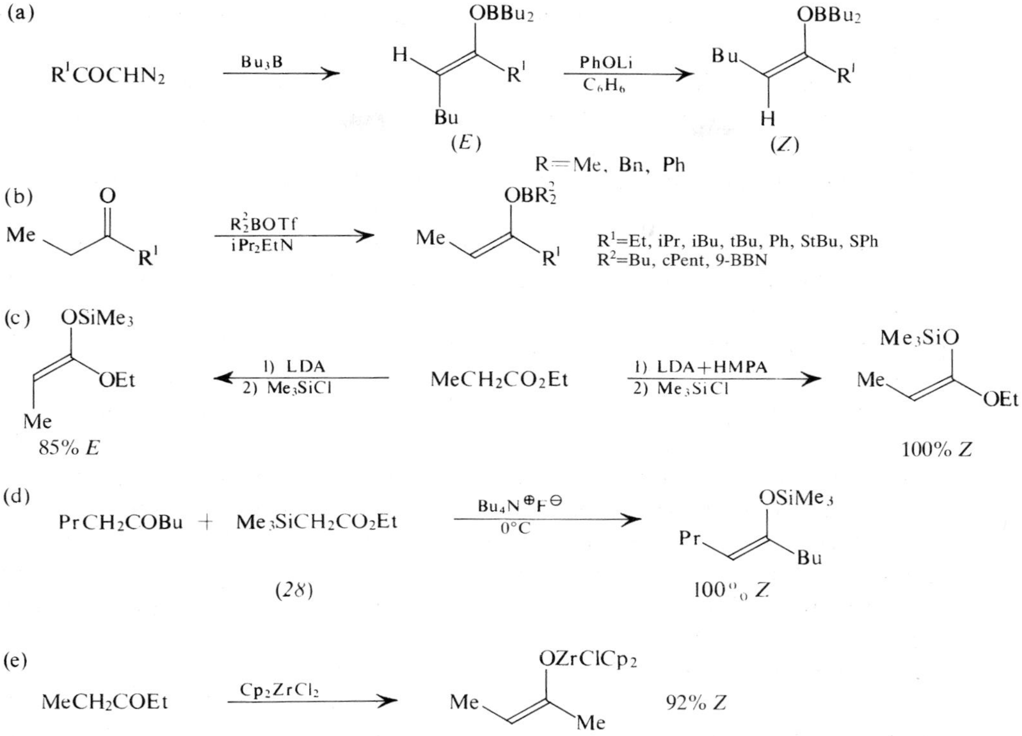

(a) R^1COCHN_2 —Bu_3B→ (*E*)-H(Bu)C=C($OBBu_2$)R^1 —PhOLi, C_6H_6→ (*Z*)-Bu(H)C=C($OBBu_2$)R^1

R=Me, Bn, Ph

(b) $MeCH_2COR^1$ —R^2_2BOTf, iPr_2EtN→ MeCH=C(OBR^2_2)R^1

R^1=Et, iPr, iBu, tBu, Ph, StBu, SPh
R^2=Bu, cPent, 9-BBN

(c) $MeCH_2CO_2Et$ —1) LDA 2) Me_3SiCl→ MeCH=C($OSiMe_3$)OEt 85% *E*

$MeCH_2CO_2Et$ —1) LDA+HMPA 2) Me_3SiCl→ MeCH=C($OSiMe_3$)OEt 100% *Z*

(d) $PrCH_2COBu$ + $Me_3SiCH_2CO_2Et$ (*28*) —$Bu_4N^{\oplus}F^{\ominus}$, 0°C→ PrCH=C($OSiMe_3$)Bu 100% *Z*

(e) $MeCH_2COEt$ —Cp_2ZrCl_2→ MeCH=C($OZrClCp_2$)Me 92% *Z*

Fig. 5-29. Generation and aldol reaction of enolates (M = B, Si, Zr).

presence of $BnMe_3N^+F^-$ it is possible to obtain, if R^1 = tBu, the *syn*-aldol under kinetic and the *anti*-aldol under thermodynamic control [161, 192]. When the catalyst was $(Et_3N)_3S^{\oplus}SiMe_2F_2^{\ominus}$,up to 95% of the *syn* product were formed under kinetic control, regardless of enolate configuration [227].

When mediated by trimethylsilyl triflate, enolsilanes react with aldehyde dimethylacetals and give, if under kinetic control, *syn*-aldols in excess both from (*Z*)- and (*E*)-silanes [191, 228].

Enolstannanes can be prepared by reacting the lithium enolates with a chlorostannane and may used *in situ* [193] or after isolation [194]. The reacting species is uncertain since *O*- and *C*-stannanes equilibrate in solution. Selectivity appears to depend on the ligands attached to tin. According to Yamamoto *et al.,* at low temperatures with the cyclohexanone enolate the *syn* product was slightly preferred (~70%) in the case of the triphenylstannyl derivative [193]. On the other hand Stille reported [195] that the trialkyl analogues yielded up to 92% of the *anti*-aldol at low temperatures and about 70% of the *syn*-aldol at 45°C.

Tin(II) enolates, which can be prepared *in situ* by treating an α-bromocarbonyl compound with tin [197] or more conveniently by reaction of a ketone with

$Sn(OTf)_2$ [198], were introduced by Mukaiyama and Stevens. *Syn* selectivities in excess of 90% were achieved with ketones [197, 198, 203], while selectivity was modest with esters [198, 203]. As mentioned in Section 5.3.1, the method is amenable to chiral catalysis [229, 230].

Titanium enolates of ketones (of which the triisopropoxy derivatives are of practical interest) are prepared by reacting the lithium enolates with $(iPrO)_3TiCl$ [182]. Titanate composition reflects that of the parent enolate, but this is irrelevant, since the *syn*-aldols are the preferred products from both (*E*)- and (*Z*)-enolates. Addition is under kinetic control and probably follows a non-cyclic mechanism.

Treatment of enolsilanes with $TiCl_4$ affords trichlorotitanium enolates which react with aldehydes with poor to modest *syn* preference (inclusive cyclohexanone enolates). A boat transition state was proposed by Nakamura and Kuwajima to account for the results [183].

The last metal to be discussed here, but certainly not the last to be tested for the aldol reaction, is zirconium. Zirconium enolates, which are accessible from lithium enolates [181, 208] (*e. g.* Fig. 5-29 (e)) react, irrespective of enolate configuration, with *syn* preference. Selectivity is modest with ketones [181] and good to excellent with amides [208]. With some chiral amide enolates (*cf.* p. 213) *syn* selectivity is excellent [170].

5.3.3 Addition of Enolates to Ketones (the Cross-aldol Reaction)

It follows from the cyclic stereochemical model for the aldol reaction that stereoselectivity depends, among other factors, on the difference in size of the ligands attached to the carbonyl group. Therefore much better selectivity can be expected with aldehydes than with ketones. Data on ketones are scarce and appear to confirm this point. In the Reformatsky reaction selectivity increases with increasing size of R^2 (Fig. 5-30 (a)) [205, 229–231]. Under the conditions quoted, the reaction is thermodynamically controlled, while in methylal at −70°C (Fig. 5-30 (b)) [205] it is under kinetic control and, as expected, selectivity improves with bulky esters.* The trend was similar in the Reformatsky reaction of a series of esters ($RCHBrCO_2Me$, R = Me, Et, iPr) with α- and β-aminoketones; behavior of the tert-butyl derivative was anomalous [232].

Recently, promising results were reported by Mukaiyama *et al.* with tin(II) enolates (*e. g.* Fig. 5-30 (c)) [233, 234].

---

* OR corresponds to R^1 in Fig. 5-28.

In Fig. 5-30 (d) an example for stereoselective addition to a chiral ketone is shown [235].

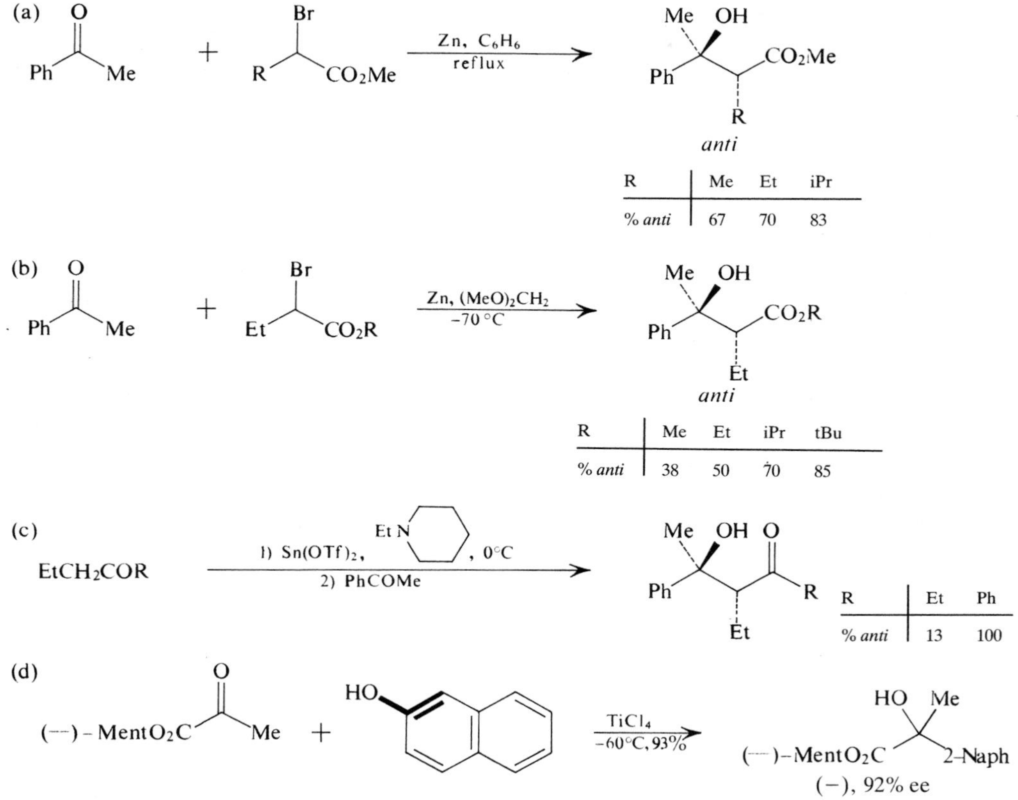

(a) PhCOMe + RCH(Br)CO_2Me — Zn, C_6H_6, reflux → Ph(Me)C(OH)–CH(R)CO_2Me *anti*

R	Me	Et	iPr
% *anti*	67	70	83

(b) PhCOMe + EtCH(Br)CO_2R — Zn, $(MeO)_2CH_2$, −70 °C → Ph(Me)C(OH)–CH(Et)CO_2R *anti*

R	Me	Et	iPr	tBu
% *anti*	38	50	70	85

(c) $EtCH_2COR$ — 1) $Sn(OTf)_2$, N-ethylpiperidine, 0 °C; 2) PhCOMe → Ph(Me)C(OH)–CH(Et)COR

R	Et	Ph
% *anti*	13	100

(d) (−)-MentO$_2$C–CO–Me + 2-naphthol — $TiCl_4$, −60 °C, 93% → (−)-MentO$_2$C–C(OH)(Me)–2-Naph (−), 92% ee

Fig. 5-30. The cross-aldol reaction.

5.3.4 Addition of Chiral Enolates to Achiral Aldehydes

The purpose of adding a chiral enolate to an achiral carbonyl compound is usually the transfer of chirality to the product, and at the end of the process the original chiral moiety is removed.

The simplest way to accomplish this is to use an ester of a chiral alcohol. Thus as early as in 1949 Reid and Turner described the enantioselective Reformatsky reaction of (−)-menthyl bromoacetate with acetophenone [ee 30% (*S*)] [236].

Despite repeated efforts (including the use of lithium enolates) the process could not be improved to provide a useful level of enantioselection* [152, 237, 224, 238].

Solladié and his coworkers prepared optically active β-hydroxy acids with the aid of magnesium enolates derived from chiral α-sulfinyl acetates (Fig. 5-31 (a)) [241, 242]. In the case when it was determined addition proceeded with complete *anti* selectivity at C-2 and C-3 [242]. Recently Annunziata *et al.* have shown that the corresponding dimethylamide gives better results with aldehydes (Fig. 5-31 (b)) [243]. The chiral inducing group could be readily removed by amalgamated sodium.

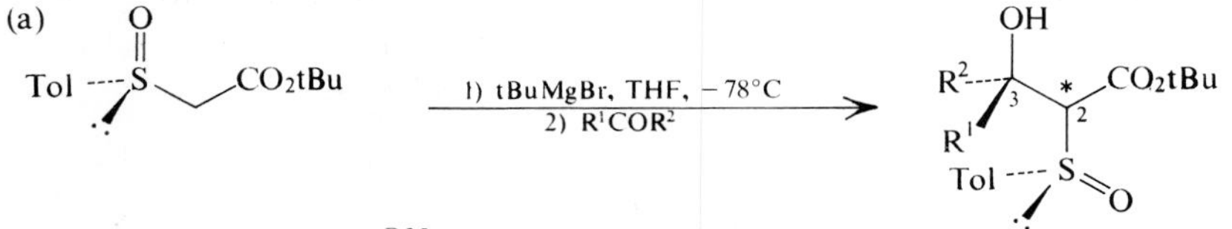

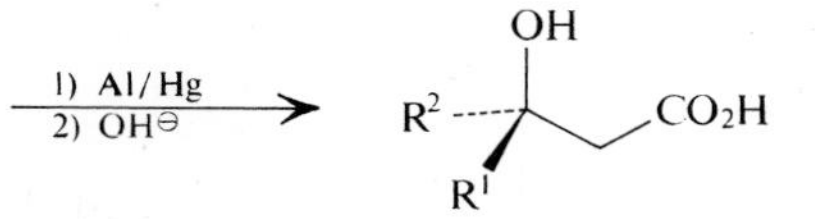

(a)

Tol–S(O)–CH₂CO₂tBu → 1) tBuMgBr, THF, −78°C; 2) R^1COR^2 → 1) Al/Hg; 2) OH⊖ → $R^2R^1C(OH)CH_2CO_2H$

R^1	cHex	Hept	Ph	Ph	CF_3	CO_2Et
R^2	H	H	H	Me	Ph	Me
% ee	95	86	91	68	20	85

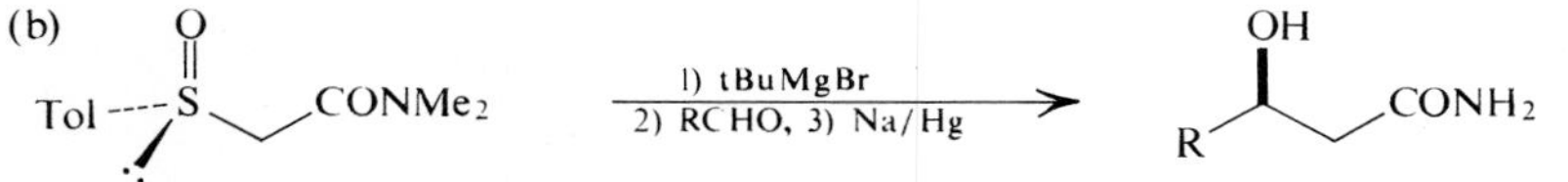

(b)

Tol–S(O)–CH₂CONMe₂ → 1) tBuMgBr; 2) RCHO, 3) Na/Hg → RCH(OH)CH₂CONH₂

R	Me	iPr	iBu	tBu
% ee	≥99	95	98	90

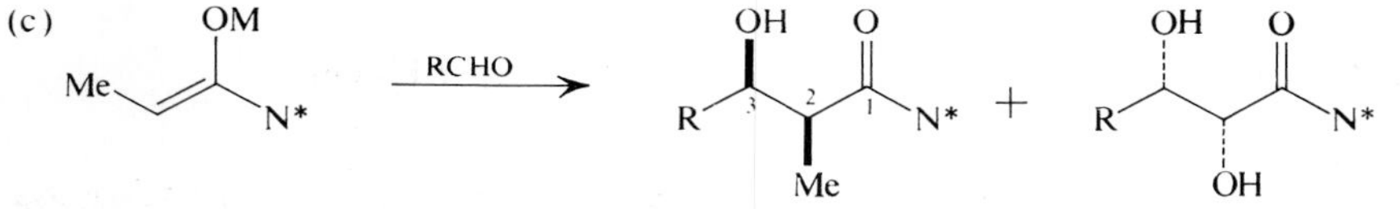

(c)

(29)–(34) → RCHO → (2S)–*syn* + (2R)–*syn*

M	N*	R, % ee of *syn*			conf.
		iPr	Bu	Ph	
$ZrCp_2Cl$	*a* (29)	96	96	98	2*R*
$ZrCp_2Cl$	*b* (30)	99	98	97	2*S*
BBu_2	*c* (31)	>99	>99	>99	2*R*
BBu_2	*d* (32)	>99	>99	>99	2*S*

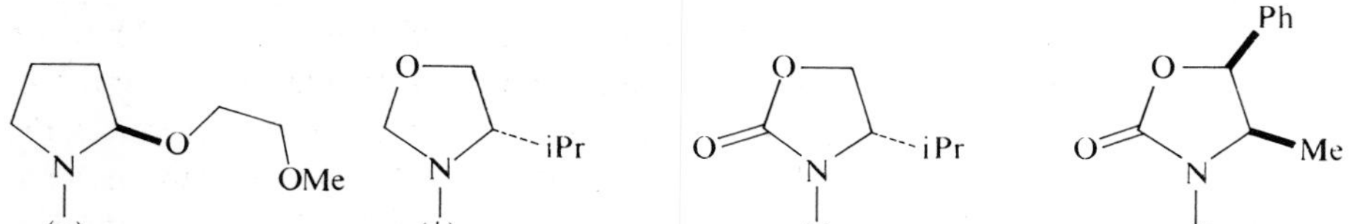

(*a*) (*b*) (*c*) (*d*)

Fig. 5-31. Aldol reaction of chiral enolates with achiral carbonyl compounds.

*Early claims for up to 93% ee [239, 240] were later shown to be incorrect [158].

Evans and his coworkers prepared the chiral amides *29, 30* [170], and *31* from amino acids, and further *32* from (1*R*,2*S*)-norephedrine [244]. They found that the zirconium enolates of *29* and *30* as well as the boron enolates of *31* and *32* added to achiral aldehydes not only with excellent *syn* selectivity ($\geqslant$96%) but also with high enantioselectivity (Fig. 5-31 (c)).

Some enolates prepared from esters or amides containing an acyclic chiral moiety showed enantio and diastereoselectivities which did not qualify for practical application [245, 246].

An interesting case for double chirality transfer, in which the chirality of the original center was temporarily "deposited", was described by Seebach and Naef (Fig. 5-32) [247]. For a similar scheme based on (*R*)-cysteine, see ref. [248].

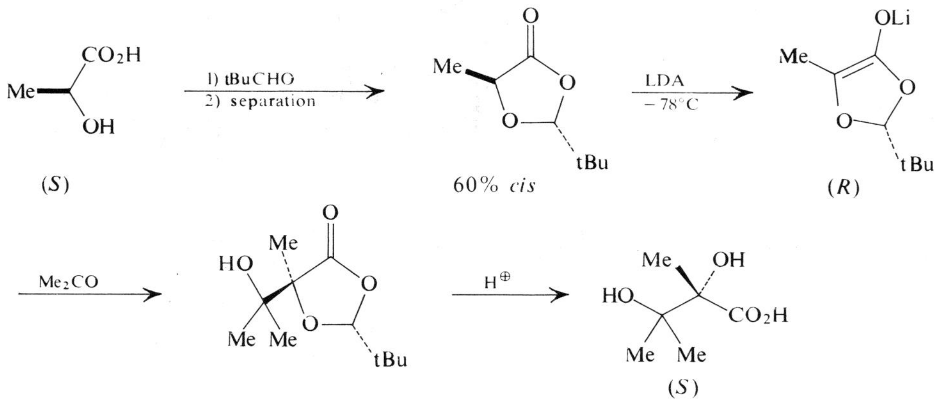

Fig. 5-32. Repeated transfer ("deposition") of chirality in aldol addition.

The use of chiral ketone enolates is only expedient when either the chiral center is destined to become an integral part of the target molecule or when it can be easily removed at a later stage. After exploratory studies by Seebach *et al.* [249] and by Evans and Taber [159], a series of chiral α-alkoxy ketones *33* [250, 251] available as both enantiomers, further *34, 35* [172], *36* [171, 252] and *37* [191] were developed in the laboratories of Masamune and Heathcock respectively (Fig. 5-33 (a)). After desilylation the aldols can be converted to α-methyl-β-hydroxycarboxylic acids by periodate cleavage of the ketol function. *33* was used as the boron enolate; asymmetric induction (diastereoface selection in respect of the enolate) increased in the series 9-BBN $<$ Bu_2B $<$ $cPent_2B$ [250]. *34-37* were reacted as the lithium enolates and showed less diastereoface selection than the boronates. *Syn* selectivity at the ends of the new bond was complete, while the degree of 1,3-induction is shown in Fig. 5-33 (a). The methylketone analogue of *36* was totally unselective in this respect [171]. For the products arising from the lithium enolates 1,3-relative configurations have not been determined, but are assumed to be *syn* by analogy with X-ray results for a product obtained with a chiral aldehyde [253].

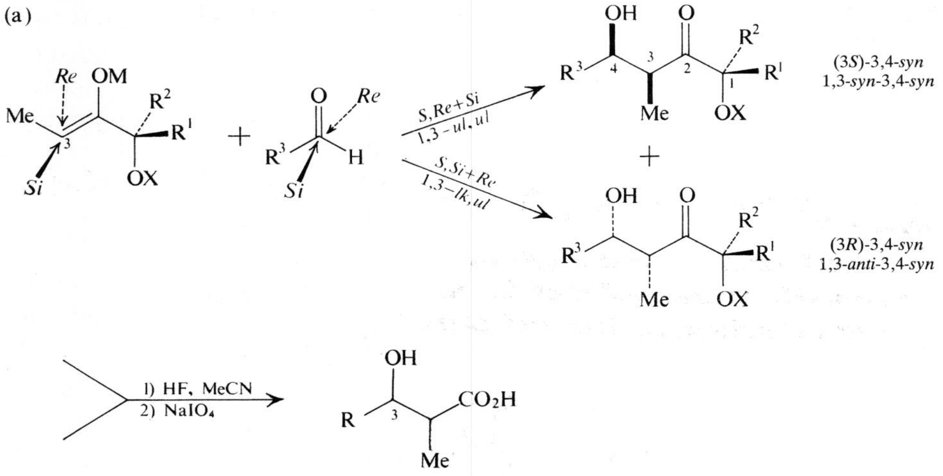

(a)

(3*S*)-3,4-*syn*
1,3-*syn*-3,4-*syn*

(3*R*)-3,4-*syn*
1,3-*anti*-3,4-*syn*

1) HF, MeCN
2) $NaIO_4$

Enolate	X	R¹	R²	M	% ee, R= Et	iPr	Ph	Bn	conf.
(*S*)−(*33*)	SitBuMe_2	H	cHex	Bu_2B	96	>98	97		3*R*
				cPent$_2$B	>98		95		3*R*
(*S*)−(*34*)	$SiMe_3$	Me	Ph	Li			56	62	3*S**
(*S*)−(*35*)			cHex					92	3*S**
(*S*)−(*36*)		H	tBu			50	50	>80	3*S**
(*S*)−(*37*)			iPr			69	34		3*S**

* Probable but not proved.

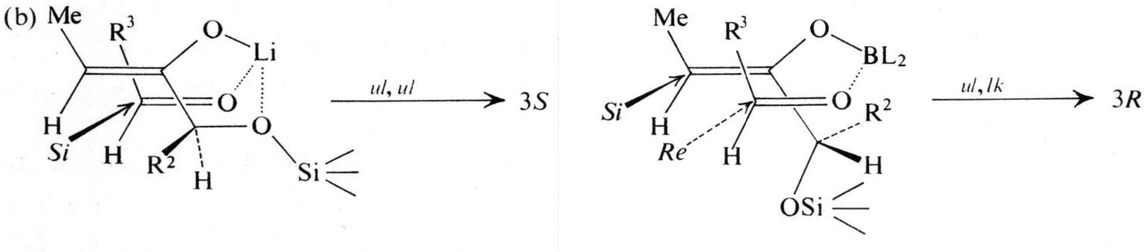

(b)

Fig. 5-33. Aldol reaction involving chiral enolates (I).

Ketones with a chiral iron center, introduced by Davies and his coworkers [254], proved their versatility in aldol addition also [255]. The configuration of the product was inverted when, instead of the aluminum enolate, the tin(II) enolate was used [256]. Removal of the inducing group was rather cumbersome.

Lithium and boron enolates differ not only in the magnitude but also in the mode of selectivity.* This has been interpreted by Heathcock by assuming complexation

* Fig. 5-33 (a)) is also an exercize in nomenclature. In the first row only the new centers are characterized. The absolute configuration at C-4 cannot be given, since this depends on the nature of the R group (*S* if Ph, *R* if iPr). In the second row the relative disposition of all three centers is defined. In the third and fourth rows configuration and mode of approach of the enolate and mode of approach of the aldehyde are expressed in two ways, the second is more general, since it also applies to (*R*)-enolates.

of the silyloxy oxygen by lithium as opposed to boron, which only complexes with the carbonyl group (Fig. 5-33 (b)).

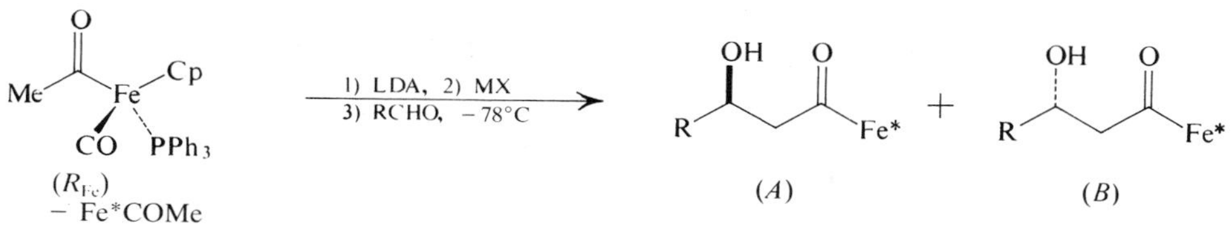

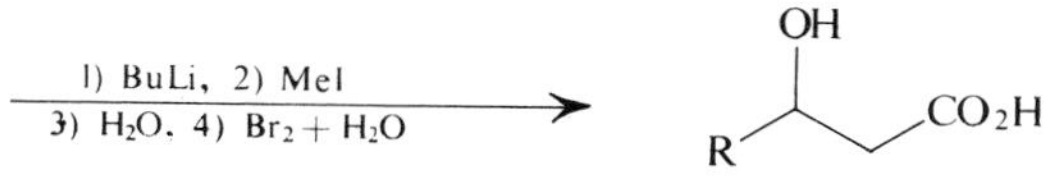

MX=Et_2AlCl

R	Me	Et	iPr	tBu	Ph
A, %de	96	>99	>99	>99	95

MX=$SnCl_2$

R	Me	Et	iPr	Ph
B, %de	68	84	84	84

Fig. 5-34. Aldol reaction involving chiral enolates (II).

The stereodirective power of chiral enolates derived from carbohydrates, though probably less expensive, did not surpass that of *33* [167, 168].

Enantiomerically pure (*R*)-3-hydroxy-2-methylpropionic acid was prepared by reacting the dicyclopentylboronate of (*S*)-*33* with formaldehyde gas followed by oxidative cleavage [251]. Shieh and Prestwich prepared the lactone *38* from (*S*)-malic acid which, with α-branched aldehydes, gave the (α,*S*)-*trans*-diols in high excess (Fig. 5-35 (a)) [257].

Lithiation of chiral 2-alkyloxazolines leads to s.c. azaenolates, which add to aldehydes with low to medium stereoselectivity [258]. Chiral boron enolates fare better [259, 260] in providing at least good *syn* or *anti* selectivity, but enantioface selection is still unsatisfactory (≤85%) (Fig. 5-35 (b)). Azaenolates obtained by lithiation of chiral hydrazones derived from methylketones were also of poor selectivity [261].

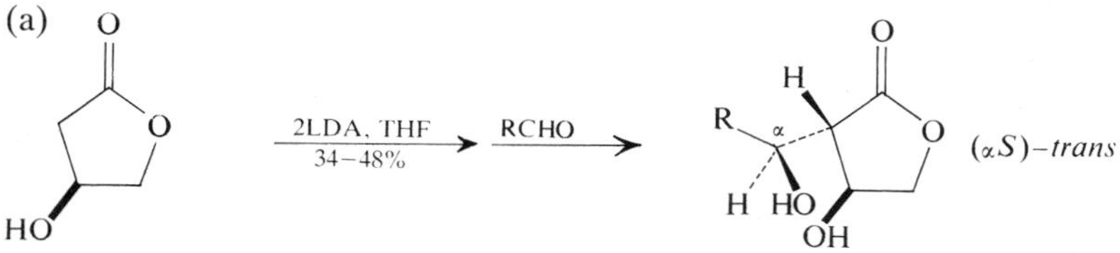

R	tBu	Ph	Bn	$C_{14}H_{25}$
% (α*S*)-*trans*	>99	91	56	50

Fig. 5-35. Aldol reaction involving chiral enolates (III).

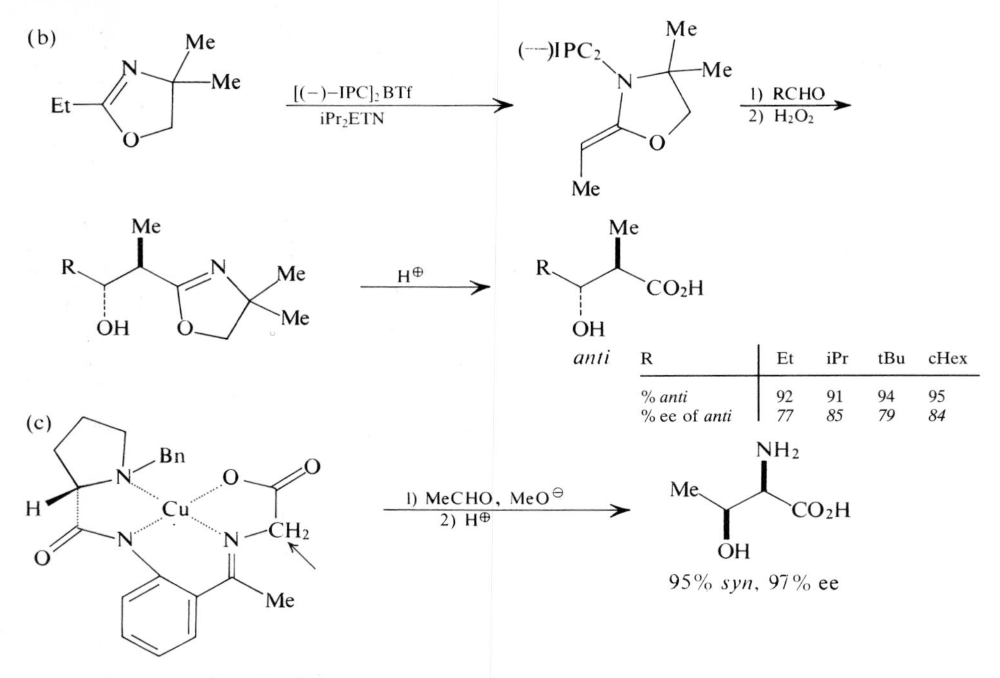

R	Et	iPr	tBu	cHex
% *anti*	92	91	94	95
% ee of *anti*	*77*	*85*	*79*	*84*

Fig. 5-35. Aldol reaction involving chiral enolates (III). (contd.)

Metal complex catalyzed addition of aldehydes to amino acid Schiff bases was studied by Japanese and Russian authors [262–264]. A recent example is shown in Fig. 5-35 (c)) [265].

5.3.5 Addition of Achiral Enolates to Chiral Aldehydes

When an enolate is added to a carbonyl compound with diastereotopic faces, usually to a chiral aldehyde, cases in which one or two new stereogenic centers are created should be distinguished (*e. g.* Fig. 5-36 (a) and (b)). In the first case the enolate is unsubstituted at the α-position and therefore only two stereoisomers are conceivable, and their ratio is dominated by the rules governing nucleophilic addition to chiral carbonyl compounds in general (*cf.* Section 3.2.1). In the second case the formation of four stereoisomers is possible, only two of which were observed in our example. Here the high stereodirective power of the aldol reaction was combined with the lower efficiency of asymmetric induction.

When there is no polar group in the aldehyde, Cram's rule is valid and consequently the *syn* product dominates. The reaction was optimized for 2-

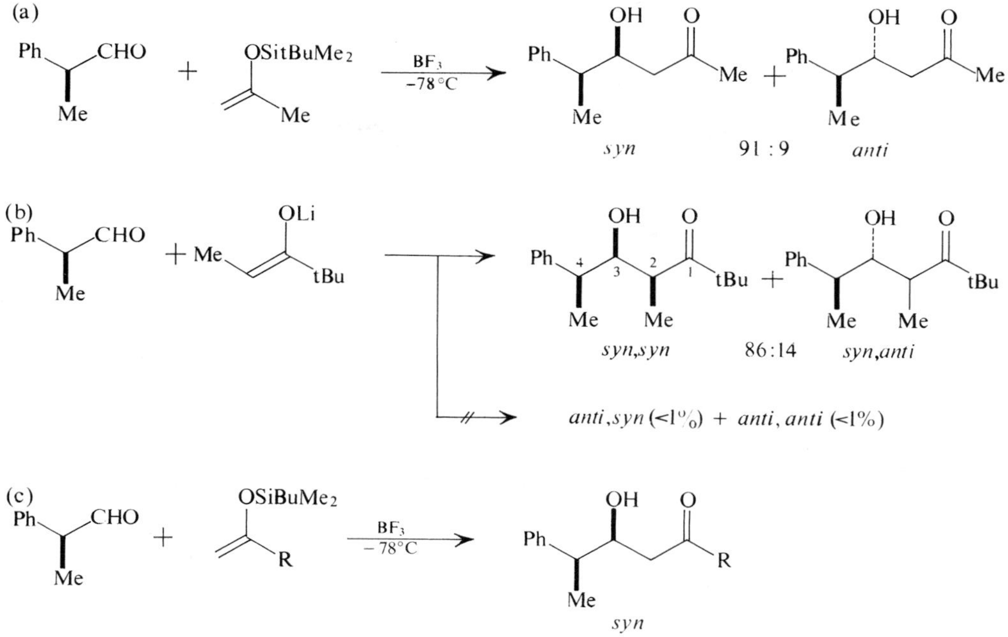

R	Me	tBu	OMe	OtBu
% *syn*	91	96	94	>97

Fig. 5-36. Aldol reaction of achiral enolates with chiral carbonyl compounds (I).

phenylpropanal by Flippin and Heathcock, and dimethyl-tert-butylsilyl enolates proved to be the most efficient (Fig. 5-36 (d)) [191]. Unfortunately, under identical conditions, addition to 2-benzylpropanol was hardly selective [191], but this is exactly what can be expected from stereochemical models.

Addition of α-substituted enolates to chiral aldehydes with no polar group follows the pattern of example (b) in Fig. 5-36. Namely, at the new bond it is strictly *syn* from (*Z*)- and *anti* from (*E*)-enolates, and the configuration of C-3 and C-4 can usually be predicted by Cram's rule [152, 161, 201, 215] (*cf.* Fig. 5-36 (b)), although some inexplicable exceptions are known [169].

Interest in polyoxygenated aliphatics prompted several studies in chiral alkoxyaldehydes [44]. Selectivity in additions to α-alkoxyaldehydes (1,2-induction) is rather unpredictable and depends on M and the structure of both substrates. Enolates derived from acetates show no selectivity [191, 253], while ketone enolates may be highly selective, and the configuration in excess depends on M and the catalyst [266] (Fig. 5-37) [47]. As another example, addition of methyl tert-butyl ketone as the lithium enolate to 2,3-isopropylidene-glyceraldehyde is 97% *anti* selective [253], whereas that of the trimethylsilyl enolate under $TiCl_4$ catalysis to 2-benzyloxypropanal gives >95% of the *syn* product [128]. Addition of various lithium

(a)

M	
M=$SiMe_3$ (+$TiCl_4$)	97% *syn, syn*
M=$SiMe_3$ (+$Bu_4N^+F^-$)	82% *syn, anti*
M=OTi $(OiPr)_3$	87% *anti, anti*

(b)

R = Ph, 85% *anti*
R = tBu, 90% *anti*

92% *syn,anti*

(c)

R	Me	Et	iPr	Ph
% ee	92	87	92	84

(d)

R	2-Me	3-Me	4-Me	4-tBu
% *cis*	>95	95	92	93

Fig. 5-37. Aldol reaction of achiral enolates with chiral carbonyl compounds (II).

enolates to both of the above mentioned aldehydes has been thoroughly investigated by Heathcock and his coworkers [152, 253].

Effective 1,3-induction in the Lewis acid catalyzed addition of enolsilanes was reported by Reetz *et al.* (Fig. 5-37 (b)) [60, 128].

Ester enolate addition to a ketone attached to a chiral auxiliary group permitted the enantioselective synthesis of α-hydroxyaldehyde esters (Fig. 5-37 (c)) [267].

High equatorial selectivity was observed in the addition of a silyl enolate to cyclohexanone acetals Fig. 5-37 (d) [268].

5.3.6 Stereoselective Aldol Condensations*

An aldol reaction can be carried out not only by transforming the nucleophilic partner to an isolable enolate but also in the presence of an amine salt, when the product is usually dehydrated to an α,β-unsaturated ketone. It was discovered that by taking a chiral amino acid as catalyst the triketones *39* could be enantioselectively transformed *via* the aldol *40* to the enones *41* [269]. The method was improved later by several groups [270–273] and became an important step in the enantioselective synthesis of steroids (Fig. 5–37(a)). Other amines were less effective, except (*S*)-phenylalanine, which was occasionally better than (*S*)-proline [274, 275]. Interestingly, with (*S*)-homoproline (*42*) as catalyst (*R*)-*41* (n = 1) was obtained in 65% ee [276]. With non-cyclic analogues both yields and enantioselectivities were poor [277, 278].

Two pathways can be envisaged for the above condensations. One involves the selective formation of an aminal (*43*) with one of the enantiotopic carbonyl groups.

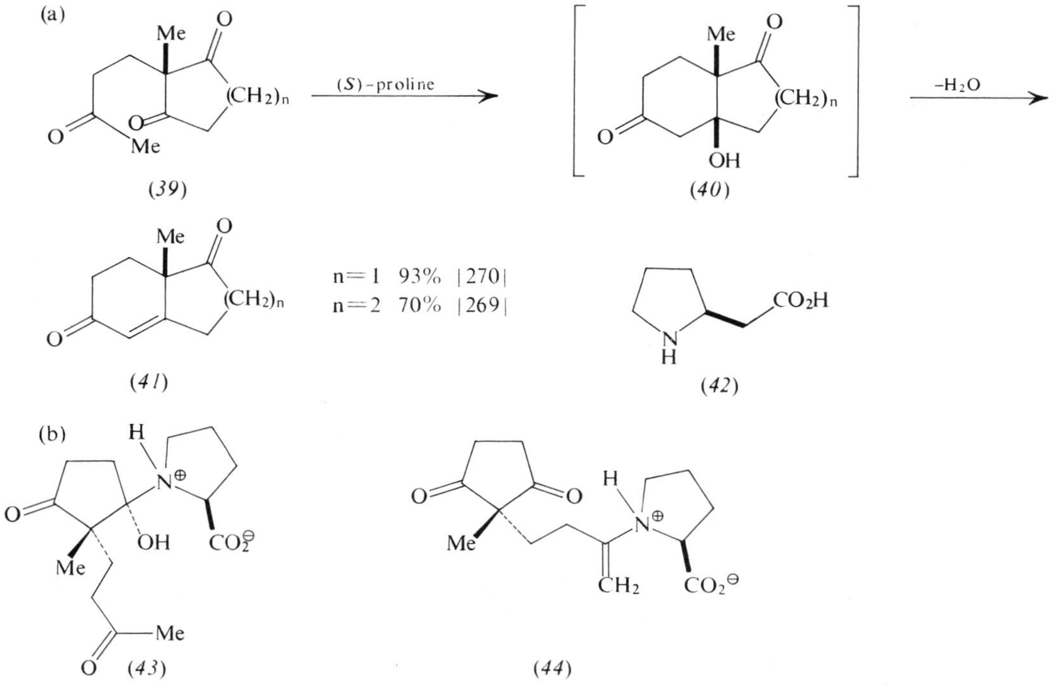

(a) (*39*) —(*S*)-proline→ [(*40*)] —$-H_2O$→ (*41*)

n = 1 93% [270]
n = 2 70% [269]

(*42*)

(b) (*43*) (*44*)

Fig. 5-38. Enantioselective aldol condensation.

* The aldol reaction, in fact an addition, is often misnamed aldol condensation (*e. g.* ref. [156]).

The other pathway goes through an enamine in which the carbonyl groups become diastereotopic (*44*) (Fig. 5-38 (b)). Recently, a careful study by Agami *et al.* gave support to the enamine pathway [144].

5.4 Miscellaneous Stereoselective Additions to Carbonyl Compounds

N-Cyclohexyl-*N*-trimethylsilyl enamines of diethylketone and cyclohexanone added, in the presence of $BF_3{-}Et_2O$ catalyst, to aldehydes giving the *syn*-aldol in excess (up to 70% *syn*) [172]. Hydrazones of methyl ketones with (*S*)-1-amino-2-methoxymethylpyrrolidine were converted by treatment with butyllithium to azaenolates, which, on addition to aldehydes and ketones gave β-hydroxyketones in up to 62% enantiomeric purity [231].

6 Stereoselective Carbon-Carbon Bond Forming Reactions

6.1 Carbon-Carbon Bond Formation Involving Olefins

6.1.1 Addition of Carbanions to Activated Olefins

This section deals with reactions covered by the following scheme:

$$>C^{\ominus} \quad + \quad >C{=}C{<}^{Y} \longrightarrow \; >C-\overset{|}{\underset{|}{C}}-\overset{|}{\underset{|}{C}}-Y$$

The carbanion may be generated from a C-acidic compound or provided by an organometallic compound, while Y is an electron attracting group such as C=O, C=N, NO_2 *etc.*

6.1.1.1 The Michael addition

In the Michael addition a C-acidic compound adds to an activated double bond in the presence of a base. Several methods have been tried out in order that the reaction should take place stereoselectively, but these efforts have rarely been successful.

Addition of C-acidic compounds to α,β-unsaturated ketones catalyzed by amines, mainly by quinine type alkaloids, has been studied by several groups [1–5] (in the homogeneous phase), [6–8] (under phase transfer conditions), and [9, 10] (with polymer bound alkaloids). Some typical optimized results are shown in Fig. 6-1 (a)*. Other additions in the presence of the chiral crown ether *2* were less

* Note that under basic conditions the chiral substrate racemizes rapidly.

selective than the one in Fig. 6-1 (a) [5] but were still higher than with other systems. Wynberg and Hermann found that the configuration of the predominant product depended on the configuration of the alkaloid at C-9 [2]. Under phase transfer conditions enantioselectivity was poor.

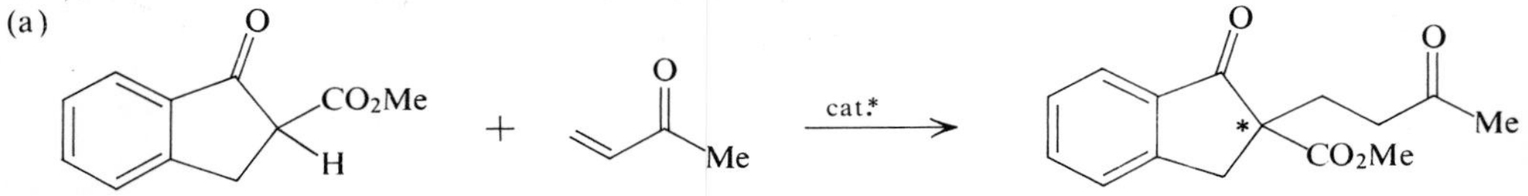

(a)

cat.*	% ee	(conf.)	ref.
quinine	76	(*S*)	2
cinchonine	68	(*R*)	1
quinine copolymer	30	(*S*)	9
$Co(acac)_2 + 1$	66	(*R*)	11
$K^+ + 2$	99	(*S*)	5

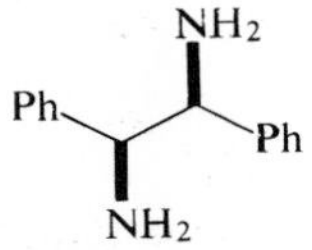

(*S*,*S*)-(*1*)

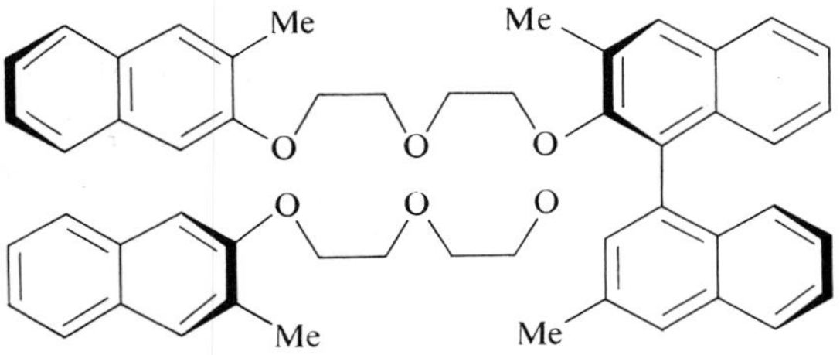

(*2*)

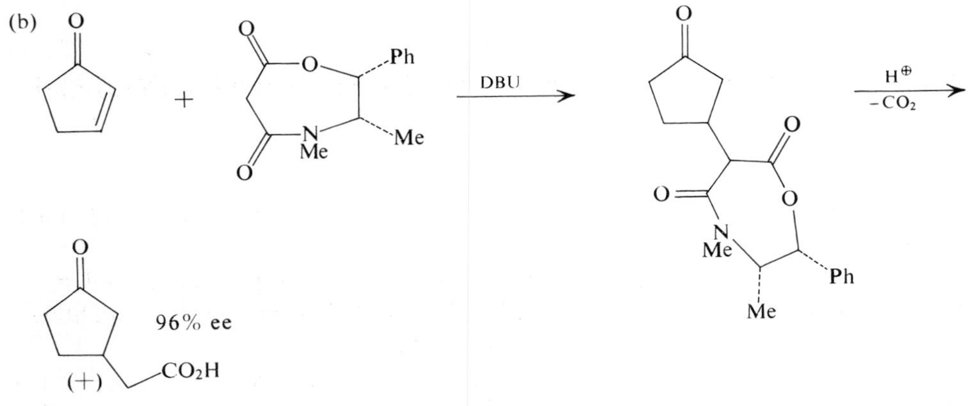

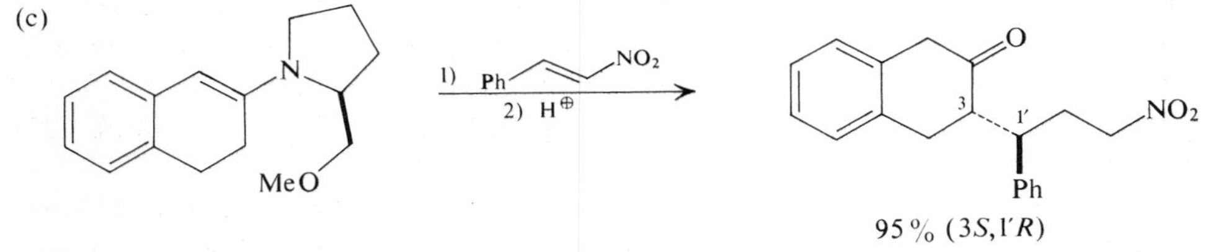

(b)

96% ee

(+)

(c)

95% (3*S*,1′*R*)

Fig. 6-1. Stereoselective Michael additions (I).

Addition of a chiral malonamide derived from (−)-ephedrine afforded high diastereoselectivities with cyclopentenone (Fig. 6-1 b)) but only 55% de with benzylidenacetone [12]. In example (c) high diastereoselectivity was somewhat offset by the appearance of 10% of a regioisomer [13].

Stork *et al.* demonstrated high diastereoselectivity in intramolecular Michael addition (Fig. 6-2 (a)) [14]. The sense of chirality transfer in example (b) could be inverted by changing the solvent and the counter ion [15].

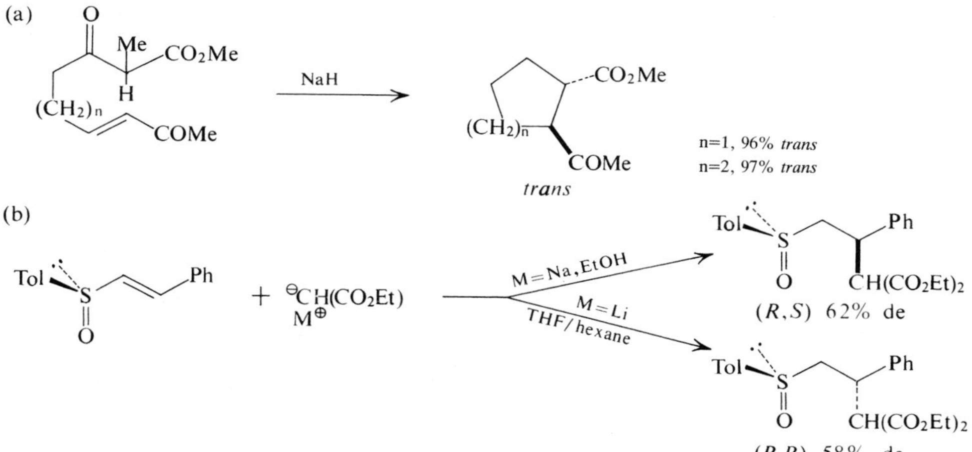

(a) O, Me, CO_2Me, H, $(CH_2)_n$, COMe — NaH → CO_2Me, $(CH_2)_n$, COMe, *trans*

n=1, 96% *trans*
n=2, 97% *trans*

(b) Tol, S, O, Ph + $^{\ominus}CH(CO_2Et)$ $M^{\oplus}$ — M=Na, EtOH → Tol, S, O, Ph, $CH(CO_2Et)_2$ (*R*,*S*) 62% de

— M=Li THF/hexane → Tol, S, O, Ph, $CH(CO_2Et)_2$ (*R*,*R*) 58% de

Fig. 6-2. Stereoselective Michael additions (II).

6.1.1.2 Addition of metalorganics to α,β-unsaturated ketones and esters

The scope of nucleophilic addition to activated double bonds has been much extended by the use of organometallic reagents.

Enantioselective addition in the presence of chiral complexants is still in its infancy. (−)-Sparteine [16] and (*S,S*)-DDB [17, 18] were tested as additives. Except for one case (58% ee), enantioselectivities were poor.

Chiral cuprate reagents capable of transferring an achiral alkyl group enantioselectively were prepared using chiral alcohols. Results are somewhat conflicting. While Imamoto and Mukaiyama reported 64% ee for the reaction shown in Fig. 6-3 (a) [19] at −30°C, only 41% ee were found by Leyendecker *et al.* at −20°C [20]. With other substrates [20], or other aminoalcohols [21] and alcohols [22] enantioselectivity was lower.

Addition of lithiumalkyls to chiral substrates concerns almost exclusively compounds with removable chiral auxiliary groups. Among them, chiral esters and sulfoxides were studied most thoroughly.

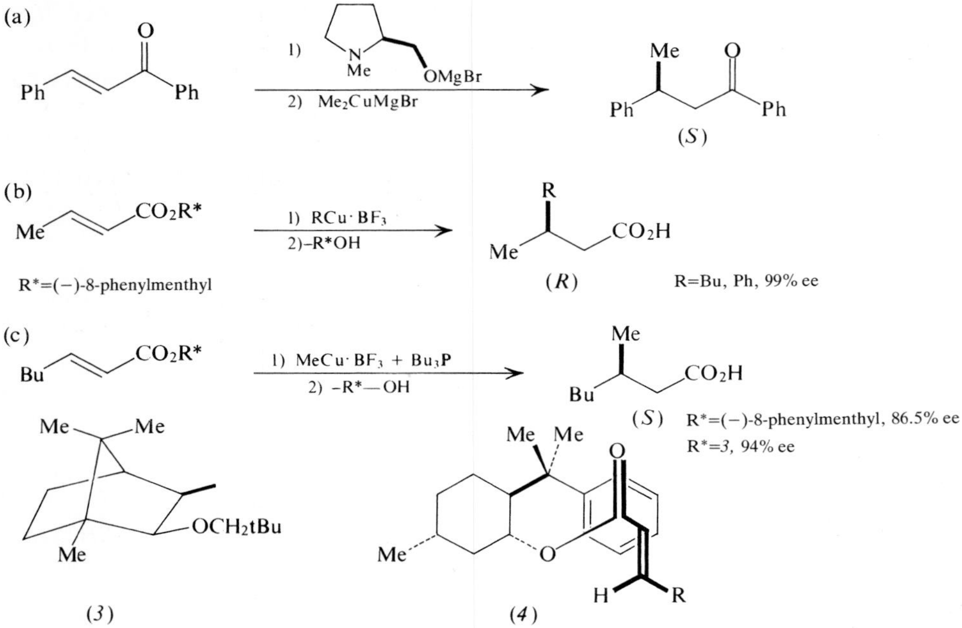

Fig. 6-3. Stereoselective addition of metalorganics to activated double bonds (I).

Chiral esters with (−)-menthol [23−25] as the alcohol component gave up to 54% ee. Claims for higher selectivity (up to 74%) with carbohydrates by Kawana and Emoto [25] were refuted by Gustafsson *et al.* [24]. 8-Phenylmenthol and other terpenic alcohols developed for enantioselective Diels-Alder synthesis by Oppolzer and his coworkers proved to be potent stereodirecting groups in conjugate addition also, as shown in Fig. 6-3 (b) [27] and Fig. 6-3 (c) [28]. With the (−)-8-phenylmenthyl esters it was assumed that the ester was in the preferred *s-trans* conformation (*4*), in which its *Si* face was effectively shielded by the phenyl ring. (*Z*)-esters gave low yields and less enantiomeric purity [27].

Addition of alkylmetals to chiral aryl α,β-ethylenic sulfoxides was exploited by Posner *et al.* [29]. A second activating group, preferably a carbonyl, is necessary, and there should be no hydrogen in α-position. Diastereoselectivity was modest with an acyclic substrate (up to 65%) [30, 31] but high with 2-arylsulfinyl cyclopentanones, especially when the proper reagent was chosen. In the case of *5*, MeMgI gave 79% ee [30], when $ZnBr_2$ was added, selectivity rose to 87% ee [32], and finally Me_2Mg in THF at −78°C gave 97% ee [33]. Selectivity depends both on the alkyl group and ring size (less with the corresponding cyclohexenones) (Fig. 6-4 (a)) [33]. Addition of alkyl lithium to the cyclohexenone analogue of *5* proceeded with 65−96% ee in 2,5-dimethyl-tetrahydrofuran [34].

Low selectivity of vinylation was eliminated by use of zinc bromide as a chelating agent (Fig. 6-4 (b)) [35]. Starting from 3-substituted analogues of *5*, construction of

(a) (S)-(5) 1) R_2Mg, DME, −78 °C; 2) Al/Hg

R	Me	Et	CH_2tBu	vinyl	Ph
% ee	97	88	91	57	>98

(b) (S)-(5) 1) $ZnBr_2$; 2) vinyl-MgBr → 1) NaH, MeI, −10 °C; 2) Al/Hg → 92% *trans*, 100% ee

(c) 1) R^2_2CuLi, THF, −78 °C; 2) Al/Hg

R^1	Tol	Tol	Me	Me
R^2	Me	Bu	Bu	Tol
% ee	78	81	88	93

Fig. 6-4. Stereoselective addition of metalorganics to activated double bonds (II).

a quaternary carbon center became possible, the configuration of which could be controlled by the order in which the groups were introduced (Fig. 6-4 (c)) [36]. Addition of a chiral bis-sulphinylmethyl carbanion to α,β-unsaturated ketones was of low enantioselectivity [37].

Mukaiyama and his coworkers used ephedrine as a very efficient chiral auxiliary for the preparation of 3-substituted-3-phenylpropionic acids [38, 39]. Product configuration can be controlled by substrate configuration (*E* or *Z*) (Fig. 6-5 (a)). Results with lithium cuprate reagents were similar. Surprisingly, open-chain amides gave even higher stereoselectivities (Fig. 6-5 (b)) [40].

Asami and Mukaiyama utilized the versatile chiral auxiliary (*S*)-2-anilinomethylpyrrolidine to prepare 3-alkylsuccinaldehyde acids of high optical purity (Fig. 6-5 (c)) [41, 42]. Only one diastereomer was obtained in the first step, and the chiral auxiliary could be recycled.

Tartaric acid *N,N*-dimethylamide proved to be a potent inducer in the addition of trialkylaluminum to its acetals with α,β-unsaturated aldehydes [43] or cyclohex-2-enone [44]. Regioselectivity was solvent dependent (Fig. 6-6 (a)).

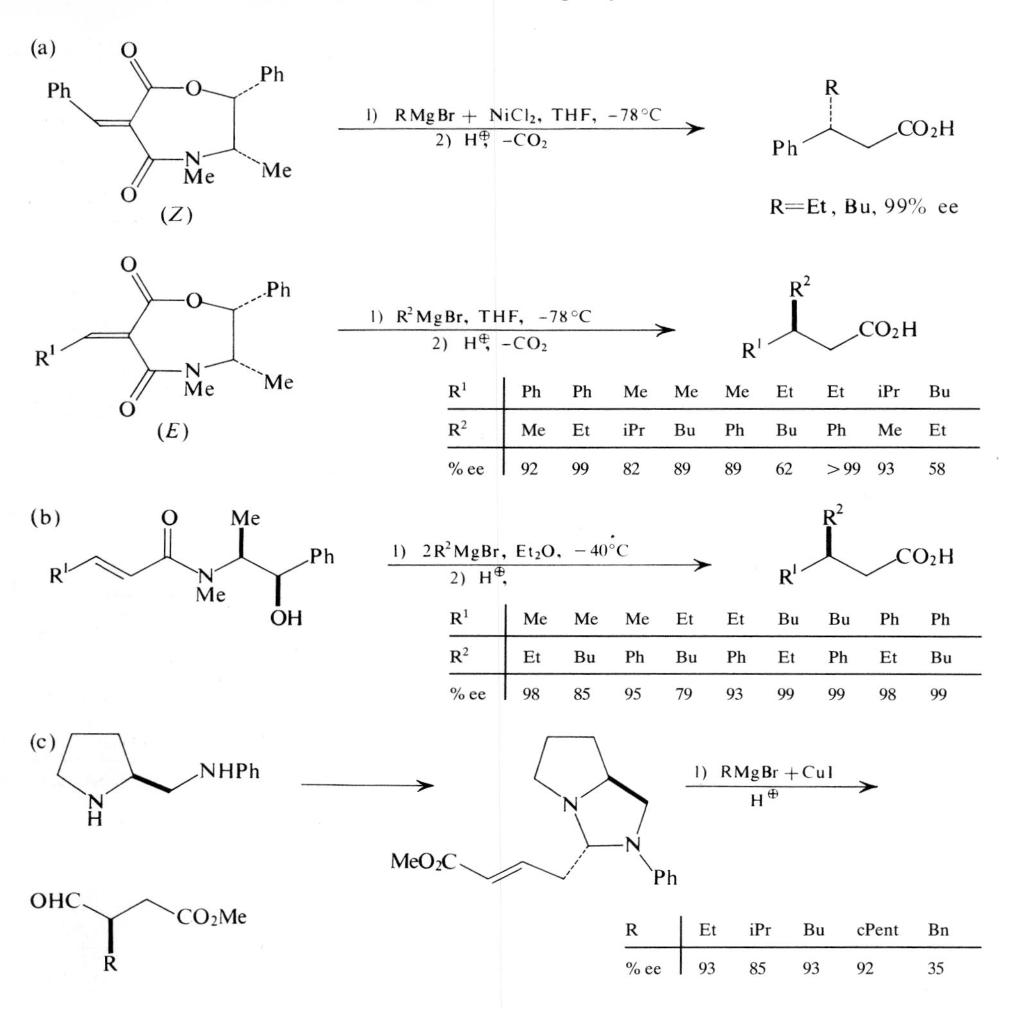

R^1	Ph	Ph	Me	Me	Me	Et	Et	iPr	Bu
R^2	Me	Et	iPr	Bu	Ph	Bu	Ph	Me	Et
% ee	92	99	82	89	89	62	>99	93	58

R^1	Me	Me	Me	Et	Et	Bu	Bu	Ph	Ph
R^2	Et	Bu	Ph	Bu	Ph	Et	Ph	Et	Bu
% ee	98	85	95	79	93	99	99	98	99

R	Et	iPr	Bu	cPent	Bn
% ee	93	85	93	92	35

Fig. 6-5. Stereoselective addition of metalorganics to activated double bonds (III).

Isobe *et al.* experienced very effective asymmetric induction by a free α-hydroxyl or by one protected with an oxygen containing group (MEM or $MeO(CH_2)_2$ in the alkylmetal addition to α,β-unsaturated sulfones (Fig. 6-5 (b)) [45–47].

Enders and Papadopoulos found that the anion generated from the chiral hydrazone *6* added to crotonates and cinnamates with excellent enantioselectivity (Fig. 6-6 (a)) [48]. House and Fischer examined the diastereoselective addition of several methylmetals to 5-methylcyclohexenone and found that Me_2CuLi was both regioselective and *trans*-selective (Fig. 6-7 (a)) [49].

Still and Galynker correlated diastereoselectivity in the addition of lithium dimethylcuprate to 8-, 9-, and 10-membered α,β-unsaturated ketones and lactones

(a)
Ph
O
O
CONMe2
CONMe2
Me3Al
CHCl3
20 °C
1) Cl(CH2)2Cl, −15 °C
2) Ac2O
HO
CONMe2
Me
Pr
1
O
CONMe2
(1R) 88% de
AcO
CONMe2
Me
Pr
3
O
CONMe2
(3S) 94% de

(b)
PhSO2
OH
Me3Si
Bn
1) 2MeLi, THF, −78 °C
2) H⊕
OH
PhSO2
Bn
Me
95% anti

(c)
N
N
OMe
R1
Me
(6)
1) LDA, THF, −78 °C
2) R2–CH=CH–CO2Me
3) H⊕
45–62%
O
R2
R1
CO2Me
>96% ee
R1=Me, Et, iPr, Ph, R2=Me, Ph

Fig. 6-6. Stereoselective addition of metalorganics to activated double bonds (IV).

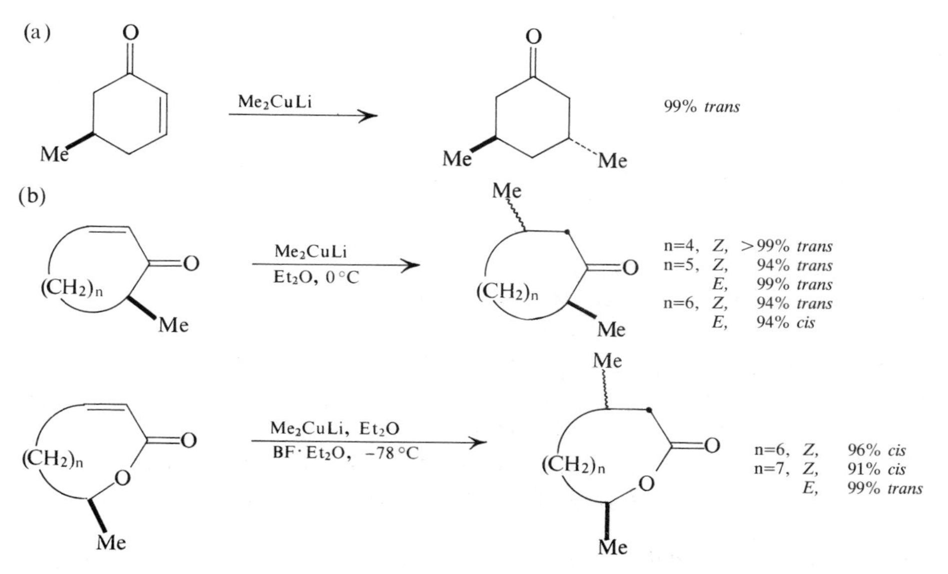

Fig. 6-7. Stereoselective addition of metalorganics to activated double bonds (V).

(c) EtO_2C–CH=CH–R^2 + R^1–CH=C(OLi)(OEt) → (THF, HMPA, −78 °C) EtO_2C–CH_2–CH(R^2)–CH(R^1)–CO_2Et

>95% *anti*

R^1=Me, Et, Oct, R^2=Me, Bu, Ph, CO_2Et

Fig. 6-7. Stereoselective addition of metalorganics to activated double bonds (V) (contd.)

with the calculated conformational composition of the substrate [50]. Some examples are quoted in Fig. 6-7 (b). Yamaguchi *et al.* reported very high *anti* selectivity in the addition of ester enolates to α,β-unsaturated esters (Fig. 6-6 (c)) [51].

6.1.1.3 Addition of metalorganics to α,β-unsaturated azomethines and 2-vinyloxazolines

Koga and his coworkers established that among several amino acid esters tested the tert-butyl ester of tert-butylglycine was the most efficient inducer in additions to azomethines derived from these amines. Examples for acyclic [52–54] and cyclic models [55–57] are presented in Fig. 6-8 (a) and (b).

(a) R^1–CH=CH–CH=N–CH(tBu)CO_2tBu → 1) R^2MgBr, THF, −55 °C; 2) $H^{\oplus}$, 41–56% → R^1–CH(R^2)–CH_2–CHO

R^1	Me	Me	Me	Ph
R^2	Bu	cHex	Ph	Et
% ee	98	96	91	95

(b) cycloalkenyl–CH=N–CH(tBu)CO_2tBu [ring $(CH_2)_n$] → 1) RMgBr, THF, −23 °C; 2) $H^{\oplus}$ → 2-R-cycloalkanecarbaldehyde [ring $(CH_2)_n$]

R	Ph	Ph	vinyl	vinyl
n	1	2	1	2
% ee	82	91	92	93*

* For the *trans*-aldehyde (*trans*/*cis* 87:13)

Fig. 6-8. Stereoselective addition of metalorganics to vinylazomethines and vinyloxazolidines.

(c)

$$\text{Me-oxazoline(Ph, CH}_2\text{OMe)} = \text{MeR}^* \xrightarrow[\text{2) (iPrO)}_2\text{POCl}]{\text{1) LDA}} (\text{iPrO})_2\overset{\text{O}}{\overset{\|}{\text{P}}}\text{CH}_2\text{R}^* \xrightarrow[\text{2) R}^1\text{CHO}]{\text{1) tBuO}^\ominus}$$

$$\text{R}^1\text{CH=CHR}^* \;(100\%\ E) \xrightarrow[\text{2) H}^\oplus]{\text{1) R}^2\text{Li, } -78\,^\circ\text{C}} \text{R}^1\text{CH(R}^2\text{)CH}_2\text{CO}_2\text{H}$$

R^1	Me	Me	Me	Et	Et	Ph	Ph
R^2	Et	Bu	Ph	Bu	Ph	Et	Bu
% ee	92	91	98	61	92	97	>95

(7) (8)

(d)

$$\text{PhCH=CH-oxazolidine(Ph, Me)} \xrightarrow[\text{2) NaBH}_4\text{, 3) H}^\oplus]{\text{1) Me}_2\text{CuLi, Et}_2\text{O}} \text{PhCH(Me)CH}_2\text{CHO} \quad 80\%\ \text{ee}\ (R)$$

Fig. 6-8. Stereoselective addition of metalorganics to vinylazomethines and vinyloxazolidines. (contd.)

One of the several areas in which the chiral oxazolines pioneered by Meyers can be used successfully is conjugate addition of metalorganics. Provided that pure (*E*)-olefins are used, (which were prepared by a special method [58],) enantiomeric purity of the end product is rarely less than 90% [59]. By interchanging R^1 and R^2, both enantiomers of the acids are accessible (Fig. 6-8 (c)).

The outcome of the reaction was rationalized by postulating double coordination with the reagent (*7*). This was supported by the fact that in the case of oxazoline *8* a product of opposite prevailing configuration was obtained with BuLi [58].

Addition of R_2CuLi to oxazolidines prepared from (−)-norephedrine and α,β-unsaturated aldehydes leads to β-alkyl aldehydes (Fig. 6-8 (d)). The reaction is fairly selective [60] and takes a solvent dependent course. In the presence of HMPA the (*S*)-aldehyde was obtained in 40% ee [61]. A similar scheme for the preparation of chiral 1,2-dialkyl-2-naphthaldehydes was elaborated by Meyers and Hoyer [62].

Addition of metalorganics to the α-position of allylic alcohols, described by Felkin *et al.* [63], belongs only formally to Section 6.1.1. While allyl and benzylmagnesium bromides add with *anti* selectivity (alkyl Grignards do not react) [64], in the presence of a cation complexant alkyl lithiums show medium to high *syn* preference [63].

RLi, TMEDA (R^1 = Me); MgBr (R^1 = H)

syn 89% *anti*

R	Et	Pr	Bn
% *syn*	86	98	75

6.1.2 Allylic Alkylation

Allylic alkylation, introduced into the synthetic arsenal by Trost [65], can be summarized by the following scheme:*

$Pd(0)(PR_3)_4$; $X^{\ominus}$; NaNu, −NaX; Pd(II), R_3P, PR_3

Either stoichiometric palldium complexes or catalytic amounts of the complex can be used, X is usually OAc or OPh, and Nu an amine, an enamine or a soft carbon nucleophile. Since it was established that the nucleophile attacks from the side opposite to palladium, then, provided that complexation takes place selectively at one face, diastereoselective alkylation can be expected, and, if the phosphine is chiral, one enantiomer of the preferred diastereomer may prevail. This is exemplified in Fig. 6-9 (a) [69]. With unsymmetrically substituted allyl systems the situation can be complicated by lack of regioselectivity.

Bosnich and McKenzie studied allylic alkylation in the presence of (*S,S*)-CHIRAPHOS (*9*) [67] and demonstrated by ^{31}P NMR that there is an equilibrium between the diastereomeric allyl complexes. Enantioselectivities obtained with sodium dimethylmalonate as nucleophile were roughly in accordance with the observed equilibrium constants (Fig. 6-9 (b), $K = 6$). A series of chiral diphosphine ligands prepared specially for allylic alkylation were tested by Hayashi *et al.* [70]. Reasonable levels of enantioselectivity could be realized with the ferrocene derivative *10* (Fig. 6-9 (c)) [71], but not with DIOP [72].

Stereoselective alkylation of an allyl ether anion with a recoverable chiral auxiliary prepared from atrolactic acid giving 3-phenylalkanals was developed by Mukaiyama *et al.* (Fig. 6-10 (a)). An aza analogue of this scheme with (*S*)-prolinol

* For reviews, see refs. [66–68].

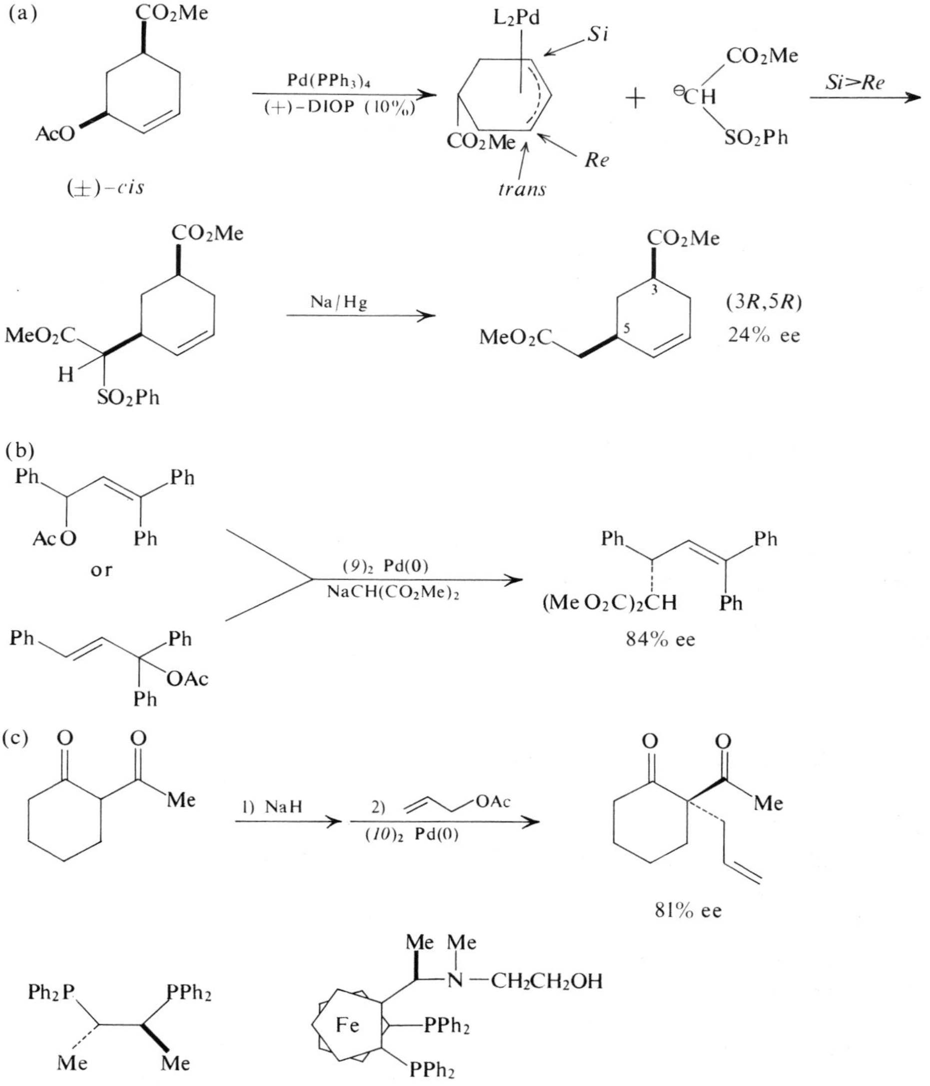

(a) (±)-*cis* —[$Pd(PPh_3)_4$, (+)-DIOP (10%)]→ L_2Pd complex (*Si*, *Re*, *trans*) + $^{\ominus}CH(CO_2Me)(SO_2Ph)$ —[*Si*>*Re*]→ —[Na/Hg]→ (3*R*,5*R*) 24% ee

(b) —[$(9)_2$ Pd(0), $NaCH(CO_2Me)_2$]→ 84% ee

(c) —[1) NaH; 2) allyl OAc, $(10)_2$ Pd(0)]→ 81% ee

(*9*) (CHIRAPHOS) (*10*)

Fig. 6-9. Stereoselectivity in allylic alkylations (I).

as the chiral auxiliary was also worked out and gave the same products but of opposite configuration and in somewhat lower optical purity (up to 65% ee) [73].

Takahashi *et al.* observed 100% transfer of chirality in intermolecular allylic alkylation (Fig. 6-10 (c)). The *Z* isomer gave an *S* product [74].

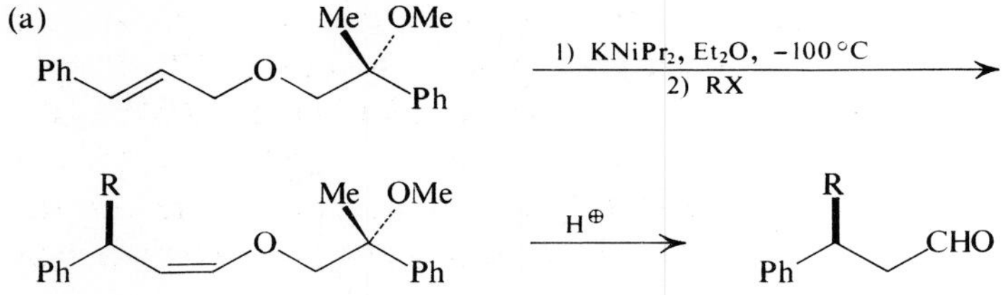

(a)

1) $KNiPr_2$, Et_2O, −100 °C
2) RX

$H^{\oplus}$

R	Me	Et	iPr	Bn	allyl
% ee	85	84	87	75	55

(b)

Pd_2L_3, $CHCl_3$
NaH, 65%

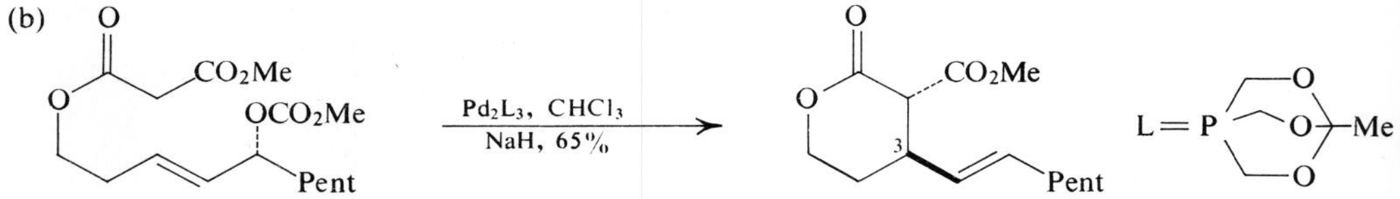

(E) (3R)

Fig. 6-10. Stereoselectivity in allylic alkylations (II).

6.1.3 Stereoselective Hydrocarbonylation*

Hydrocarbonylation, a reaction used on a very large scale in the petrochemical industry, can be briefly described by the following schemes:

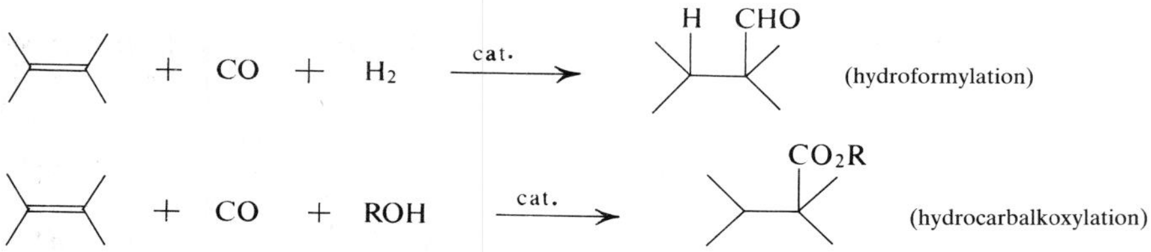

(hydroformylation)

(hydrocarbalkoxylation)

Typical situations in which a chiral center is generated by hydroformylation are shown below:

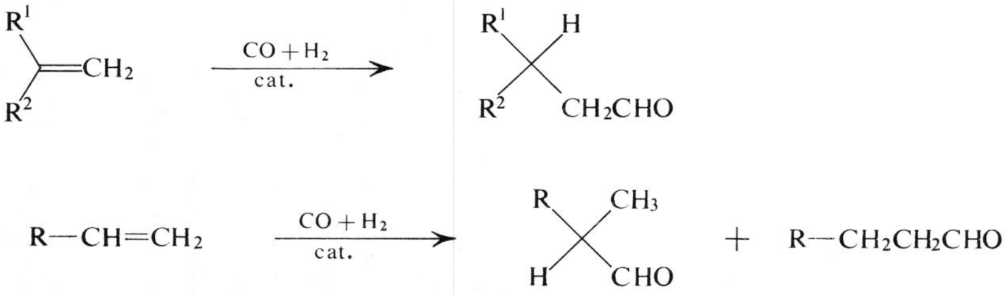

---

* For reviews, *cf.* refs. [68, 75, 76].

The first enantioselective hydroformylation, albeit in very low optical yield (<3% ee), was reported in 1972 by Botteghi, Consiglio and Pino [77]. Despite continued efforts by Pino and his coworkers, by Ogata and his group, as well as by others, enantioselectivities in hydroformylation remained modest (<50% ee). Nevertheless this reaction deserves our attention, since if a highly enantioselective catalytic system could be found, the method would certainly acquire prominent practical importance.

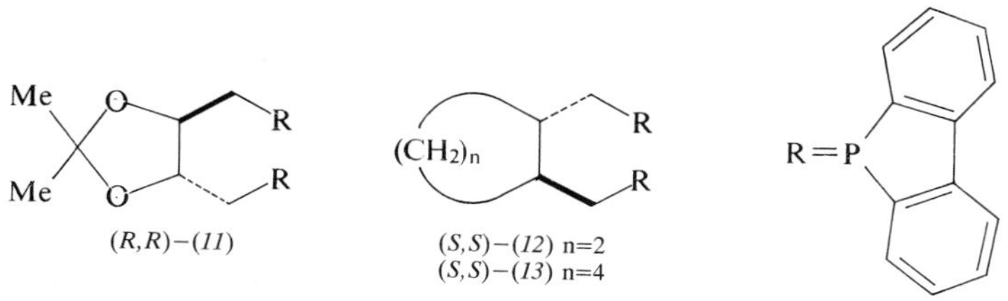

(*R,R*)-(*11*) (*S,S*)-(*12*) n=2 (*S,S*)-(*13*) n=4

Fig. 6-11. Chiral diphosphine ligands used for hydroformylation.

Three transition metal based chiral catalytic systems have been investigated: (i) cobalt complexes of chiral Schiff-base ligands, (ii) rhodium and (iii) platinum complexes containing chiral phosphine ligands. Enantioselectivity in the presence of chiral cobalt complexes was barely detectable [76, 77]. For rhodium and platinum catalysts (−)-DIOP was almost exclusively used as ligand. A few chiral ligands of diphosphol type which were synthesized by Hayashi *et al.* [78, 79] specially for use in hydroformylation (*11*−*13*) are shown in Fig. 6-11. Typical and optimized results for hydroformylations catalyzed by (−)-DIOP based systems

Table 6-1. Enantioselective Hydroformylation of Olefins Using (−)-DIOP as Chiral Phosphine Ligand.

$$R^1HC{=}CR^2R^3 \xrightarrow[\text{cat.*}]{H_2+CO\ (1\,atm)} H(R^1)C^*(CHO){-}CHR^2R^3\ (A) + R^1CH_2{-}C(CHO)R^2R^3\ (B)$$

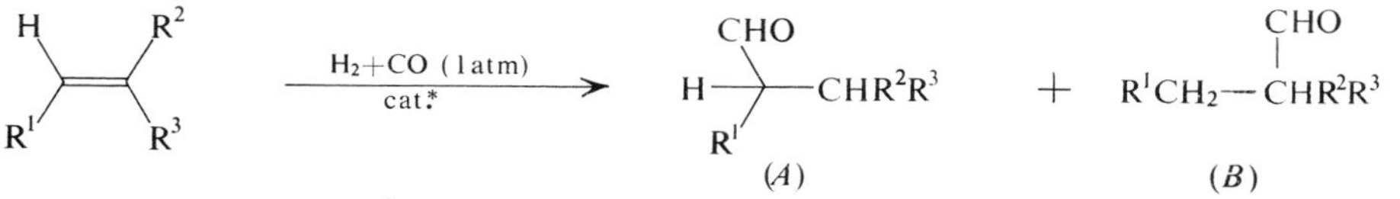

R^1	R^2	R^3	$Rh(CO)H(PPh_3)$ + (−)-DIOP %*A*	% ee	(Conf.)	Ref.	[(+)-DIOP]$PtCl_2$ + $SnCl_2$ %*A*	% ee	(Conf.)	Ref.
Et	H	H	8.5	18.8	(*R*)	76	14	2.8	(*R*)	82
Pr			7	19.7	(*R*)	76				
iPr			7	15.2	(*R*)	76				
Me		Me	> 98	27.0	(*S*)	76	56	9.9	(*S*)	82
	Me	H	> 98[a]	3.2	(*S*)	76	54	12.8	(*S*)	82
Ph	H		66	25.2	(*R*)	80	57	18.1	(*S*)	82
H	Ph	Me		1.6	(*R*)	80		9.9	(*S*)	80
		Et		1.8	(*R*)	80	52[b]	21	(*S*)	83
OAc	H	H	~100	31[c]	(*S*)	81				

[a] Total pressure 84 atm; [b] At 100°C and 250 atm; [c] With Rh(COD)[(−)-DIOP] catalyst at 80°C and 8 atm.

have been compiled in Table 6-1. It is apparent that low enantioselectivity is generally aggravated by poor regioselectivity.

The diphosphol type ligands *11*–*13* offer some advantages, and the prevailing configuration, as compared with (–)-DIOP, is inverted. Thus styrene was hydroformylated in the presence of the rhodium complex of *11* to give 89% of the branched (*S*)-aldehyde in 44% ee at 81% conversion [83]. With other models, or at higher conversion, the improvement was smaller. Decrease of optical purity with increasing conversion is due to the easy racemization of aldehydes bearing a hydrogen at the α-position. Recently Hobbs and Knowles compared the rhodium complexes of seven different DIOP-type chiral ligands in the hydroformylation of vinyl acetate, and again the diphosphol analogues proved to be the best (51% ee) [84]. The hope that asymmetric hydroformylation can be turned into a practical process may not be unjustified, since the hydroformylation of styrene could be optimized by Pittman *et al.* [85] to give, with the (*11*)–$PtCl_2$–$SnCl_2$ catalyst system, 80% of the branched product in 73% ee.*

Perhaps because there is one more parameter, *i. e.* the structure of the alcohol component, which can be varied, stereoselectivity is generally higher in hydrocarbalkoxylation than in hydroformylation. In addition, esters are less prone to racemization than aldehydes. For enantioselective hydrocarbalkoxylation, mainly the complexes of chiral phosphines with palladium have been used [75]; rhodium proved to be less satisfactory. The phosphines which provided the highest selectivities were DIOP, *11* and *12*. Some results with 1-methylstyrene as standard substrate are shown in Fig. 6-12, while data for the hydroesterification of a set of substrates with $PdCl_2$–(–)-DIOP are compiled in Table 6-2 [75].

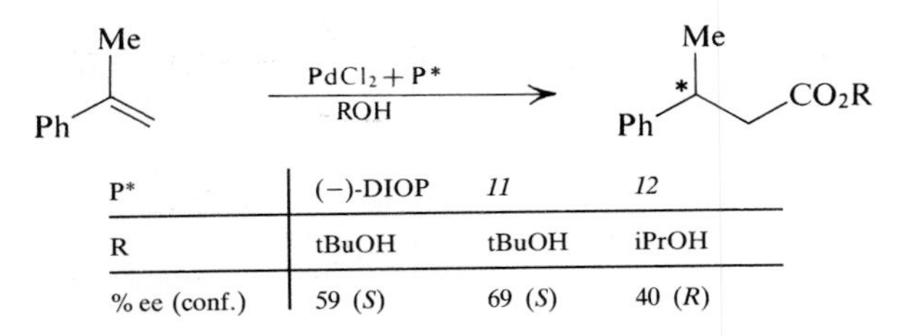

P*	(–)-DIOP	*11*	*12*
R	tBuOH	tBuOH	iPrOH
% ee (conf.)	59 (*S*)	69 (*S*)	40 (*R*)

Fig. 6-12. Enantioselective hydrocarbalkoxylations.

Recently, Cometti and Chiusoli reported a new method for enantioselective carbomethoxylation without pressure [87]. In the presence of the Pd (0) complex (+)-neomenthyldiphenylphosphine and trifluoroacetic acid, (*R*)-3-phenylbutanoic acid methylester of 52% ee was obtained from α-methylstyrene with >94% regioselectivity. Other phosphines and acids gave poor results.

Data on the influence of the individual parameters, mostly for a single substrate-catalyst pair, are available, but experience with rhodium-chiral phosphine cata-

* A claim for 94% ee was later withdrawn [86].

lyzed hydrogenation cautions that conclusions reached with one system cannot be directly applied to others.

Enantioselectivity in the hydrocarbalkoxylation of α-methylstyrene in benzene in the presence of (−)-DIOP−$PdCl_2$ catalyst increases in the order MeOH < BnOH ≅ cHexOH < iPrOH < EtOH ≅ tBuOH ≅ sBuOH, and it is hard to put any rationalization behind these facts. The order for the neat alcohols is more logical, only sBuOH is out of line (MeOH < sBuOH < EtOH < iPrOH < tBuOH). When using a chiral alcohol with an achiral catalyst enantioselectivity is negligible (<4%).

Transformation of (*E*)- and (*Z*)-2-butene gives products of opposite prevailing chirality, which means that a common intermediate can be excluded and that the formation of the palladium alkyl complex is practically irreversible.

Data in Table 6-2 reveal that it is rather difficult to forecast the major regioisomer in the case of terminal olefins. The result may be influenced by the chiral ligand, too. Thus α-ethylstyrene was attacked almost exclusively at the unsubstituted terminal in the presence of DIOP, but at the α-position when using (2-phenylbutyl)diphenylphosphine [88].

It should be noted that not too much effort was invested in order to optimize conditions with the best, *i. e.* dibenzophosphol type, catalysts [78] and therefore the hope that enantioselective hydrocarbalkoxylation can be developed to a practical method is not unreasonable.

Table 6-2. Enantioselective Hydrocarbalkoxylation of Olefins at 100°C Catalyzed by (−)-DIOP.$PdCl_2$ (cat.*).

$R^1R^2C{=}CHR^3$ $\xrightarrow[320-700\ \text{atm.}]{CO + R^4OH + cat.*}$ $R^1R^2C(CO_2R^4)CH_2R^3$ (*A*) + $R^1R^2CHCH_2CO_2R^4$ (*B*)

R^1	R^2	R^3	R^4	Type of product	% ee	(Conf.)
Et	H	H	tBu	*A*	20	(*S*)
iPr			Me	*A*	10	(*S*)
sBu				*B*	1.2	(*S*)
tBu				*A*	2	(*S*)
Ph			tBu	*A*	10	(*S*)
Et	Me			*B*	4	(*R*)
iPr				*B*	5	(*S*)
tBu				*B*	19	(*S*)
Ph				*B*	50	(*S*)
	Et			*B*	45	(*S*)
CO_2Me	Me		Me	*B*	49	(*S*)
Me	H	Me	tBu	*A*	21	(*R*)
H	Me			*A*	23	(*S*)
	Ph		Et	*A*	45	(*S*)

6.2 Stereoselective Alkylations

This section covers a large number of methods, all based on the attack of an electrophile at a nucleophilic center. Whereas the electrophilic partner is usually very simple, almost exclusively an alkyl halide, a nucleophilic center can be created in several ways, and the proper method to do this must be combined with special structural features or other means which lend stereoselectivity to the reaction. It appeared to be practical to subdivide this section according to the constitution of the nucleophile.

6.2.1 Alkylation of Enolates

Enolates prepared from esters, amides or oxazolines can be alkylated in α-position.

Enantioselective enolate alkylation is an almost untouched area. Duhamel *et al.* described the methylation of a glycine ester enolate with a chiral methanesulfonate (up to 62% ee) [89, 90], further Yamashita *et al.* reported methlation of a similar enolate generated with lithium salts of chiral amines (max. 31% ee) [91].

Diastereoselective alkylation of chiral enolates can, from a practical point of view, be divided into two groups: (i) substrates with a removable (and preferably recoverable) chiral auxiliary group and (ii) substrates in which the inducing chiral center remains in the end product.

6.2.1.1 Alkylation of enolates with a removable chiral auxiliary group

A wide range of chiral auxiliary groups has been recommended for direct alkylation of ester enolates, and the most effective ones are shown in Fig. 6-13. Selected results achieved with these groups have been compiled in Table 6-3.

In 1980 highly stereoselective alkylation of enolates derived from (*S*)-prolinol (R* = *14*) was simultaneously reported by Sonnet and Heath [92] and by Evans and Takacs [93]. The latter authors also found that the enolate was configurationally homogeneous, probably *Z*. It was suggested that allylic strain prevented the formation of the (*E*)-enolate in the corresponding transition state (Fig. 6-14 (a))

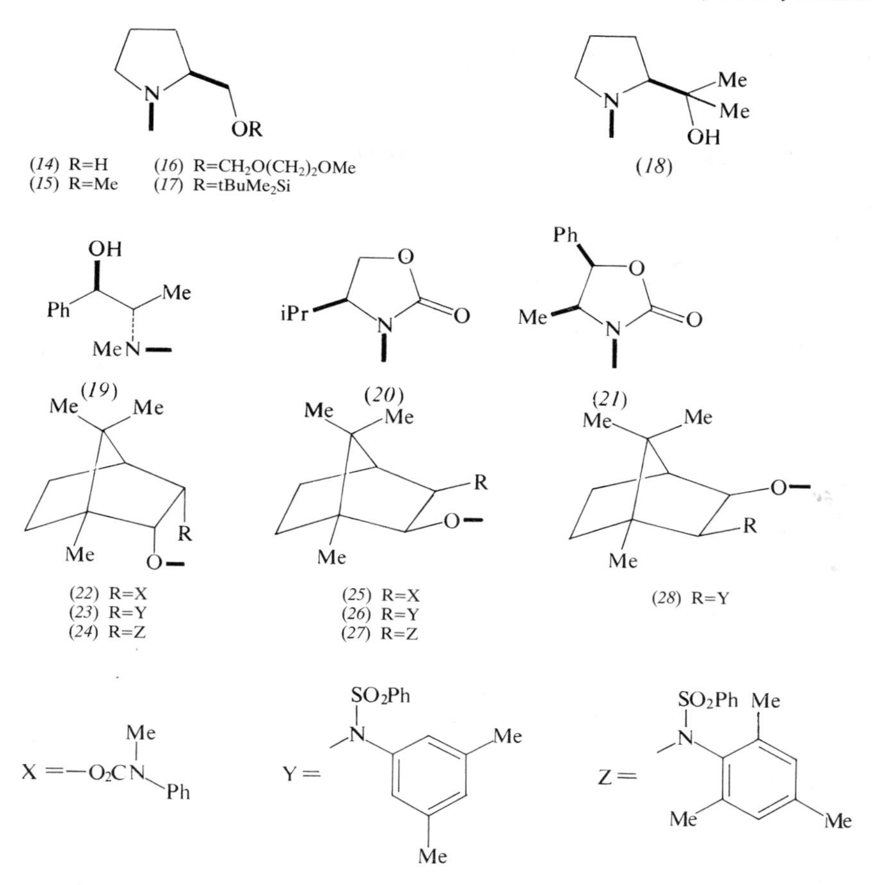

Fig. 6-13. Chiral auxiliary groups (R*) used in ester and amide enolate alkylations.

[100]. It could be anticipated that any effect wich has an influence on the rotation around the C–N bond would also be reflected in the stereoselectivity of alkylation. In fact, substitution of the hydroxyl in *14* inverted the configuration of the major product [92, 93].

Although Ireland *et al.* found that the prevailing configuration of enolates was often inverted on adding HMPA to the solvent [101], this effect was not observed with *14* [93]. In contrast, addition of HMPA changed the predominant configuration when the chiral auxiliary was *22, 24* or *27*.

A versatile series of chiral auxiliaries was prepared from camphor by Helmchen and his coworkers [97, 98, 102–104]. These compounds enable the control of configuration in several ways. Interchange of *endo* and *exo* configuration (*e. g. 22 vs. 25*), or substitution from 2,3 to 3,2 (*e. g. 26 vs. 28*), a change of solvent from THF to THF-HMPA, and, finally, changing the order of introducing R^1 and R^2 all invert the predominant configuration. It should be added that in the presence of HMPA selectivity often decreases. The best results were achieved with *23* and long

Table 6-3. Diastereoselective Alkylation of Chiral Ester and Amide Lithium Enolates. (Values shown are % de (conf. at C-2) for various R^2X.)

$$R^1CH{=}C(OLi)(OR^*) \xrightarrow{R^2X} R^1R^2CH{-}C({=}O)OR^*$$

R*	R^1	MeI	EtI	BuBr	BuI	allyl-Br	iBuI	BnBr	Ref.
(−)-Ment.	PhCH(OH)			50 (*R*)					94
14	Oct	80 (*S*)	68 (*S*)		52 (*S*)				92
	Me		84 (*R*)		88 (*R*)		94 (*R*)		92, 93
15	Oct	82 (*R*)	82 (*R*)						92
	Me		56 (*S*)						
16			56 (*S*)						93
17			54 (*S*)						93
18					87 (*R*)			87 (*R*)	99
	Bu	75 (*S*)							99
	Bn	99 (*S*)							99
	Oct	80 (*S*)							99
19	Me		>90 (*S*)		>90 (*S*)				95
	Et				>80 (*S*)			~100 (*S*)	95
20	Me		76 (*R*)			96 (*R*)		96 (*R*)	96
			88 (*R*)[a]						96
	Et	74 (*S*)							96
		86 (*S*)[a]							96
	Oct	78 (*S*)							96
		88 (*S*)[a]							96
21	Me		88 (*S*)			96 (*S*)		>98 (*S*)	96
	Et	78 (*R*)							96
	Oct	82 (*R*)							96
		86 (*R*)[a]							96
22	Me							88 (*S*)	97
	Bn	>90 (*R*)			80 (*R*)[b]			40 (*R*)[b]	97
	Bu				70 (*S*)			86 (*S*)	97
24	Me							94 (*R*)	98
								90 (*S*)[b]	98
27								96 (*S*)	98
								52 (*R*)[b]	98
28	Bn	62 (*S*)						72 (*R*)	97

[a] With the sodium enolate; [b] Enolate formed in the presence of HMPA.

chain alkyl iodides (Fig. 6-14 (b)) [97]. Recently, the method was adapted to the 2-benzyloxyacetates of *26* [104]. The reaction proceeded with remarkable *S* selectivity (>91%) but was unaffected by the addition of HMPA.

By acylation of the enolate of acetyl-*20* and -*21* followed by immediate quenching, it first became possible to prepare optically active β-dicarbonyl compounds (Fig. 6-14 (c)) [105].

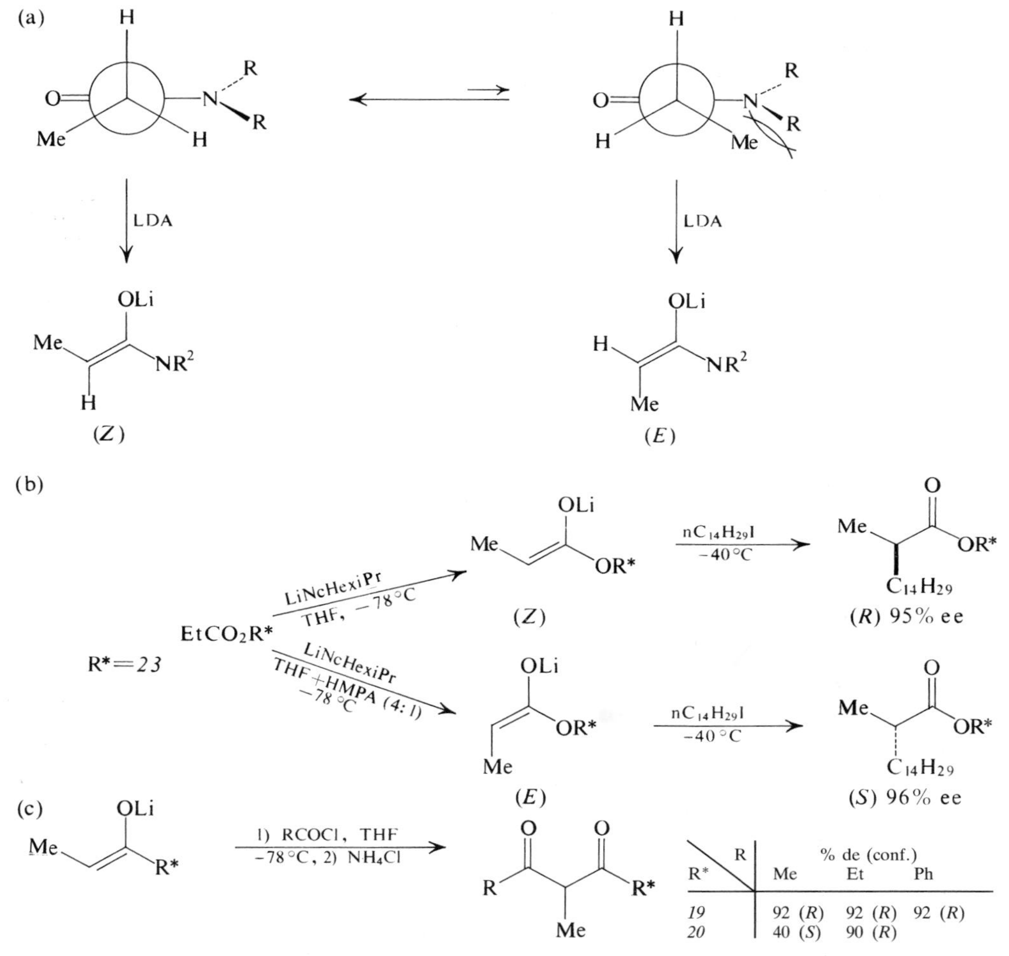

Fig. 6-14. Alkylation of chiral enolates (I).

Alkylation of enolates derived from Schiff bases of amino acids with a terpenic ketone was fairly stereoselective (Fig. 6-15 (a)) [106, 107]* **, while asymmetric induction in *N*-camphanoyl glycine was rather ineffective [108].

A special removable chiral group containing a chiral iron center was devised by Davies *et al.* and used as the enol methyl ether (Fig. 6-15 (c)) [109, 110].

The technique of "chirality deposition", *i. e.* conserving the chirality of an enolizable chiral substrate at a center not affected by enolization, was reported for mandelic acid by Fráter *et al.* [111]. The method was later extended to other

---

* Note that the chirality at the α-carbon is destroyed by enolate formation.

** The (2*R*)-ketone was used in ref. [107]. Results were changed as if obtained with the 2*S* enantiomer.

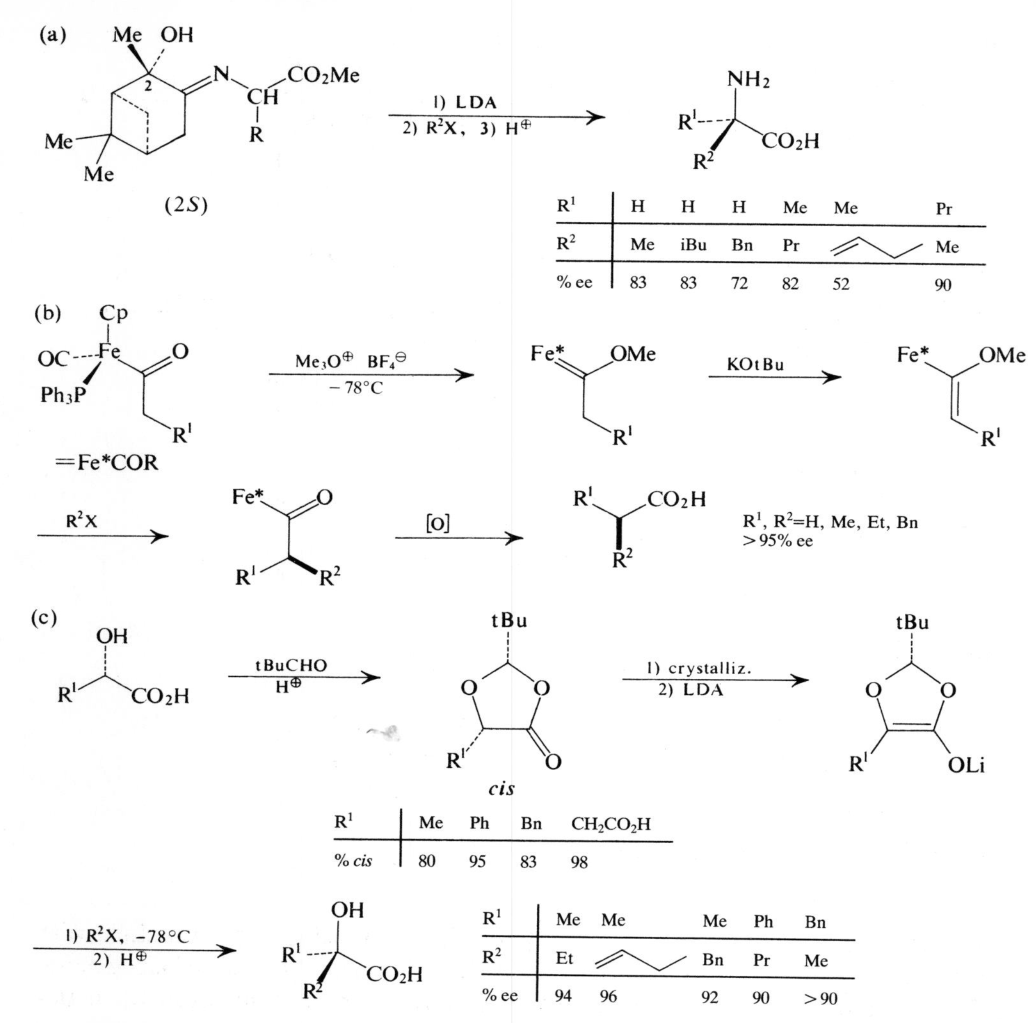

R^1	H	H	H	Me	Me	Pr
R^2	Me	iBu	Bn	Pr		Me
% ee	83	83	72	82	52	90

R^1	Me	Ph	Bn	CH_2CO_2H
% *cis*	80	95	83	98

R^1	Me	Me	Me	Ph	Bn
R^2	Et		Bn	Pr	Me
% ee	94	96	92	90	>90

Fig. 6-15. Alkylation of chiral enolates (II).

α-hydroxy acids [112] and proline [113] by Seebach and his coworkers (Fig. 6-15 (c)).*

The application of s. c. azaenolates, obtained by lithiation of chiral oxazolines, has been very successfully exploited for the stereoselective synthesis of a wide range of end products by Meyers and his coworkers [114, 115] and many other researchers. A typical example is shown in Fig. 6-16 (a) [116]. Meyers' procedure is an excellent example of "second generation" stereoselective methods which are not hinged upon steric hindrance but on directing the reagent to a specific face of the

* Fráter *et al.* reported exclusive formation of the *cis* product for R^1 = Ph [111].

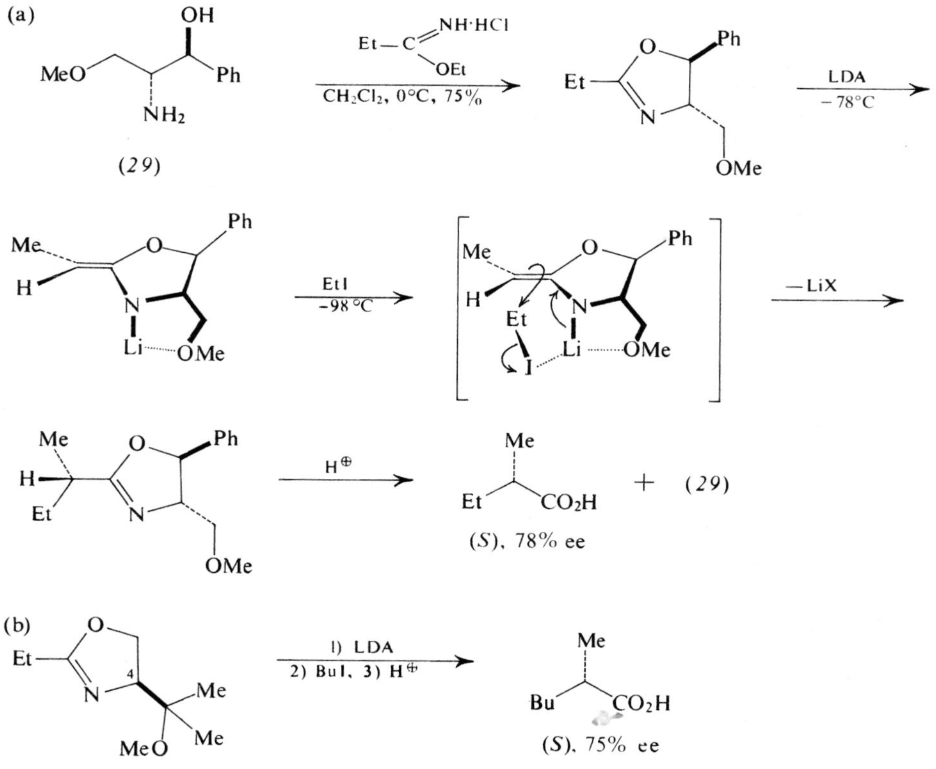

(a)

(29)

(S), 78% ee

(b)

(30)

(S), 75% ee

Fig. 6-16. Alkylation of chiral oxazolines.

substrate by coordination to a metal atom. This approach is then combined with the specific formation of one of two possible geometrical isomers of an enolate. The chiral aminoalcohol (*29*) can be prepared from commercially available material and is recoverable. As demonstrated in Fig. 1-33 (c) (p. 37), both enantiomers of a given α-alkyl-alkanoic acid are accessible by interchanging the acid and alkyl components. It is apparent from Fig. 6-16 that the methoxymethyl group is an essential part of the system, and, indeed, selectivity dropped sharply with the methyl analogue of *29* [118] and with (*S*)-4-benzyloxazolines [117]. Though without a clear rationale, the presence of the phenyl group is also essential [118]. The formation of the (*Z*)-enolate is kinetically favored, and no equilibration took place up to −30°C [119]. Nevertheless, the stereoselectivity of alkylation dropped sharply when carried out at −30°C instead of at −78°C to −90°C [114]. Unlike with other enolates (*cf.* Selections 6.2.1.1 and 7.2.1), the *Z*/*E* ratio was only somewhat diminished but not inverted by the addition of HMPT [120]. Selected results of the application of the above sequence are compiled in Table 6-4. The 2-alkyloxazolines ($R^1 \neq H$) themselves can be conveniently prepared by alkylation

Table 6-4. Enantioselective Synthesis of α-Alkylalkanoic Acids *via* Chiral Oxazolines.

R^1—CH_2–(oxazoline, 4-CH_2OMe, 5-Ph) → 1) LDA, THF, −78 °C; 2) R^2X, −78 °C−−98 °C, 3) $H^{\oplus}$ → $R^1CH(R^2)CO_2H$

R^1	R^2X	% ee	Ref.	R^1	R^2X	% ee	Ref.
Me	EtI	78	116	Ph	Me_2SO	45	116
	PrI	72	116		EtI	51	116
	BuI	75	116		BuI	52	116
	BnCl	74	116		iPrI	65	116
Et	Me_2SO_4	79	116	$Me_3SiO(CH_2)_2$	Me_2SO_4	70	122
	BnCl	85	116		EtI	72	122
Bu	Me_2SO_4	70	116		$CH_2{=}CHCH_2I$	72	122
	BnCl	82	116		BuI	60	122
	$Me_3SiO(CH_2)_2$	70	122	$Me_3SiO(CH_2)_3$	Me_2SO_4	70	122
Bn	BuI	86	116		EtI	61	122
	$Me_3SiO(CH_2)_3$	62	122		BnCl	64	122
					$CH_2{=}CHCH_2I$	72	122

of the enolate of the 2-methyl compound [116, 121]. The rather uniform results for the dialkyl products also indicate that steric effects play a subordinate role in shaping stereoselectivity.

No optically pure analogues of *29* could be prepared from serine and cysteine [123]. Hansen and Cooper experienced stereodirection in the opposite sense, as compared with *29*, in the alkylation of the enolate of *30*, (prepared from (*S*)-serine) (Fig. 6-16 (b)) [124].*

For the transfer of chirality from one amino acid to another, Schöllkopf and his coworkers elaborated two efficient schemes. The first starts from isocyanides and works only with benzyl halides (Fig. 6-16 (b)) [127].

The second method is based on mixed diketopiperazines composed of a chiral amino acid as the stereodirecting moiety and a simple amino acid, like glycine or alanine, which is alkylated. The cyclopeptide is first transformed to the bis-lactimether, (*e.g.* *31*), deprotonated with butyl lithium or LDA, and the delocalized anion is treated with an alkyl halide or added to a carbonyl compound [125–127]. As stereodirecting group, even a methyl group is quite efficient (de >90%) [128], while isopropyl and tert-butyl provide selectivities, generally beyond the detection level of the minor stereoisomer. The sequence is concluded by acidic methanolysis and separation of the two amino acid esters (a weak point in practical applications). The method is illustrated for (*S*)-valine in Fig. 6-17 (b) [129–131]. Selectivities for the tert-butyl analogue are even better [132], but unfortunately tert-leucine, though available commercially, is not a natural product.

* Though heterochiral at C-4, both give an (*S*)-acid in excess.

R	Me		iPr	Bu
% ee	>99	90	95	95

R	Hex	Bn	Ph	Ph—≡—	H—≡—	$BnOCH_2$
% ee	75–80	91–95	>95	>95	60–65	93

(33) R=Me
(34) R=iPr

Fig. 6-17. Transfer of chirality between amino acids (I).

Results were explained by coordination of $Li^{\oplus}$ with the delocalized anion at the face opposite to the substituent. Attack at this face by the electrophile is promoted both by complexation with $Li^{\oplus}$ and shielding of the other face by the bulky group at C-2. Proton abstraction seems to affect only C-5, even when this position is substituted by a methyl group [130]. Diastereoselectivity in the alkylation of a bis-lactimether (*32*) from (*S*)-3,4-dimethoxyphenyl-α-methylalanine, a drug intermediate, was sightly less (de 80–95%), which is not surprising, in view of the presence of a methyl group on the less hindered side.

3,5-Dialkyl-bis-lactimethers, such as *32* [130, 133] or *34* [134] are also amenable to alkylation, and de values in excess of 95% were obtained with the same electrophiles as were used with *31* [130, 133]. Note that the configuration at C-5 is lost on deprotonation. Not unexpectedly, the 2,5-dimethyl analogue (*35*) was alkylated less selectively [135].

Addition of bis-lactimether anions to carbonyl compounds is highly selective at the center which is part of the ring but much less so at the side chain. This is

(a) (35) 1) BuLi, 2) R^1COR^2 3) $H^{\oplus}$, MeOH

R^1	R^2	% ee at C−2	% ee (conf.) at C−3
H	H	81	
Me	Me	85	
Ph	Ph	>95	
Ph	H	88	52 (*R*)
Ph	Me	>95	41 (*R*)

(b) 1) RCHO 2) $H^{\oplus}$ MeOH → *anti*

R=Me, iPr, Ph
% ee at C−2 >95%
% ee at C−3 >97%

Fig. 6-18. Transfer of chirality between amino acids (II).

illustrated in Fig. 6-18 (a) [136, 137]. Analogous results were obtained with *31* [137] and *33* [138].

Reaction of lithiated *31* with $ClTi(NMe_2)_3$ gave a titanium compound, the addition of which was highly stereoselective with respect to both new chiral centers (Fig. 6-18 (b)) [139].

6.2.1.2 Diastereoselective alkylation of enolates

Diastereoselective alkylation of the lithium enolates of achiral amides with chiral epoxides resembles the aldol reaction and has been studied by Sauriol-Lord and Grindley (Fig. 6-19 (a)) [140, 141].

Otherwise, the scene is dominated by studies on β-oxygenated compounds. Stereoselectivities are often high, and this has been ascribed to chelate formation involving the alcoholate function [142]. Some of the more remarkable results are as follows. Alkylation of the dianion of β-hydroxyesters was found by Fráter to yield predominantly the *anti* product, both when the α-position was unsubstituted (Fig. 6-19 (b)) [143, 144] and substituted [145].

Fráter's results were later confirmed on different β-hydroxyesters by several other authors [142, 146−149]. *Anti* selectivity was high with most of the alkylating agents, and this was rationalized by the formation of a chelated intermediate (*36*) [142], which was attacked by the electrophile from the less hindered side. More than 97% of the *anti* product was obtained in bad to modest yields on alkylation of the dianion of 3-hydroxy-γ-butyrolactone in the presence of HMPA at −78°C [150].

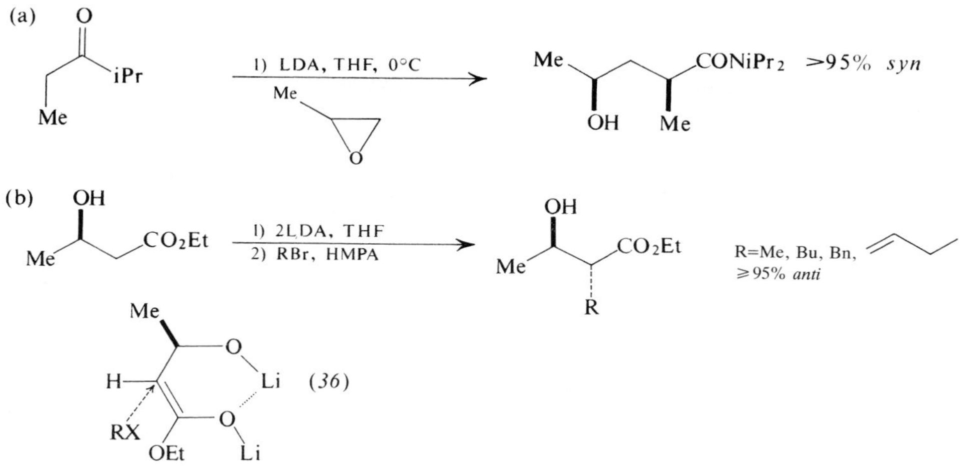

(a) 1) LDA, THF, 0°C; Me-epoxide → >95% *syn* CONiPr2

(b) 1) 2LDA, THF 2) RBr, HMPA → R=Me, Bu, Bn, ≥95% *anti*

(*36*)

Fig. 6-19. Diastereoselective alkylation of enolates.

Diastereoselectivity in the alkylation of macrocyclic ketone enolates was successfully correlated with the conformational composition of the substrate by Still and Galynker in a similar way as has been discussed for other types of reactions elsewhere (*cf.* Sections 1.5 and 6.1.1.2) [50]. Ketone enolate alkylation is, however, not an attractive reaction due to problems of regioselectivity and dialkylation.

6.2.2 Alkylation of Stabilized Carbanions

Most of the work done in this field is associated with the alkylation of carbon centers adjacent to sulfur.

Eliel and his coworkers demonstrated that under thermodynamic control 1,3-dithianes form the equatorial lithio derivatives exclusively; these can be then alkylated with retention of configuration [151–154]. An example is shown in Fig. 6-20. Preference for the equatorial metallation is so strong that even a 2-tert-butyl-1,3-dithiane was lithiated equatorially [152]. For another use of this effect *cf.* p. 167.

Asymmetric induction in α-deuteration and α-alkylation of sulfoxides is also remarkable. Methylation of the *trans*- and *cis*-thiacyclohexane-1-oxides *37* and *38* was studied by Marquet *et al.* Axial sulfoxides with two free axial α-positions gave only axial methyl-derivatives, whereas if only one of them was free axial preference was high (Fig. 6-21 (a)). *38* with an equatorially oriented oxygen, in turn, gave only equatorially methylated derivatives. As experiments with CD_3I revealed (Fig. 6-

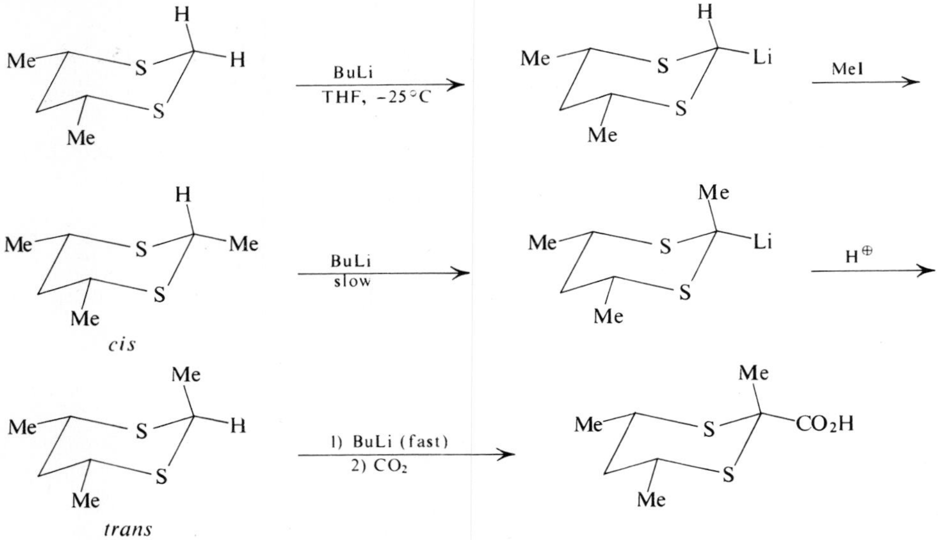

Fig. 6-20. Diastereoselektive alkylation of 1,3-dithianes.

21 (b)), the entering group can push an equatorial methyl group into an axial position [155].

^{1}H- and ^{13}C-NMR studies [156] suggest that with a planar metallated carbon the lithiated species is in a half-chair conformation (Fig. 6-21 (c)). In this case one of the faces is shielded by the metal coordinated to the sulfoxide oxygen, and this effect is combined with repulsion between the $S^+ \rightarrow O^-$ dipole and the negatively charged leaving group (path A). Indeed, for methylation with $(MeO)_3PO$, which requires assistance by $Li^{\oplus}$, selectivity was reversed (path B) (7:3 equatorial preference with *37*, 93:7 axial preference with *38*). (See Fig. 6-21.)

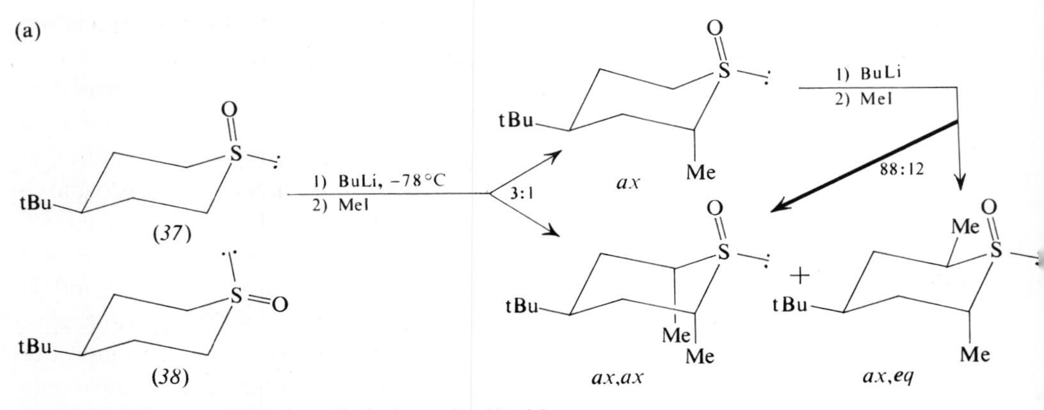

Fig. 6-21. Diastereoselective alkylation of sulfoxides.

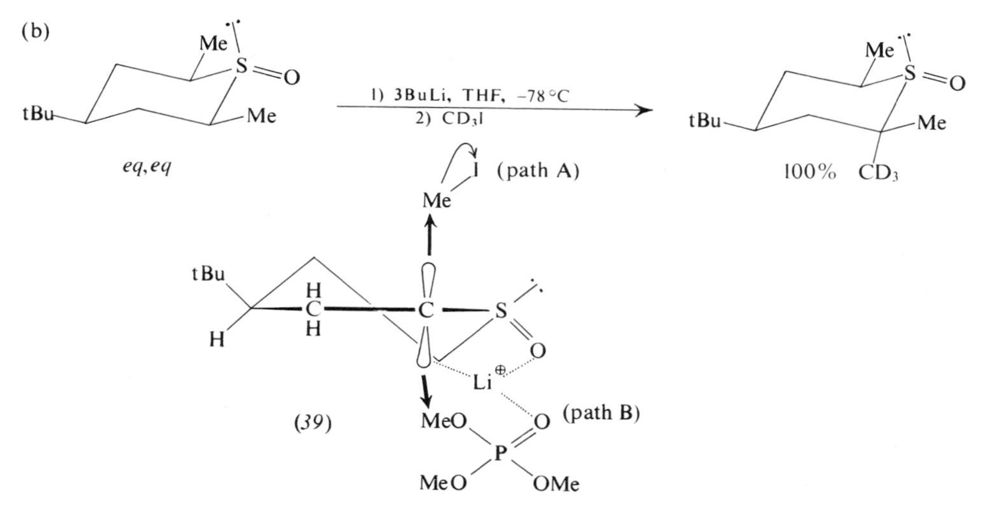

Fig. 6-21. Diastereoselective alkylation of sulfoxides. (contd.)

Inversion of selectivity in the presence of $(MeO)_3PO$ in the α-methylation of benzyl-methylsulfoxide was independently recognized by Biellmann and Vicens, too [157].

Very strong stereodirection in base catalyzed H–D exchanged was described in several interesting studies (reviewed in ref. [151]). The at first sight surprising observation, first reported by Durst *et al.* [159] and later confirmed by others [157, 158], that deuteration and methylation of lithiated benzylmethylsulfoxide occur with opposite stereoselectivity, is in accordance with an adaptation of the mechanistic picture shown in Fig. 6-21 (c) to non-cyclic models. Methyl iodide attacks from the side opposite to the coordinated $Li^{\oplus}$, whereas deuteration proceeds with retention, as with the lithiated dithianes (Fig. 6-22 (a)) [159]. The fact that ee values in deuteration and methylation are identical suggests that the results reflect the stereoselectivity of lithiation.

Methylation of the β-aminosulfones *40* was *syn* selective with tertiary amines and *anti* selective with primary and secondary amines (Fig. 6-22 (b)) [160].

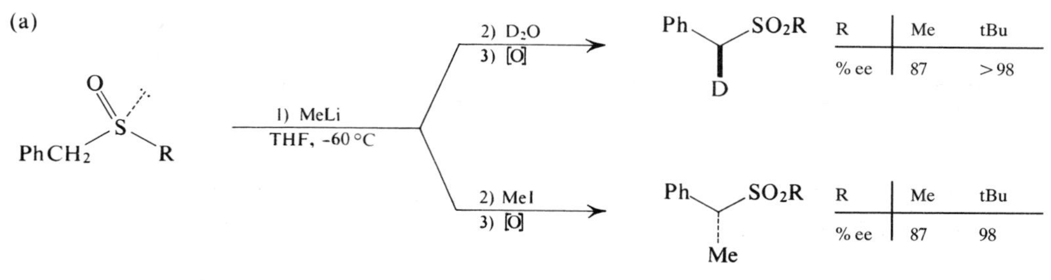

R	Me	tBu
% ee	87	>98

R	Me	tBu
% ee	87	98

Fig. 6-22. Stereoselective alkylation of carbanions.

(b)

(40) → 1) BuLi, −78°C; 2) MeI → syn + anti

X	morpholino	NMe_2	NH_2	NHMe
% syn	94	84	13	0

(c)

MeCl, PhMe, 50% aq. NaOH; N-4-$CF_3C_6H_4CH_2$-cinchoninium$^{\oplus}$ $Br^{\ominus}$

92% ee

Fig. 6-22. Stereoselective alkylation of carbanions. (contd.)

α-C-alkylation of ketones under phase transfer conditions is an established technique and recently Dolling *et al.* reported a chiral version of the process with a degree of enantioselectivity unprecedented in phase transfer catalysis (Fig. 6-22(c)) [161].

6.2.3 Alkylation of Enamines and Metallated Azomethines

Enamine alkylation is a standard method for the preparation of 2-substituted carbonyl compounds. Since chiral amines are relatively readily available, the procedure can be easily adapted to enantioselective synthesis, according to the following scheme:

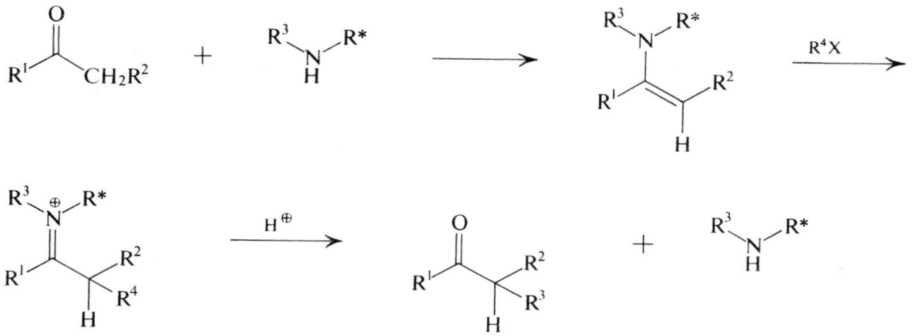

An advantage inherent in this scheme is that the chiral amine is, at least in principle, recoverable.

As amine components, the esters and amides of (*S*)-proline were first tested by Yamada and his coworkers, but enantioselectivities remained unsatisfactory (up to 60%), both with cyclic ketones and α-substituted aldehydes [162–166]. Results were no better with chiral 2-alkyl- and 2-pyrrolidinylmethyl-pyrrolidines [167–170]. Horeau reported that with methyl iodide the magnesium salt of cyclohexanone isobornyl enamine gave (*S*)-2-methylcyclohexanone in 72% ee [171], but the same method gave poor results with 2-phenylethylamine [172].

The first efficient chiral amine component, (2*S*,5*S*)-dimethylpyrrolidine, was introduced by Whitesell and Felman (Fig. 6-23(a)) [173]. High diastereoselectivities were reported by Blarer and Seebach with methylprolinol (Fig. 6-23(b)) [174].

While enamines involving secondary amines are structurally well defined, reaction of primary amines with carbonyl compounds gives a product to which either an enamine or an azomethine structure can be ascribed. Metallation shifts the

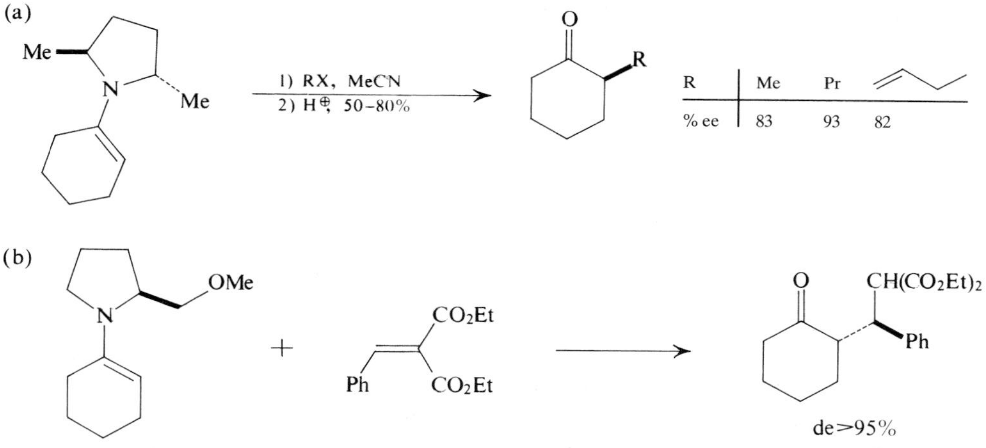

(a) 1) RX, MeCN 2) $H^{\oplus}$, 50–80%

R	Me	Pr	allyl
% ee	83	93	82

(b) de>95%

Fig. 6-23. Stereoselective alkylation of enamines (I).

equilibrium to the enamine side and permits alkylation at the α-position. Such a metallated enamine is the tin(IV) compound *41* described by De Jeso and Pommier (Fig. 6-24 (a)) [175].

In most cases, however, metallatión was carried out with LDA and, for efficient stereodirection, an ether function in the amine appeared to be essential. Thus, Meyers *et al.* tested nine different chiral β-alkoxyamines [188] and selected the one derived from phenylalanine (*42*) for further experiments. The amine *42* was an

(a) Et, O, Sn, N, (*41*); 1) $CH_2{=}CHCO_2Me$ 2) H_2O, 3) $H^{\oplus}$ → CO_2Me 86% ee

(b) N–R*; LDA, THF −20 °C → Li NR*; 1) RX, −78 °C 2) $H^{\oplus}$ → O, R

RX	MeI	EtI	PrI	$CH_2{=}CHCH_2Br$	BnBr
% ee	87	94	99	99	88

Bn, OMe, NH_2
(*42*) = R*—NH_2

(c) $(CH_2)_n$, CH_2, O; 1) *42* 2) LDA, THF, −78 °C → Li, N—R*, $(CH_2)_n$, H (*E*); reflux → Li, N—R*, $(CH_2)_n$, H (*Z*)

(*E*): 1) MeI 2) $H^{\oplus}$ → $(CH_2)_n$, O, Me (*S*)

(*Z*): 1) MeI 2) $H^{\oplus}$ → $(CH_2)_n$, O, Me (*R*)

n	8	10	13
% ee	30	59	37

n	10	13
% ee	81	81

Fig. 6-24. Stereoselective alkylation of enamines (II).

efficient auxiliary in the alkylation of cyclohexanone (Fig. 6-24 (b)) [176, 177]. The Whitesells, in turn, methylated, at −100°C, the bromomagnesium azaenolate of cyclohexanone formed with the *O*-butyl analogue of *42* and obtained the *S* product in 85% ee [178]. Enantioselectivity with *42* was slightly less with cycloheptanone and poor with cyclooctanone. An interesting way of influencing product configuration by azaenolate configuration was possible with macrocyclic ketones (Fig. 6-24 (c)). Under kinetic conditions mainly the (*E*)-enolates were formed from the C_{12} and C_{15} ketones which inverted on refluxing in THF to the (*Z*)-enolates. Cyclodecanone, however, having a medium size ring, gave an *S* product under both conditions [179].

Azomethines of cyclohexanone and 2-phenylethylamine [180] or (1*S*,2*R*)-2-methoxy-1,2-diphenylethylamine [180] were alkylated as the lithium, magnesium, or zinc azaenolates, but enantioselectivity was not better than in earlier studies. Selectivity was, however, remarkable with (*S*)-2-amino-1-propanol linked through the oxygen to a Merrifield-type resin (95 and 60% ee with MeI and PrI resp.) [181] or with a polymer of 2-(4-vinylphenyl)-ethylamine [182]. To conclude, examples with cyclic ketones, the work of Koga *et al.*, should be mentioned. They found that the amino acid esters *43* and *44* were powerful chiral auxiliaries for lithium azaenolate alkylations (up to 98% ee with *44*) [183, 184]. Recently it was observed that in the presence of an ester group in α-position the direction of alkylation could be influenced by additives (Fig. 6-25) [185, 186].

Enantioselective α-alkylation of acyclic ketones *via* azomethines proved to be a much more demanding task than that of cyclic ketones. Meyers *et al.* reported high stereoselectivities with symmetric dialkyl ketones using *42* as chiral auxiliary, provided that the enolates were converted, prior to alkylation, to the *E* form by refluxing (Fig. 6-26 (a)) [179, 187]. Note that no inversion of configuration, only

RX	cosolvent	% ee (conf.)
MeI	HMPT	>99 *B*
	THF	92 *A*
allyl bromide	HMPT	76 *B*
	THF	56 *A*
BnBr	HMPT	>99 *B*
	dioxolan	71 *A*

Fig. 6-25. Stereoselective alkylation of enamines (III).

enhancement of enantioselectivity, was brought about by $Z \rightarrow E$ transformation. Azomethines of *42* and aldehydes were alkylated as the lithio azaenolates with poor selectivity (<50%) [188]. The 2-(2-methoxyethoxy)-ethyl analogue of *42* [188] performed slightly better, and the terpenic amine *45* [190], when lithiated with Li-2,2,6,6,-tetramethylpiperidine, gave up to 58 and 75% ee resp. of 2-alkylated aldehydes. Selectivities in the same range were realized by Fraser *et al.* using 2-phenylethylamine and $MgBr_2$ as additive (Fig. 6-26(b)) [191].

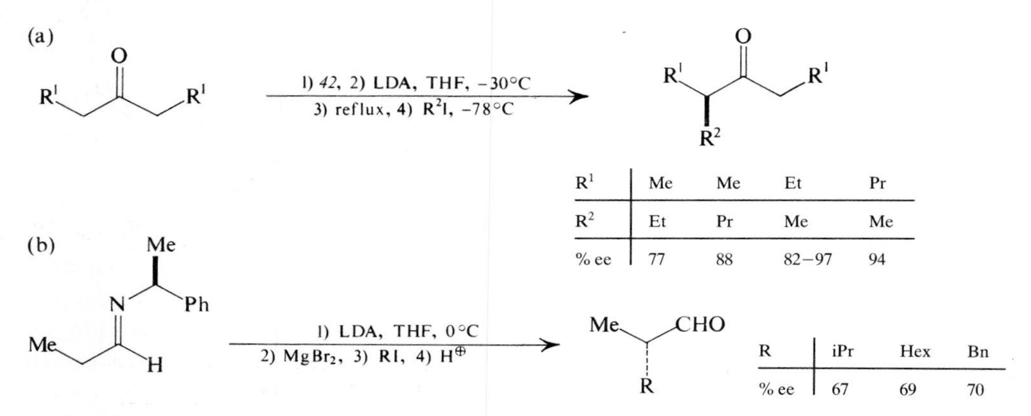

R^1	Me	Me	Et	Pr
R^2	Et	Pr	Me	Me
% ee	77	88	82–97	94

R	iPr	Hex	Bn
% ee	67	69	70

Fig. 6-26. Stereoselective alkylation of enamines (IV).

6.2.4 Alkylation of Metallated Hydrazones and Amidines*

Stereoselective α-alkylation of lithiated chiral hydrazones was first reported by Enders and Eichenauer in 1976 [195], and later developments in this field were also associated with Enders and his coworkers. The key to success was the synthesis of a chiral hydrazine *46*, available as both enantiomers (SAMP and RAMP). SAMP was prepared from (*S*)-proline in 5 steps in 50% overall yield [195, 196], while RAMP was obtained from the relatively expensive, but commercially available, (*R*)-glutamic acid in five steps in 30% overall yield [197, 198]. The general scheme of metallated hydrazone alkylation is shown in Fig. 6-27 (a). Removal of the chiral auxiliary is also feasible by quaternization with methyl iodide, followed by two-

* For reviews, see refs. [192–194].

(a)

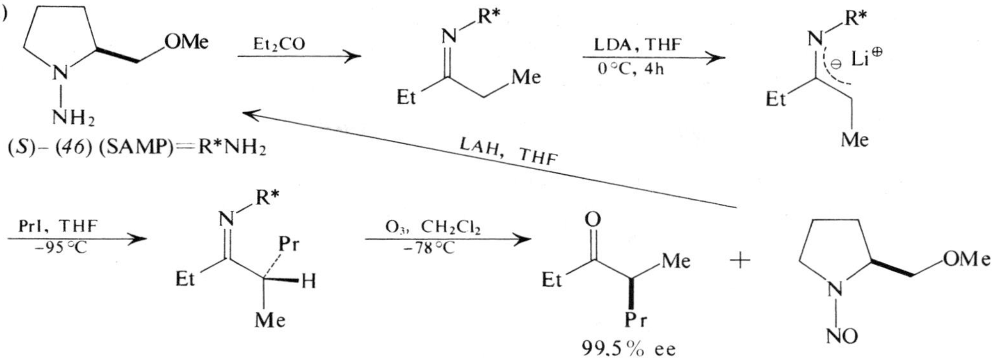

(b)

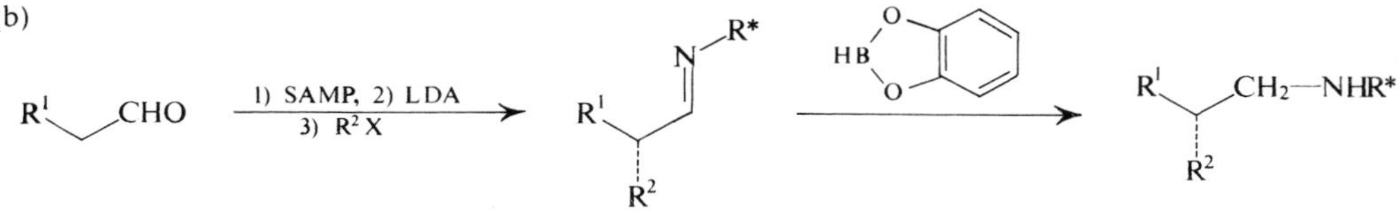

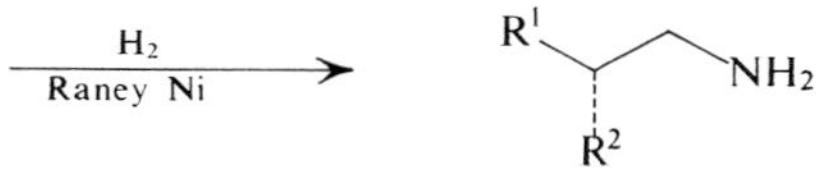

R^1	Me	Me	Hex	Ph	Bn
R^2	Hex	Bn	Me	Me	Me
% ee	>90	>90	>90	95	>95

(c)

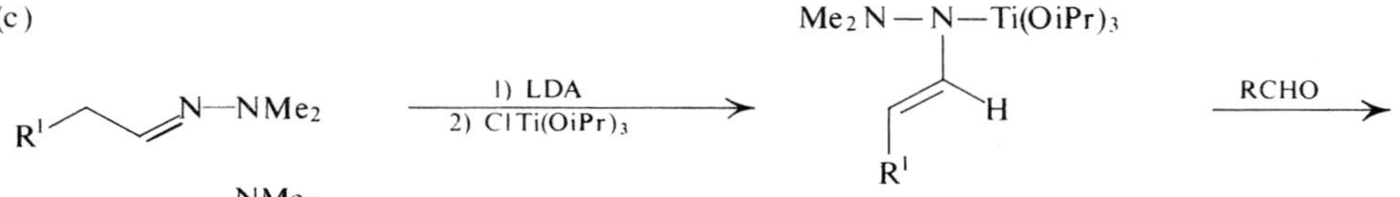

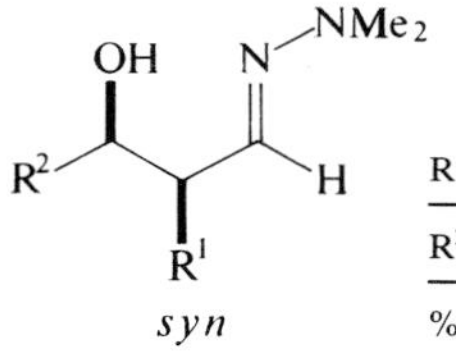

R^1	Me	Me	Me	iPr	Ph	Ph	Ph
R^2	Me	iPr	Ph	Ph	Me	iPr	Ph
% *syn*	90	86	82	88	92	96	96

Fig. 6-27. Stereoselective alkylation of hydrazones.

phase acidic hydrolysis, but in this case only a part of *46* can be regained [192]. *Z* configuration of the metallated hydrazone was proved [200, 201]. The conformation around the N–N-bond is a subject of speculation, but it is reasonable to assume that $Li^{\oplus}$ is coordinated to both the allylic type anion and the ether oxygen, and this feature is the clue to the exceptional stereoselectivity of the alkylations. So far the method has undoubtedly been the best approach to chiral α-alkyl carbonyl compounds. The most important results are compiled in Table 6-5.

Table 6-5. Stereoselective α-Alkylation of Lithiated Hydrazones of Carbonyl Compounds (R^1COR^2) with (*S*)-1-Amino-2-methoxymethylpyrrolidine (*45*, SAMP).

R^1	R^2	R^3X	% ee	(Conf.)	Ref.
H	Et	EtI	77	(*S*)	199
	Pr	MeI	62	(*R*)	199
	Et	BnBr	82–95	(*S*)	199
	$BnCH_2$	MeI	>90	(*R*)	199
	Et	HexI	>95	(*S*)	199
	Hept	MeI	95	(*R*)	199
	iBu	MeI	57	(*R*)	199
	Bn	Me_2SO_4	31	(*R*)	199
Et	Et	EtI	94	(*S*)	199
		PrI	99.5	(*S*)	199
Pr	Pr	MeI	≥98	(*R*)	199
Bu	Bu	MeI	≥98	(*R*)	199
		EtI	≥98	(*R*)	199
Me	Bn	MeI	20	(*R*)	199
Bn	Bn	MeI	30	(*S*)	199
Ph	Bn	MeI	10	(*R*)	195
Et	Et	$tBuO_2CCH_2Br$	≥95	(*S*)	199
$-(CH_2)_4-$		Me_2CO_4	86	(*R*)	199
$-(CH_2)_5-$		Me_2SO_4	99	(*R*)	199
		EtI	>98	(*R*)	199
		PrI	>86	(*R*)	199
		⁄⁄∼ Br	73	(*S*)	199
$-(CH_2)_6-$		MeI	94	(*R*)	199
$-CH{=}CH-(CH_2)_3-$		MeI	61–75	(*R*)	199
$-CH{=}CH-(CH_2)_3-$		⁄⁄∼ $(CH_2)_2Br$	89	(*R*)	202

The hydrazone method is flexible in two ways:
The configuration of the end product can be controlled, both by that of the auxiliary and by interchanging the roles of R^1 and R^2. A third method which was successfully employed with *O*-enolates, namely, altering the configuration of the enolate by the addition of HMPA (*cf.* pp. 239 and 285) was not practicable in the present case. Although inversion of configuration from *E* (93%) to *Z* (70%) at the C=C bond of the anion was demonstrated, and alkylation led to opposite prevailing configuration, selectivity was very poor (10% ee) [200].

The method was recently exploited for the preparation of β-chiral primary amines also, as shown in (Fig. 6-27 (b)) [203].

While the addition of aldehydes to *O*-enolates obtained from aldehydes is unselective, high *syn* selectivity, presumably *via* a non-cyclic transition state, was reported by Reetz *et al.* for titanium enolates of dimethylhydrazones (Fig. 6-27 (c)) [204].

Kolb and Barth have shown that, similarly to hydrazones, amidines can also be lithiated in α-position to the C=N bond, and if the sp^2-nitrogen is part of a chiral moiety, such as prolinol, alkylation proceeds stereoselectively. Enantioselectivity was 87% at best [26], but usually much less [205]. An adaptation of the method by Meyers *et al.* proved to be more successful (Fig. 6-28) [206, 207].

1) LDA, 2) RX
3) H_2NNH_2

R	Me	Bu		Bn	$Ph(CH_2)_2$
% ee	99	96	96	98	93

Fig. 6-28. Alkylation of metallated amidines.

6.3 Carbon-Carbon Bond Forming Reactions Involving C=N Bonds

Certain nucleophiles, such as the cyanide ion and Grignard reagents, readily attack the carbon atom of azomethines. The method can be made enantioselective and provide amines chiral at the α-position, when the azomethine comprises a chiral amine component. Prospects for the recovery of the chiral auxiliary are poor, since the original C–N bond has to be severed in order to liberate the product.

Attempts to utilize 2-phenylethylamine as chiral aid in the addition of cyanide to azomethines (Strecker synthesis) met with varying success. While Harada and Okawara [208] reported up to 53% ee with aliphatic aldehydes, Stout *et al.* [209] reported 75–78% ee for aliphatic, and 69–85% ee for aromatic aldehydes. An earlier claim for 100% ee did not reflect kinetic selectivity, because the primary product was enriched by chromatographic purification of the raw product [210].

Weinges *et al.* described complete diastereoselectivity of cyanide addition to azomethines prepared from 3-aryl-2-propanones and 4-aryl-2-butanones as the carbonyl component and 2-phenylethylamine [211] or the amine *47* [212, 213] as chiral auxiliaries. *47* can be prepared from an intermediate of chloramphenicol synthesis. The key to high selectivity is that the reaction is reversible, and one of the diastereomeric products crystallizes out. This becomes evident from the fact that with (*S*)-2-phenylethylamine product configuration was *R* in the case of phenylacetone and *S* with 4-methoxyphenylacetone, and at the same time an extra CH_2 group in the ketone had no influence on stereoselectivity. The method, as applied to the synthesis of α-methyl-3,4-dihydroxyphenylalanin, an important anti-hypertensive agent, is shown in Fig. 6-29 (a) [212]. It works well with aliphatic

(a) Ar–CH2–CO–Me + R*–NH2 —(NaCN, AcOH, MeOH)→ R*NH, CN, Ar, Me —(HCl)→ morpholinone (Pr, HO, Me, Ar, N⊕H2) —(Raney Ni, OH⊖, [O])→ Ar–CH2–C(Me)(NH2)–CO2H, 100% ee → α-Me-DOPA

R*–NH2 = (47) (Ph, Me, Me, O, O, NH2); Ar = 3,4-dimethoxyphenyl (MeO, MeO)

(b) Ph–CH=N–CH(iPr)–CH2OH —(RM)→ Ph–CH(R)–NH–CH(iPr)–CH2OH, *anti*

RM	EtMgBr	BnMgBr	PhLi
% anti	77	>99	>98

(c) Ph–CH=N–N(Me)–CH(Me)–CH(OH)–Ph —(1) RM; 2) H_2/Pd)→ Ph–CH(R)–NH2 + (−)-ephedrine

RM	MeMgI	EtMgI	iPrMgBr	BnMgCl	PhLi
% ee	>96	>96	76	44	91

(d) Cp(OC)(Ph3P)Fe–CO–Me = Fe*COMe —(1) LDA, THF, −42 °C; 2) Et_2AlCl)→ Fe*–C(OAlEt2)=CH2 (48) —($R^1CH{=}NR^2$)→ Fe*–CO–CH2–CH(R^1)–NHR² —([O])→ β-lactam (O, N–R², R¹)

R^1	Et	iPr	Ph	Ph–CH=CH–
R^2	Bn	Pr	Pr	Pr
% ee	86	>90	>90	43

Fig. 6-29. Stereoselective additions to C = N bonds.

methyl ketones [214, 215] and also with anisaldehyde [216]. The results confirm that the stereoselective step is crystallization: the ethyl, butyl and isopropyl ketones give the (*S*)-, the propyl, pentyl and 3,3-dimethylallyl ketones give (*R*)-amino acids in >95% ee [214]. Analogues of *47* with aldehydes or other ketones as acetalizing partner were equally selective but less accessible [217]. In a patent, synthesis of (*R*)-phenylglycine in 95% ee with (*S*)-valine tert-butyl ester as chiral auxiliary was reported [218].

Grignard reagents add to −HC=N− bonds to from secondary amines. Suzuki *et al.* used (*S*)-valinol [219], its methyl ether [220] and other chiral β-aminoalcohols [221] as chiral auxiliaries (Fig. 6-29 (b)). The method is of limited applicability since the removal of the inducing group has not been solved. The same applies to procedures using a chiral chromium−arene complex [222, 223], or (*S*)-2-methoxymethyl pyrrolidine as inducing group [224]. Takahashi *et al.* solved the problem by using a hydrazine prepared from ephedrine (Fig. 6-29 (c)) [225, 226].

In Fig. 6-29 (d) an example involving, instead of a chiral amine, the rather exotic alkylating agent *48* is shown [227].

6.4 Enantioselective Grignard Cross-Coupling and Other Stereoselective Catalytic Carbon-Carbon Bond Forming Reactions

6.4.1 Grignard Cross-Coupling*

Grignard cross-coupling is the reaction of a Grignard reagent with a halogen compound in the presence of a transition metal catalyst, whereby a carbon-carbon bond is fromed with the formal elimination of magnesium halide. In practice, only vinyl (occasionally aromatic) halides and nickel(II)- or palladium(II)-phosphine complexes are eligible as substrates and catalysts, respectively. If the Grignard

* For reviews, *cf.* refs. [67 and 228].

reagent is prepared from a chiral but racemic secondary alkyl halide and the catalyst contains a chiral phosphine (P*), kinetic resolution may take place, and an optically active product is obtained. Since chiral Grignard reagents undergo fast racemization, in principle there is no limitation to the yield and optical purity of the product. Isomerization of the secondary Grignard reagent to a primary one may, however, diminish the yield of the chiral product.

In 1973, the first asymmetric Grignard cross-coupling was reported by Consiglio and Botteghi, who used [(−)-DIOP.Ni]Cl_2 as catalyst. Enantioselectivities were less than 17% [229], (*R*)-1,2-bis(diphenylphosphino)propane and (−)-NORPHOS proved to be better ligands, giving up to 40 and 67% ee respectively [230, 231], but diphosphines were later superseded by the more efficient aminophosphines. An interesting feature of the above study was that the configuration of the major product changed, when instead of bromo compounds the chloro analogues were used as substrates. Despite careful optimization, diphosphines are not competitive with aminophosphines [231].

The reaction has been thoroughly investigated by Hayashi, Kumada and their coworkers, who prepared for this very purpose a large number of chiral ferrocenylphosphines. Some of them were also valuable, as are the rhodium complexes, for enantioselective hydrogenation (*cf.* Section 2.1.1.1) [232]. They tested the ligands on the model system shown in Fig. 6-30 (a). The structures of the ligands and the enantioselectivities recorded in the model reaction [232] are shown in Fig. 6-30 (b). From these and further data reported in the original papers the following conclusions can be reached:

(i) Enantioselectivity is primarily controlled by the configuration around iron; that of the side chain has only a modifying role (compare *49* with *60*, and both with *58*, which has an opposite ferrocenyl configuration).

(ii) The presence of a second coordinating group (NR_2 or OR) is essential (compare *56* with *e. g. 49*).

(iii) Bulky substituents at nitrogen seem to diminish or even invert enantioselectivity, but comparison of ligands *51, 53, 54,* and *55* reveals that the effect of a third coordination site may be more important than the steric requirement of the *N*-substituents.

(iv) Variation of the aryl groups attached to phosphorus had no effect on enantioselectivity, except for 2-methylphenyl (33% ee), indicating that, unlike with catalytic hydrogenation, the environment around phosphorus was not decisive.

It turned out later that much simpler compounds bearing the essential features of the best ferrocenyl type catalysts, but which form a five-membered chelate ring with the metal, are even better ligands. Eight different β-dimethylaminophosphines and a pyrrolidine analogue were prepared and tested in the model system by Hayashi *et al.* [233, 234]; *59* was the most selective.*

* *S* ligands gave *S* products in excess.

(a) Ph(Me)CHMgCl + $CH_2{=}CHBr$ —[P*₂Ni]Cl₂, 0°C→ Ph(Me)C(H)–CH=CH₂

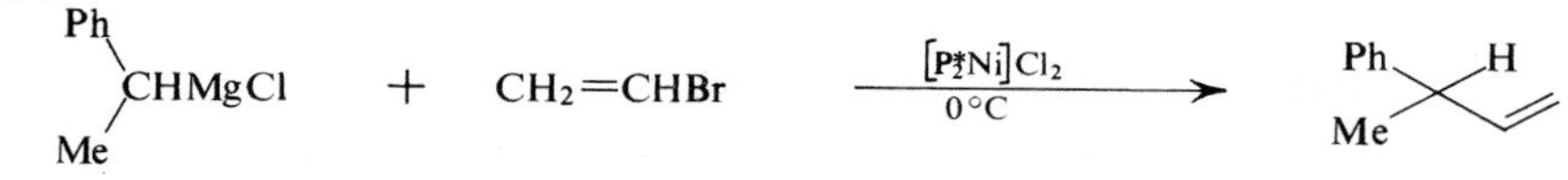

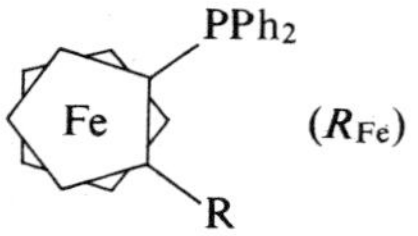

(b) (R_Fe) ferrocenyl ligand with PPh₂ and R

P*	R	% ee (conf.)	
49	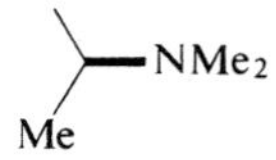 CH(Me)NMe₂	66	(*R*)
50	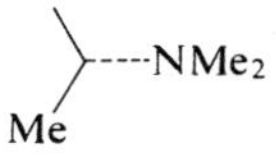 CH(Me)----NMe₂	54	(*R*)
51	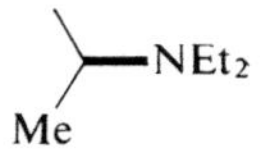 CH(Me)NEt₂	35	(*R*)
52	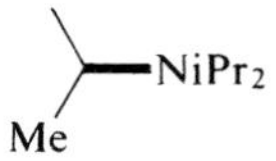 CH(Me)NiPr₂	7	(*S*)
53	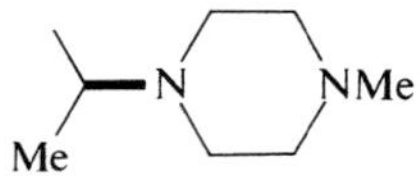 CH(Me)N(piperazine)NMe	65	(*R*)
54	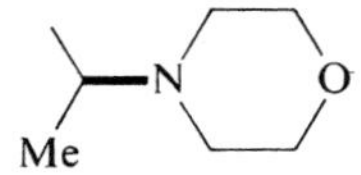 CH(Me)N(morpholine)O	17	(*R*)
55	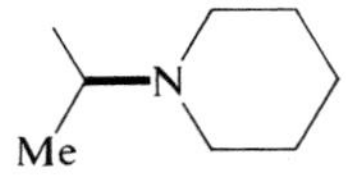 CH(Me)N(piperidine)	42	(*R*)
56	—Et	5	(*S*)
57	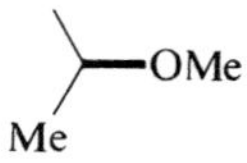 CH(Me)OMe	57	(*R*)

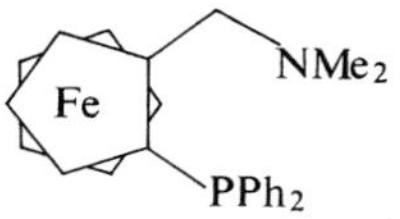

(*58*) (S_{Fe}) 65% (*S*)

(*59*)

R	iPr	tBu	Bn	Ph
% ee	81	94	71	70

Fig. 6-30. Enantioselective Grignard cross-coupling (I).

A tentative model of the process drawn up for an *S* ligand, which is in accordance with the above findings but not yet supported by direct evidence, is shown in Fig. 6-31 (a). It is assumed that one of the enantiomers of the Grignard reagent is selectively coordinated and retains its configurational integrity in the course of the following two steps. As already mentioned, the stock of the kinetically favored enantiomer is rapidly replenished by racemization of the unreacted reagent.

The model reaction has direct practical relevance, since on oxidation 2-arylpropionic acids are obtained which are important antiinflammatory agents. Thus the 4-isobutylphenyl analogue, the precursor for the well known drug Ibuprofen could be prepared in 80% ee [234].

R	H	Me	Ph
% ee	95	85	95

Fig. 6-31. Enantioselective Grignard cross-coupling (II).

Substrate oriented studies on Grignard cross-coupling have been scarce and incomplete as yet. Early investigations by Consiglio *et al.* also involved aryl halides [229, 230], but selectivities with such substrates were insufficient, too. In the presence of the nickel or palladium complexes of PPFA (*49*), the Grignard compounds RMeCHMgCl (R=Ph, Hex and Et) were reacted with vinyl bromides [CH_2=CH−Br, PhCH=CHBr, and CH_2=C(Me)Br] and bromobenzene to give products with 12−52% ee [232]. Useful selectivities could be realized when vinyl bromide was coupled with 1-arylethylmagnesium chlorides in the presence of the aminophosphine *59* (R = iPr) (Ph 81%, Tol 83%, 2-Naph 72% ee). In contrast, enantioselectivity with 2-octylmagnesium chloride was negligible (6%) [234].

Hayashi *et al.* extended the scope of Grignard cross-coupling to α-silylated Grignard reagents [235]. No *E*−*Z* isomerization takes place during coupling, and the products, optically active silanes, can be transformed stereoselectively to useful intermediates by Lewis acid catalyzed S_E1-type electrophilic substitution (Fig. 6-31 (b)). Enantioselectivity with (*Z*)-vinyl bromides was poor (<24%).

Alkylzinc halides are also amenable to the asymmetric cross-coupling reaction, but no special advantage is gained by this modification [236].

6.4.2 Stereoselective Codimerization of Olefins*

It had been recognized by Wilke and his coworkers in 1963 that phosphines were excellent cocatalysts in the codimerization of olefins catalyzed by π-allylnickel halides and aluminum halides. It was a logical development that this reaction should be tried out with chiral phosphines. Indeed, codimerization of 2-butene and propene in the presence of a chiral phosphine gave a product with detectable optical activity, and this discovery was claimed to have been made in 1967 [237], one year before the first results on enantioselective hydrogenation with rhodium-chiral phosphine complexes were made public.

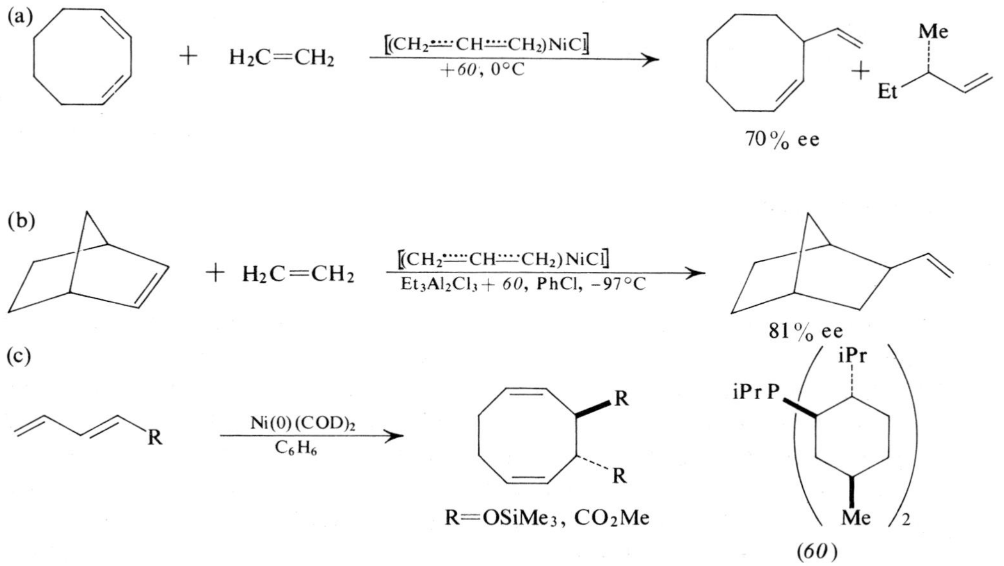

(a) + $H_2C{=}CH_2$ $\xrightarrow[+60,\ 0°C]{[(CH_2\cdots CH\cdots CH_2)NiCl]}$ + Me, Et

70% ee

(b) + $H_2C{=}CH_2$ $\xrightarrow[Et_3Al_2Cl_3 + 60,\ PhCl,\ -97°C]{[(CH_2\cdots CH\cdots CH_2)NiCl]}$

81% ee

(c) R $\xrightarrow[C_6H_6]{Ni(0)(COD)_2}$ R, R

R=$OSiMe_3$, CO_2Me

iPr P, iPr, Me, 2

(*60*)

Fig. 6-32. Stereoselective codimerization of olefins.

* For a review, see ref. [237].

The codimerization of cyclooctadiene and ethene (Fig. 6-32 (a)) was studied in detail, and by a systematic variation of the phosphine component, of the Ni/P ratio, and of other parameters it was possible to raise the optical purity of the products to a useful level [238]. The maximum ee of the by-product, 3-methylpentene, was 64%, but under different conditions and with another catalyst [237]. Optical purity of 3-vinylcyclooctene increased on lowering the temperature, and the same was observed in the codimerization of norbornene and ethene (Fig. 6-32 (b)). With norbornadiene the analogous product was obtained in 77% ee [238].

Despite this encouraging start, further interest in the method seems to be very small. Buono *et al.* carried out the codimerization of cyclohexadiene and ethene using a somewhat different nickel-chiral phosphine catalyst system and obtained (*S*)-3-vinylcyclohexene in 73% ee [239].

Recently, the diastereoselective dimerization of butadiene derivatives was described (Fig. 6-32 (c)) [240].

6.5 Miscellaneous Stereoselective Carbon-Carbon Bond Forming Reactions

This section is a collection of stereoselective carbon-carbon bond forming reactions which did not fit into any of the types discussed in other sections of Chapters 5 and 6.

In the course of their pioneering investigations on polyene cyclizations (*cf.* Section 7.3), Johnson and his group exploited stereodirection by chiral 2,4-pentanediol in its acetals. Later it was shown that such acetals can be attacked very selectively at one of their diastereotopic C–O bonds by silylacetylenes [241], trimethylsilyl cyanide [242, 243], allyltrimethylsilanes [244] and Grignard reagents [245]. Removal of the chiral auxiliary is performed by oxidation to a ketolether, followed by base or acid catalyzed elimination (Fig. 6-33).

The following reactions concern the synthesis of chiral binaphthyls. Miyano *et al.* found that chirality of 2,2′-binaphthyl was completely transferred to the product in the Ullmann coupling (Fig. 6-34 (a)) [246, 247]. With other carboxylic acid components selectivity was less [248].

Wilson and Cram provided the first example for asymmetric induction by a chiral leaving group in the reaction shown in Fig. 6-34 (b) [249]. Meyers and Lutowski carried out the same coupling with a substrate in which the leaving group was

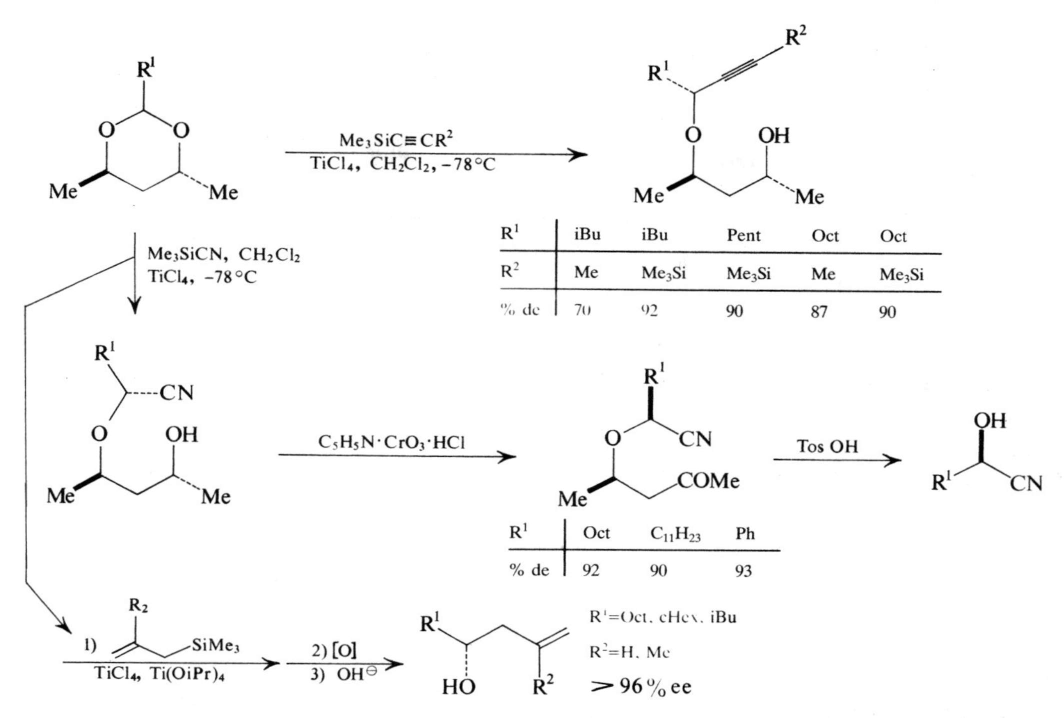

R^1	iBu	iBu	Pent	Oct	Oct
R^2	Me	Me_3Si	Me_3Si	Me	Me_3Si
% de	70	92	90	87	90

R^1	Oct	$C_{11}H_{23}$	Ph
% de	92	90	93

Fig. 6-33. Stereoselective ring opening of chiral acetals.

achiral but the oxazoline moiety was chiral. Stereoselectivity was somewhat lower [250].

Complete transfer of chirality was observed in an alkyl migration resembling the pinacol rearrangement by Suzuki *et al.* (Fig. 6-35 (a)) [251, 252]. If the groups at the quaternary center are methyl and substituted vinyl, it is the latter which migrates (ee 95–99%) [253]. A rhodium catalyzed cyclization of a chiral α-

(a)

O-OC Br Br O-OC

1) Cu, Δ, DMF
2) $LiAlH_4$, 36%

HO HO

(*S*) (*S*) 100% ee

Fig. 6-34. Stereoselective Ullmann coupling.

Fig. 6-34. Stereoselective Ullmann coupling. (contd.)

diazoester, in which the nearest chiral center of the auxiliary alcohol was three and seven bonds away from the reacting center, was described by Taber and Raman (Fig. 6-35 (b)) [189].

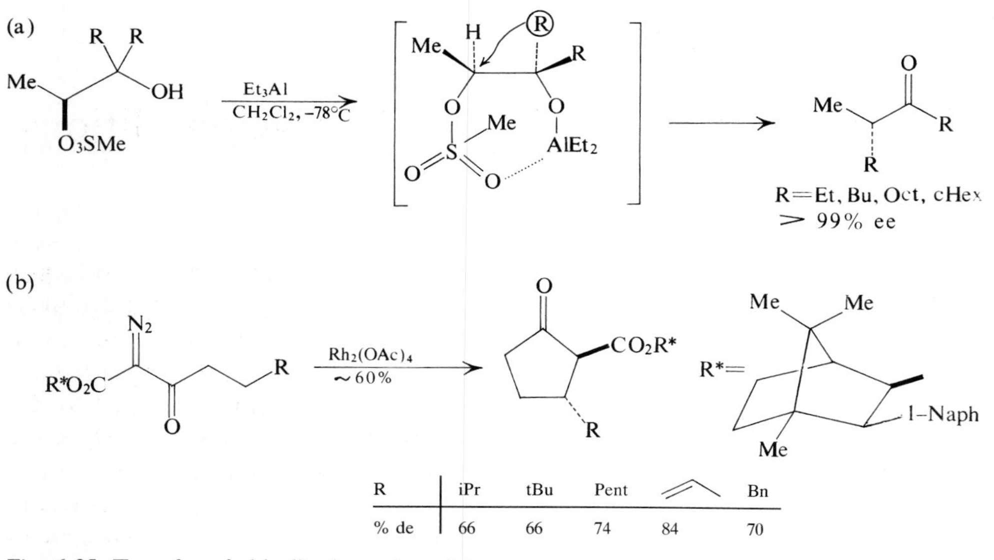

Fig. 6-35. Transfer of chirality in a pinacol-type rearrangement.

7 Stereoselective Carbon-Carbon Bond Formation by Pericyclic Reactions

According to Woodward and Hoffmann [1] pericyclic "is a reaction in which all first order changes in bonding relationships take place in concert on a closed curve". Because of the constraints inherent to cyclic transition states, pericyclic reactions are good candidates for stereoselective transformations. In fact most of the pericyclic reactions were found to be highly or even totally diastereoselective. Stereoselective pericyclic reactions are adequately covered by reviews [*e. g.* 1, 2]. From this vast area two topics were selected for more detailed discussion in this book, namely, asymmetric induction in cycloadditions and transfer of chirality in sigmatropic rearrangements.

7.1 Asymmetric Induction in Cycloadditions

Cycloaddition is a process in which two or more reactants combine to form a stable cyclic molecule, during which no small fragments are eliminated and σ bonds are formed but not broken [3]. The Diels-Alder reaction is a typical cycloaddition, while cyclizations involving a single molecule (*e. g.* electrocyclic reactions) are excluded from the definition. Cycloadditions are conveniently classified by the number of ring atoms contributed by each component to the new ring [3], or the number of electrons (usually π electrons) involved in each component [1]. The two notations coincide in the case of the Diels-Alder reaction ($4+2$ in both cases) but differ, for example, with 1,3-dipolar cycloadditions, in which the dipolar component contains four π electrons over a chain of three atoms. The number of electrons involved may be indexed according to their character (σ, π or n) and the mode of approach of the components (supra- or antarafacial, s, or a) [1]. For example, the Diels-Alder reaction is a $\pi_s^4 + \pi_s^2$ cycloaddition.

In the following sections asymmetric induction in the Diels-Alder reaction, stereoselective 1,3-dipolar cycloadditions and carbene additions will be described.

7.1.1 Asymmetric Induction in the Diels-Alder Reaction*

The Diels-Alder cycloaddition was perhaps the first reaction in which complete stereoselectivity was recognized. In Diels-Alder type cycloadditions stereoselectivity is manifested at three levels and is often combined with problems of regioselectivity.

(1) The reaction is a concerted process and therefore its basic stereochemistry is dictated by the Woodward-Hoffmann rules: the orbitals of both partners interact in a suprafacial mode, resulting in exclusive *syn* addition.**

In other words, the initial diastereomeric configuration of the double bonds of the partners is retained in the product. Thus, addition of an (*E*)-olefin to a diene gives rise to a *trans* disubstituted product exclusively (*cf.* Fig. 7-1 (a)).

(2) Two different relative orientations of the approaching reactants can be envisaged, and it was found that the one leading to the *endo* product was generally, though not exclusively, preferred (Fig. 7-1 (b)). This is the s. c. *endo* rule. The explanation for a preference of the usually more hindered *endo* product is based on secondary orbital interactions [2, 3].

(3) A further complicating factor is regioselectivity, which arises when both components are asymmetrically substituted (Fig. 7-1 (c)). The Diels-Alder reaction is often remarkably selective even in this respect. A rationalization of this regioselectivity invoking orbital interactions was proposed by Houk [6].

(4) Asymmetrically substituted dienes and dienophiles have enantiotopic faces. Therefore, if one of the partners has enantiotopic faces, even in the case of exclusive *cis-endo* addition and total regioselectivity, the product occurs as a pair of enantiomers. Thus, in the reaction of acrylates with cyclopentadiene (Fig. 7-1 (d)), the ester may react on either its *Re* or its *Si* face yielding the (2*R*)- and (2*S*)-esters, respectively. Provided that the *endo* rule is operative and addition is regioselective, no additional stereoisomers emerge when both partners are asymmetrically substituted.

From the two theoretically possible ways to obtain one of the enantiomers in excess, *i. e.* chiral catalysis and attaching a chiral auxiliary to one of the components, only the latter has met with real success so far.

The first, asymmetric Diels-Alder reaction was reported in 1948 by Korolev and Mur [7], who reacted di-(−)-menthyl fumarate with butadiene, but the observed enantioselectivity was infinitesimal. However, in 1963 Walborsky *et al.* [8] discovered that enantioselectivity increased dramatically when the reaction was cataly-

---

* For a recent review see ref. [4].

** For the characterization of the stereochemistry of addition the terms *syn* and *anti* were recommended by IUPAC Rules for Organic Chemical Transformations [5].

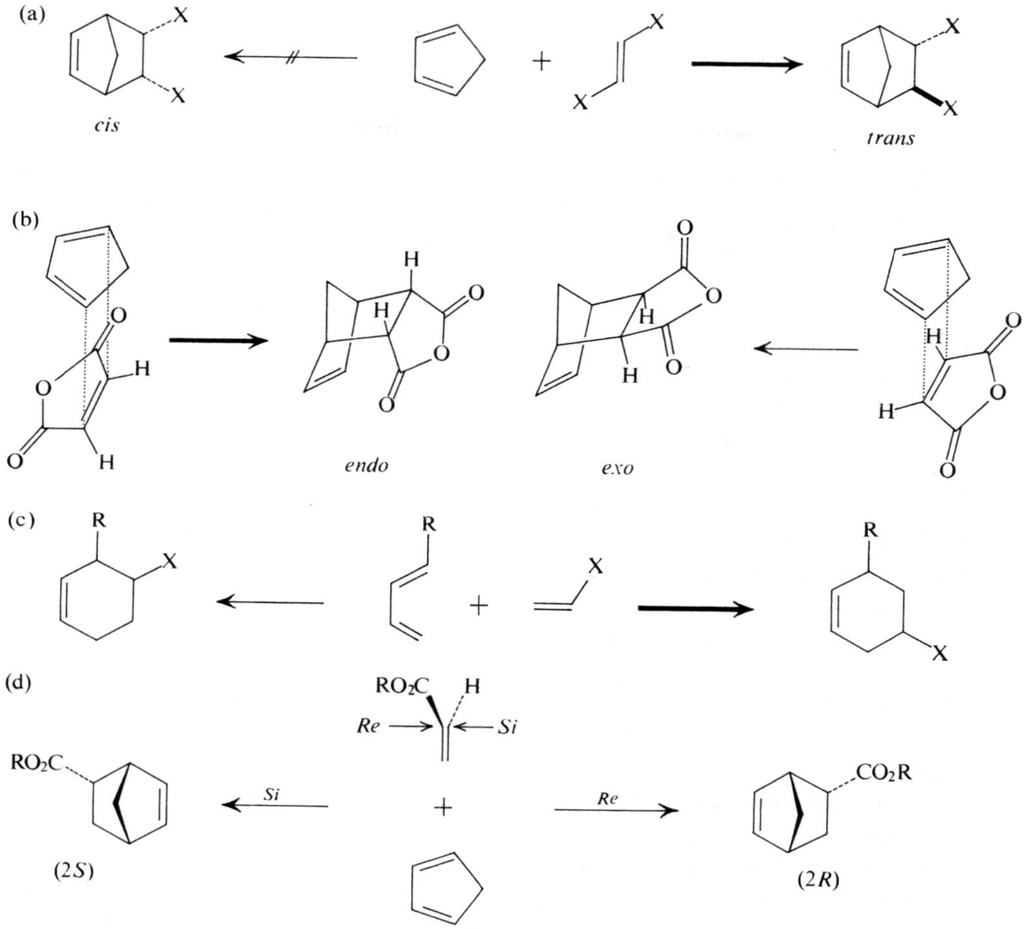

Fig. 7-1. Stereo- and regioselectivity in the Diels-Alder reaction.

zed by a Lewis acid. $TiCl_4$ in toluene gave the best results (78% de of *R,R* product). In the presence of a Lewis acid it was possible to carry out the reaction at an acceptable rate at room temperature and even at −70°C.

Addition of chiral acrylates to cyclopentadiene became a popular model for the exploration of asymmetric induction in the Diels-Alder reaction, although incomplete *endo* selectivity made the evaluation of the results more difficult. Product analysis was usually performed on the mixture of diols obtained after removal of the chiral auxiliary group by $LiAlH_4$ reduction. The high enantioselectivities (88% ee with 3,3-dimethyl-2-butanol as the chiral alcohol [9]) achieved by this method were almost unprecedented at that time, but faded in the light of later discoveries. Corey and Ensley reported virtually complete (99% de) stereoselectivity when (−)-8-phenylmenthol (*1*), prepared from pulegone, was used as a chiral

Table 7-1. Stereoselectivity in the Diels-Alder Reaction of Cyclopentadiene with Chiral Acrylates (CH_2=CH–CO_2R^*).

R*	Catalyst	Temp. °C	% *endo*[a]	% ee of *endo* (Conf.)	% ee of *exo* (Conf.)	Ref.
(−)-menthyl	none	35	69	7 (2*R*)	3 (2*S*)	16
	$AlCl_3 \cdot Et_2O$	0		49 (2*R*)	36 (2*R*)	16
	$AlCl_3 \cdot Et_2O$	−70		67 (2*R*)		16
	$BF_3 \cdot Et_2O$	−70		85 (2*R*)		9
(*S*)-2-octyl	$BF_3 \cdot Et_2O$	0		23 (2*S*)	16 (2*S*)	9
(*S*)-lactoyl	$TiCl_4$	−64	97	86 (2*S*)		17
(*S*)-3,3-dimethyl-2-butyl	$BF_3 \cdot Et_2O$	−70		88 (2*S*)		9
(*S*)-2-butyl	$SnCl_4$	4−8		24 (2*S*)		18
(−)-8-phenylmenthyl	$TiCl_4$	−20		90 (2*R*)		12
2	$TiCl_4$	−20		63 (2*R*)		12
3	$TiCl_4$	−20		85 (2*R*)		12
4	$TiCl_4$	−20		84 (2*S*)		12
5	$TiCl_4$	−20		74 (2*R*)		12
6	$TiCl_4$	−20		81 (2*R*)		12
7	$TiCl_2(OiPr)_2$	0	86	64 (2*S*)		15
8	$TiCl_2(OiPr)_2$	−20	96	97 (2*S*)		15
9	$TiCl_2(OiPr)_2$	−20	95	91 (2*R*)		15
10	$TiCl_2(OiPr)_2$	−20	96	99 (2*S*)		19
11	$TiCl_2(OiPr)_2$	−20	97	88 (2*R*)		14
12	$TiCl_2(OiPr)_2$	−20	96	93 (2*R*)		20
15	Et_2AlCl	−100	>99	86 (2*R*)		21
16	Et_2AlCl	−100	>99	90 (2*R*)		21
17	Et_2AlCl	−100	>99	90 (2*S*)		21

[a] Normalized to %*endo* + %*exo* = 100.

aid [10, 11].* By conversion of natural (*R*)-pulegone to (*S*)-pulegone [13] both enantiomers of *1* became available. Oppolzer and his coworkers undertook a very extensive study of the reaction to find, on the one hand, the best experimental conditions, and, on the other, chiral alcohols which were both highly stereodirecting and easily available [12, 14, 15]. In Table 7-1 selected data for the Diels-Alder reaction of chiral acrylates and cyclopentadiene are presented, while the less well known alcohol components are depicted in Fig. 7-2.

It is apparent from Table 7-1 that among the monooxygenated alcohols (*1*−*6*) 8-phenylmenthol (*1*) provides highest selectivity. Unfortunately, this substance is an oil, which is not readily available and has to be purified by HPLC. The ether-alcohol type auxiliaries *6*−*12* can be prepared from inexpensive (+)-camphor and mostly offer excellent selectivities, but they decompose on exposure to $TiCl_4$.

* Only 90% ee was found later by Oppolzer *et al.* [12].

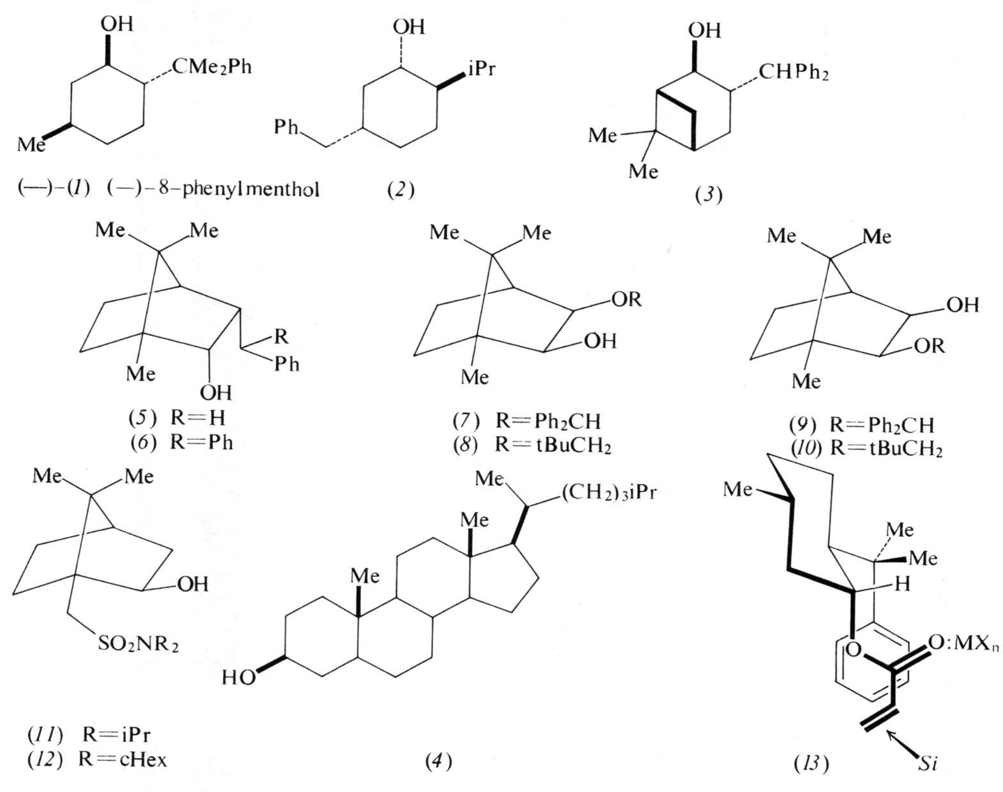

Fig. 7-2. Chiral alcohols used as auxiliary groups in enantioselective Diels-Alder synthesis.

This difficulty was circumvented by applying, as catalyst, a mild Lewis acid, $TiCl_2(OiPr)_2$, at low temperature. Homologues of *8* and *10* were less effective [22].

A plausible model, still not supported by direct experimental evidence, which may explain the stereoselectivity of 8-phenylmenthol and its congeners is *13*, in which the dienophile is in the *s-trans* conformation. For simple acrylates this has been shown to be the most stable arrangement. The fact that enantioselectivity improved on lowering the temperature supported the proposition that selectivity was controlled by a conformational equilibrium involving relatively small energy differences. Although this model, as well as analogous ones constructed for other substrates, was based on the assailable assumption that the preferred ground state conformation was retained in the transition state, it not only predicted the predominant configuration surprisingly well, but was also useful for the design of more selective chiral auxiliaries. Thus it could be anticipated that chiral alcohols which provided for a parallel positioning of the acrylate moiety and had a bulky shielding group would be hopeful candidates as highly selective auxiliaries. The bornane derivatives *5*–*12* fulfilled this requirement particularly well, and those in

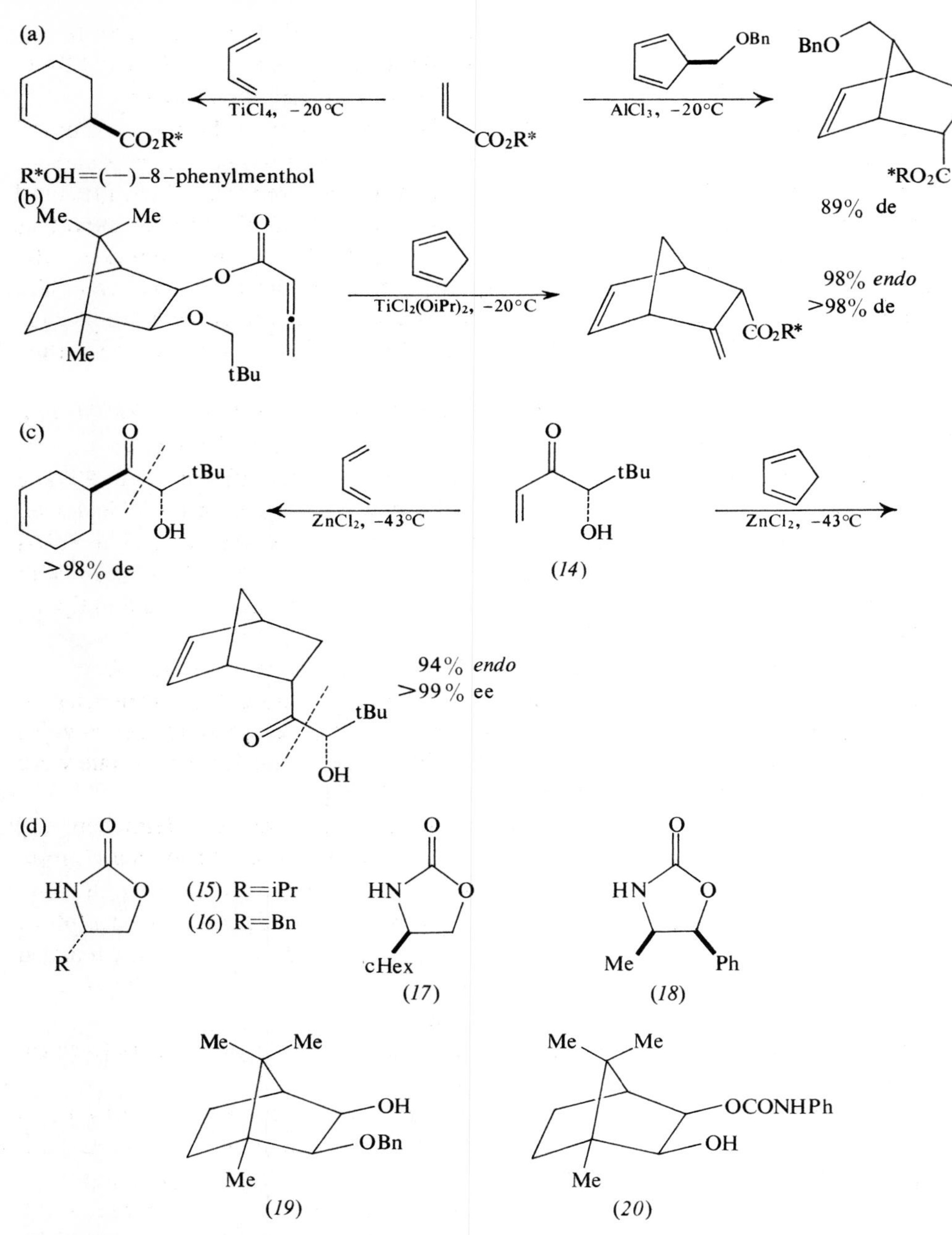

Fig. 7-3. Examples for the use of chiral auxiliary groups in the Diels-Alder reaction.

which an angular methyl group adjacent to the ester function enhanced constraints in the transition state (*9* and *10*) were especially selective. The pairs *7*–*8* and *9*–*10* are quasi-enantiomeric and therefore give rise to enantiomeric products.

Addition of chiral acrylates to cyclopentadiene is just one of the many examples in which the Diels-Alder reaction was exploited to produce chiral compounds of high optical purity (Fig. 7-3). (−)-8-Phenylmenthyl acrylate was, in fact, first used as a partner to be reacted with 5-benzyloxycyclopentadiene; the product served as a chiral starting material for a prostaglandin synthesis [10, 11].* With butadiene the same acrylate gave rise to a cyclohexene used for the synthesis of sarcomycin [23] (Fig. 7-3 (a)). The neopentyl compound *10* proved to be a powerful directing group when it was linked with allene-carboxylic acid (Fig. 7-3 (b)). The product was then transformed to (−)-β-santalene [24].

In chiral acrylates the directing chiral moiety is rather remote from the reaction center. Indeed, the enone *14* and its congeners, in which this distance is smaller, are very selective dienophiles which produce, with cyclopentadiene, up to 90% ee in an uncatalyzed addition at −20°C [25]. Addition of zinc chloride increases stereoselectivity beyond the detection limit of the minor component. The chiral inducing group can be removed oxidatively (Fig. 7-3 (c)). Reaction of *14* with butadiene is also highly selective and gives, after oxidative cleavage, the antipode of the product in example (a) [26].

Recently, *N*-acrylates and crotylates of chiral oxazolines, prepared from (*S*)-valine (*15*), (*S*)-phenylalane (*16*) and (1*S*,2*R*)-norephedrine (*18*) were used as chiral dienophiles (see Table 7-1 for the acrylates, for the crotylates selectivities were slightly better). Diastereoselectivities with isoprene and 1,4-pentadiene were also high when *16* was used as chiral auxiliary [21, 27].

Analysis of the molecular geometry of chiral auxiliaries led Helmchen and Schmierer to the conclusion that for effective stereodirection the inducing group should form a concave face around the reacting functional group [28]. The monosubstituted diols *14* and *20* (Fig. 7-3 (d)), both prepared from (+)-camphor, are examples of such convex directing groups, and in fact the $AlCl_3$-catalyzed Diels-Alder addition of the corresponding methyl fumarates to anthracene at −30°C proceeded with complete stereoselectivity [29].

The uncatalyzed hetero Diels-Alder reaction of 2,3-dimethylbutadiene to chiral sulfines produced a single diastereomer [30].

Very few studies have been reported in which a chiral diene was reacted with an achiral dienophile. Addition of butylglyoxylate to enolether-type dienes with sugars as the inducing group exhibited high diastereoface selectivity but poor *endo*-*exo* selectivity [29]. Selectivity of the enolester *21* was modest towards acrolein (de 60%) [31] but excellent when the dienophile was a naphthoquinone (Fig. 7-1 (a)) [32].

---

* Note that the diene component has diastereotopic faces, one of which (*ul* approach) is highly preferred.

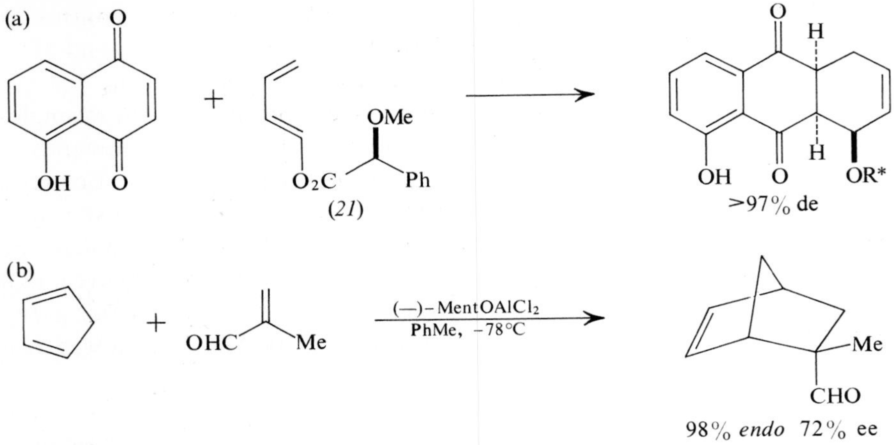

(a) OH O O + OMe O2C Ph (21) → O H H OH O OR* >97% de

(b) + OHC Me —(–)-MentOAlCl2, PhMe, –78°C→ Me CHO 98% *endo* 72% ee

Fig. 7-4. Examples for the Diels-Alder addition involving a chiral diene and a chiral catalyst.

Diels-Alder reaction under asymmetric catalysis was pioneered by Guseinov *et al.* [33]. Selectivities are generally poor, except for the example shown in Fig. 7-4(b) [34].

In the intramolecular version of chirally biassed Diels-Alder cycloaddition the reacting groups are linked by a chain of at least three atoms (*e. g.* Fig. 7-5 (a)) [35, 36]. Here and in example (b) [37] constraints in the substrate only permit an *exo* approach, which involves less favorable orbital interactions and leads to a *trans* anellated product which in the case of (a) is rather strained. When it is sterically

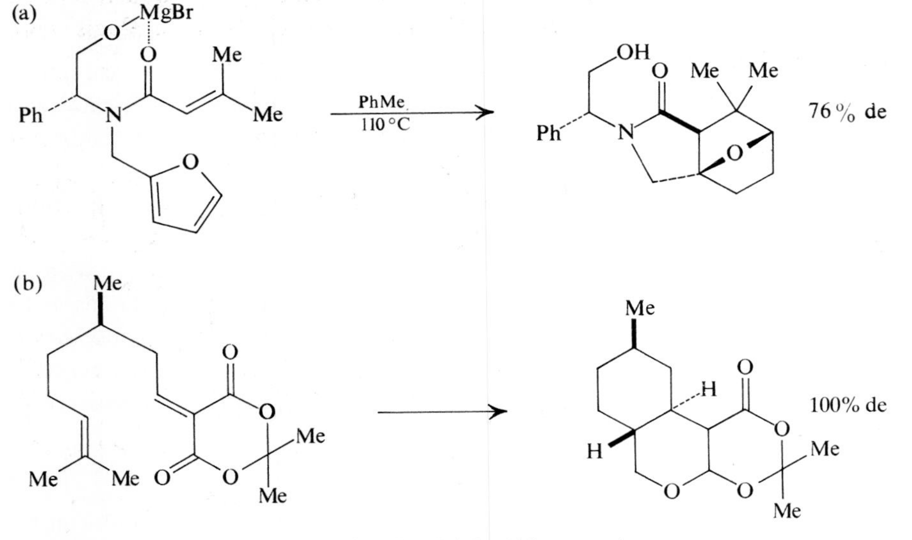

(a) MgBr O O Me Ph N Me O —PhMe, 110 °C→ OH O Me Me Ph N O 76% de

(b) Me O O Me Me O O Me → Me H O O Me H O O Me 100% de

Fig. 7-5. Intramolecular stereoselective Diels-Alder reactions.

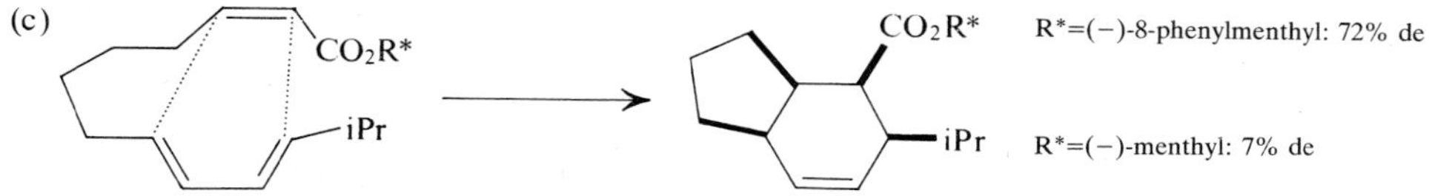
(c)
CO2R*
iPr
CO2R*
iPr
R*=(−)-8-phenylmenthyl: 72% de
R*=(−)-menthyl: 7% de

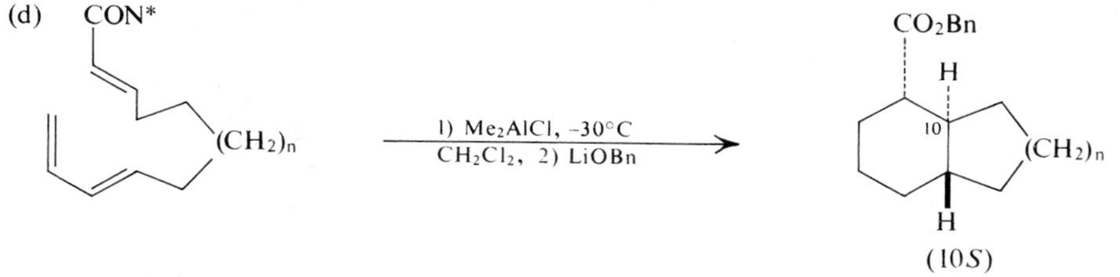
(d)
CON*
(CH2)n
1) Me2AlCl, -30°C
CH2Cl2, 2) LiOBn
CO2Bn
H
10
(CH2)n
H
(10S)

N*	% ee n=0	n=1	conf.
15	66	84	10*S*
16	90	94	10*S*
17	94	88	10*R*
18	70	82	10*R*

Fig. 7-5. Intramolecular stereoselective Diels-Alder reactions. (contd.)

possible, the *endo* rule prevails, as in example (c), in which a removable chiral auxiliary group also imparts enantioselectivity to the addition [130]. As shown by Evans *et al.*, enantioselectivity is even higher when amides prepared from *15–18* are cyclized (Fig. 7-5 (d)) [38].

o-Quinodimethanes generated from benzocyclobutanes or 1,3-dihydroisothionaphthene-2,2-dioxides are very useful dienes for the stereoselective construction of complex polycyclic molecules. The general scheme [39, 40] and two of its applications are shown in Figs. 7-6 (a), (b) and (c) [41]. Both products were converted to (+)-estradiol.

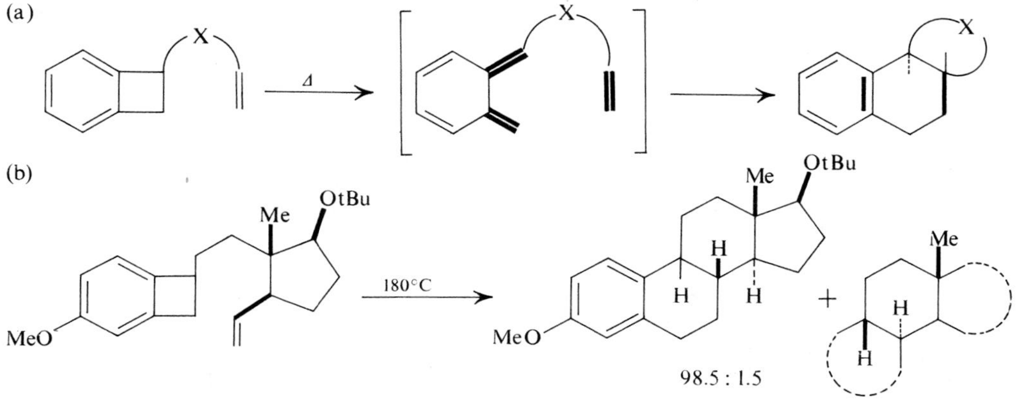
(a)
X
Δ
X
X
(b)
Me
OtBu
MeO
180°C
Me
OtBu
H
H
H
MeO
+
Me
H
H
98.5 : 1.5

Fig. 7-6. Intramolecular stereoselective Diels-Alder reactions *via o*-chinodimethanes.

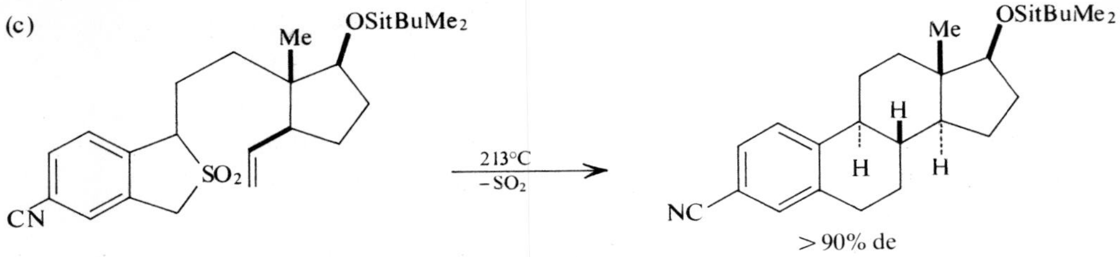

(c)

$>90\%$ de

Fig. 7-6. Intramolecular stereoselective Diels-Alder reactions *via* *o*-chinodimethanes. (contd.)

7.1.2 Asymmetric Induction in [3 + 2] Cycloadditions

[3 + 2] cycloadditions are mainly represented by 1,3-dipolar cycloadditions. These involve, as partners, an olefin and a linear system of three atoms over which four π electrons are distributed. In the Woodward-Hoffmann notation this is a $\pi_s^4 + \pi_s^2$ cycloaddition, which is allowed in the ground state in the suprafacial-suprafacial mode. The stereochemistry of this reaction has not been so exhaustively studied as that of the Diels-Alder reaction, and examples for asymmetric induction are rare.

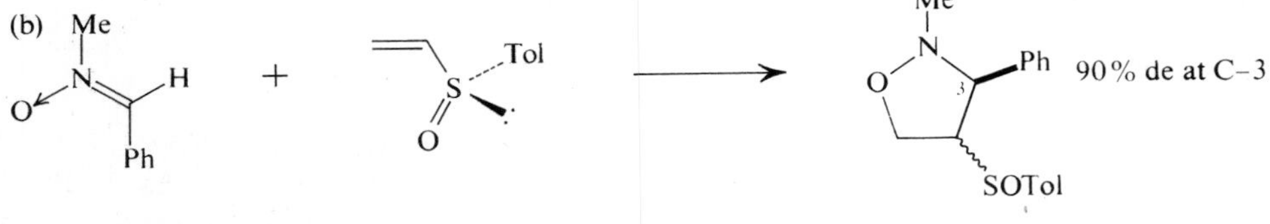

(a)

(3*R*,5*S*)–*cis* (38%) 100% de

(3*R*,5*R*)–*trans* (62%) 100% de

(b)

90% de at C-3

Fig. 7-7. Asymmetric induction in [3 + 2] cycloadditions.

In the addition of olefins to nitrones either partner may be chiral. In example (a) in Fig. 7-7 the nitrone [42], while in example (b) the olefin is chiral [43]. In example (a) total *Re*/*Si* discrimination is accompanied by veryl low *endo*/*exo* selectivity. Diastereoselectivity with some nitrones substituted at the nitrogen with carbohydrates was not outstanding either [44–46].

7.1.3 Asymmetric Induction in [2 + 2] Cycloadditions

[2 + 2] cycloadditions can proceed in a concerted manner under photochemical activation or in a non-concerted way by a stepwise ionic mechanism [2]. Both modes are subject to asymmetric induction.

A classic of $[\pi_s^2 + \pi_s^2]$ cycloadditions, the dimerization of cinnamic esters to stereoisomeric diphenylcyclobutane-dicarboxylic esters (truxinates), has also been realized in an enantioselective and regioselective way, by photolysis of the threitol derivative *22* (Fig. 7-8 (a)). Again, enantioselectivity was higher than diastereoselectivity [46]. The complementary reaction, *i. e.* the cycloaddition of stilbene to a series of chiral fumarates again to yield truxinates, was also studied [47], and the best result is shown in Fig. 7-8 (b). Di-(−)-bornyl fumarate only gave δ-truxinate in but 20% ee, suggesting that in the selectivity-determining step (probably the recombination of a 1,4-biradical intermediate) only one of the ester groups is involved.

Photoaddition of (−)-menthyl phenylglyoxylate to 2,2-dimethyl-2-butene [48] or furan [49] to form oxetanes proceeded with low stereoselectivity, while the esters with 8-phenylmenthol and the neopentylether *10* both ensured high stereoselectivity (Fig. 7-8 (c)) [50].

[2 + 2] reactions proceeding by an ionic or other non-conterted mechanism have acquired great importance in the preparation of four-membered heterocycles, for

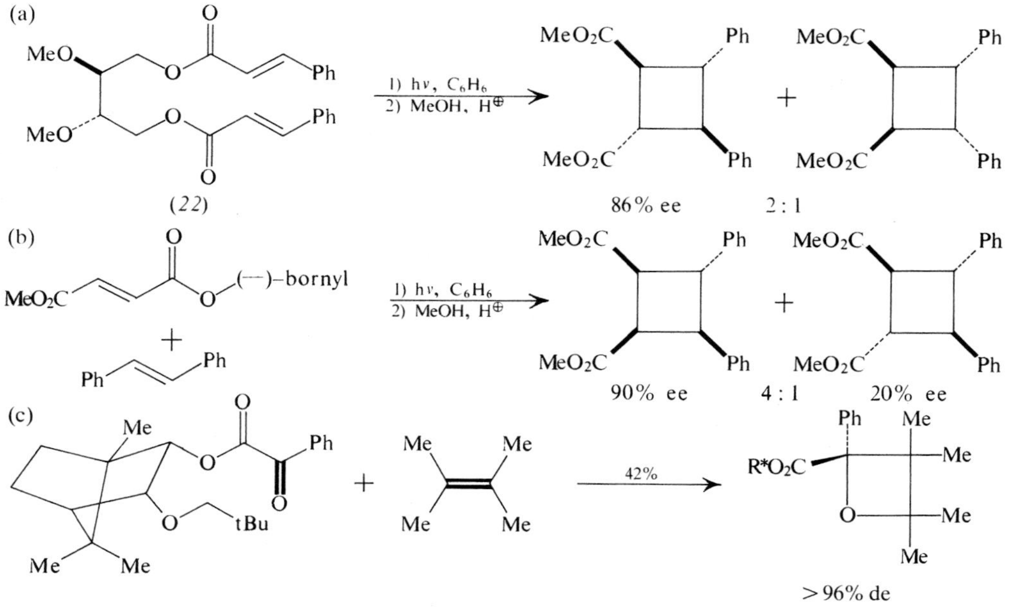

Fig. 7-8. Asymmetric induction in photoinduced [2 + 2] cycloadditions.

example, β-lactams. Owing to the prominent importance of β-lactam antibiotics, the pertinant literature is very extensive, only a few examples of which are quoted here.

Azomethines and ester enolates undergo [2 + 2] cycloaddition, and, when one of the partners is chiral, a product non-racemic in the β-lactam ring may be obtained. In example (a) in Fig. 7-9 the amine component of the azomethine [51, 52], whereas in example (b) the ester component, is chiral [53]. In the $TiCl_4$-catalyzed

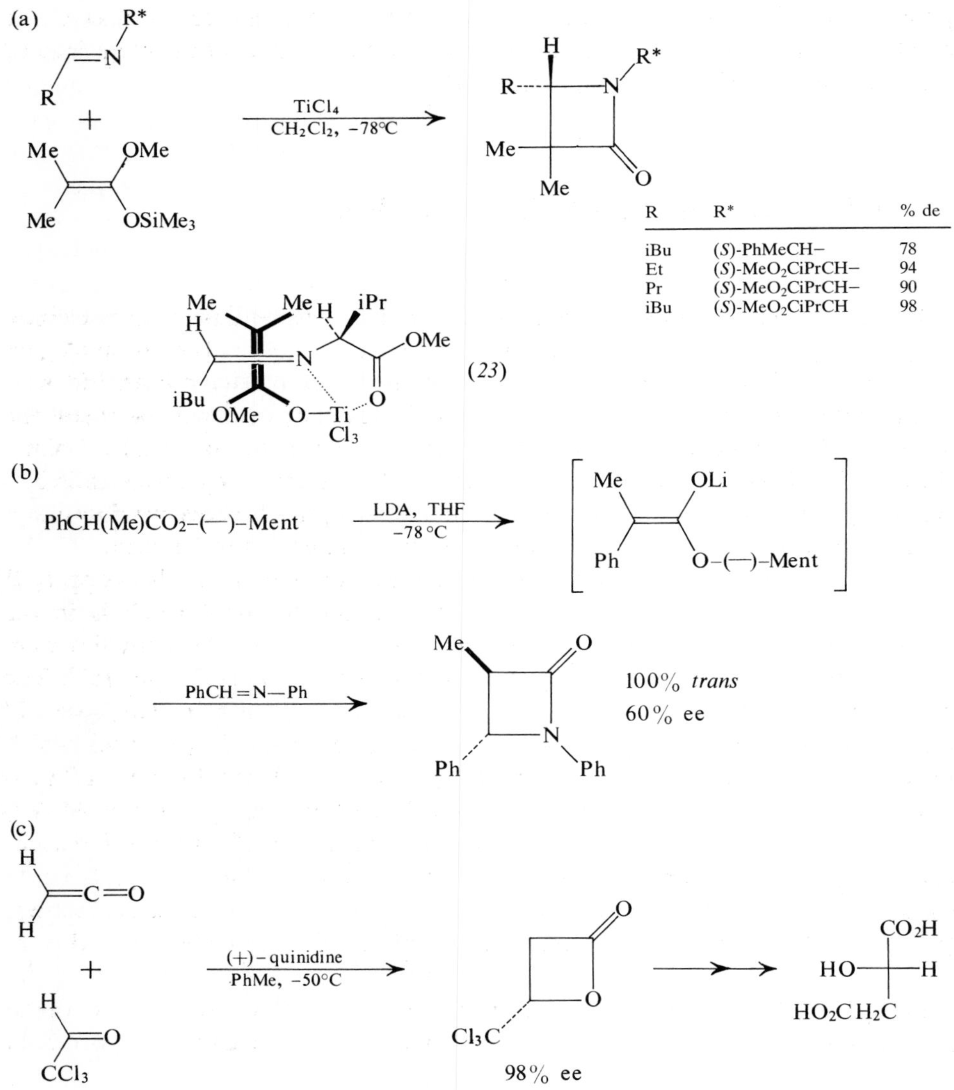

R	R*	% de
iBu	(*S*)-PhMeCH–	78
Et	(*S*)-MeO_2CiPrCH–	94
Pr	(*S*)-MeO_2CiPrCH–	90
iBu	(*S*)-MeO_2CiPrCH	98

(*23*)

100% *trans*
60% ee

98% ee

Fig. 7-9. Stereoselective formation of four-membered heterocycles by [2 + 2] cycloaddition.

reaction the favored transition state can be envisaged as in *23*. *E* and *Z* isomers of the azomethine are at equilibrium at room temperature, while at −78°C the *E* form becomes exclusive [52].

Ester enolates derived from (−)-menthyl mandelates combine with diaryl azomethines with complete or very high *trans* selectivity, but with quite poor enantioface selection. The latter is much higher with the corresponding 2-phenylpropionates (Fig. 7-9 (b)).*

[2+2] cycloaddition under chiral catalysis is exemplified in Fig. 7-9 (c). With quinine, the (*R*)-lactone was obtained in 76% ee and with (*S*)-*N,N*-dimethyl-1-phenylethylamine the (*S*)-lactone in 77% ee. By comparing a series of cinchona alkaloids, it was found that the chirality of the product in excess was determined by the configuration of the catalyst at C-8 [54].

7.1.4 Stereoselective Carbene Additions

Interest in the stereoselective additions of carbenes to olefins to form a cyclopropane ring was stimulated by a demand for chrysanthemic acid and its analogues which are building blocks of environmentally safe pyrethroid-type insecticides.

The only industrially viable method to synthesize these compounds is still the addition of a carbene generated from ethyl diazoacetate to (*Z,Z*)-2,5-dimethylhexa-2,4-dienes. This reaction is catalyzed, for example, by salycylaldimine-type copper(II) complexes, and in order to obtain an optically active product either a chiral ester, or a catalyst prepared from a chiral amine has to be used.

The use of 2-phenylethylamine as the amine component in the copper(II) complex [55] or of diazoacetates prepared with simple chiral alcohols in the presence of copper(I) chloride [56] gave discouraging results. Selectivities were much better in a study involving various amino sugars and aldehydes [57]. The optimum was attained with the Schiff base prepared from 2-pyridinealdehyde and the rather inaccessible 2-amino-4,6-benzylidene-2-deoxy-α-D-allopyranoside which provided, in the addition of ethyl diazoacetate to 1,1-dichloro-4-methyl-1,3-pentadiene, 80% of the *cis* product in 44% ee and 20% of the *trans* product in 38% ee.

In a systematic study Aratani *et al.* first investigated a series of chiral binuclear copper(II) complexes and selected *24* (prepared from (*R*)-alanine) as the best catalyst [58]. This gave appreciable enantioselectivities for both diastereomers (62−68% ee) but practically no diastereoselectivity. Other catalysts tested were more diastereoselective but less enantioselective.

In experiments using the same catalyst (*24*), it was found that the structure of the alcohol component of the diazoester played an important role in the control of both

* The absolute configuration of the product is unkown.

dia- and enantioselectivity [59]. Bulky alkoxy groups increased the predominance of the *trans* product and, out of the ten different alcohols tried out, diisopropylmethylcarbinol gave the best result. The chirality of the alcohol did not influence the absolute configuration of the predominant product. Application of the method

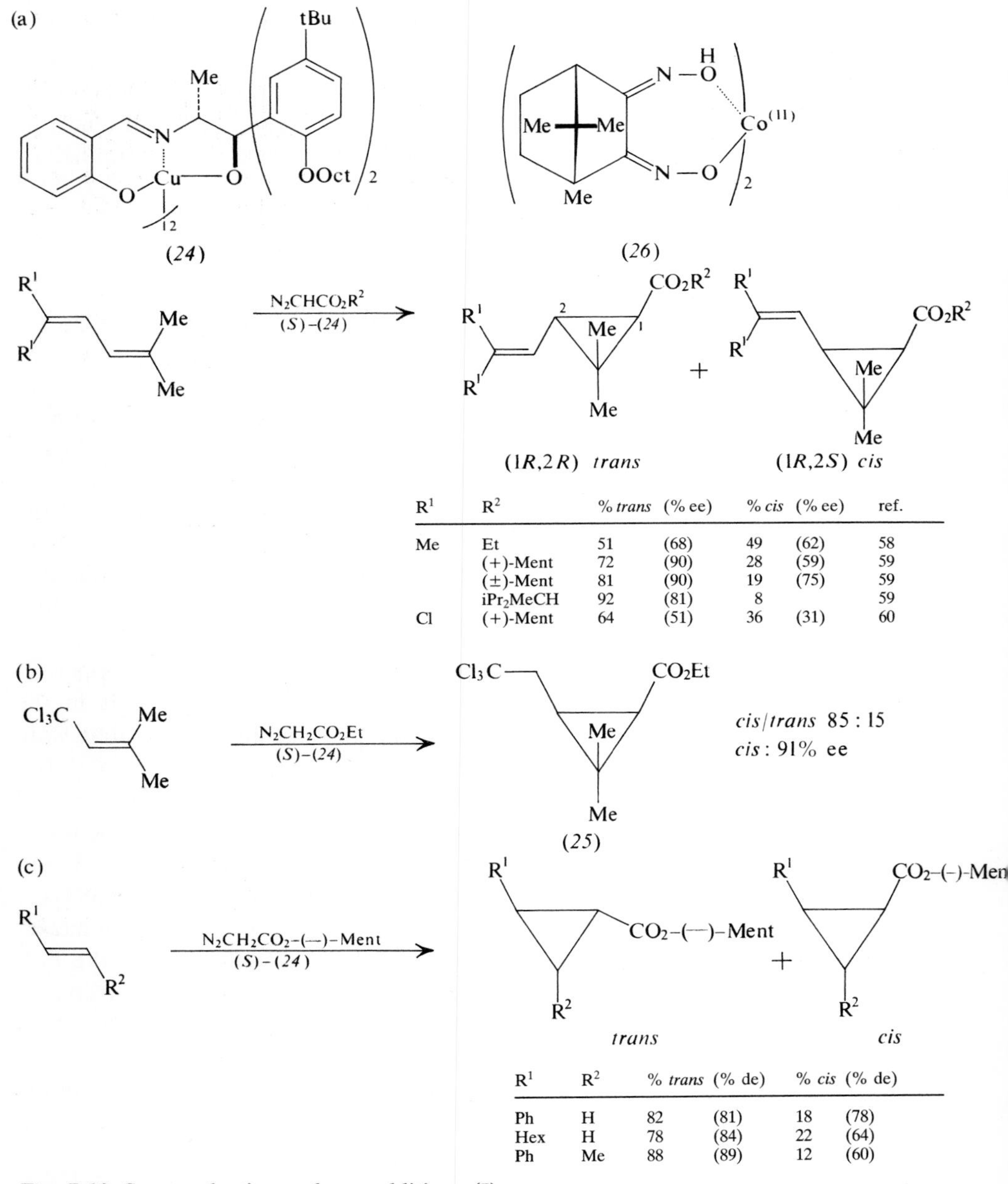

R^1	R^2	% *trans*	(% ee)	% *cis*	(% ee)	ref.
Me	Et	51	(68)	49	(62)	58
	(+)-Ment	72	(90)	28	(59)	59
	(±)-Ment	81	(90)	19	(75)	59
	iPr_2MeCH	92	(81)	8		59
Cl	(+)-Ment	64	(51)	36	(31)	60

R^1	R^2	% *trans*	(% de)	% *cis*	(% de)
Ph	H	82	(81)	18	(78)
Hex	H	78	(84)	22	(64)
Ph	Me	88	(89)	12	(60)

Fig. 7-10. Stereoselective carbene additions (I).

to the synthesis of the commercially important *cis*-permethric acid (R^1=Cl) was less successful [60]. *25*, in turn, could be obtained rather selectively (Fig. 7-10 (b)) and was smoothly converted to *cis*-permethric acid by dehydrochlorination [60].

Addition of (−)-menthyl diazoacetate to some simple olefins was also rather selective, especially when (*S*)-*24* was used as catalyst (Fig. 7-10 (c)) [60].

The cobalt complex *26* derived from (−)-camphorquinone is a more readily available chiral catalyst for the diazoacetate addition to olefins than *24*, but it gives inferior results [61−63].

Cyclopropane aldehydes of high optical purity (>90%) were prepared *via* the sequence of reactions exemplified in Fig. 7-11 (a) [64]. Note that oxazolidine formation was also highly diastereoselective. Addition of $N_2CHCH(OMe)_2$ to the same substrate, followed by photodecomposition of the isolable pyrazoline-type intermediate and hydrolysis gave the corresponding cyclopropane dialdehyde derivative in more than 90% ee.

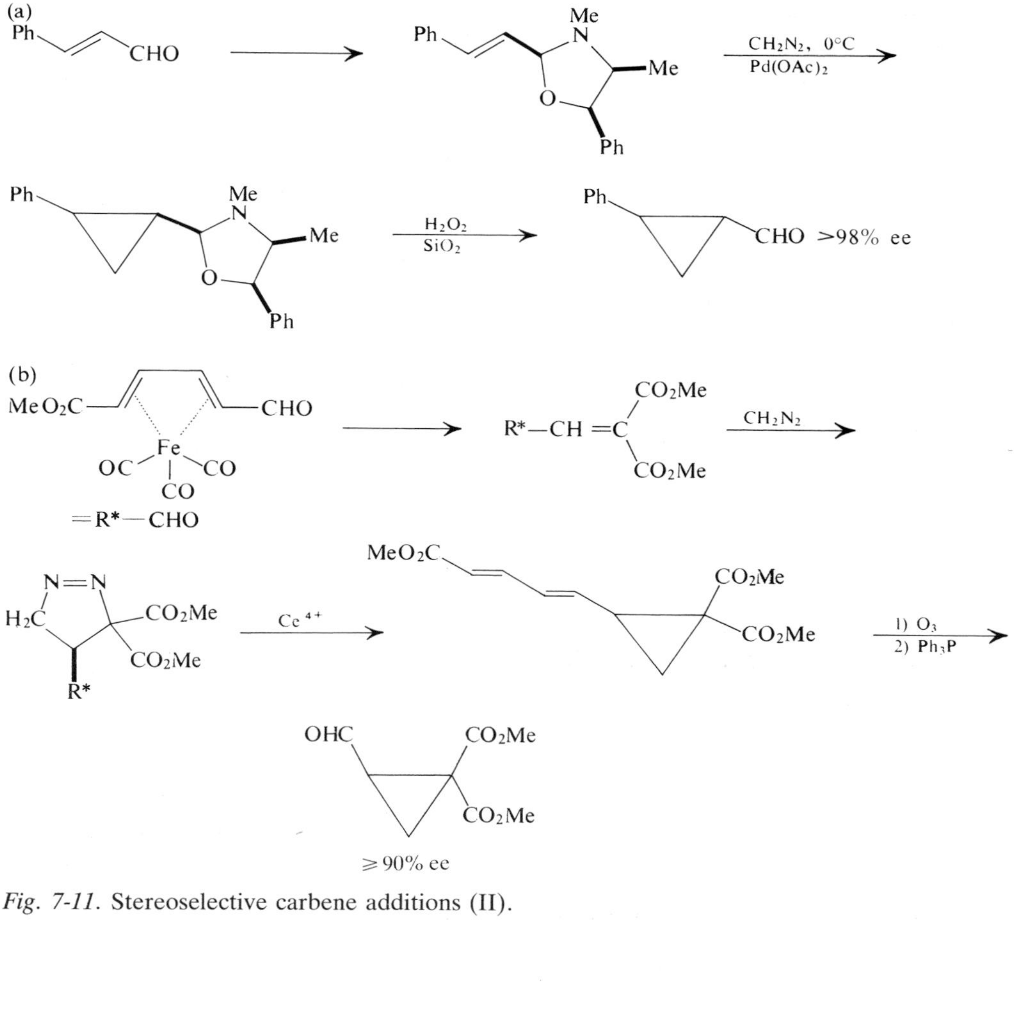

Fig. 7-11. Stereoselective carbene additions (II).

(c)

CH_2Br / BrH_2C + $H_2C(CO_2R^*)_2$ $\xrightarrow[50\%\ NaOH,\ H_2O]{Oct_3MeN^{\oplus}Cl^{\ominus},\ hexane}$ CO_2R^*, CO_2R^*

R* = (−)-8-phenylmenthyl

96% de

(d)

CO_2-(−)-Ment / (−)-MentO_2C + Ph_3P=C(Me)Me $\xrightarrow[-78°C--20°C]{THF}$ CO_2-(−)-Ment, Me, CO_2-(−)-Ment, Me

(1*S*,2*S*) 74% de

Fig. 7-11. Stereoselective carbene additions (II). (contd.)

A less practicable, but very interesting scheme, also leading to a cyclopropane aldehyde is shown in Fig. 7-11 (b) [65]. The starting material was resolved to antipodes *via* an oxazoline formed with (+)-ephedrine.

A rather special reaction leading to a vinylcyclopropane with high stereoselectivity was reported by Quinkert *et al.* [66], who used the product as starting material for the total synthesis of 19-nor-steroids (Fig. 7-11 (c)).

Phosphoranes can transfer a C_1 unit to activated olefins, and relatively high diastereoselectivities could be realized with chiral fumarates (Fig. 7-11 (d)) [67]. Using 8-phenylmenthol as the chiral auxiliary, the 1*R*,2*R* diastereomer was obtained in 82% de.

7.2 Chirality Transfer in Sigmatropic Rearrangements

The concerted transposition of a σ bond from an atom next to the end of a π system to the other end of the π system coupled with the transposition of the π bonds is a sigmatropic rearrangement. The rearrangement may involve a single π system, as in example (a) in Fig. 7-12, or two π systems simultaneously (b). The figures also illustrate the nomenclature used to classify sigmatropic rearrangements. The ene reaction, which is a hybrid of a Diels-Alder reaction and a [1,5] hydrogen shift (d), will also be discussed here, as well as a family of concerted multiple cyclization reactions involving polyenes.

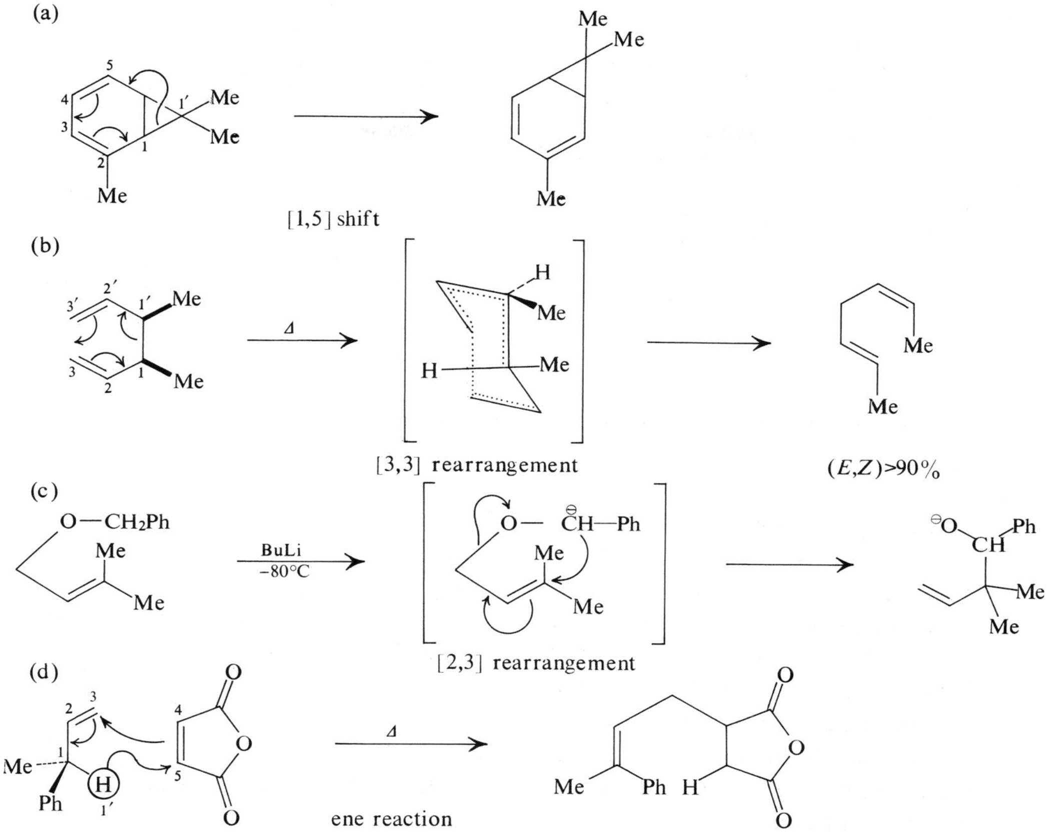

Fig. 7-12. Examples for different types of sigmatropic rearrangements.

For stereoselective synthesis [2,3] and [2,5] rearrangements and the ene reaction have been exploited. An essential feature of these transformations is the intramolecular transfer of chirality from one center to the other. Except for chiral sulfur centers, the moiety donating chirality is usually not removed from the molecule afterwards.

7.2.1 [3,3] Sigmatropic Rearrangements

[3,3] sigmatropic rearrangements have been shown to be generally concerted, involving a chair-like transition state, as shown in Fig. 7-13 (b) [68]. A boat-shaped transition state would lead either to (*Z,Z*)- or to (*E,E*)-2,4-octadiene [69]. A chair-

like transition state only provides for diastereoselectivity but not for enantioselectivity, since with chiral substrates two competing diastereomeric chair-like transition states can be envisaged. Efficient transfer of chirality from one center to the other was first demonstrated by Hill and Gilman for the Cope rearrangement (Fig. 7-13 (a), and the favored pathway was distinguished by an equatorial phenyl group [70]. Note that if we take the chair-like transition state for granted, an experiment with racemic starting material would yield the same information, since an *E* configuration of the double bond is a marker for transition state *A* and *Z* for *B*.

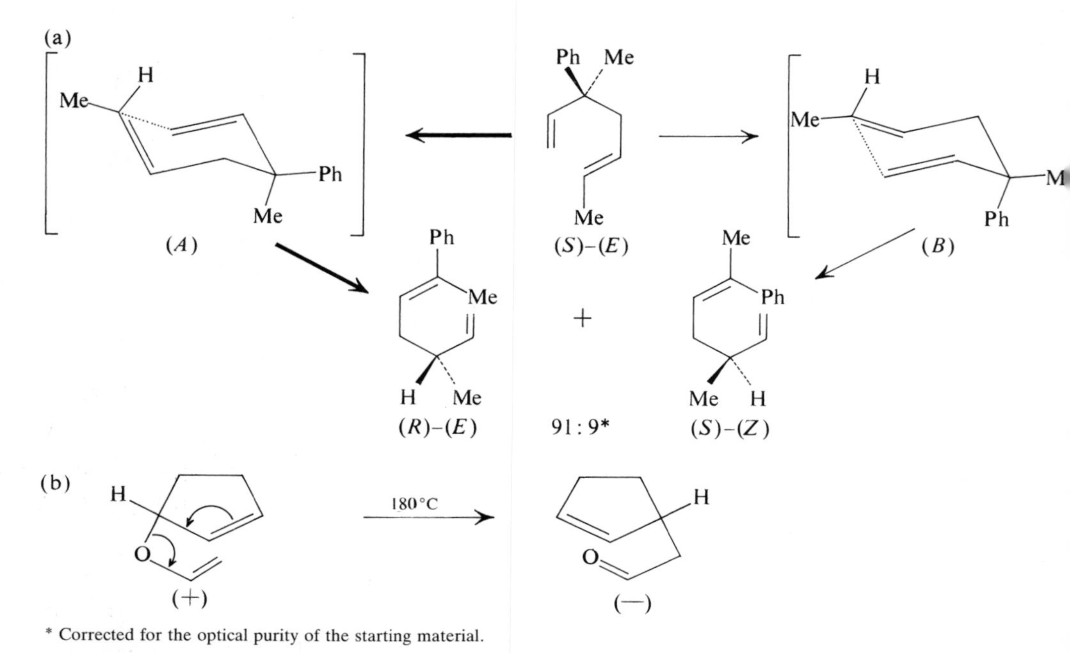

* Corrected for the optical purity of the starting material.

Fig. 7-13. The stereochemistry of [3,3] sigmatropic rearrangements.

When at least one of the π systems is part of a ring, only suprafacial transposition of the bonds is possible, and, as far the cyclic part of the system is concerned, complete transfer of chirality can be anticipated when the reaction is concerted. This was shown to be valid for the Claisen rearrangement and its thio and aza analogues (Fig. 7-13 (b)) [71].

Among [3,3] sigmatropic changes, the Claisen rearrangement has the most relevance to practical stereoselective synthesis.* In the original version an allyl vinyl ether was transformed thermally to a γ,δ-unsaturated carbonyl compound. The reaction was shown to be concerted and to proceed through a chair-like

---

* For reviews, *cf.* refs. [72] and [73].

transition state. The observed stereoselectivities can usually be adequately rationalized by assuming preference for such transition states which minimize 1,3-diaxial interactions [73].

Efficient conservation of chirality in *O*-Claisen rearrangements was demonstrated for simple models by Hill and coworkers. Interestingly, *E* selectivity was higher than enantiomeric purity, which may indicate that the chair-like transition state was not exclusive. When the allylic alcohol was reacted with triethylorthoace-

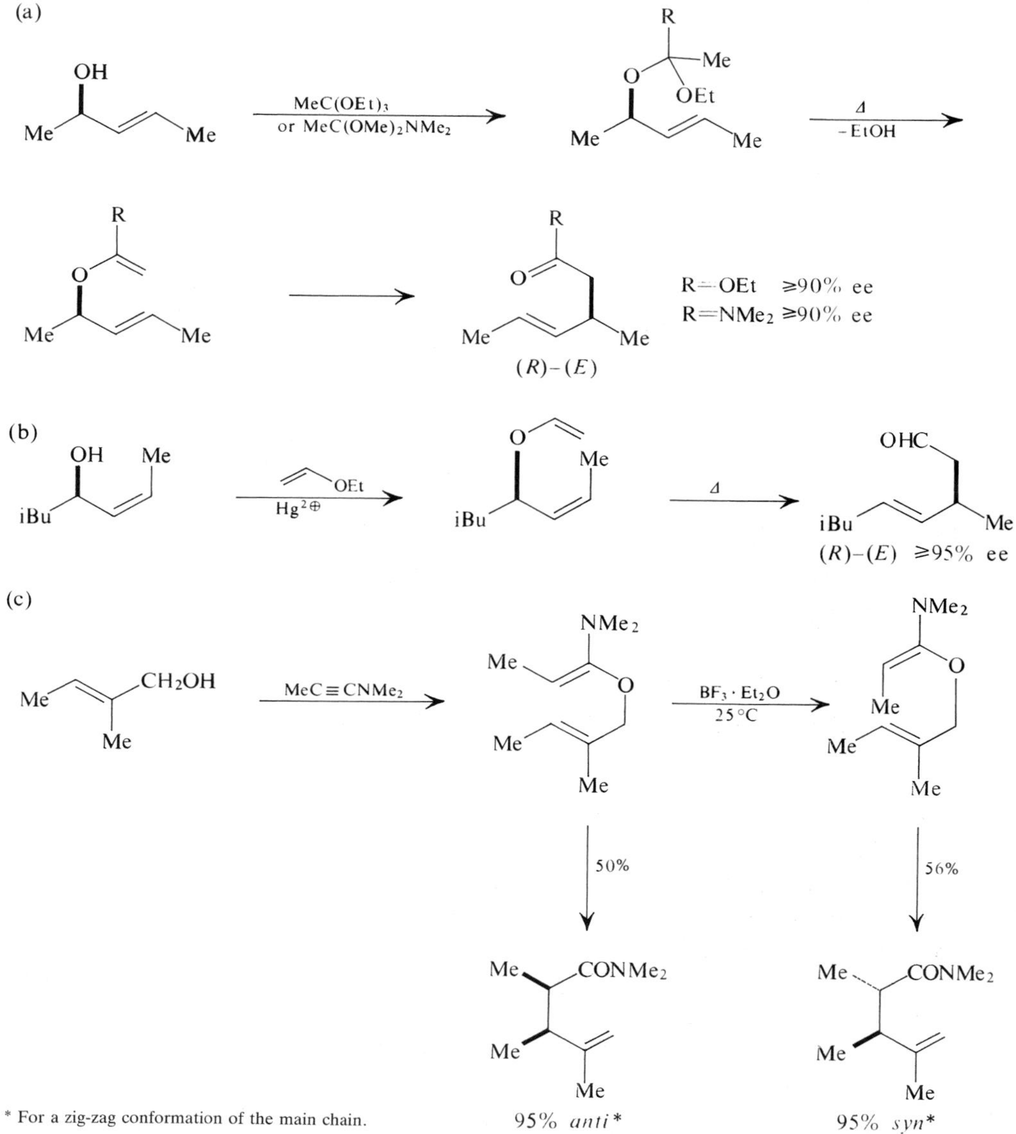

* For a zig-zag conformation of the main chain.

Fig. 7-14. Stereoselectivity in the Claisen rearrangement (I).

tate or *N,N*-dimethylacetamide dimethylacetal, the rearrangement product was an ester or an amide, respectively, (Fig. 7-14 (a)) [74]. Using a vinyl ether as the two-carbon fragment, an aldehyde was obtained (Fig. 7-13 (b)) [75].

Chirality transfer by the Claisen rearrangement has been utilized by Saucy, Cohen and their coworkers for the enantioselective and enantioconvergent synthesis of the side chain of tocopherol by a combination of enantioselective ketone reduction, diastereoselective triple bond reduction, and repeated Claisen rearrangement (*cf.* Fig. 1-32) [75–79]. A version of this procedure which so far has only been tested for diastereoselectivity is shown in Fig. 7-14 (c) [80].

Under the action of a strong base esters can be converted to ester enolates, and when this is done with an allyl ester a system amenable to Claisen rearrangement is

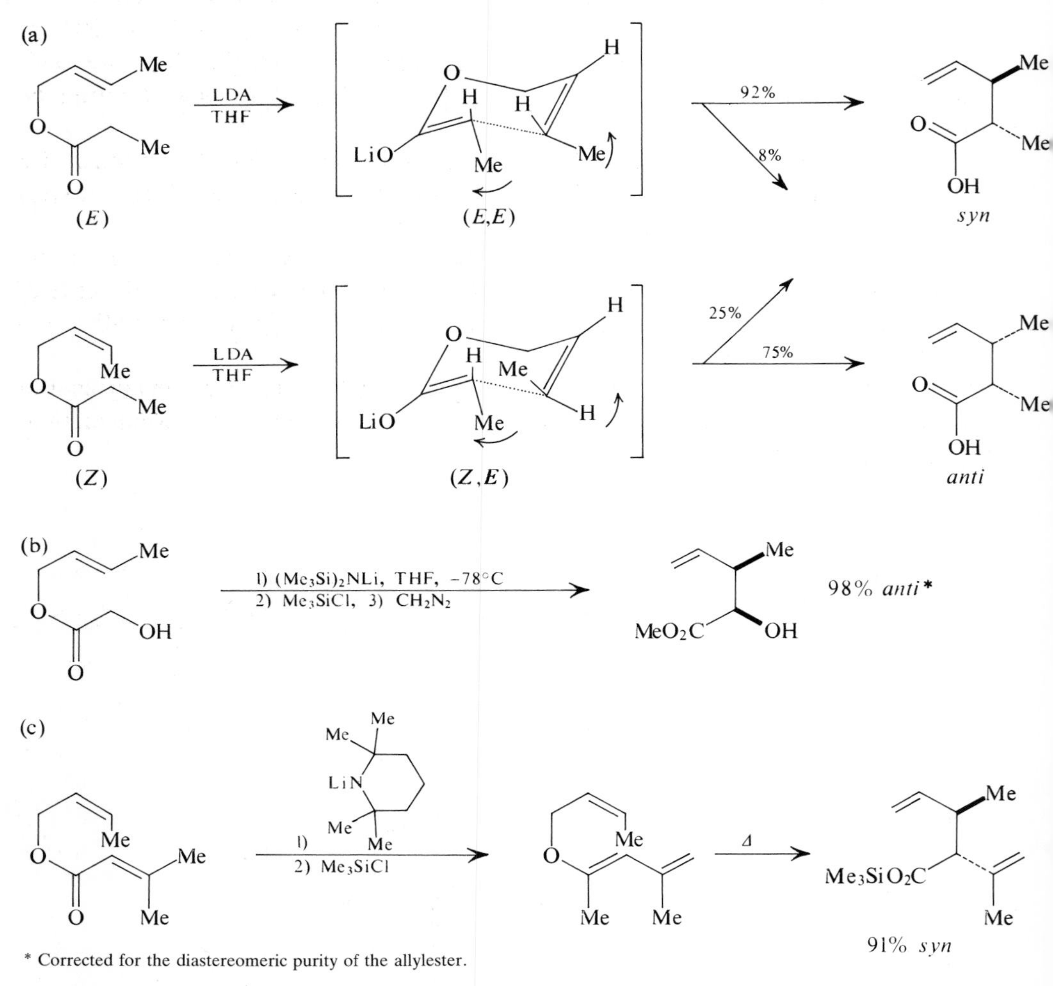

Fig. 7-15. Stereoselectivity in the Claisen rearrangement (II).

formed. In fact Ireland *et al.* found [81] that on treatment with LDA allyl esters readily rearranged to γ,δ-unsaturated acids. The lithium enolates could be stabilized by silylation, and the resulting ketene acetals reacted in the same way. Both double bond formation and single bond formation are stereoselective, and product configurations can be predicted on the basis of a chair-like transition state model (Fig. 7-15(a)). Interestingly, on conversion of the lithium enolates to the $tBuMe_2Si$-ethers selectivity decreased with the *E* substrate (87% *anti*) and increased with the *Z* substrate (89% *anti*).

Note that since the allylic double bond is fixed, *syn/anti* selectivity is governed by the configuration of the enolate bond, and this was found to be solvent dependent: by adding to THF 23% of HMPA, a solvent coordinating with Li^+, selectivity was reversed and 87% of *syn* product was obtained from the (*E*)-allyl ester. Except for the anion derived from the (*E*)-allyl ester in pure THF, yields are much better with the silylated ketene acetals, as also proved for other esters (Fig. 7-15 (b)) (98% of the *syn* diastereomer were obtained from the (*Z*)-allyl ester) [82]. The predominant diastereomer can again be changed by adding HMPA to the solvent during the formation of the enolate anion [83].

With allyl esters of α,β-unsaturated acids enolate formation may be coupled with the migration of the double bond, as shown by Wilson and Myers (Fig. 7-15 (c)) (the (*E*)-allyl ester yielded 86% of the *anti* product) [84].

Excellent to complete transfer of chirality at C-1 was experienced in the rearrangement of 1-substituted ester enolates by Ireland and Varney in the case of the 1-silyl [85], and by Nagatsuma *et al.* [85], as well as by Fujisawa *et al.* with the 1-methyl derivatives (Fig. 7-16 (a)) [86].

The Claisen rearrangement can also be triggered by forming an anion adjacent to the ether oxygen, but this approach offers little advantage over the thermal method (Fig. 7-16 (b)) [87].

(a) (*R*) (91% ee) — 1) $3(Me_3Si)_2NLi$, THF, −78 °C 2) Me_3SiCl, 3) 25 °C → 98% *syn*, 91% ee

(b) NaH → 98% *syn*

Fig. 7-16. Stereoselectivity in the Claisen rearrangement (III).

Usually allyl vinyl thioethers derived from thioamides cannot be isolated because they undergo a spontaneous and highly stereoselective thio-Claisen rearrangement. The procedure is illustrated in Fig. 7-17 (a) for thioanilides. With tertiary thioamides the *N*-silylation step drops out, but diastereoselectivity is less complete [88]. Alkylation with (*Z*)-crotyl tosylate gives *syn* products with 99% selectivity. Tamaru *et al.* have shown that stereoselectivity depends on the diastereomeric purity of the thioamide anion. Thus the thiopyrrolidone *28* can only form an *E* anion and accordingly gives a single product, while *29* gives a mixture of *syn* and *anti* diastereomers in a ratio corresponding to the *E*/*Z* ratio of the substrate (Fig. 7-17 (b) and (c)) [89].

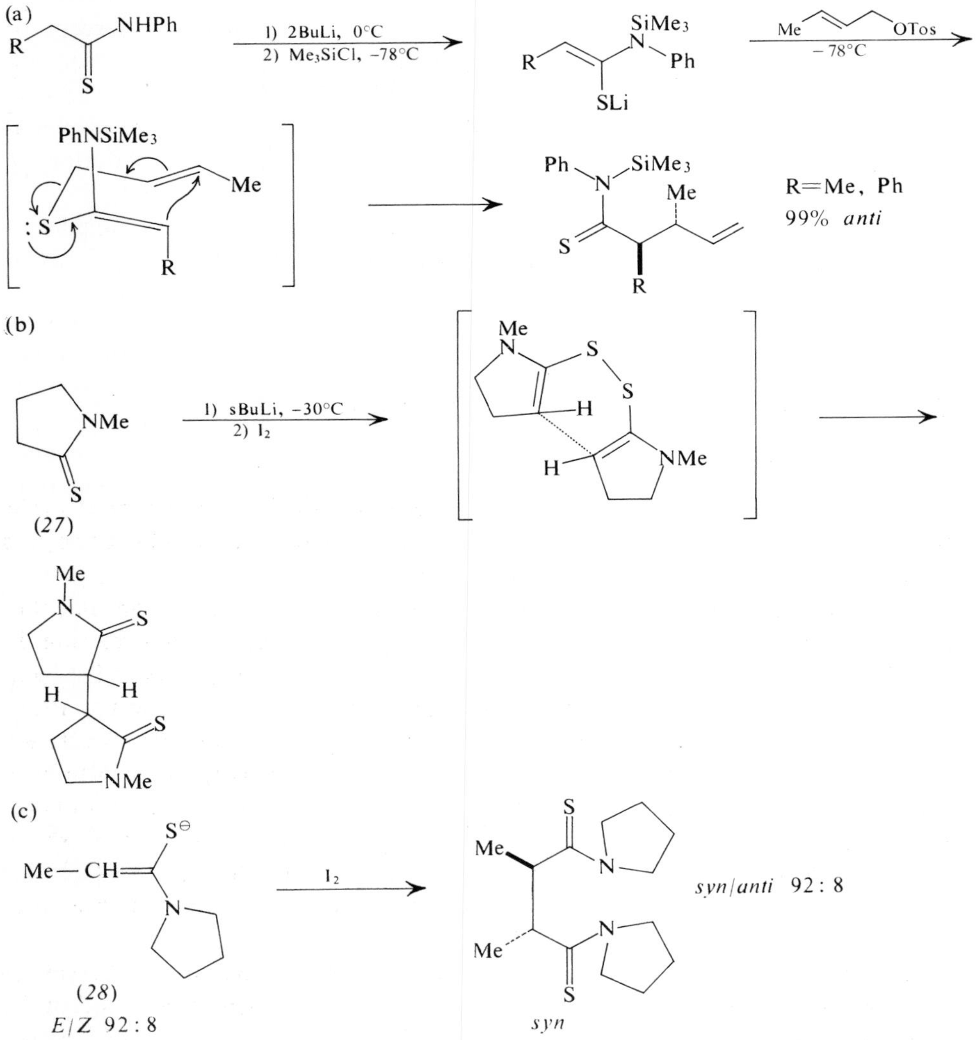

(a) 1) 2BuLi, 0°C 2) Me_3SiCl, −78°C; Me–CH=CH–CH$_2$OTos, −78°C

R=Me, Ph 99% *anti*

(b) (27) 1) sBuLi, −30°C 2) I_2

(c) (28) *E*/*Z* 92 : 8 → I_2 → *syn*; *syn*/*anti* 92 : 8

Fig. 7-17. Stereoselectivity in the thio-Claisen rearrangement.

An aza-Claisen rearrangement proceeding with moderate stereoselectivity was reported by Kurth and Decker (Fig. 7-18) [90].

R^1	R^2	R^3	% de
Me	H	H	74
H	Me	H	74
H	H	Me	72

Fig. 7-18. Stereoselectivity in the aza-Claisen rearrangement.

7.2.2 [2,3] Sigmatropic Rearrangements

[2,3] sigmatropic changes involve six electrons and a five-membered cyclic transition state. Of the six electrons two each are provided by a double bond, a σ bond, a non-bonding pair of a carbanion or a heteroatom.

A [2,3] sigmatropic change involving a carbanion is the Witting rearrangement of allyl ethers of the type *29*, where R is a group facilitating carbanion formation. In Fig. 7-19 (a) the reaction is illustrated for some simple crotyl ethers [91–94]. Stereoselectivity has been rationalized by considering envelope-shaped five-membered cyclic transition states, of which those leading to the minor products are destabilized by a "pseudo-1,3-diaxial" interaction (Fig. 7-19 (b)). It has also been suggested that in the case of bulky R groups, this preference diminishes due to R $\leftrightarrow$ Me gauche interactions for the *E* but not for the *Z* substrates [92]. This was, however, not borne out in the case of an oxazine-substituted model, which was more selective in the *E* form. Removal of the blocking group gave rise to 2-hydroxy-3-methyl-4-pentenoic acids [94].

While the above examples only demonstrated diastereoselectivity in [2,3] sigmatropic rearrangements, moderately efficient transfer of chirality was found in the transformation of a macrocyclic allyl ether (Fig. 7-20 (a)) [95] and of a chiral ammonium ylide derived from (*S*)-prolinol [96].

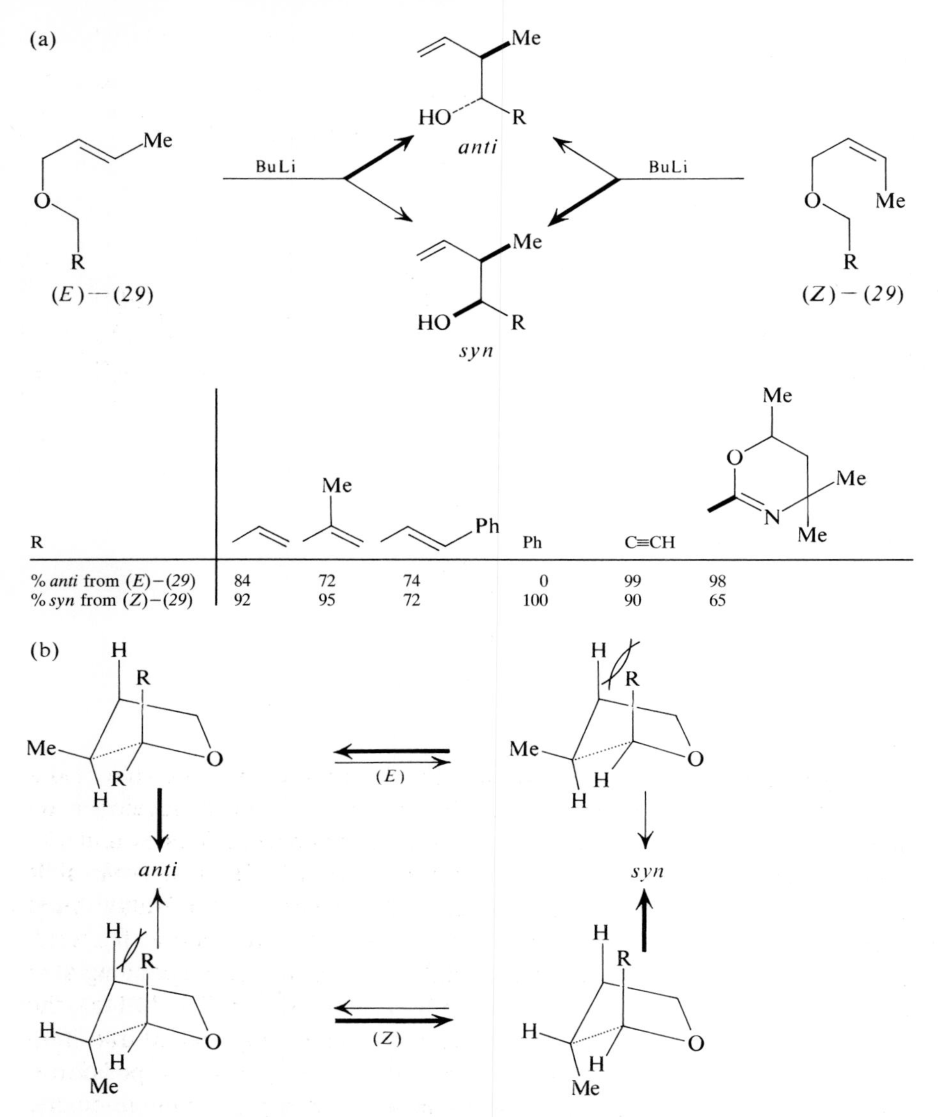

R	(structure)	(structure)	(structure)	Ph	C≡CH	(structure)
% *anti* from (*E*)−(*29*)	84	72	74	0	99	98
% *syn* from (*Z*)−(*29*)	92	95	72	100	90	65

Fig. 7-19. Stereoselectivity in the Witting rearrangement (I).

A [2,3] sigmatropic rearrangement which was supposed to proceed through a carbene intermediate was used by Chan and Saucy to prepare chiral isoprenoid synthons (Fig. 7-20 (b)) [97].

Transfer of chirality in the allylic rearrangement of amine oxides was demonstrated by Inouye and his coworkers [98–100]. As long as there are no efficient methods available for the preparation of enantiomerically pure amine oxides the method has no practical significance.

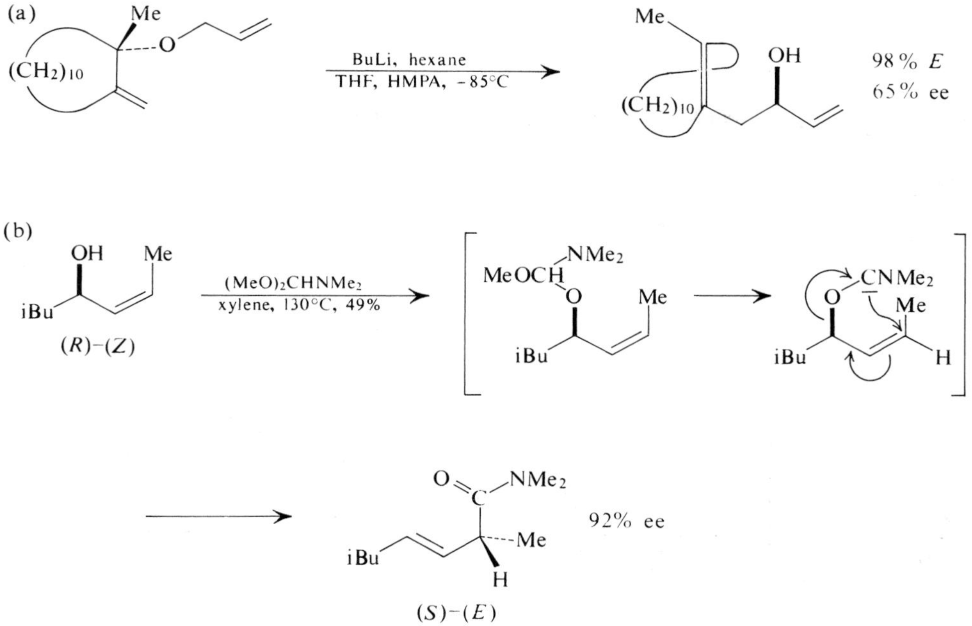

Fig. 7-20. Stereoselectivity in the Witting rearrangement (II).

Aryl vinyl sulfoxides of high optical purity are relatively readily available, and a method for their transformation to chiral allyl alcohols has been developed by Hofmann and his coworkers [101]. It is known that allyl aryl sulfoxides undergo rapid racemization at room temperature *via* a concerted [2,3] sigmatropic shift involving a sulfinate-type intermediate (Fig. 7-21 (a)) [102]. Aryl 2-alkyl vinyl sulfoxides can be converted by treatment with a strong base and then with a weak acid to an aryl allyl sulfoxide which is at equilibrium with the corresponding allyl sulfenate. If the allyl group is unsubstituted at the β-position, as in Fig. 7-21 (a), the sulfenate is achiral and is reached through two enantiomerically related transition states, and thus the chirality of the sulfoxide gets lost. If, however, this position is substituted, the transition states are diastereomeric and a pair of enantiomeric sulfenates are formed at different rates, each of which may revert to the original sulfoxide or to its antipode, again with unequal probability. If the kinetically favored sulfenate can be trapped by fast desulfuration, an optically active allyl alcohol is obtained. This situation can be best understood by considering the transformations shown in Fig. 7-21 (b). The optical purity of the product depends on K and the ratio of k_1, k_2 and k_3. In our example the (R)-sulfoxide gave the (S)-alcohol in 75% yield and 60% ee [101], with the cyclopentene analogue ee was only 44% [103]. This result is in accordance with a preference for the *exo* conformation of the (R)-allyl sulfoxide intermediate. Enantioselectivity with substituents in β-position other than tetramethylene was poor [104].

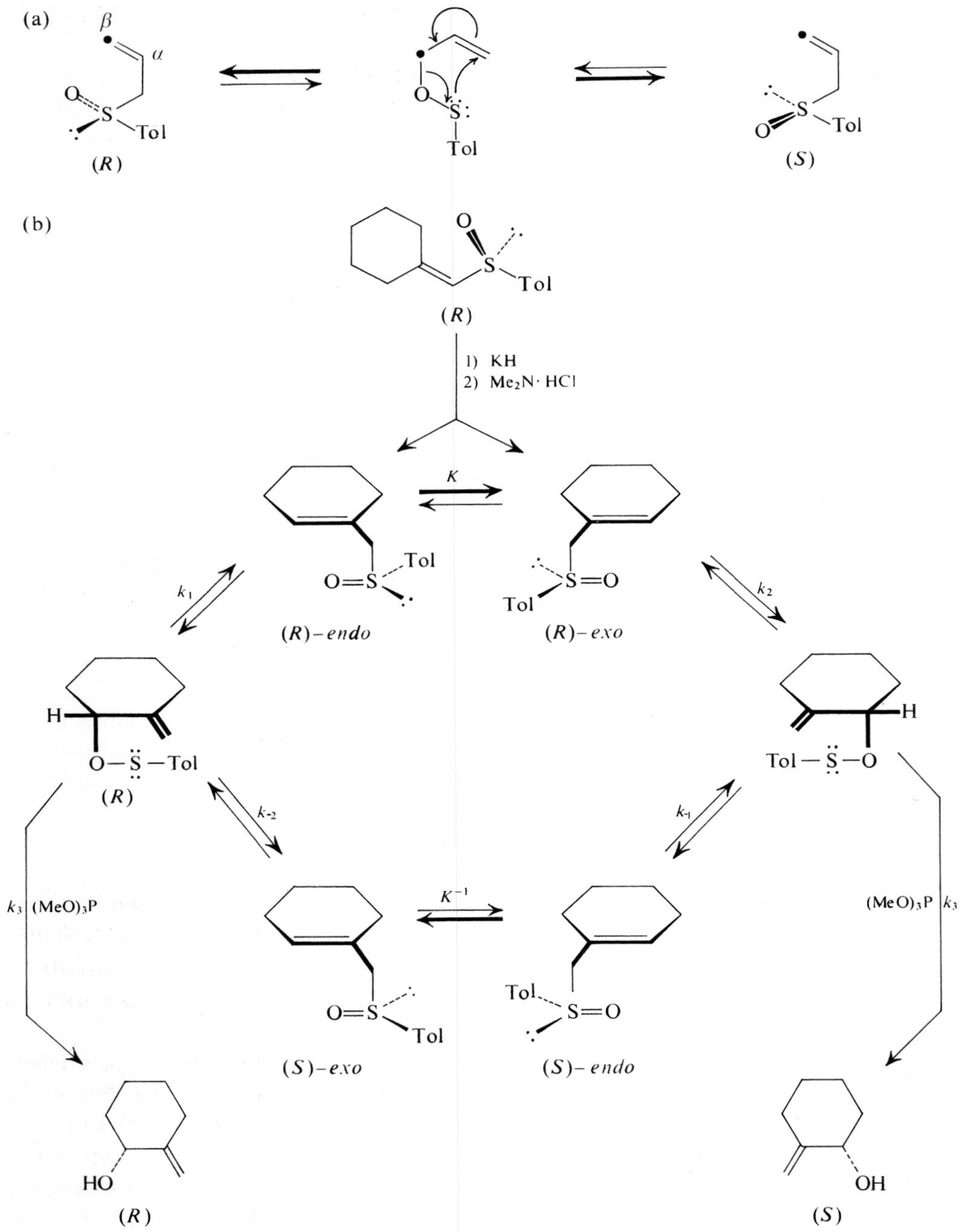
(a)
β
α
Tol
(R)
Tol
Tol
(S)
(b)
Tol
(R)
1) KH
2) Me2N·HCl
K
Tol
(R)-endo
(R)-exo
k1
k2
H
O—S—Tol
(R)
Tol—S—O
k-2
k-1
k3 (MeO)3P
(MeO)3P k3
K^-1
(S)-exo
(S)-endo
HO
(R)
OH
(S)

Fig. 7-21. Stereoselectivity in the rearrangement of allylsulfenates (I).

The use of trimethyl phosphite can be obviated by using an internal thiophilic group (Fig. 7-22 (a)) [103].

Transfer of chirality in the thermal rearrangement of sulfinates to sulphones was studied by Hiroi *et al.* [105], who established that the preferred transition state was again the one in which the bulkiest group was quasi-equatorially disposed (Fig. 7-22 (b)).

(a) NMe2 S O KOtBu HO >90% ee

(b) O S O Tol R DMF 90–120 °C Me TolSO2

R	(*E*)–Me	(*E*)–Bu	(*Z*)–Me	(*Z*)–Bu
% ee (conf.)	87 (*S*)	84 (*R*)	89 (*S*)	82 (*R*)

Fig. 7-22. Stereoselectivity in the rearrangement of allylsulfenates (II).

7.2.3 Ene Reactions

The ene reaction is a hybrid between a Diels-Alder reaction and a [1,5] sigmatropic hydrogen shift comprising the addition of an olefin containing an allylic hydrogen atom to a π bond, as shown in Fig. 7-23 (a). It can be regarded as a $(\sigma_s^2 + \pi_s^2 + \pi_s^2)$ reaction and is thermally allowed in a concerted suprafacial fashion [2].

Asymmetric induction in the ene reaction was first demonstrated by Hill and Rabinovitz [106], and the experiment supported its concerted nature. As the Diels-Alder reaction, it can be accelerated by Lewis-acid catalysis. In the uncatalyzed reaction of chloral with (−)-β-pinene the diastereomer with an *R* configuration in the side chain was obtained in 64% ee. Stereoselectivity was both enhanced and inverted on addition of $TiCl_4$ (Fig. 7-23 (b)) [107]. On complexation with the catalyst the bulky CCl_3 group was forced into the unfavorable *endo* disposition.

The intermolecular ene reaction was successfully exploited by Dauben and Brookhart for the stereocontrolled construction of steroidal side chains (Fig. 7-23 (c) [108]. The *E* olefin afforded the 20*S* diastereomer [109].

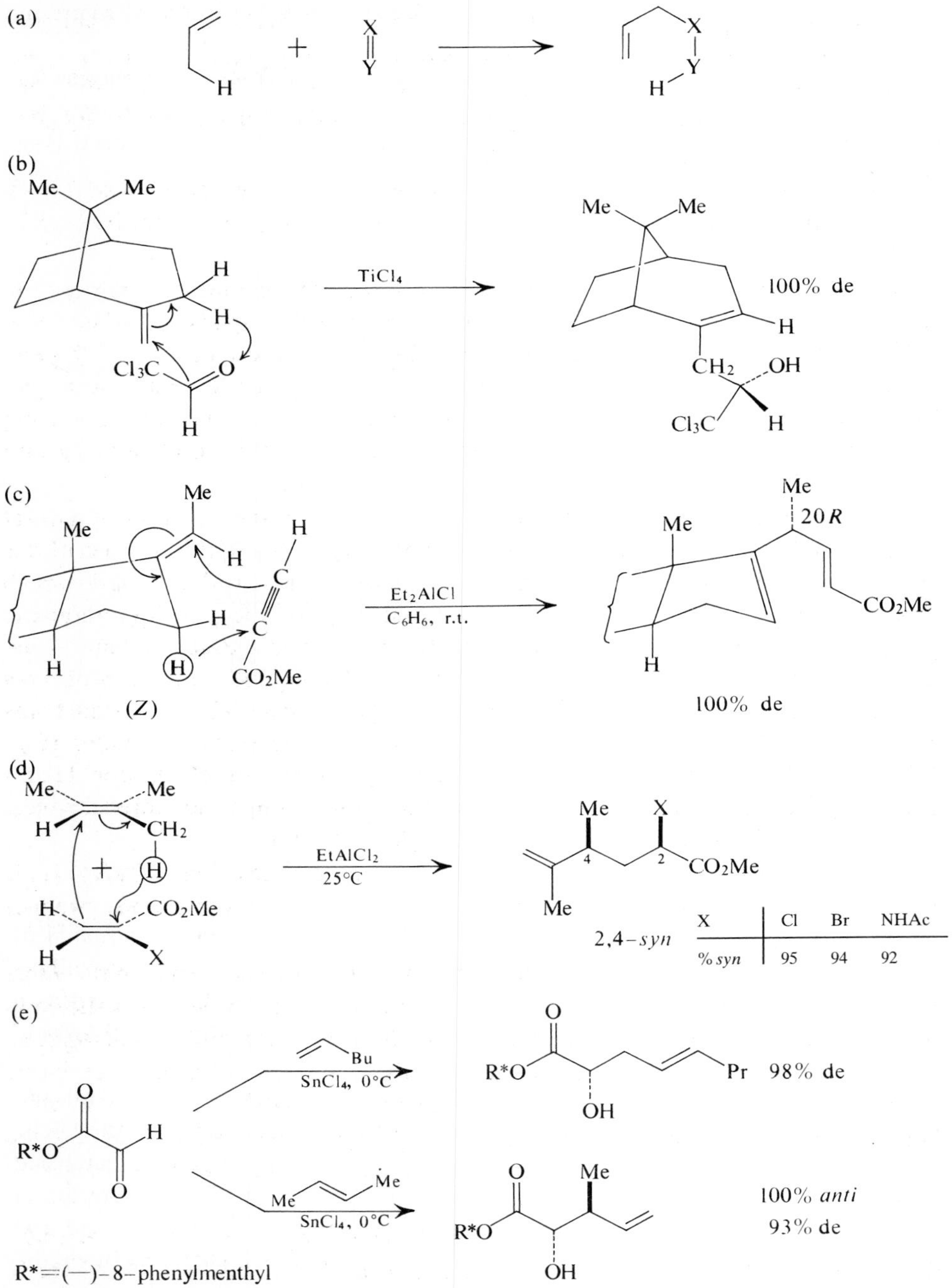

(a)

(b)

TiCl$_4$ 100% de

(c)

Et_2AlCl
C_6H_6, r.t.

(Z) 100% de

(d)

$EtAlCl_2$
25°C

2,4-*syn*

X	Cl	Br	NHAc
% *syn*	95	94	92

(e)

$SnCl_4$, 0°C 98% de

$SnCl_4$, 0°C 100% *anti* 93% de

R*=(−)-8-phenylmenthyl

Fig. 7-23. Stereoselectivity in intramolecular ene reactions (I).

Appreciable diastereoselectivity has been achieved in the Lewis acid catalyzed ene reaction of α-substituted acrylates, but asymmetric induction using chiral acrylates is poor (Fig. 7-23 (d)) [110]. Note that the reaction was regioselective for attachment to the less substituted terminal of the olefin. With (*E*)-2-butene as partner, diastereoselectivities were similar.

Efficient asymmetric induction was reported in the ene reaction of (−)-8-phenylmenthylglyoxylate with olefins by Whitesell *et al.* [27], as exemplified in Fig. 7-23 (e).

One of the early examples for the utilization of an intramolecular ene reaction was the ingenious synthesis of chiral acetic acid by Arigoni and his coworkers (Fig. 7-24 (a)) [111]. Here, in the first step, only the *Z* diastereoselectivity of the ene reaction was exploited, the configuration at C(3) was not checked and was in fact irrelevant. Chirality transfer in the second step, a retro-ene reaction involving suprafacial deuterium migration, completed the formation of a chiral methyl group of predictable configuration.

Intramolecular ene reactions became a powerful tool in the synthesis of natural products in the hands of Oppolzer and his coworkers [15, 40]. The basic idea of this approach is to link the allyl group and the acceptor double bond by a chain which should be long enough to allow the proper mutual disposition of the interacting groups, but not too long in order to lend sufficient constraint to the system. When the chain contained three atoms and a (*Z*)-allyl enophile, the reaction was completely diastereoselective, while with (*E*)-allyl enophiles the *cis* product was not exclusive due to competition of two diastereomeric transition states (Fig. 7-24 (b)) [40]. The method was employed for the synthesis of cyclopentanoid sesquiterpenes [112, 113]. With 1,2-disubstituted ene groups the stereochemical result was indifferent to olefin configuration (*E* or *Z*) [40].

Lewis acid catalysis opened up the way to the *trans* series (Fig. 7-24 (c)) [114, 115]. The corresponding (*E*)-1,6-hexadiene also gave mainly the *trans* product (*trans*/*cis* 89:11).

Finally, the problem of enantioselectivity was solved by subjecting chiral esters to Lewis-acid-catalyzed cyclization. The (−)-menthyl esters were not sufficiently selective, but the (−)-8-phenylmenthyl esters afforded good enantioselectivities, the (*Z*)-ester giving the 3*S*,4*R* enantiomer in 80% ee and the (*E*)-ester its antipode in 70% ee. The former was transformed to the natural (+)-α-allokainic acid. Stereochemical results could be predicted by considering the shielding effect of the phenyl group of the chiral auxiliary in a six-membered chair-like transition state [116, 117].

A chiral center adjacent to the ene function can also effectively control enantioselection, as demonstrated by the cyclization of *30* derived from (*S*)-glutamic acid to a precursor of (−)-kainic acid (Fig. 7-25 (a)) [118].

The scope of the stereoselective intramolecular ene reaction can be extended to substrates in which, instead of a hydrogen atom, magnesium halide is transferred from one end of the system to the other. The intermolecular version of this reaction

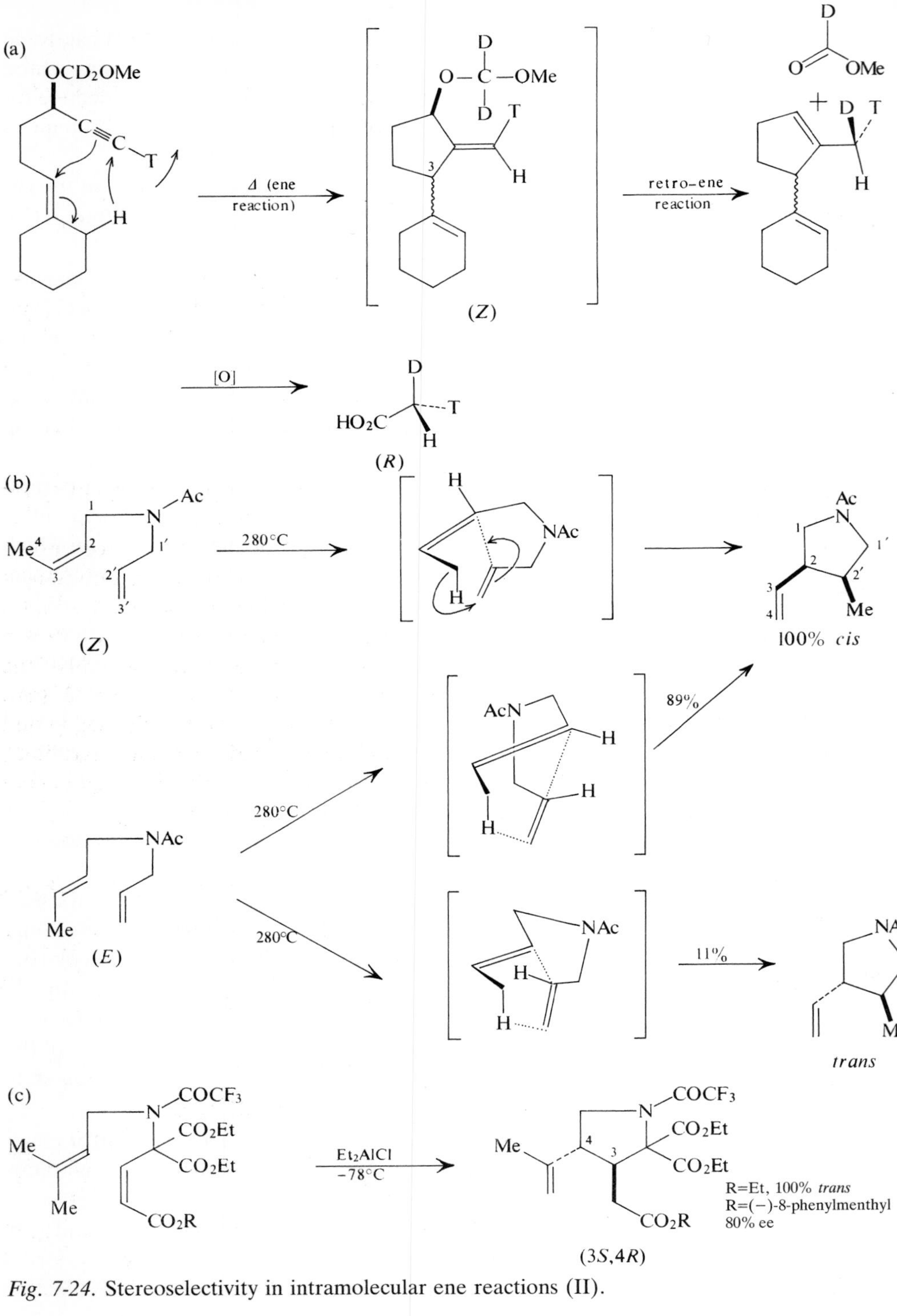

(a)

OCD$_2$OMe

Δ (ene reaction)

O—C—OMe D D T H

(Z)

retro-ene reaction

D O OMe + D T H

[O]

HO_2C D T H

(R)

(b)

Ac N Me 1 2 3 4 1′ 2′ 3′

(Z)

280°C

H NAc H

Ac N 1 1′ 2 2′ 3 4 Me

100% *cis*

NAc Me

(E)

280°C

AcN H H H

89%

280°C

NAc H H

11%

NAc Me

trans

(c)

N COCF$_3$ CO$_2$Et CO$_2$Et Me Me CO$_2$R

Et$_2$AlCl −78°C

COCF$_3$ N CO$_2$Et CO$_2$Et Me 4 3 CO$_2$R

R=Et, 100% *trans*
R=(−)-8-phenylmenthyl 80% ee

(3S,4R)

Fig. 7-24. Stereoselectivity in intramolecular ene reactions (II).

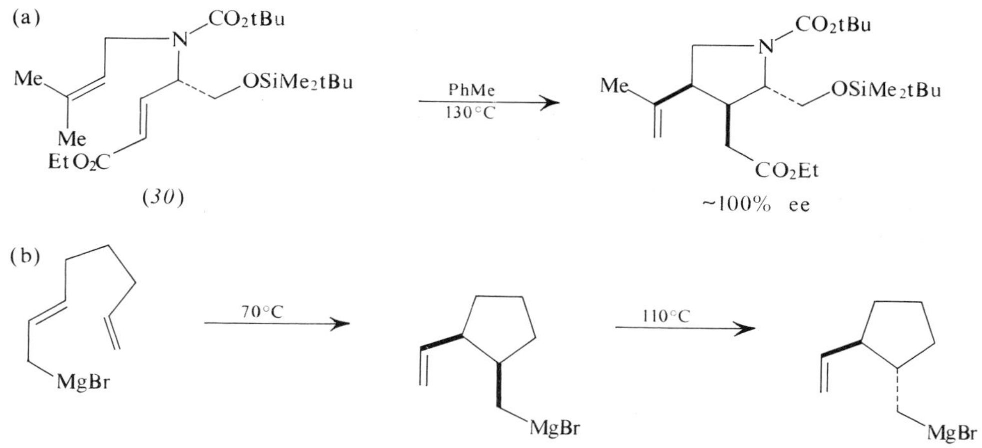

(a) (30) PhMe 130 °C → ~100% ee

(b) 70 °C → 110 °C →

Fig. 7-25. Stereoselectivity in intramolecular ene reactions (III).

suffers from poor regioselectivity, but Felkin *et al.* found surprising selectivity for an intramolecular magnesium-ene reaction (Fig. 7-25 (b)) [119].

So far only such ene reactions have been considered in which the ene component is linked to the migrating group in γ-position. Compounds in which the same is attached to the β-position can also undergo an ene reaction, and several examples for this involving the transposition of magnesium have been reported. The reaction is regioselective inasmuch as magnesium always migrates to the remote end of the double bond, and the smaller of the two possible rings is formed (Fig. 7-26 (a)). Further C–C coupling occurs with the more substituted end of the allylmagnesium halide system. (The allylmagnesium moiety undergoes rapid 1,3-metal migration.) The reaction was found to be remarkably stereoselective, which is in agreement with a concerted reaction, as shown in Fig. 7-26 (b) [120].

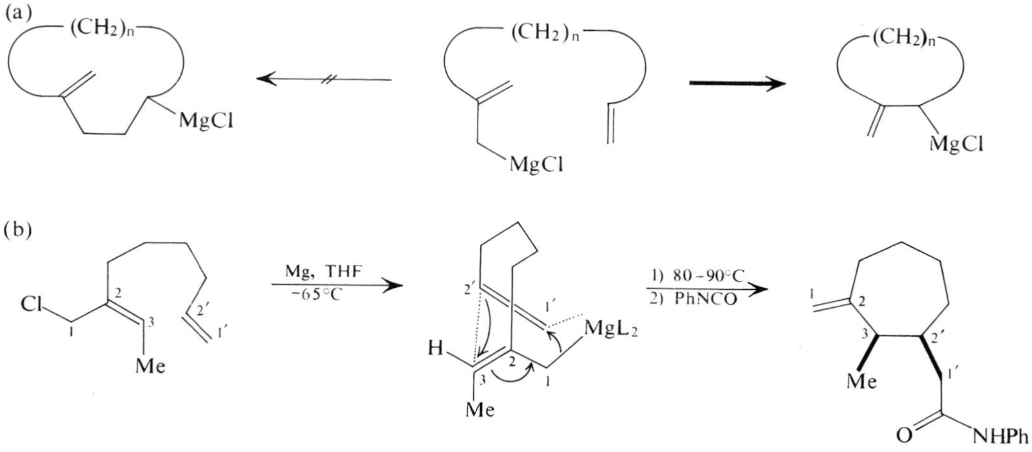

(a)

(b) Mg, THF −65 °C → 1) 80–90 °C 2) PhNCO →

Fig. 7-26. Stereoselectivity in intramolecular ene reactions (IV).

7.3 Biomimetic Polyene Cyclizations

It is known that in nature the pentacyclic ring system of steroids is formed, not by a series of distinct transformations, but in one step from squalene oxide. The process may be envisaged as a sequence of antiperiplanar electrophilic additions to the olefinic bonds, triggered by protonation of the epoxide oxygen (Fig. 7-27 (a)).

Several researchers set out to elaborate a non-enzymatic synthetic scheme which would imitate the polycyclization of squalene oxide. Among them, Johnson and his team played a prominent role [121].* Violating our self-imposed limitation that we shall be primarily concerned with basic stereoselective methodology and not its application to natural product synthesis, we nevertheless discuss this topic, although, we have to restrict ourselve to some highlights of these pioneering studies.

The first indication that biomimetic polyene cyclization was a realistic proposition was an experiment in which the dienic acetal (*E*)-*31* was converted to the *trans*-octalin *32* as the major product, while, in accordance with the supposed mechanism, (*Z*)-*31* gave only *cis*-decalin derivatives. With the careful optimization of conditions, the scope of Lewis-acid-catalyzed cyclization could be extended to the construction of all-*trans* tri- and tetracyclic systems (*e. g.* Fig. 7-27 (c)).

When the terminal double bond was replaced by an acetylenic function the linear shape of the latter precluded cyclization at its outer carbon atom and a *trans*-fused cyclopentane ring, conveniently functionalized at the side chain, was formed (Fig. 7-27 (d)).

So far only diastereoselectivity in ring junction formation has been discussed, but methods have been developed which permit one to conduct the cyclization enantioselectively. Thus, use of a dienic acetal derived from (*R,R*)-2,3-butanediol gave, after the removal of the diol moiety, a mixture of diastereomers and, surprisingly, both had the same chirality at the oxygenated center (Fig. 7-28 (a)) [122]. Oxidation of the mixture gave an octalinone with the unnatural configuration in 84% ee. A variant of this approach, the $TiCl_4$-catalyzed cyclization of an acetal formed with (2*S*, 4*S*)-2,4-pentanediol, gave a steroid C + D ring fragment with the natural configuration in excess [123]. Note again the identical configuration at the ether oxygen.

A substituent at the future C-11 position exhibited an astonishingly strong influence on the direction of cyclization. The racemic substrates *33* (R = Me or OH) gave the 11α-substituted diastereomers exclusively, *i. e.* the reaction is a 1,7-*ul* process (Fig. 7-28 (c)) [124, 125].

---

* Results up to 1976 have been summarized in this review.

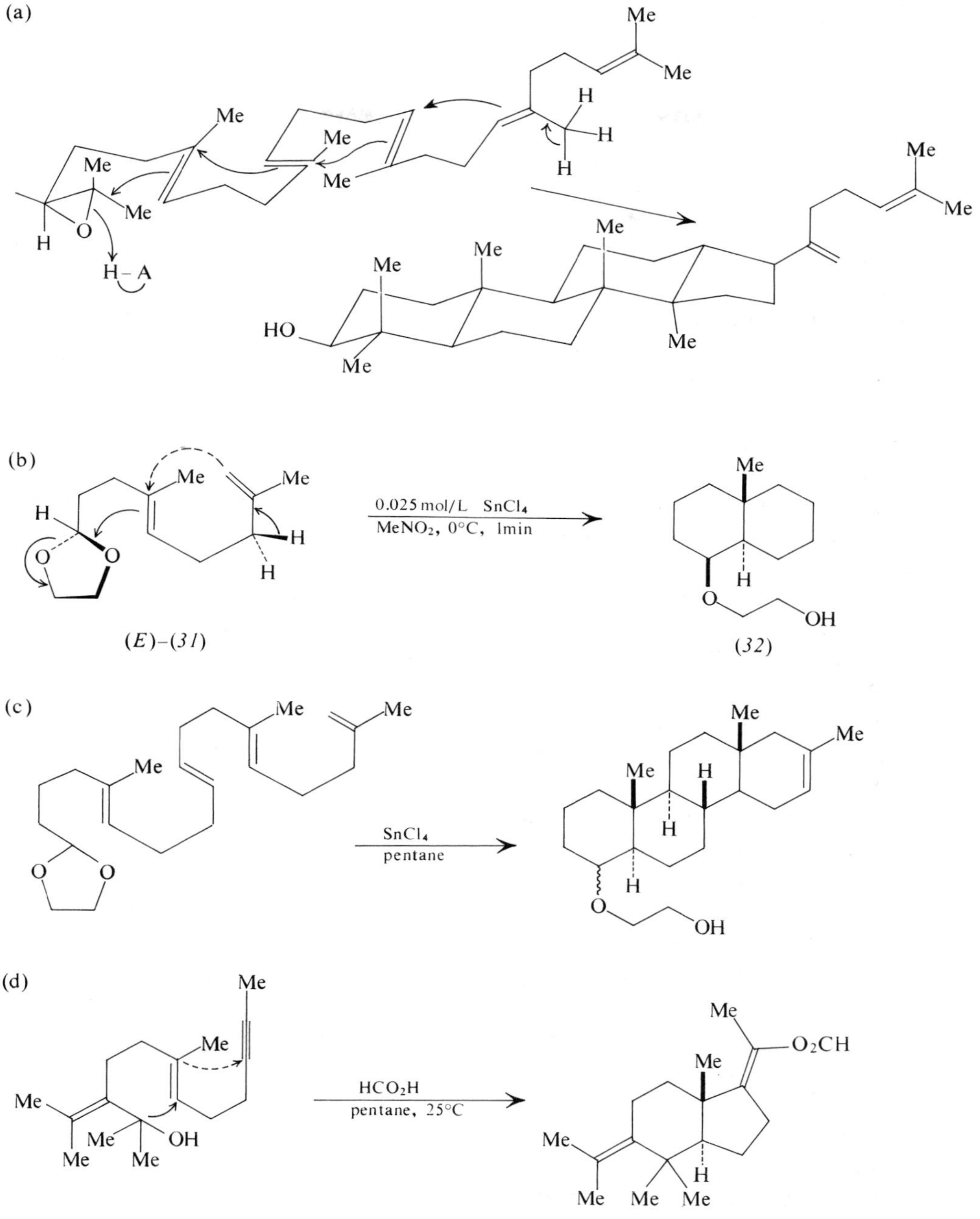

(a)

(b)

0.025 mol/L $SnCl_4$
$MeNO_2$, 0°C, 1min

(E)-(31)

(32)

(c)

$SnCl_4$
pentane

(d)

HCO_2H
pentane, 25°C

Fig. 7-27. Biomimetic polyene cyclizations (I).

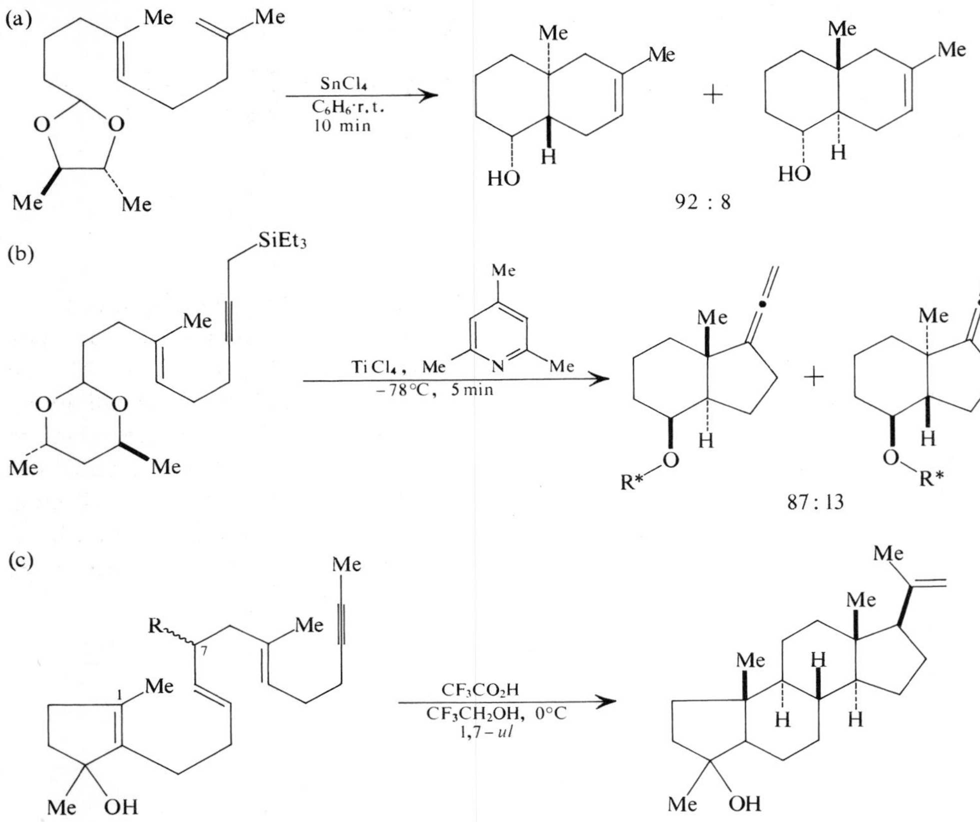

Fig. 7-28. Biomimetic polyene cyclizations (II).

An experiment with the optically active alcohol [126] confirmed this result. The directing influence of a substituent at position C-6 or C-7 was also demonstrated [127–129].

Biomimetic polyene cyclization is a very elegant approach to steroid total synthesis but a key to its practical application, *i. e.* a resonably simple preparation of the polyene precursors, has yet to be found.

8 Stereoselective Formation of Carbon-Heteroatom Bonds

In this chapter the stereoselective, mainly enantioselective, formation of C–N, C–P, C–O and C–S bonds, as well as stereoselective protonation will be briefly discussed. Although the stereoselective formation of carbon-metal bonds is an important field of research, owing to its special methodology this topic will not be covered. Some aspects of the stereochemistry of organic silicon, tin and titanium compounds have already been mentioned in Chapters 5 and 6. Despite their importance, stereoselective formation of halides, as well as the ring opening reactions of epoxides will be omitted also, since these can be readily interpreted and devised on the basis of the well known principles of classical mechanistic organic chemistry.

8.1 Stereoselective Formation of Carbon-Nitrogen Bonds

Except for the trivial case of S_N2 substitution, there are very few simple examples and even fewer general methods known for the stereoselective linking of nitrogen to carbon.

Asymmetric induction in the addition of achiral amines to chiral activated olefins or of chiral amines to achiral activated olefins is generally inefficient [1–3]. This is true even for intramolecular addition [4]. On the other hand, Panunzi *et al.* found that amine addition to olefins activated by a chiral platinum complex was highly enantioselective (Fig. 8-1 (a)). The related palladium-assisted oxyamination was shown by Bäckvall and Björkman to be highly diastereoselective with (*E*)- and (*Z*)-butene and moderately so with 1-phenyl-1-propene (Fig. 8-1 (b)) [5]. When a chiral tert-amine was used as a ligand the reaction gave up to 60% ee in the product [6].

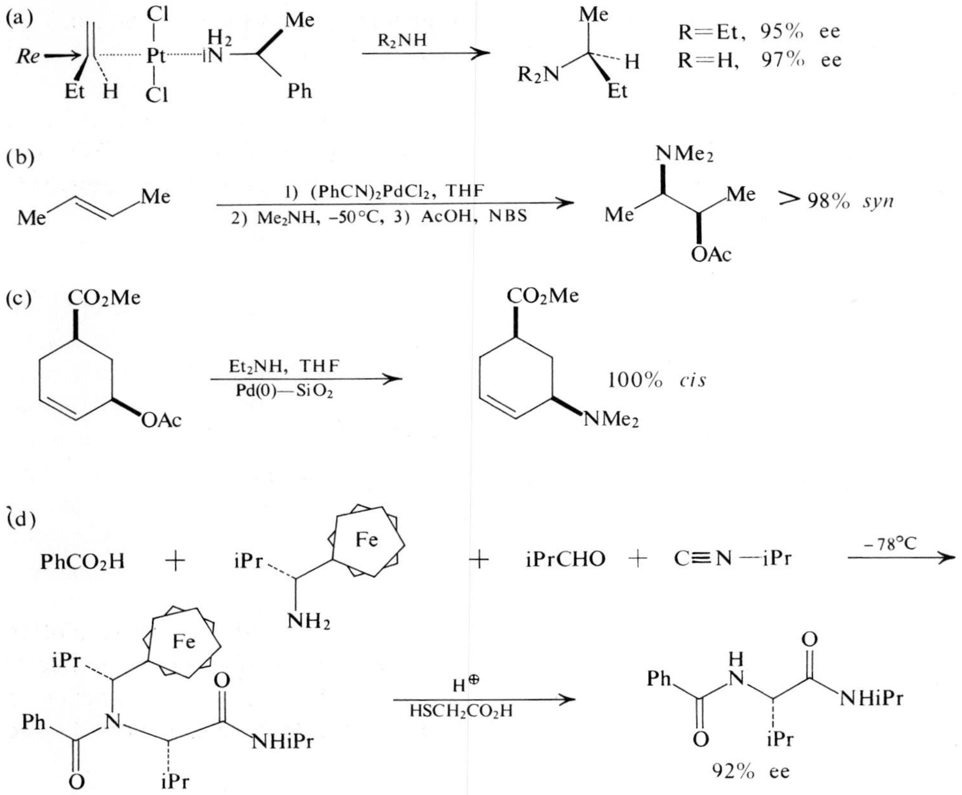

Fig. 8-1. Stereoselective formation of carbon-nitrogen bonds (I).

Transformation of an allylic acetate into an amine proceeded with complete retention of configuration when it was catalyzed by a Pd(0) complex bound to silica gel (Fig. 8-1 (c)) [7].

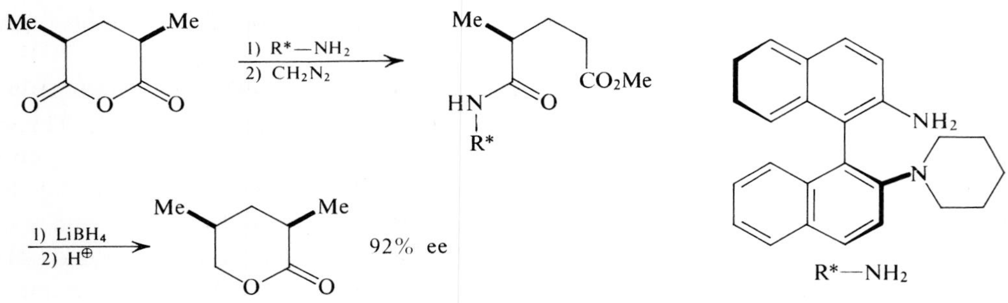

Fig. 8-2. Stereoselective formation of carbon-nitrogen bonds (II).

A rather original method in which a ferrocenylamine was used as chiral auxiliary was developed by Urban, Ugi and others for the synthesis of certain special peptides (Fig. 8-1 (d)) [8, 9].

Recently, preferential amide formation with one of the enantiotopic carbonyl groups of prochiral cyclic anhydrides was reported by Kawakami *et al.* (Fig. 8-2) [10]. A similar sequence starting from 3-hydroxy-3-methyl-pentandioic anhydride gave (*R*)-mevalolactone in 58% ee.

8.2 Stereoselective Formation of Carbon-Phosphorus Bonds

Optically active phosphorus compounds are usually obtained either by resolution or from chiral starting materials using transformations proceeding with complete or nearly complete retention and inversion, respectively [11]. Of the few examples, in which asymmetric induction and/or transfer of chirality have been exploited some

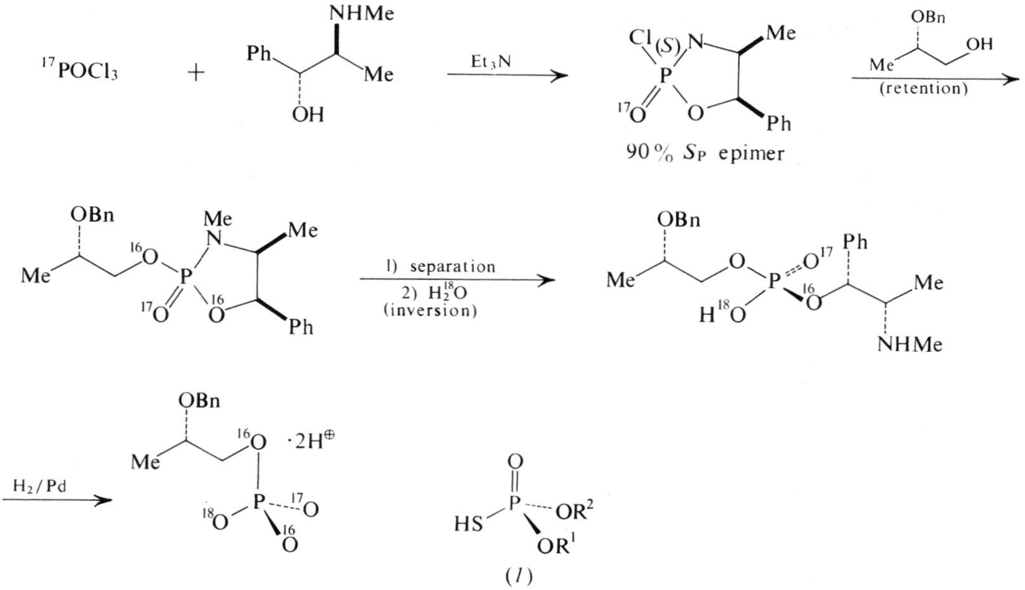

Fig. 8-3. Synthesis of a monophosphate ester chiral at phosphorus.

will be discussed here. Abbott *et al.* used ephedrine as a chiral auxiliary in the preparation of a monophosphate ester chiral at phosphorus due to the presence of three different oxygen isotopes (Fig. 8-3) [12]. In a similar way, chiral esters of thiophosphoric acid of the type *1* were prepared by Hall and Inch [13]. 2-Phenylethylamine was used as a chiral auxiliary in the synthesis of aminophosphonic acids [14] and cyclophosphamides [15] with moderate success.

8.3 Stereoselective Formation of Carbon-Oxygen Bonds

Among the many methods which can be utilized for stereoselective C–O bond formation, only hydroboration and halolactonization will be discussed in detail.

One of the most important methods for the enantioselective formation of secondary alcohols is the hydroboration of non-terminal olefins followed by oxidation. The method should have been classified as a stereoselective carbon-boron bond forming reaction, since stereoselection takes place in the first step, but the intermediate boranes are of little practical interest as compared with the end products, *i. e.* the alcohols.

In 1961 Brown and Zweifel discovered that on reacting diborane with an excess of α-pinene a dialkylborane, diisopinocampheylborane $(IPC)_2BH$ (*2*) was obtained in a completely diastereoselective reaction. *2* proved to be a highly enantioselective

B_2H_6

(*2*)

hydroborating reagent [16].* $(IPC)_2BH$ was found to be most useful with unhindered (*Z*)-olefins. Reaction with (*E*)-olefins was sluggish and less selective [18]. The primary products were usually not isolated but oxidized to secondary alcohols with complete retention of configuration. Typical results obtained with optically pure $[(-)\text{-IPC}]_2BH$ [19] or recalculated to such conditions [20–23] are shown in Fig. 8-4.**

* For a review, see ref. [17].

** The optical purity of natural α-pinene is ~95%.

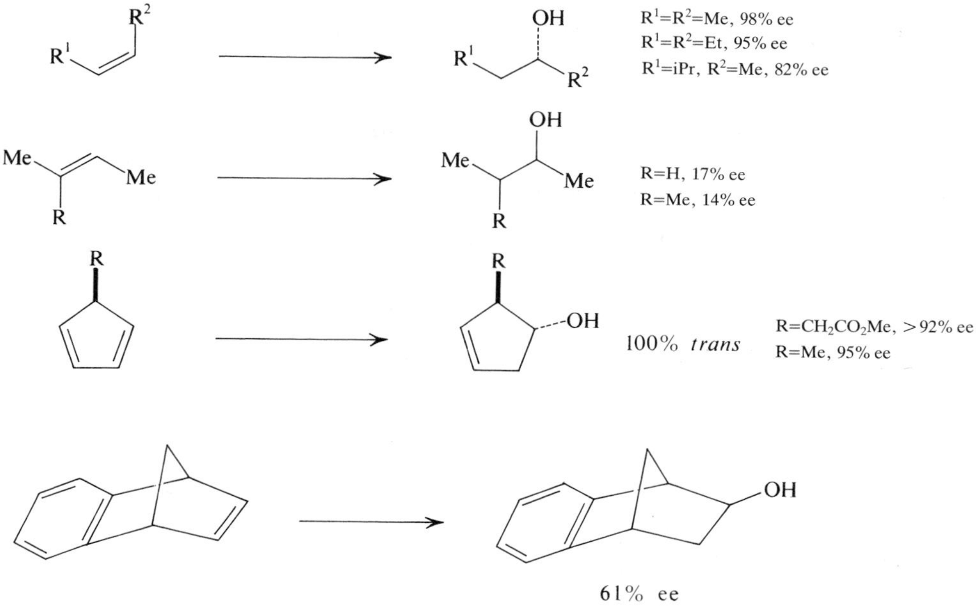

R^1=R^2=Me, 98% ee
R^1=R^2=Et, 95% ee
R^1=iPr, R^2=Me, 82% ee

R=H, 17% ee
R=Me, 14% ee

100% *trans*
R=CH_2CO_2Me, >92% ee
R=Me, 95% ee

61% ee

Fig. 8-4. Stereoselective hydroboration with diisopinocampheylborane.

Monoisopinocampheylborane ($IPCBH_2$) is a more reactive but less convenient reagent than $(IPC)_2BH$, because it is more difficult to prepare and consists of a mixture of isomeric dimers, the ratio of which changes on standing [24]. $IPCBH_2$ is not competitive with $(IPC)_2BH$ in the case of reactive olefins but exhibits moderate to excellent enantioselection with (*E*)-olefins (Fig. 8-5) [25–27].

When, instead of being oxidized, the primary adducts were reacted with acetaldehyde, chiral boranic acids were obtained in 70–100% ee [29, 30].

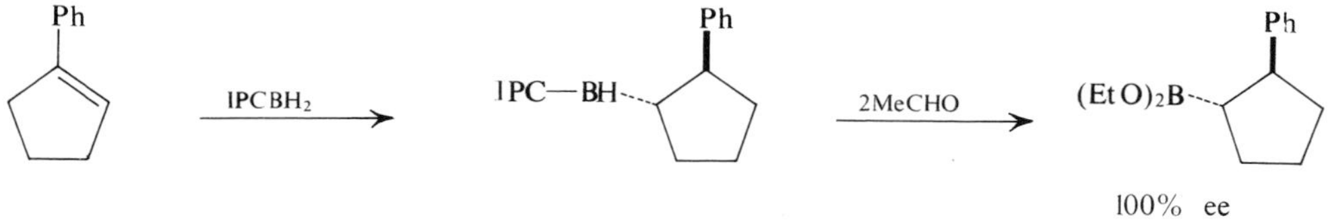

Ph — $IPCBH_2$ → IPC—BH (Ph) — 2MeCHO → $(EtO)_2B$ (Ph)

100% ee

Dilongifolinylborane, prepared from longifolene, has the advantage of being a relatively stable solid, but is less enantioselective than $(IPC)_2BH$ [28].

Hydroboration of olefins by diborane is generally assumed to involve *syn* addition and to proceed with high regiospecificity for the less substituted carbon atom. A study by Zioudrou *et al.* on a series of simple olefins has shown that regioselectivity is 80–100%, and the ratio of *syn* to *anti* addition is 87:3–97:3 [31].

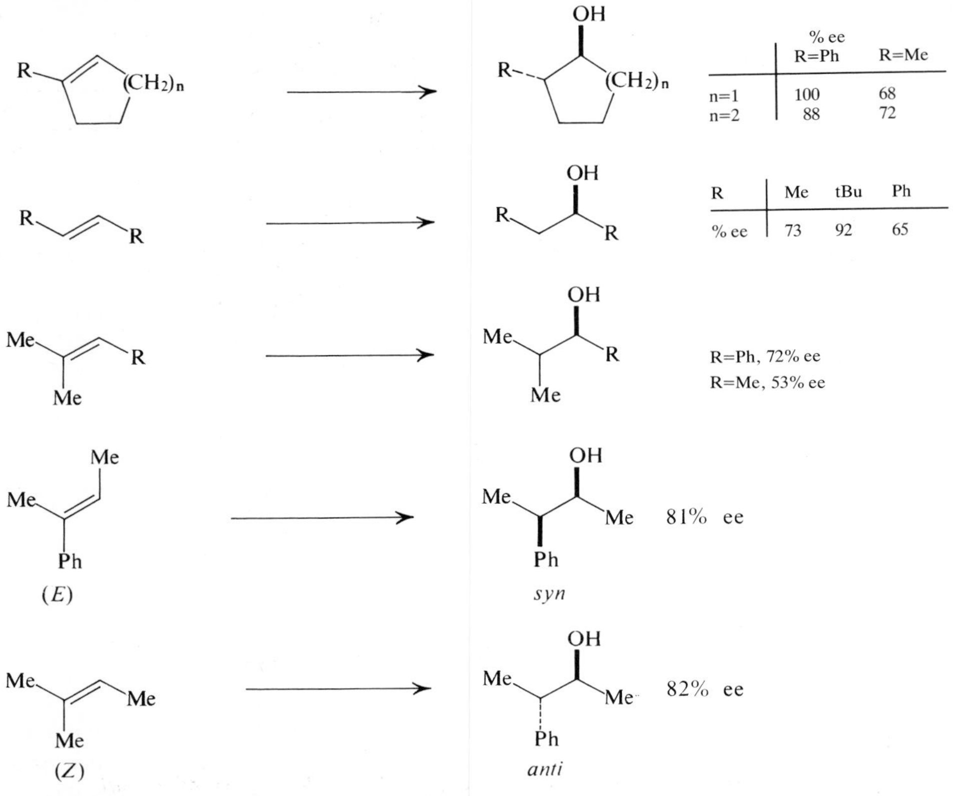

Fig. 8-5. Stereoselective hydroboration with monoisopinocampheylborane.

Asymmetric induction in hydroboration is only efficient when hindered boranes, such as 1,1,2-trimethylpropylborane (thexylborane), are used (Fig. 8-6(a)) [32]. Concurrently, Still and Shaw [33] and Morgens [34] exploited asymmetric induction in the hydroboration of a 1,5-diene for the synthesis of the Prelog-Djerassi lactone. The procedure is illustrated on simpler models (Fig. 8-6(b)). The concept of attaining acyclic stereoselection with the aid of a cyclic intermediate was successfully extended to 1,4-dienes, but failed with 1,6-dienes [35].

A computational study of stereoselective hydroboration was published by Houk *et al.* [36].

In halolactonization reactions a halogen atom and an intermolecular carboxy group are added to a double bond in an *anti* manner to give an α-halogenolactone. Terashima, Jew and Koga recognized that application of this transformation to the proline amides of α,β-unsaturated acids permitted the highly enantioselective preparation of certain α-hydroxycarboxylic acids [37–39]. The procedure is illustrated in Fig. 8-7(a). Unfortunately, no reaction takes place with the crotyl amide,

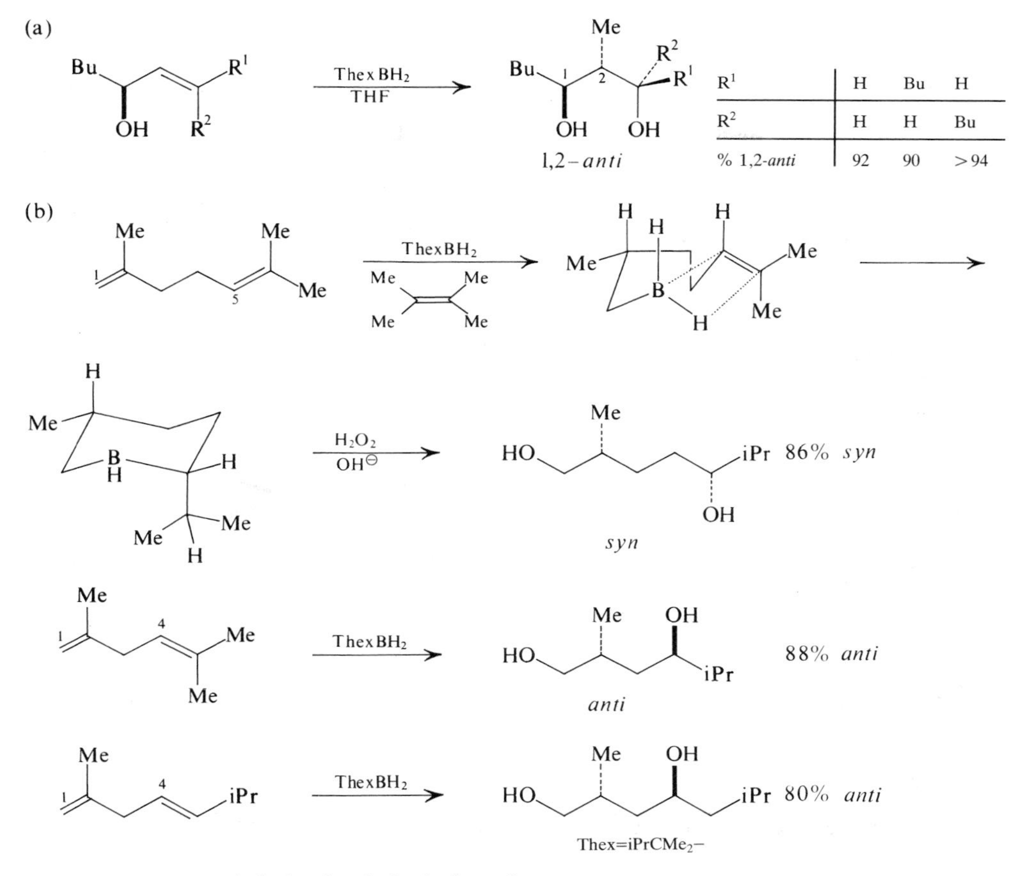

Fig. 8-6. Asymmetric induction in hydroboration.

and a 7-membered bromolactone is formed from the 3,3-dimethylacrylamide [39]. Treatment of the bromolactones with sodium methoxide and reduction with bis-(2-methoxyethoxy)-aluminum hydride yields chiral α,β-epoxyaldehydes [39].

Stereoselectivity of the halolactonization of chiral 4,5- and 5,6-unsaturated carboxylic acids is sensitive both to reaction conditions and the constitution of the substrate. From acids having a substituent adjacent to the double bond the *trans*-halolactones are formed in excess under equilibrium conditions (no base added), while the *cis*-products arise under kinetic control in the presence of $KHCO_3$ or $NaHCO_3$ (*e. g.* Fig. 8-7 (b)) [40, 41]. When in the 5,6-unsaturated acid *3* the methyl group was moved to C-3 and C-2, the predominant product became the *cis*- and the *trans*-lactone, respectively [41]. Similarly, from 2-substituted 4,5-unsaturated amides and thioamides, *trans*-lactones and -thiolactones were obtained after reduction (Fig. 8-7 (c)) [42].

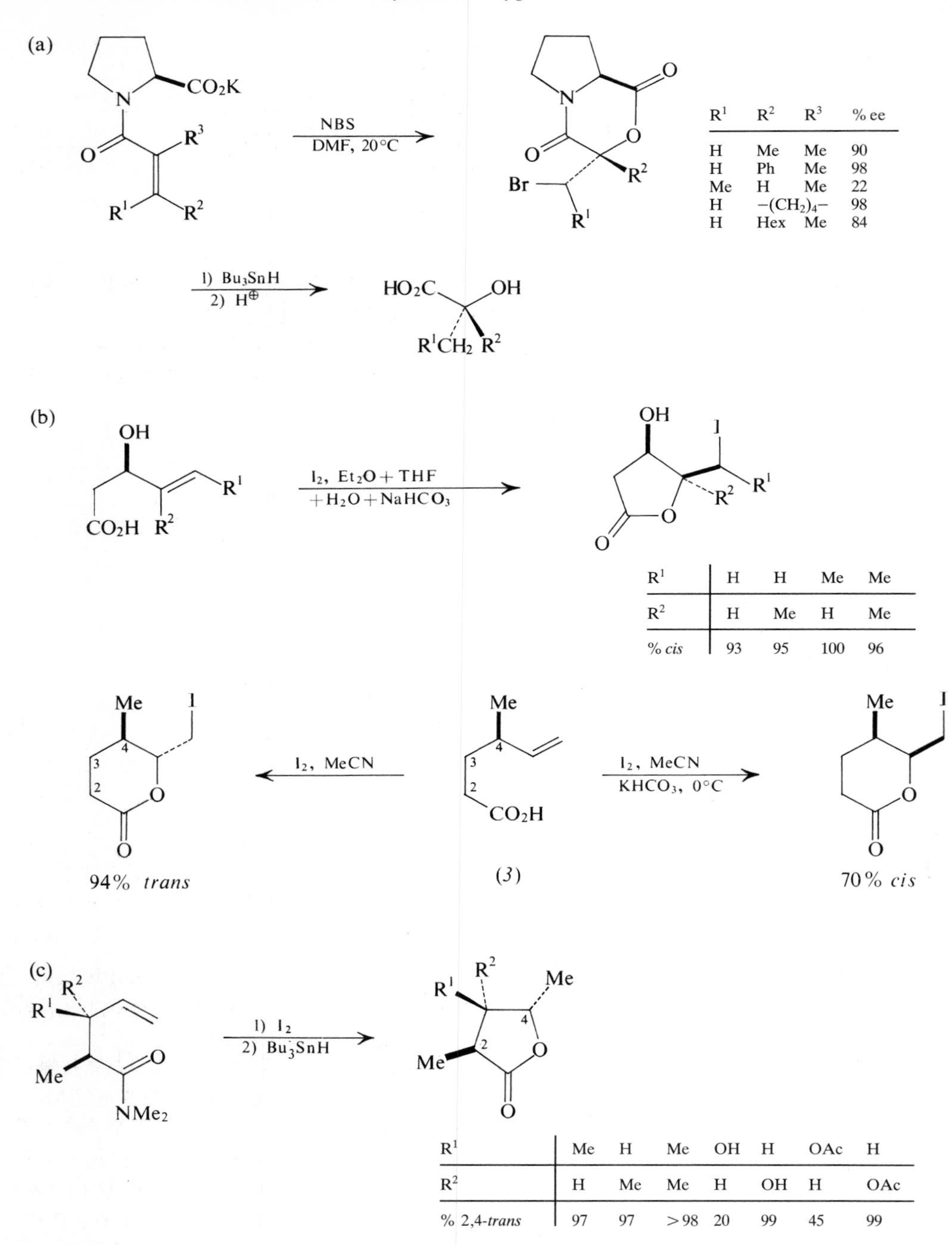

R^1	R^2	R^3	% ee
H	Me	Me	90
H	Ph	Me	98
Me	H	Me	22
H	$-(CH_2)_4-$		98
H	Hex	Me	84

R^1	H	H	Me	Me
R^2	H	Me	H	Me
% *cis*	93	95	100	96

R^1	Me	H	Me	OH	H	OAc	H
R^2	H	Me	Me	H	OH	H	OAc
% 2,4-*trans*	97	97	>98	20	99	45	99

Fig. 8-7. Enantioselective synthesis of α-hydroxycarboxylic acids by halolactonization.

OH R^3 R^1 R^2 → (I_2, H_2O) → OH OH R^1 R^2 R^3 I *syn, syn*

R^1	R^1	R^3	% *syn, syn*
Me	Bu	H	93
	H	Bu	97
Bu	Me	H	99
	H	Me	77
tBu	Me	H	97

Fig. 8-8. Asymmetric induction in iodohydrin formation.

Asymmetric induction by a hydroxyl group of allylic alcohols was found to be surprisingly effective in idodohydrin formation (Fig. 8-8) [43].

Acylation of alcohols with methylphenylketene in the presence of chiral catalysts has been thoroughly studied (*e. g.* [44]), but no practical method has emerged from

TosO OH OH → (ClSn, Me) →

TosO O SnCl O (*R*) ⇌ TosO O SnCl O (*S*)

1) N Me N O
2) PhCOCl, 0°C

TosO OH O_2CPh (*S*) 20% 80% ee

R OH R OH → (Bu_2SnO, PhMe) → R O $SnBu_2$ R O

Me Me COCl O = R*Cl

C_6H_5N, 0°C → R OH R OR*

R	MeO_2C	Ph
% de	89	46

Fig. 8-9. Stereoselective acylation of symmetrical diols.

these studies. Recently, Mukaiyama and his coworkers discovered that tin(II) and tin(IV) esters derived from diols having enantiotopic hydroxy groups can be acylated with considerable selectivity, either with benzoyl chloride in the presence of a chiral base [45] or with a chiral acid chloride [46] (Fig. 8-9). The method may be valuable in lipid synthesis.

8.4 Stereoselective Formation of Carbon-Sulfur Bonds

Enantioselective Michael addition of thiols in the presence of chiral catalysts has been extensively studied, not so much for the sake of the products, but rather to collect information about chiral catalyst systems.

Up to 67% ee were realized by Wynberg and his coworkers in the addition of thiolesters [47] and thiophenols [48] to cyclohexen-2-one in the presence of alkaloids. In the latter reaction, as shown by Mukaiyama *et al.*, hydroxyproline derivatives were more efficient (Fig. 8-10) [49].

Thiol addition to open-chain enones and nitroolefins was much investigated but remained only of theoretical interest [*e.g.* 50, 51].

Since chiral sulfur compounds, primarily sulfoxides, are useful as chiral auxiliaries [52], their enantioselective synthesis is important.

The *p*-tolylsulfinylester of (−)-menthol, one diastereomer of which can be obtained pure relatively easily [53], is a useful starting material since it can be transformed with complete inversion to compounds amenable to versatile transformations (Fig. 8-11 (a)) [54].

An efficient scheme for the transfer of chirality form an amine to a sulfoxide based on a cyclic intermediate was first reported by Wudle and Lee [55] and later

OH
N
Et
NHPh
SH
R
O
PhMe, −5 °C, 5 days
S

R	H	Cl	tBu
% ee	77	47	88

Fig. 8-10. Enantioselective Michael addition of thiols.

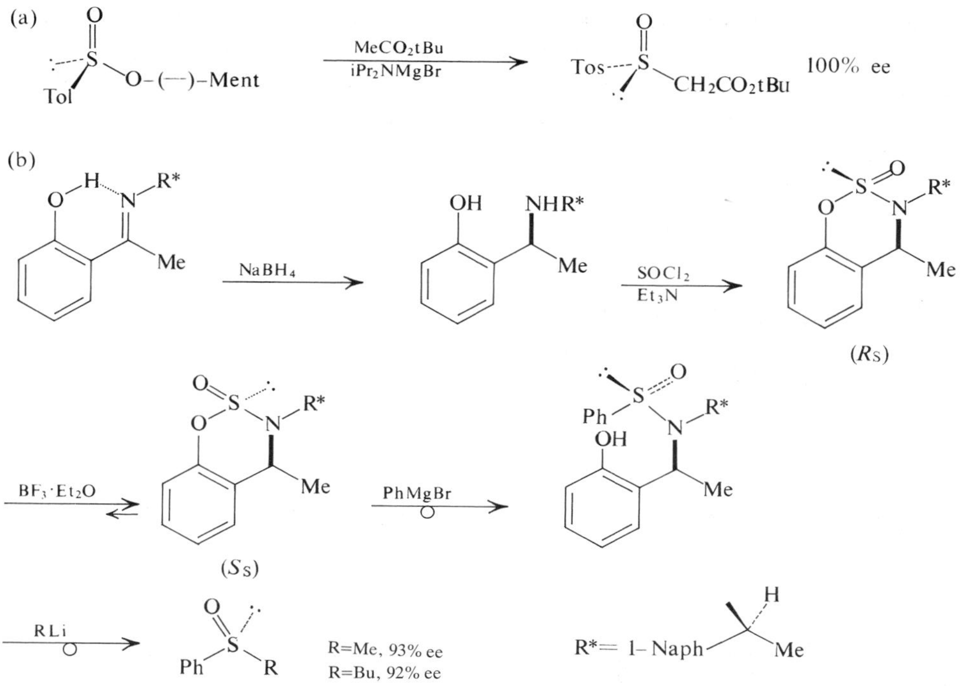

(a)

Tol–S(O)–O-(–)-Ment $\xrightarrow[iPr_2NMgBr]{MeCO_2tBu}$ Tos–S(O)–CH_2CO_2tBu 100% ee

(b)

$NaBH_4$ → ; $SOCl_2$, Et_3N → (R_S)

$BF_3 \cdot Et_2O$ ⇄ (S_S)

PhMgBr →

RLi → Ph–S(O)–R R=Me, 93% ee; R=Bu, 92% ee

R* = 1-Naph–CH(Me)

Fig. 8-11. Synthesis of optically active sulfoxides.

improved by Hiroi *et al.* [56] (Fig. 8-11 (b)). The original 2:1 mixture of (R_S)- and (S_S)-sulfinamides can either be separated, or their equilibrium shifted by the addition of boron trifluoride to obtain the pure S_S diastereomer.

8.5 Stereoselective Protonation and Hydrogen Migration

Although hydrogen is not regarded as a heteroatom, the discussion of protonation and hydrogen migration seems to be more in place in this chapter than in those dealing with reductions.

Chiral centers in which one of the ligands is an acidic hydrogen can be racemized by deprotonation-protonation, since carbanions are configurationally unstable. If this process interconverts diastereomers (epimerization), product distribution is almost invariably under thermodynamic control, since even a hydrated proton is very small, and therefore it is difficult to direct its approach to a specific face of the molecule. Despite their importance, such protonation reactions will not be discussed here.

Intermolecular enantioselective protonation of enamines [57, 58] and enolates [59] with di-*O*-acyltartaric acids has been described by Duhamel and Plaquevent. Enantioselectivities are low, but nevertheless remarkable. When combined with deprotonation by a chiral lithium base, enantioselectivity improved (Fig. 8-12 (a)).

Deuterium-hydrogen exchange in α-deuterated cyclopentanone affected the pro-*S* deuterium preferentially when catalyzed by a chiral diamine (Fig. 8-12 (b)) [61].

Decarboxylation of a prochiral malonate was found to be enantioselective in the presence of a chiral cobalt(III) complex. In fact enantioselectivity was due to preferential coordination of one of the enantiotopic carboxyl groups (Fig. 8-12 (c)) [60].

The concerted [1,2n + 1] sigmatropic shift of a hydrogen atom from one end of a π electron system to the other follows the Woodward-Hoffmann rules and is

(a)
tBu, CH—N(morpholine), Br → (2*R*,3*R*)-di-*O*-benzoyl-tartaric acid, −60 °C → tBu, CHO, Br — 32% ee

Ph, CO_2Me, N=CHPh (±) → 1) RLi, −70 °C; 2) (2*R*,3*R*)-di-*O*-adamantoyltartaric acid, −70 °C → Ph, CO_2Me, N=CHPh
R=$NiPr_2$, 53% ee
R= EtN–CH(Me)Ph, 70% ee

(b)
2,2,5,5-tetradeuterocyclopentanone → (chiral diamine: Me, Me, NMe_2, NH_2) → (2*S*,5*S*) 2,5-dideuterocyclopentanone — > 94% ee

(c)
Me, H_2N, CO_2H, CO_2H → 1) $[COCl_2L^*]^{\oplus}Cl^{\ominus}$; 2) MeOH, 65 °C, $-CO_2$ → NH_2, Me, CO_2H — 96% ee

L* = (pyrrolidin-2-yl)CH_2–NH–$(CH_2)_3$–NH–CH_2(pyrrolidin-2-yl)

Fig. 8-12. Enantioselective protonation.

(a) Me, O=S, OH, iPr — 50 °C, $-SO_2$ → Me, H, iPr

(b) Me, 4, $SiMe_3$, 2, Me, CO_2H — Δ → Me, Me, CO_2H — > 89% *anti*

anti

Me, 4, $SiMe_3$, 1, OH, R (*4*) — $BF_3 \cdot AcOH$, CH_2Cl_2 → Me, OH, R

anti

R	Me	Bu	tBu	Me Me
% *anti*	66	75	87	90

(c) Me, Me, Me, NEt_2 — Rh, P*, Δ → Me, Me, Me, NEt_2

— $H^{\oplus}$ → Me, Me, Me, CHO

P*=(*S*)−(*5*), 99% ee
P*=(*R*)-BINAP (*76*, Fig. 2-4), 96% ee

Me, Me, PPh_2, PPh_2 (*S*)−(*5*)

Fig. 8-13. Stereoselective intramolecular hydrogen transfer.

completely diastereoselective, provided that the formation of only one of the diastereomers is symmetry allowed (*cf.* Chapter 7). Some interesting examples of hydrogen migration in allylic systems, the mechanism of which has not yet been clarified, will be discussed below.

Corey and Engler prepared the less stable *trans* diastereomer of a hydrindane by exploiting intramolecular hydrogen (or proton) transfer (Fig. 8-13 (a)) [62]. With allylsilanes, Wilson and Price observed effective 1,3- and 1,4-induction in hydrogen transfer (Fig. 8-13 (b)) [63, 64]. The acetates of the alcohols *4* gave predominantly the *syn* products (uniformly 66–67%), which supported the participation of the hydroxy group.

Enantioselective hydrogen migration in an allylamine-enamine transformation catalyzed by a cobalt–DIOP complex was first reported by Kumobayashi and Akutagawa (ee up to 33%) [65]. A dramatic increase in selectivity when the rhodium complex of a chiral biphenyl or binaphthyl was applied as catalyst was simultaneously reported by Hansen *et al.* [66, 67] and by Noyori and his coworkers [68] (Fig. 8-13 (c)). The (*E*)-allylamine gave the (*S*)-enamine in excess [68].

Appendix

Relative prices of some chiral reagents per mole as compared with that of (*R,R*)-tartaric acid (taken as 1.00).*

D-Alanine	233
L-Alanine	11
L-Alanine-t-butylester · HCl	2140
(*R*)-2-Amino-1-butanol	356
(*S*)-2-Amino-1-butanol	88
(1*S*, 2*S*)-2-Amino-1-phenyl-1,3-propanediol	531
(*S*)-2-Amino-3-phenyl-1-propanol (phenylalaninol)	513
(*R*)-1-Amino-2-propanol	549
(*S*)-1-Amino-2-propanol	693
(*S*)-2-Amino-1-propanol (alaninol)	30.3
(−)-Borneol	32.3
(*R,R*)-2,3-Butanediol	630
(+)-Camphor	0.10
(−)-Camphor	1.00
(+)-Camphoric acid	0.10
(+)-Carvone	39
(−)-Carvone	14
(−)-Chinchonidine	53
(+)-Cinchonine	45
L-Cysteine	15
(*R,R*)-2,3-Dimethoxy-1,4-bis(dimethylamino)butane (DDB)	618
(*S,S*)-2,3-Dimethoxy-1,4-bis(dimethylamino)butane (DDB)	563
(*R*)- and (*S*)-*N,N*-Dimethyl-1-phenylethylamine	1490
(+)-Ephedrine	25
(−)-Ephedrine	18
trans-4-Hydroxy-L-proline	62
trans-4-Hydroxy-D-proline	1780
(*R*)-Glyceraldehyde	1292
(*S*)-Glyceraldehyde	3276
(+)- and (−)-Isopinocampheyl-9-borabicyclo[3.3.1]nonane (Alpine-borane®)	550
(*R,R*)- and (*S,S*)-2,3-Isopropylidene-2,3-dihydroxy-1,4-bis(diphenylphosphino)butane (DIOP)	20000
(+)-Isomenthol	11
Lithium-B-isopinocampheyl-9-borabicyclo[3.3.1]nonyl hydride	10110
(*S*)-Lactic acid	0.72
(+)-Longifolene	867
(*R*)-Malic acid	360
(*S*)-Malic acid	27
(−)-Menthol	16
(−)-Menthone	37
(−)-Menthyl chloroformate	50

* Average price for 1985 US $ 1.00.

(4*S*, 5*S*)-4-Methoxymethyl-2-methyl-5-phenyl-2-oxazoline	754
(*R*)- and (*S*)-α-Methoxy-α-trifluoromethylphenylacetic acid	2470
(−)-*N*-Methylephedrine	662
(−)-*cis*-Myrtanylamine	40
(*R*)- and (*S*)-1-(1-Naphthyl)ethylamine	1580
(−)-Neomenthol	500
L-Phenylalanine	24
D-Phenylalanine	38
(*R*)- and (*S*)-1-Phenylethanol	4700
(*R*)-1-Phenylethylamine	97
(*S*)-1-Phenylethylamine	92
(*R*)-Phenylglycine	42
(*S*)-Phenylglycine	254
(*S*)-3-Phenyllactic acid	1030
(+)- and (−)-Pinanemethylamine · HCl	57
(+)-α-Pinene	10
(−)-α-Pinene	152
(−)-β-Pinene	0.12
D-Proline	2300
L-Proline	21
(+)-Pulegone	27
(*R*)-2-Pyrrolidinemethanol (prolinol)	27
(−)-Quinine	4.0
D-Serine	76
L-Serine	24
(*R*,*R*)-Tartaric acid	1.00
(*S*,*S*)-Tartaric acid	50
L-Threonine	27
D-Valine	272
L-Valine	17

Literature

Chapter 1

[1] *Nomenclature of Organic Chemistry, Part E.* Pure Appl. Chem. **45**, 11 (1976).
[2] a) R. S. Cahn, C. Ingold and V. Prelog. Angew. Chem. **78**, 413 (1966); Int. Ed. Engl. **5**, 385 (1966).
b) V. Prelog and G. Helmchen, Angew. Chem. **94**, 614 (1982); Int. Ed. Engl. **21**, 567 (1982).
[3] C. H. Heathcock, C. T. Buse, W. A. Kleschick, M. C. Pirrung, J. E. Sohn and J. Lampe, J. Org. Chem. **45**, 1066 (1980).
[4] S. Masamune, S. A. Ali, D. L. Snitman and D. S. Garrey, Angew. Chem. **92**, 573 (1980); Int. Ed. Engl. **19**, 557 (1980).
[5] F. A. Carey and M. E. Kuehne, J. Org. Chem. **17**, 3811 (1982).
[6] R. Noyori, I. Nishida and J. Sakata, J. Am. Chem. Soc. **103**, 2108 (1981).
[7] D. Seebach and V. Prelog, Angew. Chem. **94**, 696 (1982); Int. Ed. Engl. **21**, 654 (1982).
[8] *Nomenclature for Straightforward Transformations,* Pure Appl. Chem. **53**, 305 (1981).
[9] C. A. Heathcock, in *Asymmetric Synthesis,* (J. D. Morrison, ed.). Academic Press, New York, 1984, p. 111.
[10] K. Mislow and M. Raban, Top. Stereochem. **1**, 1 (1967).
[11] M. Raban, Tetrahedron Lett. 3105 (1966).
[12] K. R. Hanson, J. Am. Chem. Soc. **88**, 2771 (1966).
[13] V. Prelog and S. Helmchen, Helv. Chim. Acta **55**, 2581 (1972).
[14] P. Müller and J. C. Perlberger, Helv. Chim. Acta **59**, 2335 (1976).
[15] S. Winstein, D. Pressman and W. G. Young, J. Am. Chem. Soc. **61**, 1645 (1939).
[16] P. Pfeiffer, Ber. dtsch. chem. Ges. **45**, 1816 (1912).
[17] W. Kirmse and R. Jendralla, Chem. Ber. **111**, 1858 (1978).
[18] E.-O. Renth, Angew. Chem. **87**, 379 (1975); Int. Ed. Engl. **14**, 361 (1975).
[19] P. H. Schippers and H. P. J. M. Dekkers, J. C. S. Perkin **2**, 1429 (1982).
[20] W. H. Pirkle and P. L. Rinaldi, J. Am. Chem. Soc. **99**, 3510 (1977).
[21] F. Cramer and W. Dietsche, Chem. Ber. **92**, 1739 (1959).
[22] W. Mikolajczyk, Chem. Ind. 2059 (1966).
[23] S. Miyano, L. D. L. Lu, S. M. Viti and K. B. Sharpless, J. Org. Chem. **48**, 3608 (1983).
[24] G. Balavoine, S. Juce and H. B. Kagan, Tetrahedron Lett. 4159 (1973).
[25] K. Liu, Y. Tong and Y. Chang, Tetrahedron Lett. 2725 (1980).
[26] K. Mislow, R. R. O'Brien and R. Schaeser, J. Am. Chem. Soc. **84**, 1940 (1982).
[27] W. Marckwald and A. McKenzie, Ber. dtsch. chem. Ges. **32**, 2130 (1899).
[28] V. S. Martin, S. S. Woodard, T. Katsuki, Y. Yamada, M. Ikeda and K. B. Sharpless, J. Am. Chem. Soc. **103**, 6237 (1981).
[29] J. Jacques, A. Collet and S. H. Wilen, *Enantiomers, Racemates and Resolution.* Wiley-Interscience, New York, 1981, p. 81.
[30] D. A. Evans, J. V. Nelson and T. R. Taber, Top Stereochem. **13**, 1 (1981).
[31] J. F. Nicoud and H. B. Kagan, Israel J. Chem. **15**, 78 (1976/77).
[32] C. Quannes, R. Beugelmans and G. Rossi, J. Am. Chem. Soc. **95**, 8272 (1973).
[33] J. Hagenbuch and P. Vogel, J. C. S. Chem. Comm. 1062 (1980).

[34] R. Noyori, I. Tomino and Y. Tanimoto, J. Am. Chem. Soc. **101**, 3129 (1979).
[35] Y. Izumi, Angew. Chem. **83**, 956 (1971); Int. Ed. Engl. **10**, 871 (1971).
[36] Y. Izumi and A. Tai, *Stereodifferentiating Reactions*. Academic Press, New York, 1977.
[37] D. Seebach, R. Dörr, B. Bastani and V. Ehrig, Angew. Chem. **81**, 1002 (1969).
[38] Y. Inoue, Y. Kunitomi, S. Takumaku and H. Sakurai, J. C. S. Chem. Comm. 1024 (1978).
[39] R. Hazard, S. Jaouannat and A. Tallec, Tetrahedron **38**, 93 (1982).
[40] Z. G. Hajós and D. R. Parrish, J. Org. Chem. **39**, 1615 (1974).
[41] S. Juliá, A. Ginebreda, J. Huixer, J. Masana, A. Tomás and S. Colonna, J. C. S. Perkin 1 575 (1981).
[42] A. Tai, M. Nakahata, T. Harada and Y. Izumi, Chem. Lett. 1125 (1980).
[43] I. Ojima, T. Kogure and T. Terasaki, J. Org. Chem. **43**, 3444 (1978).
[44] K. Osakada, M. Obana, T. Ikariya, M. Saburi and S. Yoshikawa, Tetrahedron Lett. 4297 (1981).
[45] N. Baba, J. Oda and Y. Inouye, J. C. S. Chem. Comm. 815 (1980).
[46] K. Mislow, *Introduction to Stereochemistry*. Benjamin, New York, 1966, p. 131.
[47] D. R. Boyd and R. Graham, J. Chem. Soc. C 2648 (1969).
[48] S. Hashimoto and K. Koga, Chem. Pharm. Bull. **27**, 2760 (1979).
[49] R. Meric and J.-P. Vigneron, Tetrahedron Lett. 2059 (1974).
[50] K. Weinges, K.-L. Klotz and H. Droste, Chem. Ber. **113**, 710 (1980).
[51] J. C. Fiaud and H. Horeau, Tetrahedron Lett. 2565 (1972).
[52] A. McKenzie, J. Chem. Soc. **101**, 1196 (1912).
[53] E. L. Eliel and Y. Senda, Tetrahedron **26**, 2411 (1970).
[54] E. Fischer, Ber. dtsch. chem. Ges. **27**, 3189 (1894).
[55] V. Prelog, Helv. Chim. Acta **36**, 308 (1953).
[56] D. J. Cram and F. A. Abd Elhafez, J. Am. Chem. Soc. **74**, 5828 and 5951 (1952).
[57] E. Ruch and I. Ugi, Top. Stereochem. **4**, 99 (1969).
[58] E. Anders, E. Ruch and I. Ugi, Angew. Chem. **85**, 16 (1973); Int. Ed. Engl. **12**, 25 (1973).
[59] M. Raban and E. M. Carlson, Israel J. Chem. **15**, 106 (1976/1977).
[60] J.-P. Vigneron, M. Dhaenens and A. Horeau, Tetrahedron **33**, 497 (1977).
[61] W. T. Wipke and P. Gund, J. Am. Chem. Soc. **98**, 8107 (1976).
[62] M. Hirota, K. Abe, H. Tashiro and M. Nishio, Chem. Lett. 777 (1982).
[63] W. Marckwald, Ber. dtsch. chem. Ges. **37**, 1368 (1904).
[64] J. D. Morrison and H. S. Mosher, *Asymmetric Organic Reactions*. Prentice-Hall, Engelwood Cliffs, N.J. U.S.A., 1971.
[65] J.-P. Guetté and A. Horeau, Bull. Soc. Chim. France, 1747 (1967).
[66] H. Handel and J.-L. Pierre, Tetrahedron Lett. 2029 (1976).
[67] R. K. Hill and N. W. Gilman, J. C. S. Chem. Comm 619 (1967).
[68] T. Nakai, K. Mikami and S. Taya, J. Am. Chem. Soc. **103**, 6492 (1981).
[69] A. I. Meyers and D. G. Wettlaufer, J. Am. Chem. Soc. **106**, 1135 (1984).
[70] B. M. Trost, J. M. Timko and J. L. Stanton, J.C.S. Chem. Comm. 436 (1978).
[71] A. Fischli, M. Klaus, H. Mayer, P. Schönholzer and R. Rüegg, Helv. Chim. Acta **58**, 564 (1975).
[72] S. Terashima and S. Yamada, Tetrahedron Lett. 1001 (1977).
[73] K.-K. Chan, N. Cohen, J. P. DeNoble, A. C. Specian, Jr. and G. Saucy, J. Org. Chem. **41**, 3505 (1976).
[74] T. Koizumi, Y. Kobayashi, H. Amitani and E. Yoshii, J. Org. Chem. **42**, 3459 (1977).
[75] A. I. Meyers, G. Knaus, K. Kamata and M. E. Ford, J. Am. Chem. Soc. **98**, 567 (1976).
[76] For details see p. 237 in E. L. Eliel, *Stereochemistry of Carbon Compounds*. McGraw-Hill, 1962.
[77] W. C. Still and I. Galynker, Tetrahedron **23**, 3981 (1981).

Chapter 2

[1] J. A. Osborne, F. H. Jardine, J. F. Young and G. Wilkinson, J. Chem. Soc. A, 1711 (1966).
[2] L. Horner, H. Winkler, A. Rapp, A. Mentrup, H. Hoffmann and H. Beck, Tetrahedron Lett. 161 (1961).
[3] For a review, see N. J. Gallagher and I. D. Jenkins, in *Topics in Stereochem.* Vol. 3, Chap. 1, (N. L. Allinger and E. L. Eliel, eds.). Wiley, New York, 1968.
[4] O. Korpium, R. A. Lewis, J. Chickos and K. Mislow, J. Am. Chem. Soc. **90**, 4842 (1968).
[5] W. S. Knowles and M. J. Sabacky, J.C.S. Chem. Comm. 1445 (1968).
[6] B. D. Vineyard, W. S. Knowles and M. J. Sabacky, J. Mol. Catal. **19**, 159 (1983).
[7] L. Horner, H. Siegel and H. Büthe, Angew. Chem. **80**, 1034 (1968); Int. Ed. Engl. **7**, 942 (1968).
[8] P. Abley and F. J. McQuillin, J.C.S. Chem. Comm. 477 (1969).
[9] P. Abley and F. J. McQuillin, J. Chem. Soc. C. 844 (1971).
[10] J. D. Morrison, R. E. Burnett, A. M. Aguiar, C. J. Morrow and C. Phillips, J. Am. Chem. Soc. **93**, 1301 (1971).
[11] W. S. Knowles, M. J. Sabacky and B. D. Vineyard, J.C.S. Chem. Comm. 10 (1972); U.S. Pat. 4.005.127 (25. 1. 1977). C.A. **86**, 190463 (1977).
[12] B. D. Vineyard, W. S. Knowles, M. J. Sabacky, G. L. Bachman and D. J. Weinkauff, J. Am. Chem. Soc. **99**, 5946 (1977).
[13] H. B. Kagan and T.-P. Dang, J.C.S. Chem. Comm. 481 (1971).
[14] H. B. Kagan and T.-P. Dang, J. Am. Chem. Soc. **94**, 6429 (1972).
[15] V. Čaplar, G. Comisso and V. Šunjic, Synthesis 85 (1981).
[16] J. D. Morrison, R. E. Burnett, A. M. Aguiar, C. J. Morrow and C. Phillips, J. Am. Chem. Soc. **93**, 1301 (1971).
[17] F. Joó and E. Trócsányi, J. Organomet. Chem. **231**, 63 (1982).
[18] F. Joó and E. Trócsányi, J. Organomet. Chem. **231**, 63 (1982).
[19] T. Hayashi, K. Kanehira and M. Kumada, Tetrahedron Lett. 4417 (1981).
[20] I. Ojima, T. Kogure, N. Yoda, T. Suzuki, M. Yatabe and T. Tanaka, J. Org. Chem. **47**, 1329 (1982).
[21] K. Achiwa, Tetrahedron Lett. 3735 (1977).
[22] Ger. Offen. DE 3.302.697 (18. 8. 1983). C.A. **100**, 22797 (1984).
[23] T. Hayashi, M. Tanaka and I. Ogata, Tetrahedron Lett. 295 (1977).
[24] C. Fisher and H. S. Mosher, Tetrahedron Lett. 2487 (1977).
[25] K. Yamamoto, A. Tomita and J. Tsuji, Chem. Lett. 3 (1978).
[26] W. R. Cullen, F. W. B. Einstein, C. Huang, A. C. Willis and E.-S. Yeh, J. Am. Chem. Soc. **102**, 988 (1980).
[27] U. Nagel, H. Menzel, P. W. Lednor, W. Beck, A. Guyot and M. Barholin, Naturf. **36**, 578 (1981).
[28] M. Yamada, M. Yamashita and S. Inokawa, Carbohydr. Res. **95**, C9 (1981).
[29] Y. Nakamura, S. Saito and Y. Morita, Chem. Lett. 7 (1980).
[30] B. D. Vineyard, W. S. Knowles, M. J. Sabacky, G. L. Bachman and D. J. Weinkauff, J. Am. Chem. Soc. **99**, 5946 (1977).
[31] W. S. Knowles, M. J. Sabacky, B. D. Vineyard and D. J. Weinkauff, J. Am. Chem. Soc. **97**, 2567 (1975).
[32] T. Yoshikuni and J. C. Bailar, J. Inorg. Chem. **21**, 2129 (1982).
[33] R. B. King, J. Bakos, C. D. Hoff and L. Markó, J. Org. Chem. **44**, 3095 (1979).
[34] M. D. Fryzuk and B. Bosnich, J. Am. Chem. Soc. **99**, 6262 (1977).
[35] J. Köttner and G. Greber, Chem. Ber. **113**, 2323 (1980).
[36] P. A. MacNeil, N. K. Roberts and B. Bosnich, J. Am. Chem. Soc. **103**, 2273 (1981).
[37] J. M. Brown and B. A. Murrer, J.C.S. Perkin Trans. 2, 489 (1982).

[38] M. Fiorini and G. M. Giongo, J. Mol. Catal. **7**, 411 (1980).
[39] Rhone Poulenc S. A. French. Pat. 2.230.654 (20. 6. 1974).
[40] R. Glaser, M. Twaik, S. Geresh and J. Blumenfeld, Tetrahedron Lett. **52**, 4635 (1977).
[41] R. Glaser, J. Blumenfeld and M. Twaik, Tetrahedron Lett. **52**, 4639 (1977).
[42] P. Aviron-Violet, Y. Collenille and J. Varagnal, J. Mol. Catal. **5**, 41 (1979).
[43] T.-P. Dang, J. C. Poulin and H. B. Kagan, J. Organomet. Chem. **91**, 105 (1975).
[44] R. Glaser, S. Geresh, J. Blumenfeld and M. Twaik, Tetrahedron **34**, 2405 (1978).
[45] A. Miyashita, A. Yasuda, H. Takaya, K. Toriumi, T. Ito, T. Souchi and R. Noyori, J. Am. Chem. Soc. **102**, 7932 (1980).
[46] K. J. Brown, M. S. Berg, K. K. Waterman, D. Lingenfelter and J. R. Murdoch, J. Am. Chem. Soc. **106**, 4717 (1984).
[47] A. Miyashita, H. Takaya, T. Souchi and R. Noyori, Tetrahedron **40**, 1245 (1984).
[48] D. L. Allen, V. C. Gibson, M. H. L. Green, J. F. Skinner, J. Bashkin and P. D. Grebenik, J.C.S. Chem. Comm. 895 (1983).
[49] H. Brunner, W. Pieronczyk, B. Schönhammer, K. Streng, I. Bernal and J. Korp, Chem. Ber. **114**, 1137 (1981).
[50] H.-J. Kreuzfeld and C. Döbler, React. Kinet. Catal. Lett. **16**, 229 (1981).
[51] T.-P. Dang and H. Kagan, J.C.S. Chem. Comm. 481 (1971).
[52] M. D. Fryzuk and B. Bosnich, J. Am. Chem. Soc. **101**, 3043 (1979).
[53] W. Bergstein, A. Kleemann and J. Martens, Synthesis 76 (1981).
[54] R. B. King, J. Bakos, C. D. Hoff and L. Markó, J. Org. Chem. **44**, 1729 (1979).
[55] D. P. Riley, J. Organomet. Chem. **234**, 85 (1982).
[56] U.S. Pat. 4.393.240 (12. 7. 1983). C.A. **99**, 212720 (1983).
[57] T. H. Johnson, D. K. Pretzer, S. Thomen, V. J. K. Chaffin and G. Rangarajan, J. Org. Chem. **44**, 1878 (1979).
[58] V. Caplar, G. Comisso and V. Sunjic, Synthesis 85 (1981).
[59] O. Samuel, R. Couffignal, M. Lauer, S. Y. Zhang and H. B. Kagan, Nouv. J. Chim. **5**, 15 (1981).
[60] B. A. Murrer, J. M. Brown, P. A. Chaloner, P. N. Nicholson and D. Parker, Synthesis 350 (1979).
[61] S. Y. Zhang, S. Yemul, H. B. Kagan, R. Stern, D. Commereuc and Y. Chauvin, Tetrahedron Lett. 3955 (1981).
[62] D. Lafont, D. Sinou and G. Descotes, J. Organomet. Chem. **169**, 87 (1979).
[63] D. Lafont, D. Sinou, G. Descotes, R. Glaser and R. Geresh, J. Mol. Catal. **10**, 305 (1981).
[64] D. Sinou, Tetrahedron Lett. 2987 (1981).
[65] G. Descotes, D. Lafont and D. Sinou, J. Organomet. Chem. **150**, C14–C16 (1978).
[66] J. M. Brown and B. A. Murrer, Tetrahedron Lett. 581 (1980).
[67] T. H. Johnson and G. Rangarajan, J. Org. Chem. **45**, 62 (1980).
[68] T. Hayashi, T. Mise, M. Fukushima, M. Kogatani, N. Nagashima, Y. Hamada, A. Matsumoto, S. Kawarami, M. Konishi, K. Yamamoto and M. Kumada, Bull. Chem. Soc. Jap. **53**, 1138 (1980).
[69] T. Hayashi, T. Mise, S. Mitachi, K. Yamamoto and M. Kumada, Tetrahedron Lett. 1133 (1976).
[70] H. Brunner, B. Schönhammer and C. Steinberger, Chem. Ber. **116**, 3529 (1983).
[71] K. Achiwa, Chem. Lett. 777 (1977).
[72] K. Achiwa and T. Soga, Tetrahedron Lett. 1119 (1978).
[73] K. Achiwa, Tetrahedron Lett. **17**, 1475 (1978).
[74] K. Achiwa, Chem. Lett. 905 (1978).
[75] B. L. Baker, S. J. Fritschel, J. R. Stille and J. K. Stille, J. Org. Chem. **47**, 2954 (1981).
[76] K. Achiwa, J. Am. Chem. Soc. **98**, 8265 (1976).
[77] K. Achiwa, Chem. Lett. 561 (1978).
[78] I. Ojima and T. Kogure, Chem. Lett. 567 (1978).
[79] I. Ojima, T. Kogure and N. Yoda, J. Org. Chem. **45**, 4728 (1980).
[80] I. Ojima and N. Yoda, Tetrahedron Lett. **21**, 1051 (1980).
[81] U. Nagel, Angew. Chem. **96**, 425 (1984); Int. Ed. Engl. **23**, 435 (1984).

[82] K. Hanaki, K. Kashiwabara and J. Fujita, Chem. Lett. 489 (1978).
[83] M. Yamashita, K. Hiramatsu, M. Yamada, N. Suzuki and S. Inakawa, Bull. Chem. Soc. Jap. **55**, 2917 (1982).
[84] J. Brunner and W. Pieronczyk, J. Chem. Res. (S) 74 (1980).
[85] M. Tanaka and I. Ogata, J.C.S. Chem. Comm. 735 (1975).
[86] J. Bakos, I. Tóth and B. Heil, Tetrahedron Lett. **25**, 4965 (1984).
[87] R. H. Grubbs and R. A. DeVries, Tetrahedron Lett. 1879 (1977).
[88] J. Bourson and L. Oliveros, L. Organomet. Chem. **229**, 77 (1982).
[89] R. Selke, React. Kinet. Catal. Lett. **10**, 135 (1979).
[90] W. R. Cullen and Y. Sugi, Tetrahedron Lett. 1635 (1978).
[91] D. Sinou and G. Descotes, React. Kinet. Catal. Lett. **14**, 463 (1980).
[92] R. Jackson and D. J. Thompson, J. Organomet. Chem. **159**, C29 (1978).
[93] E. Cesarotti, A. Chiesa and G. D. Alfonso, Tetrahedron Lett. 2995 (1982).
[94] E. Cesarotti, A. Chiesa, G. Ciani and A. Sironi, J. Organomet. Chem. **251**, 79 (1983).
[95] M. Yatagai, M. Zama, T. Yamagishi and M. Hida, Chem. Lett. 1203 (1983).
[96] M. Fiorini, F. Marcati and G. M. Giongo, J. Mol. Catal. **4**, 125 (1978).
[97] E. Kashiwabara, K. Hanaki and J. Fujita, Bull. Chem. Soc. Jap. **53**, 2275 (1980).
[98] K. Onuma, T. Ito and A. Nakamura, Chem. Lett. 905 (1979).
[99] K. Onuma, T. Ito and A. Nakamura, Bull. Chem. Soc. Jap. **53**, 2016 (1980).
[100] A. Uehara, T. Kubota and R. Tsuchiya, Chem. Lett. 441 (1983).
[101] S. Miyano, M. Nawa and H. Hashimoto, Chem. Lett. 729 (1980).
[102] S. Miyano, M. Nawa, A. Mori and H. Hahimoto, Bull. Chem. Soc. Jap. **57**, 2178 (1984).
[103] G. Pracejus and H. Pracejus, Tetrahedron Lett. 3497 (1977).
[104] M. Fiorini and G. M. Giongo, J. Mol. Catal. **5**, 303 (1979).
[105] B. A. Murrer, J. M. Brown, P. A. Chaloner, P. N. Nicholson and D. Parker, Synthesis 350 (1979).
[106] O. Červinka and K. Gajewski, Czechosl. Pat. 194.386 (1979).
[107] T. Hayashi and M. Kumada, Acc. Chem. Res. **15**, 395 (1982).
[108] H. W. Krause, React. Kinet. Catal. Lett. **10**, 243 (1979).
[109] K. Ohkubo, K. Fujimori and K. Yoshinaga, Inorg. Nucl. Chem. Letters **15**, 231 (1979).
[110] W. Dumont, J.-C. Poulin, T.-P. Dang and H. B. Kagan, J. Am. Chem. Soc. **95**, 8295 (1973).
[111] K. Ohkubo, M. Haga, K. Yoshinaga and Y. Motazato, Inorg. Nucl. Chem. Letters **17**, 215 (1981).
[112] T. Masuda and J. K. Stille, J. Am. Chem. Soc. **100**, 268 (1978).
[113] N. Takaishi, H. Imai, C. A. Bertelo and J. K. Stille, J. Am. Chem. Soc. **100**, 264 (1978).
[114] G. L. Baker, S. J. Fritschel and J. K. Stille, J. Org. Chem. **46**, 2960 (1981).
[115] K. Achiwa, Heterocycles **9**, 1539 (1978).
[116] M. Inoue, K. Ohta, N. Ishizuka and S. Enomoto, Chem. Pharm. Bull. **31**, 3371 (1983).
[117] R. Glaser, S. Geresh and J. Blumenfeld, J. Organomet. Chem. **112**, 355 (1976).
[118] I. Ojima and T. Kogure, Chem. Lett. 567 (1978).
[119] M. D. Fryzuk and B. Bosnich, J. Am. Chem. Soc. **100**, 5491 (1978).
[120] J. M. Brown, P. A. Chaloner, B. A. Murrer and D. Parker, A.C.S. Symposium Ser. 119, 1731 (1980).
[121] G. Descotes, D. Lafont, D. Sinou, J. M. Brown, P. A. Chaloner and D. Parker, Nouv., J. Chim. **5**, 167 (1981).
[122] K. Achiwa and T. Soga, Tetrahedron Lett. **13**, 1119 (1978).
[123] Y. Ohgo, Y. Natori, S. Takenchi, J. Yoshimura, Chem. Lett. **33**, 709 (1974).
[124] K. Achiwa, Tetrahedron Lett. **50**, 4431 (1977).
[125] I. Ojima and T. Kogure, J. Organomet. Chem. **195**, 239 (1980).
[126] Sz. Tőrös, B. Heil, L. Kollár and L. Markó, J. Organomet. Chem. **197**, 85 (1980).
[127] T.-P. Dang and H. Kagan, J. Organomet. Chem. **91**, 105 (1975).
[128] W. S. Knowles, W. S. Christophel, K. E. Koenig and C. F. Hobbs, Advances in Chemistry Series **196**, 325 (1982).

[129] W. S. Knowles, Acc. Chem. Res. **16**, 106 (1983).
[130] M. Capka, J. Hetflejs and R. Selke, React. Kinet. Catal. Lett. **10**, 225 (1979).
[131] J. Halpern, Pure Appl. Chem. **55**, 99 (1983).
[132] Y. Izumi, Angew. Chem. **83**, 956 (1971); Int. Ed. Eng. **10**, 871 (1971).
[133] C. Cativiela, J. A. Mayoral, E. Melendez, R. Ulson, L. A. Oro and M. T. Pinillos, React. Kinet. Catal. Lett. **21**, 173 (1982).
[134] C. Fisher and H. S. Mosher, Tetrahedron Lett. **42**, 2487 (1977).
[135] J. M. Brown and D. Parker, J.C.S. Chem. Comm 342 (1980).
[136] K. E. Koenig, G. L. Bachmann and B. D. Vineyard, J. Org. Chem. **45**, 2362 (1980).
[137] K. Achiwa, Tetrahedron Lett. 2583 (1978).
[138] W. C. Christophel and B. D. Vineyard, J. Am. Chem. Soc. **101**, 4406 (1979).
[139] U.S. Pat. 4.409.397 (11. 10. 1983). C.A. **100**, 51098 (1984).
[140] H. B. Kagan, N. Langlois and T.-P. Dang, J. Organomet. Chem. **90**, 353 (1975).
[141] W. S. Knowles, M. J. Sabacky and B. D. Vineyard, Advan. Chem. Ser. **132**, 274 (1974).
[142] R. Glaser and S. Geresh, Tetrahedron Lett. **29**, 2527 (1977).
[143] R. Glaser and B. Vainas, J. Organomet. **121**, 249 (1976).
[144] R. Glaser, S. Geresh, J. Blumenfeld, B. Vainas and M. Twaik, Israel J. Chem. **15**, 17 (1976/77).
[145] J. W. Scott, D. D. Keith, G. Nix, Jr., D. R. Parrish, S. Remington, G. P. Roth, J. M. Townsend, D. Valentine, Jr. and R. Yang, J. Org. Chem. **46**, 5086 (1981).
[146] J. M. Brown and P. A. Chaloner, J.C.S. Chem. Comm. 613 (1979).
[147] K. E. Koenig and W. S. Knowles, J. Am. Chem. Soc. **100**, 7561 (1978).
[148] C. Detellier, G. Gelhard and H. B. Kagan, J. Am. Chem. Soc. **100**, 7556 (1978).
[149] A. Kleemann, J. Martens, M. Samson und W. Bergstein, Synthesis 740 (1981).
[150] J. C. Poulin and H. B. Kagan, J.C.S. Chem. Comm. 1261 (1982).
[151] I. Ojima and T. Suzuki, Tetrahedron Lett. 1239 (1980).
[152] D. Sinou, D. Lafont, G. Descotes and A. G. Kent, J. Organomet. Chem. **217**, 119 (1981).
[153] D. Meyer, J.-C. Poulin and H. B. Kagan, J. Org. Chem. **45**, 4680 (1980).
[154] K. Onuma, T. Ito and A. Nakamura, Chem. Lett. 481 (1980).
[155] I. Ojima and N. Yoda, Tetrahedron Lett. 3913 (1982).
[156] J. M. Brown and S. A. Hall, Tetrahedron Lett. **25**, 1393 (1984).
[157] D. A. Evans and M. M. Morrisey, J. Am. Chem. Soc. **106**, 3866 (1984).
[158] I. Ojima, N. Yoda, M. Yatabe, T. Tanaka and T. Kogure, Tetrahedron **40**, 1255 (1984).
[159] J. M. Brown and P. A. Chaloner, Tetrahedron Lett. **21**, 1877 (1978).
[160] A. S. C. Chan, J. J. Pluth and J. Halpern, J. Am. Chem. Soc. **102**, 5952 (1980).
[161] P. S. Chua, N. K. Roberts, B. Bosnich, S. J. Okrasinski and J. Halpern, J.C.S. Chem. Comm. 1278 (1981).
[162] J. M. Brown and P. A. Chaloner, J.C.S. Chem. Comm. 344 (1980).
[163] J. M. Brown and P. A. Chaloner, J.C.S. Chem. Comm. 321 (1978).
[164] K. Achiwa, P. A. Chaloner and D. Parker, Organomet. Chem. **218**, 249 (1981).
[165] K. Achiwa, Y. Ohga, Y. Itaka and H. Saito, Tetrahedron Lett. 4683 (1978).
[166] K. Achiwa, P. A. Chaloner and D. Parker, J. Org. Chem. **218**, 249 (1981).
[167] J. Halpern, D. P. Riley, A. S. C. Chan and J. J. Pluth, J. Am. Chem. Soc. **99**, 8055 (1977).
[168] A. S. C. Chan and J. Halpern, J. Am. Chem. Soc. **102**, 838 (1980).
[169] A. S. C. Chan, J. J. Pluth and J. Halpern, Inorg. Chim. Acta **37**, 2477 (1979).
[170] J. M. Brown and B. A. Murrer, Tetrahedron Lett. 4859 (1979).
[171] J. M. Brown and P. A. Chaloner, J. Am. Chem. Soc. **102**, 3040 (1980).
[172] J. M. Brown and D. Parker, J. Org. Chem. **47**, 2722 (1982).
[173] J. M. Brown, P. A. Chaloner, G. Descotes, R. Glaser, D. Lafont and D. Sinou, J.C.S. Chem. Comm. 611 (1979).
[174] M. Bianchi, F. Piacenti, P. Frediani, U. Matteoli, C. Botteghi, S. Gladiali and E. Benedetti, J. Organomet. Chem. **141**, 107 (1977).
[175] B. R. James, D. K. W. Wang and R. F. Voigt, J.C.S. Chem. Comm. 574 (1975).

[176] C. Botteghi, S. Gladiali, M. Bianchi, U. Matteoli, P. Frediani, P. G. Vergamini and E. Benedetti, J. Organomet. Chem. **140**, 221 (1977).
[177] G. Balavoine, T. P. Dang, C. Eskenazi and H. B. Kagan, J. Mol. Catal. **7**, 531 (1980).
[178] T. H. Johnson, L. A. Siegle and V. J. K. Chaffin, J. Mol. Catal. **9**, 307 (1980).
[179] E. R. James and R. S. McMillan, Can. J. Chem. **55**, 3927 (1977).
[180] U. Matteoli, P. Frediani, M. Bianchi, C. Botteghi and S. Gladiali, J. Mol. Catal. **12**, 265 (1981).
[181] K. Ohkubo, I. Terada and K. Yoshinaga, Inorg. Nucl. Chem. Lett. **15**, 421 (1979).
[182] K. Ohkubo, I. Terada, K. Sugahara and K. Yoshinaga, J. Mol. Catal. **7**, 421 (1980).
[183] A. Fischli and D. Süss, Helv. Chim. Acta **62**, 2361 (1979).
[184] A. Fischli and J. J. Daly, Helv. Chim. Acta **63**, 1628 (1980).
[185] Y. Ohgo, K. Kobayashi, S. Takeuchi and J. Yoshimura, Bull. Chem. Soc. Jap. **45**, 933 (1972).
[186] S. Takeuchi and Y. Ohgo, Bull. Chem. Soc. Jap. **54**, 2136 (1981).
[187] E. Cesarotti, R. Ugo and H. B. Kagan, Angew. Chem. **91**, 842 (1979); Int. Ed. Engl. **18**, 779 (1979).
[188] E. Cesarotti, R. Ugo and R. Vitiello, J. Mol. Catal. **12**, 63 (1981).
[189] J. Solodar, Chem. Techn. 421 (1975).
[190] M. Fiorini, F. Marcati and G. M. Giongo, J. Mol. Catal. **3**, 385 (1977/78).
[191] I. Ojima, T. Kogure and K. Achiwa, J.C.S. Chem. Comm. 428 (1977).
[192] G. Zassinovich and F. Grisoni, J. Organomet. Chem. **247**, C24 (1983).
[193] K. Achiwa, T. Kogure and I. Ojima, Chem. Lett. 297 (1978).
[194] T. Hayashi, A. Katsumura, M. Konishi and M. Kumada, Tetrahedron Lett. 425 (1979).
[195] Sz. Tőrös, B. Heil and L. Markó, J. Organomet. Chem. **159**, 401 (1978).
[196] B. Heil, Sz. Tőrös, J. Bakos and L. Markó, J. Organomet. Chem. **175**, 229 (1979).
[197] G. Strukul, M. Bonivento, M. Graziani, E. Cernia and N. Palladino, Inorg. Chim. Acta **12**, 15 (1975).
[198] K. Ohkubo, M. Setoguchi and K. Yoshinaga, Inorg. Nucl. Chem. Letters **15**, 235 (1979).
[199] G. Zassinovich, A. Camus and G. Mestroni, J. Mol. Catal. **9**, 345 (1980).
[200] G. Zassinovich, C. del Bianco and G. Mestroni, J. Org. Chem. **222**, 323 (1981).
[201] K. Osakada, M. Obana, T. Ikariya, M. Saburi and S. Yoshikawa, Tetrahedron Lett. **22**, 4297 (1981).
[202] Y. Ohgo, S. Takeuchi and J. Yoshumura, Bull. Chem. Soc. Jap. **44**, 583 (1971).
[203] Y. Ohgo, S. Takeuchi, Y. Natori and J. Yoshimura, Bull. Chem. Soc. Jap. **54**, 2124 (1981).
[204] M. M. Kucharska and S. Tyrlik, React. Kinet. Catal. Lett. **15**, 145 (1980).
[205] H. Pracejus and R. Rennau, React. Kinet. Catal. Lett. **15**, 203 (1980).
[206] S. Vastag, J. Bakos, S. Tőrös, N. E. Takach, R. B. King, B. Heil and L. Markó, J. Mol. Catal. **22**, 283 (1984).
[207] C. Botteghi, M. Bianchi, E. Benedetti and U. Matteoli, Chimia **29**, 256 (1975).
[208] I. Ojima, K. Yamamoto and M. Kumada, in *Aspects of Homogeneous Catalysis*, Vol. 3. D. Reidel, Dorndrecht, 1977, p. 186.
[209] H. Brunner, Angew. Chem. **95**, 921 (1983); Int. Ed. Engl. **22**, 897 (1983).
[210] B. Bosnich and M. D. Fryzuk, Top. Stereochem. **12**, 119 (1980).
[211] R. N. Meals, Pure Appl. Chem. **13**, 148 (1966).
[212] C. Eaborn and R. W. Bott, in *Organometallic Compounds of the Group IV Elements*, Vol. 1. Part 1, (A. G. MacDiarmid ed.). Marcel Dekker, New York, 1968.
[213] I. Ojima, M. Nihonyanagi and Y. Nagai, Chem. Commun. 938 (1972).
[214] I. Ojima, T. Kogure, M. Nihonyanagi and Y. Nagai, Bull. Chem. Soc. Jap. **45**, 3506 (1972).
[215] K. Yamamoto, Y. Uramoto and M. Kumada, J. Organomet. Chem. **31**, C9 (1971).
[216] T. Hayashi, K. Tamao, Y. Katsuro, I. Nakae and M. Kumada, Tetrahedron Lett. 1871 (1980).
[217] T. Hayashi, K. Yamamoto and M. Kumada, Tetrahedron Lett. 4405 (1974).
[218] H. Brunner und G. Riepl, Angew. Chem. **94**, 369 (1982); Int. Ed. Engl. **21**, 377 (1982).
[219] H. Brunner, G. Riepl and H. Weitzer, Angew. Chem. **95**, 326 (1983); Int. Ed. Engl. **22**, 331 (1983).

[220] H. Brunner, B. Reiter and G. Riepl, Chem. Ber. **117**, 1330 (1984).
[221] H. Brunner, R. Becker and G. Riepl, Organometallics **3**, 1354 (1983).
[222] K. Yamamoto, T. Hayashi and M. Kumada, J. Organomet. Chem. **46**, C 65 (1972); *ibid.* **113**, 127 (1976).
[223] I. Ojima, T. Kogure and Y. Nagai, Chem. Lett. 541 (1973); I. Ojima and Y. Nagai, *ibid.* 223 (1974).
[224] W. Dumont, J. C. Poulin, T.-P. Dang and H. B. Kagan, J. Am. Chem. Soc. **95**, 8295 (1973).
[225] I. Ojima, T. Kogure and M. Kumagai, J. Org. Chem. **42**, 1671 (1977).
[226] R. Glaser, Tetrahedron Lett. **25**, 2127 (1975).
[227] I. Ojima, T. Kogure and Y. Nagai, Tetrahedron Lett. 5035 (1972).
[228] T. Hayashi, K. Yamamoto and M. Kumada, Tetrahedron Lett. 3 (1975).
[229] T. Kogure and I. Ojima, J. Organomet. Chem. **234**, 249 (1982).
[230] I. Ojima, T. Kogure and Y. Nagai, Tetrahedron Lett. 1889 (1974).
[231] I. Ojima, T. Kogure and Y. Nagai, Tetrahedron Lett. 2475 (1973).
[232] N. Langlois, T.-P. Dang and H. B. Kagan, Tetrahedron Lett. 4865 (1973).
[233] H. Brunner and R. Becker, Angew. Chem. **96**, 221 (1984); Int. Ed. Engl. **23**, 222 (1984).
[234] R. J. P. Corrin and J. J. E. Moreau, J. Organomet. Chem. **64**, C 51 (1974); *ibid. 85*, 19 (1975).
[235] T. Onoda and S. Tomita, Japan Kokai **74**, 110.631, quoted in ref. [272].
[236] W. Dumont, J. C. Poulin, T.-P. Dang and H. B. Kagan, J. Am. Chem. Soc. **95**, 8295 (1973).
[237] T. H. Johnson, K. C. Klein and S. Tomen, J. Mol. Catal. **12**, 37 (1981).
[238] I. Kolb and J. Hetflejš, Coll. Czechoslov. Chem. Comm. **45**, 2808 (1980).
[239] I. Kolb and J. Hetflejš, Coll. Czechoslov. Chem. Comm. **45**, 2224 (1980).
[240] I. Ojima, M. Nihonyanagi and Y. Nagai, Bull. Chem. Soc. Jap. **45**, 3722 (1972).
[241] I. Ojima, T. Tanaka and T. Kogure, Chem. Lett. 823 (1981).
[242] T. Harada and Y. Izumi, Chem. Lett 1195 (1978).
[243] A. Tai, M. Nakahata, T. Harada and Y. Izumi, Chem. Lett. 1125 (1980).
[244] A. Tai, T. Harada, Y. Hiraki and S. Murakami, Bull. Chem. Soc. Jap. **56**, 1414 (1983).
[245] S. Murakami, T. Harada and A. Tai, Bull. Chem. Soc. Jap. **53**, 1356 (1980).
[246] T. Osawa and T. Harada, Bull. Chem. Soc. Jap. **57**, 1518 (1984).
[247] Y. Hiraki, K. Ito, T. Harada and A. Tai, Chem. Lett. 131 (1981).
[248] K. Ito, T. Harada, A. Tai and Y. Izumi, Chem. Lett. 1049 (1979).
[249] K. Ito, T. Harada and A. Tai, Bull. Chem. Soc. Jap. **53**, 3367 (1980).
[250] A. Tai, H. Watanabe and T. Harada, Bull. Chem. Soc. Jap. **52**, 1468 (1979).
[251] A. Tai, K. Ito and T. Harada, Bull. Chem. Soc. Jap. **54**, 223 (1981).
[252] E. I. Klabunovskii, Izv. Akad. Nauk SSSR, Ser. Khim. 505 (1984).
[253] H. Poisel and U. Schmidt, Chem. Ber. **106**, 3408 (1973).
[254] B. W. Bycroft and G. R. Lee, J.C.S. Chem. Comm. 988 (1975).
[255] N. Izumiya, S. Lee, T. Kanmera and H. Aoyagi, J. Am. Chem. Soc. **99**, 8346 (1977).
[256] T. Kanmera, H. Aoyagi, S. Lee and N. Izumiya, Pept. Chem. 59 (1979).
[257] J. P. Vigneron, H. Kagan and A. Horeau, Tetrahedron Lett. 5681 (1968).
[258] J. P. Vigneron, H. Kagan and A. Horeau, Tetrahedron Lett. 5681 (1968).
[259] J. Vasilevkis, J. A. Gualtieri, S. D. Hutsching, R. C. West, J. W. Scott, D. R. Parrish, F. T. Bizzarro and G. F. Field, J. Am. Chem. Soc. **100**, 7423 (1978).
[260] M. Furukawa, T. Okawara, Y. Noguchi and Y. Terawaki, Chem. Pharm. Bull. **27**, 2223 (1979).
[261] I. Ojima and M. Yatabe, Chem. Lett. 1335 (1982).
[262] C. Gallina, G. Lucente and F. Pinnen, Tetrahedron **34**, 2361 (1978).
[263] M. Tamura and K. Harada, Bull. Chem. Soc. Jap. **53**, 561 (1980).
[264] K. Harada, T. Munegumi and S. Nomoto, Tetrahedron Lett. **22**, 111 (1981).
[265] For its stereochemical aspects, see J. Rètey and J. A. Robinson, *Stereospecificity in Organic Chemistry and Enzymology*. Verlag Chemie, Weinheim, 1982, p. 169.
[266] Y. Tachibana, M. Ando and H. Kuzuhara, Bull. Chem. Soc. Jap. **56**, 3652 (1983).
[267] K. Bernauer, R. Deschenaux and T. Taura, Helv. Chim. Acta **66**, 2049 (1983).

[268] R. T. Standridge, H. G. Howell, J. A. Gylys, R. A. Partyka and A. T. Shulgin, J. Med. Chem. **19**, 1400 (1976).
[269] D. E. Nichols, C. F. Barfknecht, D. R. Rusterholz, F. Benington and R. D. Morin, J. Med. Chem. **16**, 480 (1973).
[270] S. Yamada, N. Ikota and K. Achiwa, Tetrahedron Lett. 1001 (1976).
[271] A. W. Frahm and G. Knupp, Tetrahedron Lett. 2633 (1981).
[272] G. Knupp and A. W. Rahm, Chem. Ber. **117**, 2076 (1984).
[273] K. Harada and K. Matsumoto, J. Org. Chem. **32**, 1794 (1967).
[274] K. Harada and K. Matsumoto, J. Org. Chem. **33**, 4467 (1968).
[275] K. Harada and T. Yoshida, Bull. Chem. Soc. Jap. **43**, 921 (1970).
[276] T. Yoshida and K. Harada, Bull. Chem. Soc. Jap. **45**, 3706 (1972).
[277] K. Harada and T. Yoshida, J.C.S. Chem. Comm. 1071 (1970).
[278] K. Harada and Y. Kataoka, Chem. Lett. 791 (1978).
[279] K. Harada, T. Iwasaki and T. Okawara, Bull. Chem. Soc. Jap. **46**, 1901 (1973).
[280] K. Harada and M. Tamura, Bull. Chem. Soc. Jap. **52**, 1227 (1979).
[281] K. Harada and Y. Kataoka, Tetrahedron Lett. 2103 (1978).
[282] K. Harada and K. Matsumoto, Bull. Chem. Soc. Jap. **44**, 1068 (1971).
[283] E. J. Corey, H. S. Sachdev, J. Z. Gougoutas and W. Saenger, J. Am. Chem. Soc. **92**, 2488 (1970).
[284] E. J. Corey, R. J. McCaully and H. S. Sachder, J. Am. Chem. Soc. **92**, 2476 (1970).
[285] A. N. Kost, R. S. Sagitullin and M. A. Yurovskaja, Chem. and Ind. 1496 (1966).
[286] S. Kiyooka, K. Takeshima, H. Yamamoto and K. Suzuki, Bull. Chem. Soc. Jap. **49**, 1897 (1976).
[287] J. Peyronel, J. Fiaud and H. B. Kagan, J. Chem. Res. (S) 320, (M) 4057 (1980).
[288] K. Yamamoto, T. Hayashi and M. Kumada, J. Organomet. Chem. **54**, C45 (1973).
[289] G. Descotes and D. Sinou, Tetrahedron Lett. 4083 (1976).

Chapter 3

[1] H. Haubenstock, Top. Stereochem. **14**, 231 (1982).
[2] A. A. Bothner-By, J. Am. Chem. Soc. **73**, 846 (1951).
[3] P. S. Portoghese, J. Org. Chem. **27**, 3359 (1962).
[4] S. R. Landor, B. J. Miller and A. R. Tatchell, J. Chem. Soc. C, 1922 (1966).
[5] S. Yamaguchi and H. S. Mosher, J. Org. Chem. **38**, 1870 (1973).
[6] Y. Minoura and H. Yamaguchi, J. Polym. Sc. **6**, A-1, 2012 (1968).
[7] A. Horeau, H. Kagan and J. Vigneron, Bull. Soc. Chim. France 3795 (1968).
[8] O. Červinka, Coll. Czech. Chem. Commun. **26**, 673 (1961).
[9] O. Červinka, Coll. Czech. Chem. Commun. **30**, 2403 (1965).
[10] O. Červinka, V. Suchan, O. Kotynek and V. Dudek, Coll. Czech. Chem. Commun. **30**, 2484 (1965).
[11] S. Yamaguchi and K. Kabuto, Bull. Chem. Soc. Jap. **50**, 3033 (1977).
[12] R. J. D. Evans, S. R. Landor and J. P. Regan, J. C. S. Perkin 1 552 (1974).
[13] R. Andrisano, S. R. Angeloni and S. Marzocchi, Tetrahedron **29**, 913 (1973).
[14] S. R. Landor, B. J. Miller and A. R. Tatchell, J. Chem. Soc. C 2280 (1966).
[15] S. R. Landor, B. J. Miller and A. R. Tatchell, J. Chem. Soc. C 197 (1967).
[16] J. Hutton, M. Senior and N. C. A. Wright, Synth. Comm. **9**, 799 (1979).
[17] O. Červinka and A. Fabryova, Tetrahedron Lett. 1179 (1967).

[18] N. Baggett and P. Stribblehill, J. C. S. Perkin 1 1123 (1977).
[19] E. D. Lund and P. E. Shaw, J. Org. Chem. **42**, 2073 (1977).
[20] T. H. Johnson and K. C. Klein, J. Org. Chem. **44**, 461 (1979).
[21] H. J. Schneider and R. Haller, Liebigs Ann. Chem. 187 (1971).
[22] R. Haller and H. J. Schneider, Chem. Ber. **106**, 1312 (1973).
[23] R. Noyori, I. Tomino and Y. Tanimoto, J. Am. Chem. Soc. **101**, 3129 (1979).
[24] R. Noyori, Pure Appl. Chem. **53**, 2316 (1981).
[25] R. Noyori, I. Tomino, Y. Tanimoto and M. Nishizawa, J. Am. Chem. Soc. **106**, 6709 (1984).
[26] M. Nishizawa, M. Yamada and R. Noyori, Tetrahedron Lett. 247 (1981).
[27] R. Noyori, I. Tomino, M. Yamada and M. Nishizawa, J. Am. Chem. Soc. **106**, 6717 (1984).
[28] M. Nishizawa and R. Noyori, Tetrahedron Lett. 247 (1981).
[29] R. Noyori, I. Tomino and M. Nishizawa, J. Am. Chem. Soc. **101**, 5843 (1979).
[30] H. Suda, S. Kanoh, N. Umeda, M. Ikka and M. Motoi, Chem. Lett. 899 (1984).
[31] K. Yamamoto, H. Fukushima and M. Nakazaki, J. C. S. Chem. Comm. 1490 (1984).
[32] O. Červinka and O. Bělovsky, Coll. Czech. Chem. Comm. **32**, 3897 (1965).
[33] J. D. Morrison and H. S. Mosher, *Asymmetric Organic Reactions*. Prentice-Hall, Englewood Cliffs, New Jersey, 1971.
[34] O. Červinka, P. Malon and H. Procházková, Coll. Czech. Chem. Comm. **39**, 1869 (1974).
[35] I. Jacquet and J.-P. Vigneron, Tetrahedron Lett. 2065 (1974).
[36] J.-P. Vigneron and I. Jacquet, Tetrahedron **32**, 939 (1976).
[37] J.-P. Vigneron and V. Bloy, Tetrahedron Lett. 2683 (1979).
[38] S. Yamada, M. Kitamoto and S. Terashima, Tetrahedron Lett. 3165 (1976).
[39] M. Kitamoto, K. Kameo, S. Terasjo and S. Yamada, Chem. Pharm. Bull. **25**, 1273 (1977).
[40] S. Terashima, N. Tanno and K. Koga, Chem. Lett. 981 (1980).
[41] M. Kawasaki, Y. Suzuki and S. Terashima, Chem. Lett. 239 (1984).
[42] A. Pohland and H. R. Sullivan, J. Am. Chem. Soc. **75**, 4453 (1953).
[43] S. Yamaguchi, H. S. Mosher and A. Pohland, J. Am. Chem. Soc. **94**, 9254 (1972).
[44] R. S. Brinkmeyer and V. M. Kapoor, J. Am. Chem. Soc. **99**, 8341 (1977).
[45] N. Cohen, R. J. Lopresti, C. Neukon and G. Saucy, J. Org. Chem. **45**, 582 (1980).
[46] A. I. Meyers and P. M. Kenda, Tetrahedron Lett. 1357 (1974).
[47] S. Yamaguchi, F. Yasuhara and K. Kabuto, J. Org. Chem. **42**, 1578 (1977).
[48] J. M. Hawkins and K. B. Sharpless, J. Org. Chem. **49**, 3861 (1984).
[49] R. A. Kretchmer, J. Org. Chem. **37**, 801 (1972).
[50] M. Asami, H. Ohno, S. Kobayashi and T. Mukaiyama, Bull. Chem. Soc. Jap. **51**, 1869 (1978).
[51] M. Asami and T. Mukaiyama, Heterocycles 499 (1979).
[52] T. Sato, Y. Goto and T. Fujisawa, Tetrahedron Lett. 4111 (1982).
[53] T. Sato, Y. Goto, M. Watanabe and T. Fujisawa, Chem. Lett. 1533 (1983).
[54] D. Seebach and H. Daum, Chem. Ber. **107**, 1748 (1974).
[55] M. Schmidt, R. Amstutz, G. Grass and D. Seebach, Chem. Ber. **113**, 1691 (1980).
[56] J. D. Morrison, E. R. Grandbois, S. I. Howard and G. R. Weisman, Tetrahedron Lett. **22**, 2619 (1981).
[57] O. Červinka, Coll. Czech. Chem. Comm. **26**, 673 (1961).
[58] S. R. Landor, O. O. Sonola and A. R. Tatchell, J. C. S. Perkin 1 605 (1978).
[59] S. R. Landor, Y. M. Chan, O. O. Sonola and A. R. Tatchell, J. C. S. Perkin 1 493 (1983).
[60] S. R. Landor, O. O. Sonola and A. R. Tatchell, J. C. S. Perkin 1 1902 (1974).
[61] A. Hirao, H. Mochizuki, S. Nakahama and N. Yamazaki, J. Org. Chem. **44**, 1720 (1979).
[62] A. Hirao, S. Nakahama, D. Mochizuki, S. Itsuno, M. Ohowa and N. Yamazaki, J. C. S. Chem. Comm. 807 (1979).
[63] A. Hirao, M. Ohwa, S. Itsuno, H. Mochizuki, S. Nakahara and N. Yamazaki, Bull. Chem. Soc. Jap. **54**, 1424 (1981).
[64] A. Hirao, S. Nakahama, H. Mochizuki, S. Itsuno and N. Yamazaki, J. Org. Chem. **45**, 4231 (1980).

[65] A. Hirao, S. Itsuno, M. Owa, S. Nagami, H. Mochizuki, H. H. A. Zoorov, S. Niakahama and N. Yamazaki, J. C. S. Perkin 1 900 (1981).
[66] J. D. Morrison, E. R. Grandbois and S. I. Howard, J. Org. Chem. **45**, 4229 (1980).
[67] R. J. McMahon, K. E. Wiegers and S. G. Smith, J. Org. Chem. **46**, 99 (1981).
[68] K. Soai, H. Oyamada and T. Yamanoi, Chem. Lett. 251 (1984).
[69] K. Yamada, M. Takeda and T. Iwakuma, Tetrahedron Lett. 3869 (1981).
[70] J. P. Massé and E. R. Paryre, J. C. S. Chem. Comm. 438 (1976).
[71] L. Horner and W. Brich, Liebigs Ann. Chem. 710 (1978).
[72] J. Balcells, S. Colonna and R. Fornasier, Synthesis 266 (1976).
[73] S. Colonna and R. Fornasier, J. C. S. Perkin 1 371 (1978).
[74] R. Kinishi, N. Uchida, Y. Yamamoto, J. Oda and Y. Inouye, Agric. Biol. Chem. **44**, 643 (1980).
[75] T. Doiuchi and Y. Minoura, Isr. J. Chem. **15**, 84 (1976/77).
[76] T. Sugimoto, Y. Matsumura, S. Tanimoto and M. Okano, J. C. S. Chem. Comm. 926 (1978).
[77] H. C. Brown and D. B. Bigley, J. Am. Chem. Soc. **83**, 3166 (1961).
[78] E. Caspi and K. R. Varma, J. Org. Chem. **33**, 2181 (1968).
[79] H. C. Brown and A. K. Mandal, J. Org. Chem. **42**, 2996 (1977).
[80] H. C. Brown, J. S. Cha and B. Nazer, J. Org. Chem. **49**, 2073 (1984).
[81] H. C. Brown, P. K. Jadhav and A. K. Mandal, Tetrahedron **37**, 3547 (1981).
[82] H. C. Brown and A. K. Mandal, J. Org. Chem. **49**, 2558 (1984).
[83] S. Krishnamurthy, F. Vogel and H. C. Brown, J. Org. Chem. **42**, 2534 (1977).
[84] M. M. Midland and A. Kazubski, J. Org. Chem. **47**, 2495 (1982).
[85] J. C. Fiaud and H. B. Kagan, Bull. Soc. Chim. France 2742 (1969).
[86] M. F. Grundon, W. A. Khan, D. R. Boyd and W. R. Jackson, J. Chem. Soc. C 2557 (1971).
[87] R. F. Borch and S. R. Leviton, J. Org. Chem. **37**, 2347 (1972).
[88] M. F. Grundon, D. G. McGleery and J. W. Wilson, Tetrahedron Lett. 295 (1976).
[89] M. F. Grundon, D. G. McGleery and J. W. Wilson, J. C. S. Perkin 1 231 (1981).
[90] N. Umino, T. Iwakuma and N. Itoh, Chem. Pharm. Bull. **27**, 1479 (1979).
[91] A. Hirao, S. Itsuno, S. Nakahama and N. Yamazaki, J. C. S. Chem. Comm. 315 (1981).
[92] S. Itsuno, A. Hirao, S. Nakahama and N. Yamazaki, J. C. S. Perkin 1 1673 (1983).
[93] S. Itsuno, K. Ito, A. Hirao and S. Nakahama, J. Org. Chem. **49**, 555 (1984).
[94] S. Itsuno, K. Ito, A. Hirao and S. Nakahama, J. C. S. Chem. Comm. 469 (1983).
[95] G. M. Giongo, F. Di Gregoric, N. Palladino and W. Marconi, Tetrahedron Lett. 3195 (1973).
[96] B. M. Mikhailov, Yu. N. Bubnov and V. G. Kiselev, J. Gen. Chem. USSR **36**, 65 (1966).
[97] M. M. Midland, in *Asymmetric Synthesis*, Vol. 2, (J. D. Morrison, ed.). Academic Press, New York, 1984, p. 45.
[98] M. M. Midland, A. Tramontano and S. A. Zderic, J. Am. Chem. Soc. **99**, 5211 (1977).
[99] M. M. Midland, S. Greer, A. Tramontano and S. A. Zderic, J. Am. Chem. Soc. **101**, 2352 (1979).
[100] M. M. Midland and A. Tramontano, Tetrahedron Lett. 3549 (1980).
[101] M. M. Midland, D. C. McDowell, R. L. Hatch and A. Tramontano, J. Am. Chem. Soc. **102**, 867 (1980).
[102] M. M. Midland and N. N. Nguyen, J. Org. Chem. **46**, 4107 (1981).
[103] M. M. Midland and J. I. McLoughlin, J. Org. Chem. **49**, 4102 (1984).
[104] M. M. Midland and A. Kazubski, J. Org. Chem. **47**, 2814 (1982).
[105] M. M. Midland, A. Tramontano, A. Kazubski, R. S. Graham, D. J. S. Tsai and D. Cardin, Tetrahedron **40**, 1371 (1984).
[106] M. M. Midland and J. I. McLoughlin, J. Org. Chem. **49**, 4101 (1984).
[107] M. M. Midland and J. I. McLoughlin, J. Org. Chem. **49**, 1316 (1984).
[108] H. C. Brown and G. G. Pai, J. Org. Chem. **47**, 1606 (1982).
[109] H. C. Brown, G. G. Pai and P. K. Jadhav, J. Am. Chem. Soc. **106**, 1531 (1984).
[110] G. Giacomelli, R. Menicagli and L. Lardicci, J. Org. Chem. **39**, 1757 (1974).
[111] G. Vavon, C. Rivière and B. Angelo, Compt. rend. **222**, 959 (1946).
[112] R. MacLeod, F. J. Welsh and H. S. Mosher, J. Am. Chem. Soc. **82**, 876 (1960).

[113] Unpublished results quoted in ref. [33].
[114] J. Capillon and J. P. Guetté, Tetrahedron **35**, 1807 (1979).
[115] J. Capillon and J. P. Guetté, Tetrahedron **35**, 1817 (1979).
[116] G. Giacomelli, L. Lardicci and A. M. Caporusso, J. C. S. Perkin 1 1795 (1975).
[117] W. M. Foley, F. J. Welsh, E. M. LaCombe and H. S. Mosher, J. Am. Chem. Soc. **81**, 2779 (1959).
[118] M. F. Tatibouet, Bull. Soc. Chim. France 868 (1951).
[119] G. P. Giacomelli, R. Menicagli and L. Lardicci, Tetrahedron Lett. 4135 (1971).
[120] L. Lardicci and G. Giacomelli, J. C. S. Perkin 1 337 (1974).
[121] L. Lardicci, G. P. Giacomelli and R. Menicagli, Tetrahedron Lett. 687 (1972).
[122] G. Giacomelli, R. Menicagli and L. Lardicci, J. Org. Chem. **38**, 2370 (1973).
[123] G. P. Giacomelli, R. Menicagli and L. Lardicci, J. Am. Chem. Soc. **97**, 4009 (1975).
[124] G. Giacomelli and L. Lardicci, J. Org. Chem. **46**, 3116 (1981).
[125] G. Giacomelli, A. M. Caporusso and L. Lardicci, Tetrahedron Lett. **22**, 3663 (1981).
[126] G. Giacomelli, L. Lardicci, F. Palla and A. M. Caporusso, J. Org. Chem. **49**, 1725 (1984).
[127] L. Gruber, I. Tömösközy and L. Ötvös, Tetrahedron Lett. 811 (1973).
[128] G. Giacomelli, R. Menicagli, A. Caporusso and A. M. Lardicci, J. Org. Chem. **43**, 1790 (1978).
[129] G. Giacomelli, L. Lardicci and F. Palla, J. Org. Chem. **49**, 310 (1984).
[130] For details, see pp. 160–165 in ref. [33].
[131] D. Nasipuri, C. K. Grosh and R. J. L. Martin, J. Org. Chem. **35**, 657 (1970).
[132] For details, see pp. 171–172 in ref. [33].
[133] D. Nasipuri and P. K. Bhattacharya, J. C. S. Perkin 1 576 (1977).
[134] A. K. Samaddar, S. K. Konar and D. Nasipuri, J. C. S. Perkin 1 1449 (1983).
[135] J. D. Morrison and R. W. Ridgway, J. Org. Chem. **39**, 3107 (1974).
[136] A. Streitwieser and M. R. Granger, J. Org. Chem. **32**, 1528 (1967).
[137] A. Ohno, M. Ikeguchi, T. Kimura and S. Oka, J. Am. Chem. Soc. **101**, 7036 (1979).
[138] M. Seki, N. Baba, J. Oda and Y. Inouye, J. Am. Chem. Soc. **103**, 4613 (1981).
[139] R. M. Kellog, Biomimetic Chem., Proc. 2nd Intern. Kyoto Conf. 1983, p. 136.
[140] O. Červinka and O. Bělovsky, Coll. Czech. Chem. Comm. **32**, 3897 (1967).
[141] G. M. Giongo, F. Di Gregorio, N. Palladino and W. Marconi, Tetrahedron Lett. 3195 (1973).
[142] The literature up to 1968 has been reviewed in ref. [33] pp. 84–132.
[143] J. Canceill and J. Jacques, Bull. Soc. Chim. France 2180 (1970).
[144] M. Chérest, H. Felkin and N. Prudent, Tetrahedron Lett. 2199 (1968).
[145] C. Zioudrou, I. Moustakali-Mavridis, P. Chrysochou and G. Karabatsos, Tetrahedron **34**, 3181 (1978).
[146] Data are taken from recent reviews: refs. [147, 148, 149].
[147] J. R. Boone and A. C. Ashby, Top. Ster. **11**, 53 (1979).
[148] E. C. Ashby and J. T. Laemmle, Chem. Rev. **75**, 521 (1975).
[149] D. J. Cram and F. A. Abd Elhafez, J. Am. Chem. Soc. **74**, 5828 (1952).
[150] V. Prelog, Helv. Chim. Acta **36**, 308 (1953).
[151] J. W. Cornforth, R. H. Cornforth and K. K. Mathew, J. Chem. Soc. 112 (1959).
[152] G. J. Karabatsos, J. Am. Chem. Soc. **89**, 1367 (1967).
[153] G. J. Karabatsos and D. J. Fenoglio, Top. Stereochem. **5**, 167 (1970).
[154] D. J. Cram and K. R. Kopecky, J. Am. Chem. Soc. **81**, 2748 (1959).
[155] M. Chérest and H. Felkin, Tetrahedron Lett. 2205 (1968).
[156] G. J. Karabatsos and D. J. Fenoglio, J. Am. Chem. Soc. **91**, 1124 (1969).
[157] N. T. Anh and O. Eisenstein, Nouv. J. Chim. **1**, 61 (1976).
[158] N. T. Anh, Top. Curr. Chem. **88**, 146 (1980).
[159] H. B. Bürgi, J. D. Dunitz, J. M. Lehn and G. Wipf, Tetrahedron **30**, 1563 (1974).
[160] H. B. Bürgi, J. M. Lehn and G. Wipf, J. Am. Chem. Soc. **96**, 1956 (1974).
[161] For leading reviews about this topic *cf.* refs. [147, 148, and 163].
[162] D. H. R. Barton, J. Chem. Soc. 1027 (1953).

[163] D. C. Wigfield, Tetrahedron **35**, 449 (1979).
[164] W. G. Dauben, G. J. Tonken and D. S. Noyce, J. Am. Chem. Soc. **78**, 2579 (1956).
[165] E. L. Eliel and Y. Senda, Tetrahedron **26**, 2411 (1970).
[166] J.-C. Richer, J. Org. Chem. **30**, 324 (1965).
[167] J. Huet, Y. Maroni-Barnaud, N. Tron Anh and J. Seyden-Penne, Tetrahedron Lett. 159 (1976).
[168] W. T. Wipke and P. Gund, J. Am. Chem. Soc. **98**, 8107 (1976).
[169] M.-H. Rei, J. Org. Chem. **48**, 5386 (1983).
[170] E. C. Ashby and S. A. Noding, J. Org. Chem. **42**, 264 (1977).
[171] Y. Gault and H. Felkin, Bull. Soc. Chim. France 1342 (1960).
[172] M. M. Midland and Y. C. Kwon, J. Am. Chem. Soc. **105**, 3725 (1983).
[173] D. J. Cram, F. A. Abd Elhafez and H. L. Nyquist, J. Am. Chem. Soc. **76**, 22 (1954).
[174] D. J. Cram and J. Allinger, J. Am. Chem. Soc. **76**, 4516 (1954).
[175] D. J. Cram, F. A. Abd Elhafez and H. Weingartner, J. Am. Chem. Soc. **75**, 2293 (1953).
[176] E. C. Ashley and J. R. Boone, J. Am. Chem. Soc. **98**, 5524 (1976).
[177] D. J. Cram and F. D. Greene, J. Am. Chem. Soc. **75**, 6005 (1953).
[178] R. Guyon and P. Villa, Bull. Soc. Chim. France 152 (1977).
[179] H. Bodot, E. Dieuzeide and J. Julien, Bull. Soc. Chim. France 1086 (1960).
[180] J. H. Stocker, P. Sidismuthorn, B. M. Benjamin and C. J. Collins, J. Am. Chem. Soc. **82**, 3913 (1960).
[181] S. Yamada and K. Koga, Tetrahedron Lett. 1711 (1967).
[182] H. C. Brown and H. R. Deck, J. Am. Chem. Soc. **87**, 5620 (1965).
[183] W. J. Gensler, F. Johnson and A. D. B. Sloan, J. Am. Chem. Soc. **82**, 6074 (1960).
[184] T. Nakata and T. Oishi, Tetrahedron Lett. 1641 (1980).
[185] C. Ziodrou and P. Chrysochou, Tetrahedron **33**, 2103 (1977).
[186] J. A. Katzenellenbogen and S. B. Bowlus, J. Org. Chem. **38**, 627 (1973).
[187] T. Nakata, T. Tanaka and T. Oishi, Tetrahedron Lett. 2653 (1983).
[188] M. Proštenik and B. Alanpovič, Croat. Chem. Acta **29**, 393 (1957); C. A. **53**, 1131 (1959).
[189] H. K. Müller, I. Jarchow and G. Rieck, Liebigs Ann. Chem. **613**, 103 (1958).
[190] K. Koga and S. Yamada, Chem. Pharm. Bull. **20**, 526 (1972).
[191] M. Tramontini, Synthesis 605 (1982).
[192] S. Yamada and K. Koga, Tetrahedron Lett. 1711 (1967).
[193] T. Kametani, K. Kigasawa, M. Hiiragi, N. Wagatsuma, T. Kohagizawa and H. Inone, J. Pharm. Soc. Jap. **100**, 839 (1980).
[194] G. Guanti, L. Banfi, E. Narisano and C. Scolastico, Tetrahedron Lett. **25**, 4693 (1984).
[195] T. Nakata, Y. Tani, M. Hatozaki and T. Oishi, Chem. Pharm. Bull **32**, 1411 (1984).
[196] M. Shimagaki, T. Maeda, Y. Matsuzaki, I. Hori, T. Nakata and T. Oishi, Tetrahedron Lett. **25**, 4775 (1984).
[197] M. Shimagaki, Y. Matsuzaki, I. Hori, T. Nakata and T. Oishi, Tetrahedron Lett. **25**, 4779 (1984).
[198] K. Suzuki, E. Katayama and G. Tsuchihashi, Tetrahedron Lett. **25**, 2479 (1984).
[199] Y. Ito and M. Yamaguchi, Tetrahedron Lett. 5385 (1983).
[200] D. Kruger, A. E. Sopchik and C. A. Kingbury, J. Org. Chem. **49**, 778 (1984).
[201] S. B. Bowlus and J. A. Katzenellenbogen, J. Org. Chem. **39**, 3309 (1974).
[202] T. Oishi and T. Nakata, Acc. Chem. Res. **17**, 338 (1984).
[203] Y. Ito, T. Katsuki and M. Yamaguchi, Tetrahedron Lett. **25**, 6015 (1984).
[204] D. Hartley, Chem. Ind. 551 (1981).
[205] M. Brienne, C. Ouannis and J. Jacques, Bull. Soc. Chim. France 1036 (1968).
[206] R. Annunziata, M. Cinquini and F. Cozzi, J. C. S. Perkin 1 1109 (1981).
[207] C. R. Johnson and C. J. Stark, Jr., J. Org. Chem. **47**, 1196 (1982).
[208] G. Solladié, C. Greck, G. Demailly and A. Solladié-Cavallo, Tetrahedron Lett. 5047 (1982).
[209] A. McKenzie, J. Chem. Soc. **85**, 1249 (1904).
[210] K. Soai, K. Komiya, Y. Shigematsu, H. Hasegawa and A. Ookawa, J. C. S. Chem. Comm. 1282 (1982).

[211] B. P. Giovanni, M. Fabio, P. G. Piero, S. Daniele, B. Achille and B. Simonetta, J. C. S. Perkin Trans 1 2983 (1982).
[212] D. C. Wigfield, Can. J. Chem. **55**, 646 (1977).
[213] E. L. Eliel and S. H. Schroeter, J. Am. Chem. Soc. **87**, 5031 (1965).
[214] H. C. Brown and S. Krishnamurthy, J. Am. Chem. Soc. **94**, 7159 (1972).
[215] J. Hooz, S. Akiyama, F. J. Cedar, M. J. Bennett and R. M. Tuggle, J. Am. Chem. Soc. **96**, 274 (1974).
[216] S. Krishnamurthy and H. C. Brown, J. Am. Chem. Soc. **98**, 3383 (1976).
[217] H. C. Brown, J. L. Hubbade and B. Singaram, Tetrahedron **37**, 2359 (1981).
[218] R. O. Hutchins, J. Org. Chem. **42**, 920 (1977).
[219] S. Kim, K. H. Ahn and Y. W. Chung, J. Org. Chem. **47**, 4581 (1982).
[220] G. Kovács, G. Galambos and Z. Juvancz, Synthesis 171 (1977).
[221] E. C. Ashby, S. A. Noding and A. B. Goel, J. Org. Chem. **45**, 1028 (1980).
[222] E. C. Ashby and A. B. Goel, Inorg. Chem. **17**, 1862 (1978).

Chapter 4

[1] H. B. Henbest, Chem. Soc. Special Publ. No. 19, 83 (1965).
[2] W. H. Pirkle and P. L. Rinaldi, J. Org. Chem. **42**, 2080 (1977).
[3] M. F. Grundon and S. A. Surgenor, J. C. S. Chem. Comm. 624 (1978).
[4] U. Folli, D. Iarossi, F. Montanari and G. Torre, J. Chem. Soc. C 1317 (1968).
[5] F. Montanari, I. Moretti and G. Torre, J. C. S. Chem. Comm. 1694 (1968).
[6] D. R. Boyd and R. Graham, J. Chem. Soc. 2648 (1969).
[7] D. Boschelli and A. B. Smith, Tetrahedron Lett. 4385 (1981).
[8] F. A. Davis, M. E. Harakal and S. B. Awad, J. Am. Chem. Soc. **105**, 3123 (1983).
[9] R. Helder, J. C. Hummelen, R. W. P. M. Laane, J. S. Wiering and H. Wynberg, Tetrahedron Lett. 1831 (1976).
[10] H. Pluim and H. Wynberg, J. Org. Chem. **45**, 2498 (1980).
[11] S. Juliá, J. Guixer, J. Masana, J. Rocas, S. Colonna, R. Annunziata and H. Molinari, J. C. S. Perkin 1 1317 (1982).
[12] S. Colonna, H. Molinari, S. Banfi, S. Juliá, J. Masana and A. Alvarez, Tetrahedron **39**, 1635 (1983).
[13] T. Itoh, K. Kaneda and S. Terashini, J. C. S. Chem. Comm. 421 (1976).
[14] S. Yamada, T. Mashiko and S. Terashima, J. Am. Chem. Soc. **99**, 1988 (1977).
[15] R. C. Michaelson, R. E. Palermo and K. B. Sharpless, J. Am. Chem. Soc. **99**, 1990 (1977).
[16] K. B. Sharpless and T. R. Verhoeven, Aldrichimica Acta **12**, 63 (1979).
[17] H. B. Kagan, H. Mimoun, C. Mark and V. Schurig, Angew. Chem. **91**, 511 (1979); Int. Ed. Engl. **18**, 485 (1979).
[18] T. Katsuki and K. B. Sharpless, J. Am. Chem. Soc. **102**, 5974 (1980).
[19] B. E. Rossiter, T. Katsuki and K. B. Sharpless, J. Am. Chem. Soc. **103**, 464 (1981).
[20] K. Mori and T. Umemura, Tetrahedron Lett. 4433 (1981).
[21] K. Mori and T. Ebata, Tetrahedron Lett. 4281 (1981).
[22] E. J. Corey, S. Hashimoto and A. E. Barton, J. Am. Chem. Soc. **103**, 721 (1981).
[23] W. R. Roush and R. J. Brown, J. Org. Chem. **47**, 1373 (1982).
[24] P. Ma V. S. Martin, S. Masamune, K. B. Sharpless and S. M. Viti, J. Org. Chem. **47**, 1378 (1982).

[25] K. Mori and H. Ueda, Tetrahedron **37**, 2581 (1981).
[26] S. Takano, C. Kasahara and K. Ogasawara, Chem. Lett. 175 (1983).
[27] J. G. Hill, B. E. Rossiter and K. B. Sharpless, J. Org. Chem. **48**, 3607 (1983).
[28] K. Tani, M. Hanfusa and S. Otsuka, Tetrahedron Lett. 3017 (1979).
[29] B. E. Rossiter and K. B. Sharpless, J. Org. Chem. **49**, 3707 (1984).
[30] K. B. Sharpless, S. S. Woodard and M. G. Finn, Pure Appl. Chem. **55**, 1823 (1983).
[31] L. D.-L. Lu, R. A. Johnson, M. G. Finn and K. B. Sharpless, J. Org. Chem. **49**, 728 (1984).
[32] T. Komori and T. Nonaka, J. Am. Chem. Soc. **105**, 5691 (1983).
[33] K.-T. Liu and Y.-C. Tong, J. Chem. Res. (S), 276 (1979).
[34] A. W. Czarnik, J. Org. Chem. **49**, 924 (1984).
[35] T. Sugimoto, T. Kokubo, J. Miyazaki, S. Tanimoto and M. Okano, J. C. S. Chem. Comm. 1052 (1979).
[36] T. Sugimoto, T. Kokubo, J. Miyazaki, S. Tanimoto and M. Okano, J. C. S. Chem. Comm. 402 (1979).
[37] B. Feringa and H. Wynberg, Bioorg. Chem. 7, 397 (1978).
[38] J. Brussee and A. C. A. Janssen, Tetrahedron Lett. 3261 (1983).
[39] S. G. Hentges and K. B. Sharpless, J. Am. Chem. Soc. **102**, 4263 (1980).
[40] H. B. Henbest and R. A. L. Wilson, J. Chem. Soc. 1958 (1957).
[41] H. B. Henbest, Proc. Chem. Soc. 159 (1963).
[42] R. B. Dehnel and G. H. Whitham, J. C. S. Perkin 1 953 (1979).
[43] P. Chamberlain, M. L. Roberts and G. H. Whitham, J. Chem. Soc. B 1374 (1970).
[44] T. Itoh, K. Jitsukawa, K. Kaneda and S. Terenishi, J. Am. Chem. Soc. **101**, 159 (1979).
[45] G. Cicala, R. Curci, M. Fiorentino and O. Laricchiuta, J. Org. Chem. **47**, 2670 (1982).
[46] J.-L. Pierre, P. Chautemps and P. Arnaud, Bull. Soc. Chim. France 1317 (1969).
[47] P. Chautemps and J.-L. Pierre, Tetrahedron **32**, 549 (1976).
[48] S. Tanaka, H. Yamamoto, H. Nozaki, K. B. Sharpless, R. C. Michaelson and J. D. Cutting, J. Am. Chem. Soc. **96**, 5254 (1974).
[49] B. E. Rossiter, T. R. Verhoeven and K. B. Sharpless, Tetrahedron Lett. 4733 (1979).
[50] E. D. Mihelich, Tetrahedron Lett. 4729 (1979).
[51] K. Takai, K. Oshima and H. Nozaki, Bull. Chem. Soc. Jap. **56**, 3791 (1983).
[52] A. S. Narula, Tetrahedron Lett. 2017 (1981) and 5579 (1982).
[53] A. S. Narula, Tetrahedron Lett. 5421 (1983).
[54] N. Minami, S. S. Ko and Y. Kishi, J. Am. Chem. Soc. **104**, 1109 (1982).
[55] M. Isobe, M. Kitamura, S. Mio and T. Goto, Tetrahedron Lett. 221 (1982).
[56] V. S. Martin, S. S. Woodard, T. Katsuki, Y. Yamada, M. Ikeda and K. B. Sharpless, J. Am. Chem. Soc. **103**, 6237 (1981).
[57] K. B. Sharpless and R. C. Michaelson, J. Am. Chem. Soc. **95**, 6136 (1973).
[58] T. Fukuyama, B. Vranesic, D. P. Negri and Y. Kishi, Tetrahedron Lett. 2741 (1978).
[59] M. R. Johnson and Y. Kishi, Tetrahedron Lett. 4347 (1979).
[60] I. Hasan and Y. Kishi, Tetrahedron Lett. 4229 (1981).
[61] K. Nishihata and M. Nishio, Tetrahedron Lett. 1041 (1977).
[62] F. A. Carey, O. D. Dailey, Jr., O. Hernandez and J. R. Tucker, J. Org. Chem. **41**, 3975 (1976).
[63] J. K. Cha, W. J. Christ and Y. Kishi, Tetrahedron **40**, 2247 (1984).
[64] H. Wynberg, Chimia **30**, 445 (1976).

Chapter 5

[1] H. L. Cohen and G. F. Wright, J. Org. Chem. **18**, 432 (1953).
[2] T. D. Inch, G. J. Lewis, G. L. Sainsbury and D. J. Sellers, Tetrahedron Lett. 3657 (1969).
[3] N. Oguni and T. Omi, Tetrahedron Lett. **25**, 2823 (1984).
[4] N. Oguni, T. Omi, Y. Yamamoto and A. Nakamura, Chem. Lett. 841 (1983).
[5] D. Seebach und W. Langer, Helv. Chim. Acta **62**, 1701 (1979).
[6] D. Seebach, H. O. Kalinowski, B. Bastani, G. Crass, H. Daum, H. Dörr, N. P. DuPreez, V. Ehrig, W. Langer, C. Nüssler, H. Oei and M. Schmidt, Helv. Chim. Acta **60**, 301 (1977).
[7] D. Seebach, G. Crass, E. Wilka, D. Hilvert and E. Brunner, Helv. Chim. Acta **62**, 2695 (1979).
[8] J.-P. Mazaleyrat and D. J. Cram, J. Am. Chem. Soc. **103**, 4585 (1981).
[9] T. Mukaiyama, K. Soai and S. Kobayashi, Chem. Lett. 219 (1978).
[10] K. Soai and T. Mukaiyama, Bull. Chem. Soc. Jap. **52**, 3371 (1979).
[11] T. Mukaiyama, K. Soai, T. Sato, H. Shimizu and K. Suzuki, J. Am. Chem. Soc. **101**, 1455 (1979).
[12] L. Columbo, C. Gennari, G. Poli and C. Scolastico, Tetrahedron **38**, 2725 (1982).
[13] D. R. Williams, J. G. Phillips and J. C. Huffman, J. Org. Chem. **46**, 4101 (1981).
[14] T. Akiyama, M. Shimizu and T. Mukaiyama, Chem. Lett. 611 (1984).
[15] T. Mukaiyama, K. Suzuki, K. Soai and T. Sato, Chem. Lett. 447 (1979).
[16] T. Mukaiyama and K. Suzuki, 255 (1980).
[17] D. Abenhaim, G. Boireau and B. Sabourault, Tetrahedron Lett. 3043 (1980).
[18] A. G. Olivero, B. Weidmann and D. Seebach, Helv. Chim. Acta **64**, 2485 (1981).
[19] B. Weidmann, L. Wilder, A. G. Olivero, C. D. Maycock and D. Seebach, Helv. Chim. Acta **64**, 357 (1981).
[20] G. Bredig and P. S. Fiske, Biochem. Z. **46**, 7 (1912).
[21] V. Prelog and M. Wilhelm, Helv. Chim. Acta **37**, 1634 (1954).
[22] J. Oku and S. Inoue, J. C. S. Chem. Comm. 229 (1981).
[23] J. Oku, N. Ito and S. Inoue, Makromol. Chem. **183**, 579 (1982).
[24] J. D. Morrison and H. S. Mosher, *Asymmetric Organic Reactions*. Prentice-Hall, Englewood Cliffs, N. Y., 1971.
[25] E. L. Eliel, in *Asymmetric Synthesis,* Vol. 2, (J. D. Morrison, ed.). Academic Press, N. Y., 1983, p. 125.
[26] D. J. Cram and F. A. Abd Elhafez, J. Am. Chem. Soc. **74**, 5828 (1952).
[27] D. J. Cram and F. A. Abd Elhafez, J. Am. Chem. Soc. **76**, 22 (1954).
[28] D. J. Cram and J. Allinger, J. Am. Chem. Soc. **76**, 4516 (1954).
[29] D. J. Cram and J. D. Knight, J. Am. Chem. Soc. **74**, 5835 (1952).
[30] D. J. Cram and F. A. Abd Elhafez and H. Weingartner, J. Am. Chem. Soc. **75**, 2293 (1953).
[31] D. J. Cram and F. D. Greene, J. Am. Chem. Soc. **75**, 6005 (1953).
[32] M. T. Reetz, R. Steinbach, J. Westermann and R. Peter, Angew. Chem. **92**, 1044 (1980); Int. Ed. Engl. 19, 1011 (1980).
[33] J. H. Stocker, P. Sidisunthorn, B. M. Benjamin and C. J. Collins, J. Am. Chem. Soc. **82**, 3913 (1960).
[34] D. J. Cram and D. R. Wilson, J. Am. Chem. Soc. **85**, 1245 (1963).
[35] J. W. Cornforth, R. H. Cornforth and K. K. Mathew, J. Chem. Soc. 112 (1959).
[36] J. H. Stocker, J. Org. Chem. **29**, 3593 (1964).
[37] D. Y. Curtin, E. E. Harris and E. K. Meislich, J. Am. Chem. Soc. **74**, 2901 (1952).
[38] P. Duhamel, L. Duhamel and J. Gralak, Tetrahedron Lett. 2013 (1977).
[39] A. Gaset, P. Andoye and A. Lattes, J. Appl. Chem. Biotechnol. **25**, 13 (1975).
[40] A. Gaset, M. T. Maurette and A. Lattes, J. Appl. Chem. Biotechnol. **25**, 19 (1975).
[41] A. Gaset, M. T. Maurette and A. Lattes, C. R. Acad. Sci. Sec. C **270**, 72 and 2002 (1970).

[42] W. C. Still and J. H. McDonald, Tetrahedron Lett. 1031 (1980).
[43] W. C. Still and J. A. Schneider, Tetrahedron Lett. 1035 (1980).
[44] M. T. Reetz, Angew. Chem. **96**, 542 (1984); Int. Ed. Engl. **23**, 556 (1984).
[45] M. Tramontini, Synthesis 605 (1982).
[46] M. T. Reetz, R. Steinbach, J. Westermann, R. Urz, B. Wenderoth and R. Peter, Angew. Chem. **94**, 133 (1982); Int. Ed. Engl. **21**, 135 (1982).
[47] M. T. Reetz, K. Kesseler, S. Schmidtberger, B. Wenderoth and R. Steinbach, Angew. Chem. **95**, 1007 (1983); Int. Ed. Engl. **22**, 989 (1983).
[48] R. Méric and J. Vigneron, Bull. Soc. Chim. France 327 (1973).
[49] E. L. Eliel, J. K. Koksimies and B. Lohri, J. Am. Chem. Soc. **100**, 1614 (1978).
[50] E. L. Eliel and W. J. Frazee, J. Org. Chem. **44**, 3598 (1979).
[51] J. E. Lynch and E. L. Eliel, J. Am. Chem. Soc. **106**, 2943 (1984).
[52] E. L. Eliel and S. Morris-Natschke, J. Am. Chem. Soc. **106**, 2937 (1984).
[53] K. Ko, W. J. Frazee and E. L. Eliel, Tetrahedron **40**, 1333 (1984).
[54] T. Mukaiyama, Y. Sakito and M. Asami, Chem. Lett. 1253 (1978).
[55] T. Mukaiyama, Y. Sakito and M. Asami, Chem. Lett. 705 (1979).
[56] Y. Sakito and T. Mukaiyama, Chem. Lett. 1027 (1979).
[57] T. J. Leitereg and D. J. Cram, J. Am. Chem. Soc. **90**, 4011 (1968).
[58] T. J. Leitereg and D. J. Cram, J. Am. Chem. Soc. **90**, 4019 (1968).
[59] M. Tramontini, L. Angiolini, C. Fouquey and J. Jacques, Tetrahedron **29**, 4183 (1973).
[60] M. T. Reetz and A. Jung, J. Am. Chem. Soc. **105**, 4833 (1983).
[61] A. McKenzie, J. Chem. Soc. **85**, 1249 (1904).
[62] V. Prelog, Helv. Chim. Acta **36**, 308 (1953).
[63] J. A. Berson and M. A. Greenbaum, J. Am. Chem. Soc. **80**, 652 (1958).
[64] J. A. Berson and M. A. Greenbaum, J. Am. Chem. Soc. **79**, 2340 (1957).
[65] J. K. Whitesell, A. Bhattacharya and K. Henke, J. C. S. Chem. Comm. 988 (1982).
[66] J. K. Whitesell, D. Deyo and A. Bhattacharya, J. C. S. Chem. Comm. 802 (1983).
[67] A. I. Meyers and J. Slade, J. Org. Chem. **45**, 2785 (1980).
[68] G. Stork and J. M. Stryker, Tetrahedron Lett. 4887 (1983).
[69] E. C. Ashby and J. T. Laemmle. Chem. Rev. **75**, 521 (1975).
[70] E. C. Ashby, S. H. Yu and P. V. Roling, J. Org. Chem. **37**, 1918 (1972).
[71] T. L. MacDonald and W. C. Still, J. Am. Chem. Soc. **97**, 5280 (1975).
[72] E. C. Ashby, J. J. Lin and J. J. Watkins, Tetrahedron Lett. 1709 (1977).
[73] B. Weidmann and D. Seebach, Helv. Chim. Acta **63**, 2451 (1980).
[74] M. T. Reetz, R. Steinbach, B. Wenderoth and J. Westermann, Chem. and Ind. 541 (1981).
[75] E. C. Ashby and J. T. Laemmle, J. Org. Chem. **40**, 1469 (1975).
[76] G. Solladié, Synthesis 185 (1981).
[77] N. Kuneida, M. Kinoshita and J. Nokami, Chem. Lett. 289 (1977).
[78] M. Braun and W. Hild, Chem. Ber. **117**, 413 (1984).
[79] C. Mioskowski and G. Solladié, J. C. S. Chem. Comm. 162 (1977).
[80] G. Solladié, F. Matloubi-Moghadam, C. Luttmann and C. Mioskowski, Helv. Chim. Acta **65**, 1602 (1982).
[81] M. Asami and T. Mukaiyama, Chem. Lett. 17 (1980).
[82] A. I. Meyers, M. A. Hanagan, L. M. Trefonas and K. J. Baker, Tetrahedron **39**, 1991 (1983).
[83] E. Nakamura and I. Kuwajima, Tetrahedron Lett. 3347 (1983).
[84] G. J. McGarvey and M. Kimura, J. Org. Chem. **47**, 5420 (1982).
[85] D. Hoppe, Angew. Chem. **96**, 930 (1984); Int. Ed. Engl. **23**, 932 (1984).
[86] R. W. Hoffmann, Angew. Chem. **94**, 569 (1982); Int. Ed. Engl. **21**, 555 (1982).
[87] R. W. Hoffmann, H. J. Zeiss, W. Ladner and S. Tabche, Chem. Ber. **115**, 2357 (1982).
[88] H. C. Brown and P. K. Jadhav, J. Am. Chem. Soc. **105**, 2092 (1983).
[89] G. E. Keck and E. P. Boden, Tetrahedron Lett. **25**, 265 (1984).
[90] R. W. Hoffmann and H.-J. Zeiss, Angew. Chem. **91**, 329 (1979); Int. Ed. Engl. **18**, 306 (1979).

[91] R. W. Hoffmann and W. Ladner, Tetrahedron Lett. 4653 (1979).
[92] T. Hayashi, M. Konishi and M. Kumada, J. Am. Chem. Soc. **104**, 4963 (1982).
[93] C. T. Buse and C. H. Heathcock, Tetrahedron Lett. 1685 (1978).
[94] D. Seebach and L. Widler, Helv. Chim. Acta **65**, 1972 (1982).
[95] S. E. Denmark and E. J. Weber, Helv. Chim. Acta **66**, 1655 (1983).
[96] S. E. Denmark and E. J. Weber, J. Am. Chem. Soc. **106**, 7970 (1984).
[97] M. T. Reetz and B. Wenderoth, Tetrahedron Lett. 5259 (1982).
[98] H. C. Brown, P. K. Jadhav and P. T. Perumal, Tetrahedron Lett. **25**, 5111 (1984).
[99] H. C. Brown and P. K. Jadhav, J. Org. Chem. **49**, 4089 (1984).
[100] Y. Yamamoto, T. Komatsu and K. Maruyama, J. Am. Chem. Soc. **106**, 5031 (1984).
[101] M. Yamaguchi and T. Mukaiyama, Chem. Lett. 993 (1980).
[102] Y. Yamamoto, T. Komatsu and K. Maruyama. J. C. S. Chem. Comm. 191 (1983).
[103] Y. Yamamoto, H. Yatagai and K. Maruyama, J. Am. Chem. Soc. **103**, 3229 (1981).
[104] M. M. Midland and S. B. Preston, J. Am. Chem. Soc. **104**, 2330 (1982).
[105] R. W. Hoffmann and H.-J. Zeiss, J. Org. Chem. **46**, 1309 (1981).
[106] R. W. Hoffmann and B. Landmann, Angew. Chem. **96**, 427 (1984); Int. Ed. Engl. **23**, 437 (1984).
[107] T. Herold and R. W. Hoffmann, Angew. Chem. **90**, 822 (1978); Int. Ed. Engl. **17**, 768 (1978).
[108] T. Herold, U. Schrott and R. W. Hoffmann, Chem. Ber. **114**, 359 (1981).
[109] R. W. Hoffmann and T. Herold, Chem. Ber. **114**, 375 (1981).
[110] R. W. Hoffmann and H.-J. Zeiss, Angew. Chem. **92**, 218 (1980); Int. Ed. Engl. **19**, 218 (1980).
[111] R. W. Hoffmann, A. Endesfelder and H.-J. Zeiss, Carbohydr. Res. **123**, 320 (1983).
[112] E. Favre and M. Gaudemar, J. Organomet. Chem. **92**, 17 (1975).
[113] M. T. Reetz, Angew. Chem. **96**, 542 (1984); Int. Ed. Engl. **23**, 556 (1984).
[114] L. Widler and D. Seebach, Helv. Chim. Acta **65**, 1085 (1982).
[115] E. Sato, K. Iida, K. Ijima, S. Moriyarand and M. Sato, J. C. S. Chem. Comm. 1140 (1981).
[116] M. T. Reetz and M. Sauerwald, J. Org. Chem. **49**, 2292 (1984).
[117] F. Sato, H. Uchiyama, K. Iida, Y. Kobayashi and M. Sato, J. C. S. Chem. Comm. 921 (1983).
[118] D. Hoppe and A. Brönneke, Tetrahedron Lett. 1687 (1983).
[119] K. Furuta, M. Ishiguro, R. Haruta, N. Ikeda and H. Yamamoto, Bull. Chem. Soc. Jap. **57**, 2768 (1984).
[120] Y. Yamamoto, W. Ito and K. Maruyama, J. C. S. Chem. Comm. 1004 (1984).
[121] T. Hayashi, M. Konishi and M. Kumada, J. Org. Chem. **48**, 281 (1983).
[122] K. Mikami, T. Maeda, N. Kishi and T. Nakai, Tetrahedron Lett. **25**, 5151 (1984).
[123] T. Hayashi, K. Kabeta, I. Hamachi and M. Kumada, Tetrahedron Lett. 2865 (1983).
[124] A. Itoh, K. Oshima and H. Nozaki, Tetrahedron Lett. 1783 (1979).
[125] I. Ojima, Y. Miyazawa and M. Kumagai, J. C. S. Chem. Comm. 927 (1976).
[126] S. Kiyoka and C. H. Heathcock, Tetrahedron Lett. 4765 (1983).
[127] C. H. Heathcock, S. Kiyooka and T. A. Blumenkopf, J. Org. Chem. **49**, 4214 (1984).
[128] M. T. Reetz, K. Kesseler and A. Jung, Tetrahedron Lett. **25**, 729 (1984).
[129] A. I. Meyers and M. E. Ford, Tetrahedron Lett. 1341 (1974).
[130] H. Yatagai, Y. Yamamoto and K. Maruyama, J. Am. Chem. Soc. **102**, 4548 (1980).
[131] J. Otera, Y. Kawasaki, H. Mizuno and Y. Shimizu, Chem. Lett. 1529 (1983).
[132] Y. Yamamoto, H. Yatagai, Y. Naruta and K. Maruyama, J. Am. Chem. Soc. **102**, 7107 (1980).
[133] M. Koreeda and Y. Tanaka, Chem. Lett. 1297 (1982).
[134] Y. Yamamoto, H. Yatagai, Y. Ishihara, N. Maeda and K. Maruyama, Tetrahedron **40**, 2239 (1984).
[135] M. Koreeda and Y. Tanaka, Chem. Lett. 1299 (1982).
[136] A. J. Pratt and E. J. Thomas, J. C. S. Chem. Comm. 1115 (1982).
[137] Y. Yamamoto, N. Maeda and K. Maruyama, J. C. S. Chem. Comm. 774 (1983).
[138] G. E. Keck and E. P. Boden, Tetrahedron Lett. **25**, 265 (1984).
[139] G. E. Keck and D. A. Abbott, Tetrahedron Lett. **25**, 1883 (1984).
[140] G. E. Keck and E. P. Boden, Tetrahedron Lett. **25**, 1879 (1984).

[141] Y. Yamamoto and K. Maruyama, Heterocycles **18**, 357 (1982).
[142] K. Maruyama, Y. Ishihara and Y. Yamamoto, Tetrahedron Lett. 4235 (1981).
[143] M. Asami and T. Mukaiyama, Chem. Lett. 93 (1983).
[144] C. Agami, F. Meyner, C. Puchot, J. Guilhem and C. Pascard, Tetrahedron **40**, 1031 (1984).
[145] J. Mulzer, M. Kappert, G. Huttner and I. Jibril, Angew. Chem. **96**, 726 (1984); Int. Ed. Engl. **23**, 704 (1984).
[146] G. Zweifel and G. Hahn, J. Org. Chem. **49**, 4565 (1984).
[147] D. Hoppe and F. Lichtenberg, Angew. Chem. **94**, 378 (1982); Int. Ed. Engl. **21**, 372 (1982).
[148] R. Hanko and D. Hoppe, Angew. Chem. **94**, 378 (1982); Int. Ed. Engl. **21**, 372 (1982).
[149] Y. Okuda, S. Hirano, T. Hiyama and H. Nozaki, J. Am. Chem. Soc. **99**, 3179 (1977).
[150] M. D. Lewis and Y. Kishi, Tetrahedron Lett. 2343 (1982).
[151] Y. Yamamoto and K. Maruyama, Tetrahedron Lett. 2895 (1981).
[152] C. H. Heathcock, in *Asymmetric Synthesis,* Vol. 3., (J. D. Morrison, ed.). Academic Press, N. Y., 1984, p. 111.
[153] C. H. Heathcock, IUPAC Curr. Trends in Org. Synth., (H. Nozaki, ed.). Pergamon Press, 1983, p. 27.
[154] C. H. Heathcock, Science **214**, 395 (1981).
[155] D. A. Evans, J. M. Takacs, L. R. McGee, M. D. Ennis, D. J. Mathre and J. Bartroli, Pure Appl. Chem. **53**, 1109 (1981).
[156] D. A. Evans, J. V. Nelson and T. R. Taber, Top. Stereochem. **13**, 1 (1981).
[157] L. M. Jackmann and B. C. Lange, Tetrahedron **33**, 2737 (1977).
[158] E. B. Dongala, D. L. Dull, C. Mioskowski and G. Solladié, Tetrahedron Lett. 4983 (1973).
[159] D. A. Evans and T. R. Taber, Tetrahedron Lett. 4675 (1980).
[160] D. A. Evans, J. V. Nelson, E. Vogel and T. R. Taber, J. Am. Chem. Soc. **103**, 3099 (1981).
[161] C. H. Heathcock, C. T. Buse, W. A. Kleschick, M. C. Pirrung, J. E. Sohn and J. Lampe, J. Org. Chem. **45**, 1066 (1980).
[162] M. Guetté, J. Capillon and J.-P. Guetté, Tetrahedron **29**, 3659 (1973).
[163] T. Mukaiyama and N. Iwasawa, Chem. Lett. 753 (1984).
[164] J. Mulzer, P. deLasalle, A. Chucholowski, U. Blascheck, G. Brüntrup, I. Jibril and G. Huttner, Tetrahedron **40**, 2211 (1984).
[165] S. Masamune and W. Choy, Aldrichimica Acta **15**, 47 (1982).
[166] H. B. Kagan and J. C. Fiaud, Top. Stereochem. **10**, 175 (1978).
[167] C. H. Heathcock and C. T. White, J. Am. Chem. Soc. **101**, 7076 (1979).
[168] C. H. Heathcock and C. T. White, J. J. Morrison and D. VanDerveer, J. Org. Chem. **46**, 1296 (1981).
[169] S. Masamune, S. A. Ali, D. L. Snitman and D. S. Garvey, Angew. Chem. **92**, 573 (1980); Int. Ed. Engl. **19**, 557 (1980).
[170] D. A. Evans and L. R. McGee, J. Am. Chem. Soc. **103**, 2876 (1981).
[171] C. H. Heathcock, M. C. Pirrung, J. Lampe, C. T. Buse and S. D. Young, J. Org. Chem. **46**, 2290 (1981).
[172] W. Ando and H. Tsumaki, Chem. Lett. 1409 (1983).
[173] H. E. Zimmermann and M. D. Traxler, J. Am. Chem. Soc. **79**, 1920 (1957).
[174] P. G. M. Wuts and M. A. Walters, J. Org. Chem. **49**, 4573 (1984).
[175] J. E. Dubois and M. Dubois, Tetrahedron Lett. 4215 (1967).
[176] J. E. Dubois and P. Fellmann, Tetrahedron Lett. 1225 (1974).
[177] P. Fellmann and J. E. Dubois, Tetrahedron **34**, 1349 (1978).
[178] E. A. Jeffery, A. Meisters and T. Mole, J. Organomet. Chem. **74**, 365 (1974).
[179] H. O. House, D. S. Crumrine, A. Y. Teranishi and H. D. Olmstead, J. Am. Chem. Soc. **95**, 3310 (1973).
[180] K. K. Heng and R. A. J. Smith, Tetrahedron **34**, 425 (1979).
[181] Y. Yamamoto and K. Maruyama, Tetrahedron Lett. **21**, 4607 (1981).
[182] M. T. Reetz and R. Peter, Tetrahedron Lett. 4691 (1981).

[183] E. Nakamura and I. Kuwajima, Tetrahedron Lett. 3343 (1983).
[184] W. W. Fenzl and R. Köster, Liebigs Ann. Chem. 1322 (1975).
[185] S. Masamune, S. Mori, D. Van Horn and D. W. Brooke, Tetrahedron Lett. 1665 (1979).
[186] T. Inoue and T. Mukaiyama, Bull. Chem. Soc. Jap. **53**, 174 (1980).
[187] M. Hirama, D. S. Garvey, L. D.-L. Lu and S. Masamune, Tetrahedron Lett. 3937 (1979).
[188] I. Kuwajima, M. Kato and A. Mori, Tetrahedron Lett. 4291 (1980).
[189] T. Mukaiyama, K. Banno and K. Narasaka, J. Am. Chem. Soc. **96**, 7503 (1974).
[190] R. Hoyori, K. Yokoyama, J. Sabata, I. Kuwajima, E. Nakamura and M. Shimizu, J. Am. Chem. Soc. **99**, 1265 (1977).
[191] C. H. Heathcock and L. A. Flippin, J. Am. Chem. Soc. **105**, 1667 (1983).
[192] W. A. Kleschick, C. T. Buse and C. H. Heathcock, J. Am. Chem. Soc. **99**, 247 (1977).
[193] Y. Yamamoto, H. Yatagai and K. Maruyama, J. C. S. Chem. Comm. 162 (1981).
[194] S. S. Labadie and J. K. Stille, Tetrahedron **40**, 2329 (1984).
[195] S. Shenvi and J. K. Stille, Tetrahedron Lett. 627 (1982).
[196] K. Kobayashi, M. Kawanisi, T. Hitomi and S. Nozima, Chem. Lett. 851 (1983).
[197] T. Harada and T. Mukaiyama, Chem. Lett. 467 (1982).
[198] T. Mukaiyama, R. W. Stevens and N. Iwasawa, Chem. Lett. 353 (1982).
[199] A. I. Meyers and P. J. Reider, J. Am. Chem. Soc. **101**, 2501 (1979).
[200] C. H. Heathcock, M. C. Pirrung, S. H. Montgomery and J. Lampe, Tetrahedron **37**, 4087 (1981).
[201] M. C. Pirrung and C. H. Heathcock, J. Org. Chem. **45**, 1727 (1980).
[202] R. W. Hoffmann and B. Kemper, Tetrahedron **40**, 2219 (1984).
[203] T. Harada and T. Mukaiyama, Chem. Lett. 161 (1982).
[204] J. Canceill, J. J. Basselier and J. Jacques, Bull. Chim. France 1024 (1967).
[205] J. Canceill and J. Jacques, Bull. Soc. Chim. France 2180 (1970).
[206] A. M. Touzin, Tetrahedron Lett. 1477 (1975).
[207] A. Shanzer, L. Somekh and D. Butina, J. Org. Chem. **44**, 3967 (1979).
[208] D. A. Evans and L. R. McGhee, Tetrahedron Lett. 3975 (1980).
[209] J. Mulzer, M. Zippel, G. Brüntrup, J. Segner and J. Finke, Liebigs Ann. Chem. 1108 (1980).
[210] Y. Tamaru, T. Harada, S. Nishi, M. Mizutani, T. Hoiki and Z. Yoshida, J. Am. Chem. Soc. **102**, 7806 (1980).
[211] M. Hirama and S. Masamune, Tetrahedron Lett. 2229 (1979).
[212] C. Goasdoue, N. Goasdoue and M. Gaudemar, Tetrahedron Lett. 4001 (1983).
[213] D. M. von Schriltz, E. M. Kaiser and C. R. Hauser, J. Org. Chem. **32**, 2610 (1967).
[214] C. H. Heathcock and J. Lampe, J. Org. Chem. **48**, 4330 (1983).
[215] C. T. Buse and C. H. Heathcock, J. Am. Chem. Soc. **99**, 8109 (1977).
[216] C. T. White and C. H. Heathcock, J. Org. Chem. **46**, 191 (1981).
[217] R. W. Hoffmann and K. Ditrich, Tetrahedron Lett. **25**, 1781 (1984).
[218] H. Hamana, K. Sasakura and T. Sugasawa, Chem. Lett. 1729 (1984).
[219] C. Gennari, S. Cardani, L. Colombo and C. Scolastico, Tetrahedron Lett. **25**, 2283 (1984).
[220] Y. Tamuru, T. Hioki, S. Kawamura, H. Satomi and Z. Yoshida, J. Am. Chem. Soc. **106**, 3876 (1984).
[221] E. Nakamura, K. Hashimoto and I. Kuwajima, Tetrahedron Lett. 2079 (1978).
[222] J. Mulzer, G. Brüntrup, J. Finke and M. Zippel, J. Am. Chem. Soc. **101**, 7723 (1979).
[223] Y. Tamura, Y. Amino, Y. Furukawa, M. Kagotani and Z. Yoshida, J. Am. Chem. Soc. **104**, 4018 (1982).
[224] Y. Kudo, M. Iwasawa, M. Kobayashi, Y. Senda and S. Mitsui, Tetrahedron Lett. 2125 (1972).
[225] D. A. Evans, E. Vogel and J. V. Nelson, J. Am. Chem. Soc. **101**, 6120 (1979).
[226] T. H. Chan, T. Aida, P. W. K. Lan, V. Gorys and D. H. Harpp, Tetrahedron Lett. 4029 (1979).
[227] R. Noyori, I. Nishida, and J. Sakata, J. Am. Chem. Soc. **103**, 2106 (1981).
[228] S. Murata, M. Suzuki and R. Noyori, J. Am. Chem. Soc. **102**, 3248 (1980).
[229] N. Iwasawa and T. Mukaiyama, Chem. Lett. 1441 (1982).
[230] R. W. Stevens and T. Mukaiyama, Chem. Lett. 1799 (1983).

[231] E. Eichenauer, E. Friedrich, W. Lutz and D. Enders, Angew. Chem. **90**, 219 (1978); Int. Ed. Engl. **17**, 206 (1978).
[232] M. Lucas and J.-P. Guetté, J. Chem. Res. (S) 53 (1980).
[233] R. W. Stevens and T. Mukaiyama, Chem. Lett. 595 (1983).
[234] R. W. Stevens, N. Iwasawa and T. Mukaiyama, Chem. Lett. 1459 (1984).
[235] O. Piccolo, L. Filippini, L. Tinucci, E. Valoti and A. Citterlo, Helv. Chim. Acta **67**, 739 (1984).
[236] J. A. Reid and E. E. Turner, J. Chem. Soc. 3565 (1949); *ibid* 3694 (1950).
[237] S. Brandänge, S. Josephson and S. Vallén, Acta Chem. Scand. **27**, 1084 (1973).
[238] S. Brandänge, S. Josephson, L. Mörch and S. Vallen, Acta Chem. Scand. **B 35**, 273 (1981).
[239] S. Mitsui, K. Konno, I. Onuma and K. Shimizu, J. Chem. Soc. Jap. **85**, 437 (1964).
[240] S. Mitsui and Y. Kudo, Tetrahedron **23**, 4271 (1967).
[241] C. Mioskowski and G. Solladié, J. C. S. Chem. Comm. 162 (1977).
[242] C. Mioskowski and G. Solladié, Tetrahedron **36**, 227 (1980).
[243] R. Anunnziata, M. Cinquini, F. Cozzi, F. Montanari and A. Restelli, J. C. S. Chem. Comm. 1138 (1983).
[244] D. A. Evans, J. Bartroli and T. L. Shih, J. Am. Chem. Soc. **103**, 2127 (1981).
[245] T. Nakatsuka, T. Miwa and T. Mukaiyama, Chem. Lett. 279 (1981).
[246] M. Braun and R. Devant, Angew. Chem. **95**, 802 (1983); Int. Ed. Engl. **22**, 788 (1983).
[247] D. Seebach and R. Naef, Helv. Chim. Acta **64**, 2704 (1981).
[248] D. Seebach and T. Weber, Helv. Chim. Acta **67**, 1650 (1984).
[249] D. Seebach, V. Ehrig and M. Teschner, Liebigs Ann. Chem. 1357 (1976).
[250] S. Masamune, W. Choy, F. A. J. Kerdesky and B. Imperiali, J. Am. Chem. Soc. **103**, 1566 (1981).
[251] W. Choy, P. Ma and S. Masamune, Tetrahedron Lett. 3555 (1981).
[252] C. H. Heathcock, M. C. Pirrung, C. T. Buse, J. R. Hagen, S. D. Young and J. E. Sohn, J. Am. Chem. Soc. **101**, 7077 (1979).
[253] C. H. Heathcock, S. D. Young, J. P. Hagen, M. C. Pirrung, C. T. White and D. VanDerveer, J. Org. Chem. **45**, 3846 (1980).
[254] N. Aktogu, H. Felkin and S. G. Davies, J. C. S. Chem. Comm. 1303 (1982).
[255] S. G. Davies, I. M. Dordor and P. Warner, J. C. S. Chem. Comm. 957 (1984).
[256] L. S. Liebeskind and M. E. Welker, Tetrahedron Lett. **25**, 4341 (1984).
[257] H. M. Shieh and G. D. Prestwich, J. Org. Chem. **46**, 4319 (1981).
[258] A. I. Meyers and G. Knaus, Tetrahedron Lett. 1333 (1974).
[259] A. I. Meyers and Y. Yamamoto, J. Am. Chem. Soc. **103**, 4278 (1981).
[260] A. I. Meyers and Y. Yamamoto, Tetrahedron **40**, 2309 (1984).
[261] H. Eichenauer, W. Friedrich, E. Lutz and D. Enders, Angew. Chem. **87**, 218 (1978); Int. Ed. Engl. **17**, 206 (1978).
[262] S. Suzuki, H. Narita and K. Harada, J. C. S. Chem. Comm. 29 (1979).
[263] Yu. N. Belokon, I. E. Zel'tzer, N. M. Loim, V. A. Tsiryapkin, Z. N. Parnes, D. N. Kursanov and V. M. Belikov, J. C. S. Chem. Comm. 789 (1979).
[264] Yu. N. Belokon, I. E. Zel'tzer, N. M. Loim, V. A. Tsiryapkin, G. G. Aleksandrov, D. N. Kursanov, Z. N. Parnes, Yu. T. Struchkov and V. M. Belikov, Tetrahedron **36**, 1089 (1980).
[265] Yu. N. Belokon, L. E. Zel'tzer, M. G. Ryzhov, M. B. Saporovskaya, V. I. Bakhmutov and V. M. Belikov, J. C. S. Chem. Comm. 180 (1982).
[266] M. T. Reetz and K. Kesseler, J. C. S. Chem. Comm. 1079 (1984).
[267] Y. Sakito, M. Asami and T. Mukaiyama, Chem. Lett. 455 (1980).
[268] E. Nakamura, Y. Horiguchi, J. Shimada and I. Kuwajima, J. C. S. Chem. Comm. 796 (1983).
[269] U. Eder, G. Sauer and R. Wiechert, Angew. Chem. **83**, 492 (1971); Int. Ed. Engl. **10**, 496 (1971).
[270] Z. G. Hajos and D. R. Parrish, J. Org. Chem. **39**, 1615 (1974).
[271] J. Gutzwiller, P. Buchschacher and A. Fürst, Synthesis 167 (1977).
[272] Buchschacher, J. Casal, A. Fürst and W. Meier, Helv. Chim. Acta **60**, 2747 (1977).
[273] P. A. Grieco, N. Fukamiya and M. Miyashita, J. C. S. Chem. Comm. 573 (1976).
[274] S. Danishefsky and P. Cain, J. Am. Chem. Soc. **98**, 4975 (1976).

[275] I. Shimizu, Y. Naito and J. Tsuji, Tetrahedron Lett. 487 (1980).
[276] T. Wakabayashi, K. Watanabe and Y. Kato, Synth. Comm. **7**, 238 (1977).
[277] S. Terashima, S. Sato and K. Koga, Tetrahedron Lett. 3469 (1979).
[278] C. Agami and H. Senestre, J. C. S. Chem. Comm. 1385 (1984).

Chapter 6

[1] H. Wynberg and R. Helder, Tetrahedron Lett. 4057 (1975).
[2] K. Hermann and H. Wynberg, J. Org. Chem. **44**, 2238 (1979).
[3] W. ten Hoeve and H. Wynberg, J. Org. Chem. **45**, 1508 (1980).
[4] K. Matsumoto and T. Uchida, Chem. Lett. 1673 (1981).
[5] D. J. Cram and G. D. Y. Sogah, J. C. S. Chem. Comm. 625 (1981).
[6] R. Annunziata, M. Cinquini and S. Colonna, J. C. S. Perkin 1, 2422 (1980).
[7] S. Colonna, A. Re and H. Wynberg, J. C. S. Perkin 1, 547 (1981).
[8] S. Colonna, H. Hiemstra and H. Wynberg, J. C. S. Chem. Comm. 238 (1978).
[9] N. Kobayashi and K. Iwai, J. Am. Chem. Soc. **100**, 7071 (1978).
[10] N. Kobayashi and K. Iwai, J. Polym. Sci. Polym. Chem. Ed. **18**, 923 (1980).
[11] H. Brunner and B. Hammer, Angew. Chem. **96**, 305 (1984); Int. Ed. Engl. **23**, 312 (1984).
[12] T. Mukaiyama, Y. Hirato and T. Takeda, Chem. Lett. 461 (1978).
[13] S. J. Blarer and D. Seebach, Chem. Ber. **116**, 3086 (1983).
[14] G. Stork, J. D. Winkler and N. A. Sacconamo, Tetrahedron Lett. 465 (1983).
[15] G. Tsuchihashi, S. Mitamura and K. Ogura, Tetrahedron Lett. 855 (1976).
[16] R. A. Kretchmer, J. Org. Chem. **37**, 2744 (1972).
[17] D. Seebach, G. Crass, E. Wilka, D. Hilvert and E. Brunner, Helv. Chim. Acta **62**, 2695 (1979).
[18] W. Langer and D. Seebach, Helv. Chim. Acta **62**, 1710 (1979).
[19] T. Imamoto and T. Mukaiyama, Chem. Lett. 45 (1980).
[20] F. Leyendecker, F. Jesser and B. Ruhland, Tetrahedron Lett. 3601 (1981).
[21] M. Huché, J. Berlan, G. Pourcelot and P. Cresson, Tetrahedron Lett. 1329 (1981).
[22] B. Gustafsson, G. Hallnemo and C. Ullenius, Acta Chem. Scand. B **34**, 443 (1980).
[23] B. Gustafsson and C. Ullenius, Tetrahedron Lett. 3174 (1977).
[24] B. Gustafsson, A.-T. Hansson and C. Ullenius, Acta. Chem. Scand. B **34**, 113 (1980).
[25] M. Kawana and S. Emoto, Bull. Chem. Soc. Japan, **39**, 910 (1980).
[26] M. Kolb and J. Barth, Liebigs Ann. Chem. 1068 (1983).
[27] W. Oppolzer and H. J. Löher, Helv. Chim. Acta **64**, 2808 (1981).
[28] W. Oppolzer, R. Moretti, T. Godel, A. Meunier and H. Löher, Tetrahedron Lett. 4971 (1983).
[29] G. H. Posner, in *Asymmetric Synthesis*, Vol. 2 (J. D. Morrison, ed.). Academic Press, New York, 1983, p. 225.
[30] G. H. Posner, J. P. Mallamo and K. Miura, J. Am. Chem. Soc. **103**, 2886 (1981).
[31] G. Tsuchihashi, S. Mitamura, S. Inone and K. Ogura, Tetrahedron Lett. 323 (1973).
[32] G. H. Posner, J. P. Mallamo, M. Hulce and L. L. Frye, J. Am. Chem. Soc. **104**, 4180 (1982).
[33] G. H. Posner and M. Hulce, Tetrahedron Lett. **25**, 379 (1984).
[34] G. H. Posner and L. L. Frye, Isr. J. Chem. **24**, 8 (1984).
[35] G. H. Posner, M. Hulce, J. P. Mallamo, S. A. Drexler and J. Clardy, J. Org. Chem. **46**, 5244 (1981).
[36] G. H. Posner, T. P. Kogan and M. Hulce, Tetrahedron Lett. **25**, 383 (1984).

[37] L. Colombo, C. Gennari, G. Resnati and C. Scolastico, Synthesis 74 (1981).
[38] T. Mukaiyama, T. Takeda and M. Osaki, Chem. Lett. 1165 (1977).
[39] T. Mukaiyama, T. Takeda and K. Fujimoto, Bull. Chem. Soc. Japan **51**, 3368 (1978).
[40] T. Mukaiyama and N. Iwasawa, Chem. Lett. 913 (1981).
[41] M. Asami and T. Mukaiyama, Chem. Lett. 569 (1979).
[42] K. Soai, A. Ookawa and Y. Nohara, Synth. Comm. **13**, 27 (1983).
[43] J. Fujiwara, Y. Fukutani, M. Hasegawa, K. Maruoka and H. Yamamoto, J. Am. Chem. Soc. **106**, 5004 (1984).
[44] Y. Fukutani, K. Maruoka and H. Yamamoto, Tetrahedron Lett. **25**, 5911 (1984).
[45] M. Isobe, M. Kitamura and T. Goto, Tetrahedron Lett. 3465 (1979).
[46] M. Isobe, M. Kitamura and T. Goto, Tetrahedron Lett. 4727 (1980).
[47] M. Isobe, M. Kitamura and T. Goto, Tetrahedron Lett. 239 (1981).
[48] D. Enders and K. Papadopoulos, Tetrahedron Lett. 4967 (1983).
[49] H. O. House and W. F. Fischer, Jr., J. Org. Chem. **33**, 949 (1968).
[50] W. C. Still and I. Galynker, Tetrahedron **37**, 3981 (1981).
[51] M. Yamaguchi, M. Tsukamoto, S. Tanaka and I. Hirao, Tetrahedron Lett. **25**, 5661 (1984).
[52] S. Hashimoto, S. Yamada and K. Koga, J. Am. Chem. Soc. **98**, 7450 (1976).
[53] S. Hashimoto, N. Komeshima, S. Yamada and K. Koga, Tetrahedron Lett. 2907 (1977).
[54] S. Hashimoto, S. Yamada and K. Koga, Chem. Pharm. Bull. **27**, 771 (1979).
[55] S. Hashimoto, H. Kogen, K. Tomioka and K. Koga, Tetrahedron Lett. 3009 (1979).
[56] H. Kogen, K. Tomioka, S. Hashimoto and K. Koga, Tetrahedron **37**, 3951 (1981).
[57] K. Tomioka, F. Masumi, T. Yamashita and K. Koga, Tetrahedron Lett. **25**, 333 (1984).
[58] A. I. Meyers, R. K. Smith and C. E. Whitten, J. Org. Chem. **44**, 2250 (1979).
[59] A. I. Meyers and C. E. Whitten, J. Am. Chem. Soc. **97**, 6266 (1975).
[60] J. Berlan, Y. Besace, G. Pourcelot and P. Cresson, J. Organomet. Chem. **256**, 181 (1983).
[61] J. Berlan, Y. Besace, D. Prat and G. Pourcelot, J. Organomet. Chem. **264**, 399 (1984).
[62] A. I. Meyers and D. Hoyer, Tetrahedron Lett. **25**, 3667 (1984).
[63] H. Felkin, G. Swierczewski and A. Tambuté, Tetrahedron Lett. 707 (1969).
[64] M. Chérest, H. Felkin, C. Frajerman, C. Lion, G. Roussi and G. Swierczewski, Tetrahedron Lett. 875 (1966).
[65] B. M. Trost, Acc. Chem. Res. **13**, 385 (1980).
[66] P. Pino and G. Consiglio, Pure Appl. Chem. **55**, 1781 (1983).
[67] B. Bosnich and P. B. Mackenzie, Pure Appl. Chem. **54**, 189 (1982).
[68] B. Bosnich and M. D. Fryzuk, Top. Stereochem. **12**, 119 (1980).
[69] B. M. Trost and P. E. Strege, J. Am. Chem. Soc. **99**, 1649 (1977).
[70] T. Hayashi, K. Kanehira, H. Tsuchiya and M. Kumada, J. C. S. Chem. Comm. 1162 (1982).
[71] T. Hayashi, Abstr. ESOC Conference, Freiburg, FRG, 1984.
[72] J. C. Fiaud, A. H. De Gournay, M. Larcheveque and H. B. Kagan, J. Organomet. Chem. **154**, 175 (1978).
[73] H. Ahlbrecht, G. Bonnet, D. Enders and G. Zimmermann, Tetrahedron Lett. 3175 (1980).
[74] T. Takahashi, Y. Jinbo, K. Kitamura and J. Tsuji,Tetrahedron Lett. **25**, 5921 (1984).
[75] G. Consiglio and P. Pino, Advances in Chem. Ser. **196**, 371 (1982).
[76] P. Pino, G. Consiglio, C. Botteghi and C. Salomon, Advances in Chem. Ser. **132**, 295 (1974).
[77] C. Botteghi, G. Consiglio and P. Pino, Chimia **26**, 141 (1972).
[78] T. Hayashi, M. Tanaka and I. Ogata, Tetrahedron Lett. 3925 (1978).
[79] T. Hayashi, M. Tanaka, Y. Ikeda and I. Ogata, Bull. Chem. Soc. Japan **52**, 2605 (1979).
[80] G. Consiglio and P. Pino, Helv. Chim. Acta **59**, 642 (1976).
[81] Can. Pat. 1.027.141 (28. 2. 1978). C. A. **89**, 42440 (1978).
[82] Y. Kawabata, T. M. Suzuki and I. Ogata, Chem. Lett. 361 (1978).
[83] M. Tanaka, Y. Ikeda and I. Ogata, Chem. Lett. 1115 (1975).
[84] C. F. Hobbs and W. S. Knowles, J. Org. Chem. **46**, 4422 (1981).
[85] C. U. Pittman, Y. Kawabata and L. I. Flowers, J. C. S. Chem. Comm. 473 (1982).

[86] G. Consiglio, P. Pino, L. I. Flowers and C. U. Pittman, Jr., J. C. S. Chem. Comm. 612 (1983).
[87] G. Cometti and G. P. Chiusoli, J. Organomet. Chem. **236**, C31 (1982).
[88] G. Consiglio, Helv. Chim. Acta **59**, 124 1975.
[89] P. Duhamel, J.-Y. Valnot and J. J. Eddine, Tetrahedron Lett. 2863 (1982).
[90] P. Duhamel, J. J. Eddine and J.-J. Valnot, Tetrahedron Lett. **25**, 2355 (1984).
[91] T. Yamashita, H. Mitsui, H. Watanabe and N. Nakamura, Bull. Chem. Soc. Japan **55**, 961 (1982).
[92] P. E. Sonnet and R. R. Heath, J. Org. Chem. **45**, 3137 (1980).
[93] D. A. Evans and J. M. Takacs, Tetrahedron Lett. 4233 (1980).
[94] T. Kaneko, D. Turner, M. Newcomb and D. E. Bergbreiter, Tetrahedron Lett. 103 (1979).
[95] M. Larcheveque, E. Ignatova and T. Cuvigny, Tetrahedron Lett. 3961 (1978).
[96] D. A. Evans, M. D. Ennis and D. J. Mathre, J. Am. Chem. Soc. **104**, 1737 (1982).
[97] R. Schmierer, G. Grotemeier, G. Helmchen and A. Selim, Angew. Chem. **93**, 207 (1981); Int. Ed. Engl. *20*, 209 (1981).
[98] G. Helmchen, A. Selim, D. Dorsch and I. Taufer, Tetrahedron Lett. 3213 (1983).
[99] G. Lin, E. Hoegberg and T. Norin, Kexue Tongbao **29**, 632 (1984). C. A. **101**, 15374 (1984).
[100] D. A. Evans, J. M. Takacs, L. R. McGee, M. D. Ennis, D. J. Mathre and J. Bartroli, Pure Appl. Chem. **53**, 1109 (1981).
[101] R. E. Ireland, R. H. Mueller and A. K. Willard, J. Am. Chem. Soc. **98**, 2868 (1976).
[102] E. Ade, G. Helmchen and G. Heiligenmann, Tetrahedron Lett. 1137 (1980).
[103] G. Helmchen and R. Schmierer, Tetrahedron Lett. 1235 (1983).
[104] G. Helmchen and R. Wierzchowski, Angew. Chem. **96**, 59 (1984); Int. Ed. Engl. **23**, 60 (1984).
[105] D. A. Evans, M. D. Ennis, T. Le, N. Mandel and G. Mandel, J. Am. Chem. Soc. **106**, 1154 (1984).
[106] S. Yamada, T. Oguri and T. Shioiri, J. C. S. Chem. Comm. 136 (1976).
[107] J. A. Bajgrowicz, B. Cossec, C. Pigiere, R. Jacquier and P. Viallefont, Tetrahedron Lett. 3721 (1983).
[108] K. Piotrowska and W. Abramski, Pol. J. Chem. 2397 (1979).
[109] G. J. Baird, S. G. Davies, R. H. Jones, K. Prout and P. Warner, J. C. S. Chem. Comm. 745 (1984).
[110] P. J. Curtis and S. G. Davies, J. C. S. Chem. Comm. 747 (1984).
[111] Gy. Fráter, U. Müller and W. Günther, Tetrahedron Lett. 4221 (1981).
[112] D. Seebach, R. Naef and G. Calderari, Tetrahedron **40**, 1313 (1984).
[113] D. Seebach, M. Boes, R. Naef and W. B. Schweizer, J. Am. Chem. Soc. **105**, 5390 (1983).
[114] A. I. Meyers, Pure Appl. Chem. **51**, 1255 (1979).
[115] A. I. Meyers and E. D. Mihelich, Angew. Chem. **88**, 321 (1976); Int. Ed. Engl. **15**, 270 (1976).
[116] A. I. Meyers, G. Knaus, K. Kamata and M. E. Ford, J. Am. Chem. Soc. **98**, 567 (1976).
[117] S. Shibata, H. Matsushita, H. Kaneko, M. Noguchi, M. Saburi and S. Yoshikawa, Chem. Lett. 217 (1981).
[118] A. I. Meyers, A. Mazzu and C. E. Whitten, Heterocycles **6**, 971 (1977).
[119] A. I. Meyers, E. S. Snyder and J. J. H. Ackerman, J. Am. Chem. Soc. **100**, 8186 (1978).
[120] M. A. Hoobler, D. E. Bergbreiter and M. Newcomb, J. Am. Chem. Soc. **100**, 8182 (1978).
[121] A. I. Meyers and G. Knaus, J. Am. Chem. Soc. **96**, 6508 (1974).
[122] A. I. Meyers, Y. Yamamoto, E. D. Mihelich and R. A. Bell, J. Org. Chem. **45**, 2792 (1980).
[123] A. I. Meyers and C. E. Whitten, Heterocycles **4**, 1687 (1976).
[124] J. F. Hansen and C. S. Cooper, J. Org. Chem. **41**, 3219 (1976).
[125] U. Schöllkopf, Pure Appl. Chem. **55**, 1799 (1983).
[126] U. Schöllkopf, Top. Curr. Chem. **109**, 65 (1983).
[127] U. Schöllkopf, Tetrahedron **39**, 2085 (1983).
[128] U. Schöllkopf, W. Hartwig and U. Groth, Angew. Chem. **91**, 922 (1979); Int. Ed. Engl. **18**, 863 (1979).
[129] U. Schöllkopf, U. Groth and C. Deng, Angew. Chem. **93**, 793 (1981); Int. Ed. Engl. **20**, 798 (1981).

[130] U. Schöllkopf, U. Groth, K.-O. Westphalen and C. Deng, Synthesis 969 (1981).
[131] J. Nozulak and U. Schöllkopf, Synthesis 866 (1982).
[132] U. Schöllkopf and H. Neubauer, Synthesis 861 (1982).
[133] U. Groth and U. Schöllkopf, Synthesis 37 (1983).
[134] U. Schöllkopf, U. Busse, R. Kilger and P. Lehr, Synthesis 271 (1984).
[135] U. Schöllkopf, W. Hartwig, U. Groth and K.-O. Westphalen, Liebigs Ann. Chem. 696 (1981).
[136] U. Schöllkopf, U. Groth and W. Hartwig, Liebigs Ann. Chem. 2407 (1981).
[137] U. Groth and U. Schöllkopf, Synthesis 673 (1983).
[138] U. Groth, U. Schöllkopf and Y. Chiang, Synthesis 864 (1982).
[139] U. Schöllkopf, U. Groth, M.-R. Gull and J. Nozulak, Liebigs Ann. Chem. 1133 (1983).
[140] F. Sauriol-Lord and T. B. Grindley, J. Org. Chem. **46**, 2831 (1981).
[141] F. Sauriol-Lord and T. B. Grindley, J. Org. Chem. 2833 (1981).
[142] D. Seebach and D. Wasmuth, Helv. Chim. Acta **63**, 197 (1980).
[143] Gy. Fráter, Helv. Chim. Acta **62**, 2825 (1979).
[144] Gy. Fráter, U. Müller and W. Günther, Tetrahedron **40**, 1269 (1984).
[145] Gy. Fráter, Tetrahedron Lett. 425 (1981).
[146] A. R. Chamberlin and M. Dezube, Tetrahedron Lett. 3055 (1982).
[147] K. Tomioka, Y. Cho, F. Sato and K. Koga, Chem. Lett. 1621 (1981).
[148] S. Takano, K. Chiba, M. Yonaga and K. Ogasawara, J. C. S. Chem. Comm. 616 (1980).
[149] D. Seebach and J. D. Aebi, Tetrahedron Lett. 3311 (1983).
[150] H. M. Shieh and G. D. Prestwich, J. Org. Chem. **46**, 4319 (1981).
[151] E. L. Eliel, Tetrahedron **30**, 1503 (1974).
[152] E. L. Eliel, A. A. Hartmann and A. G. Abatjoglou, J. Am. Chem. Soc. **96**, 1807 (1974).
[153] A. A. Hartmann and E. L. Eliel, J. Am. Chem. Soc. **93**, 2572 (1971).
[154] E. L. Eliel, A. Abatjoglou and A. A. Hartmann, J. Am. Chem. Soc. **94**, 4786 (1972).
[155] S. Bory and A. Marquet, Tetrahedron Lett. 4155 (1973).
[156(a)] R. Lett and A. Marquet, Tetrahedron Lett. 1579 (1975).
(b) G. Chassaing, R. Lett and A. Marquet, Tetrahedron Lett. 471 (1978).
[157] J. F. Biellmann and J. J. Vicens, Tetrahedron Lett. 467 (1978).
[158] K. Nishihata and M. Nishio, Tetrahedron Lett. 1695 (1976).
[159] T. Durst, R. Viau and M. R. McClory, J. Am. Chem. Soc. **93**, 3077 (1971).
[160] J. J. Eisch and J. E. Galle, J. Org. Chem. **45**, 4534 (1980).
[161] U. Dolling, P. Davis and E. J. J. Grabowski, J. Am. Chem. Soc. **106**, 446 (1984).
[162] S. Yamada, K. Hiroi and K. Achiwa, Tetrahedron Lett. 4233 (1969).
[163] S. Yamada and G. Otani, Tetrahedron Lett. 4237 (1969).
[164] K. Hiroi, K. Achiwa and S. Yamada, Chem. Pharm. Bull. **20**, 246 (1972).
[165] G. Otani and S. Yamada, Chem. Pharm. Bull. **21**, 2112 (1973).
[166] G. Otani and S. Yamada, Chem. Pharm. Bull. **21**, 2125 (1973).
[167] T. Sone, K. Hiroi and S. Yamada, Chem. Pharm. Bull. **21**, 2331 (1973).
[168] T. Sone, S. Terashima and S. Yamada, Chem. Pharm. Bull. **24**, 1288 (1976).
[169] C. C. Tseng, S. Terashima and S. Yamada, Chem. Pharm. Bull. **25**, 29 (1977).
[170] S. Yamada, M. Shibasaki and S. Terashima, Tetrahedron Lett. 381 (1973).
[171] D. Méa-Jacheet and A. Horeau, Bull. Soc. Chim. France 4571 (1968).
[172] M. Kitamoto, K. Hiroi, S. Terashima and S. Yamada, Chem. Pharm. Bull. **22**, 459 (1974).
[173] J. K. Whitesell and S. W. Felman, J. Org. Chem. **42**, 1663 (1977).
[174] S. J. Blarer and D. Seebach, Chem. Ber. **116**, 2250 (1983).
[175] B. De Jeso and J.-C. Pommier, Tetrahedron Lett. 4511 (1980).
[176] A. I. Meyers, D. R. Williams and M. Druelinger, J. Am. Chem. Soc. **98**, 3032 (1976).
[177] A. I. Meyers, D. R. Williams, G. W. Erickson, S. White and M. Druelinger, J. Am. Chem. Soc. **103**, 3081 (1981).
[178] J. K. Whitesell and M. A. Whitesell, J. Org. Chem. **42**, 377 (1977).
[179] A. I. Meyers, D. R. Williams, S. White and G. W. Erickson, J. Am. Chem. Soc. **103**, 3088 (1981).

[180] K. Saigo, A. Kasahara, S. Ogawa and H. Nohira, Tetrahedron Lett. 511 (1983).
[181] P. M. Worster, C. R. McArthur and C. C. Leznoff, Angew. Chem. **91**, 255 (1979); Int. Ed. Engl. **18**, 221 (1979).
[182] J. M. J. Frechet, React. Polym., Ion Exch., Sorbents **1**, 227 (1983). C. A. **100**, 5900 (1984).
[183] S. Hashimoto and K. Koga, Tetrahedron Lett. 573 (1978).
[184] S. Hashimoto and K. Koga, Chem. Pharm. Bull. **27**, 2760 (1979).
[185] K. Tomioka, K. Ando, T. Takemasa and K. Koga, J. Am. Chem. Soc. **106**, 2718 (1984).
[186] K. Tomioka, K. Ando, T. Takemasa and K. Koga, Tetrahedron Lett. 5677 (1984).
[187] A. I. Meyers and D. R. Williams, J. Org. Chem. **43**, 3245 (1978).
[188] A. I. Meyers, G. S. Pointdexter and Z. Brich, J. Org. Chem. **43**, 892 (1978).
[189] D. F. Taber and K. Raman, J. Am. Chem. Soc. **105**, 5935 (1983).
[190] A. I. Meyers, Z. Brich, G. W. Erickson and S. G. Traynor, J. C. S. Chem. Comm. 566 (1979).
[191] R. R. Fraser, F. Akiyama and J. Banwille, Tetrahedron Lett. 3929 (1979).
[192] D. Enders, in *Asymmetric Synthesis*, Vol. 3, (J. D. Morrison, ed.). Academic Press, New York, 1984, p. 275.
[193] D. Enders, Chemtech., 504 (1981).
[194] D. Enders, in *Curr. Trends in Org. Synth.*, (H. Nozaki, ed.). Pergamon Press, Oxford, 1983, p. 151.
[195] D. Enders and H. Eichenauer, Angew. Chem. **88**, 579 (1976); Int. Ed. Engl. **15**, 549 (1976).
[196] D. Enders and H. Eichenauer, Chem. Ber. **112**, 2933 (1979).
[197] D. Enders, H. Eichenauer and R. Pieter, Chem. Ber. **112**, 3703 (1979).
[198] P. M. Hardy, Synthesis 290 (1978).
[199] D. Enders, H. Eichenauer, U. Baus, H. Schubert and K. A. M. Kremer, Tetrahedron **40**, 1345 (1984).
[200] K. G. Davenport, H. Eichenauer, D. Enders, M. Newcomb and D. E. Bergbreiter, J. Am. Chem. Soc. **101**, 5654 (1979).
[201] D. Enders and H. Eichenauer, Angew. Chem. **91**, 425 (1979); Int. Ed. Engl. **18**, 397 (1979).
[202] S. I. Pennanen, Acta Chem. Scand. B **35**, 555 (1981).
[203] D. Enders and H. Schuber, Angew. Chem. **96**, 368 (1984); Int. Ed. Engl. **23**, 365 (1984).
[204] M. T. Reetz, R. Steinbach and K. Kesseler, Angew. Chem. **95**, 872 (1982); Int. Ed. Engl. **21**, 864 (1982).
[205] M. Kolb and J. Barth, Tetrahedron Lett. 2999 (1979).
[206] A. I. Meyers, M. Boes and D. A. Dickman, Angew. Chem. **96**, 448 (1984); Int. Ed. Engl. **23**, 458 (1984).
[207] A. I. Meyers, L. M. Fuentes and Y. Kubota, Tetrahedron **40**, 1361 (1984).
[208] K. Harada and T. Okawara, J. Org. Chem. **38**, 707 (1973).
[209] D. M. Stout, L. A. Black and W. L. Matier, J. Org. Chem. **48**, 5369 (1983).
[210] M. Opplinger and R. Schwyzer, Helv. Chim. Acta **60**, 43 (1977).
[211] K. Weinges, K. Gries, B. Stemmle and W. Schrank, Chem. Ber. **110**, 2098 (1977).
[212] K. Weinges, G. Graab, D. Nagel and B. Stemmle, Chem. Ber. **104**, 3594 (1971).
[213] K. Weinges, G. Brune and H. Droste, Liebigs Ann. Chem. 212 (1980).
[214] K. Weinges and B. Stemmle, Chem. Ber. **106**, 2291 (1973).
[215] K. Weinges and H. Blackholm, Chem. Ber. **113**, 3098 (1980).
[216] K. Weinges, G. Brune and H. Droste, Liebigs Ann. Chem. 212 (1980).
[217] K. Weinges, K.-P. Klotz and H. Droste, Chem. Ber. **113**, 710 (1980).
[218] Japan Kokai 76,108,002 (25. 9. 1976); C. A. **86**, 155965 (1977).
[219] H. Takahashi, Y. Suzuki and H. Inagaki, Chem. Pharm. Bull. **30**, 3160 (1982).
[220] Y. Suzuki and H. Takahashi, Chem. Pharm. Bull. **31**, 2895 (1983).
[221] Y. Suzuki and H. Takahashi, Chem. Pharm. Bull. **31**, 31 (1983).
[222] A. Solladié-Cavallo, J. Suffert and J.-L. Haesslein, Angew. Chem. **92**, 1038 (1980); Int. Ed. Engl. **19**, 1005 (1980).
[223] A. Solladié-Cavallo and E. Tsamo, J. Organomet. Chem. **172**, 165 (1979).

[224] N. Maigrot, J.-P. Mazaleyrat and Z. Welvart, J. C. S. Chem. Comm. 40 (1984).
[225] H. Takahashi, K. Tomita and H. Otamasu, J. C. S. Chem. Comm. 668 (1979).
[226] H. Takahashi and H. Inagaki, Chem. Pharm. Bull. **30**, 922 (1982).
[227] L. S. Liebeskind, M. E. Welker and V. Goedken, J. Am. Chem. Soc. **106**, 441 (1984).
[228] M. Kumada, Pure Appl. Chem. **52**, 669 (1980).
[229] G. Consiglio and C. Botteghi, Helv. Chim. Acta **56**, 460 (1973).
[230] G. Consiglio, O. Piccolo and F. Morandini, J. Organomet. Chem. **177**, C13 (1979).
[231] G. Consiglio, F. Morandini and O. Piccolo, Tetrahedron **39**, 2699 (1983).
[232] T. Hayashi, M. Konishi, M. Fukushima, T. Mise, M. Kagotani, M. Tajika and M. Kumada, J. Am. Chem. Soc. **104**, 180 (1982).
[233] T. Hayashi, M. Fukushima, M. Konishi and M. Kumada, Tetrahedron Lett. 79 (1980).
[234] T. Hayashi, M. Konishi, M. Fukushima, K. Kanehira, T. Hioki and M. Kumada, J. Org. Chem. **48**, 2195 (1983).
[235] T. Hayashi, M. Konishi, H. Ito and M. Kumada, J. Am. Chem. Soc. **104**, 4962 (1982).
[236] T. Hayashi, T. Hagihara, Y. Katsuro and M. Kumada, Bull. Chem. Soc. Jap. **56**, 363 (1983).
[237] B. Bogdanovic, Angew. Chem. **85**, 1013 (1973); Int. Ed. Engl. **12** , 954 (1973).
[238] B. Bogdanovic, B. Henc, B. Meister, H. Pauling and G. Wilke, Angew. Chem. **84**, 1070 (1972); Int. Ed. Engl. **11**, 1023 (1972).
[239] G. Buono, G. Pfeiffer, A. Montreux and F. Petit, J. C. S. Chem. Comm. 937 (1980).
[240] P. Brun, A. Tenaglia and B. Waegell, Tetrahedron Lett. 385 (1983).
[241] W. S. Johnson, R. Elliot and J. D. Elliot, J. Am. Chem. Soc. **105**, 2904 (1983).
[242] J. D. Elliott, W. M. F. Choi and W. S. Johnson, J. Org. Chem. **48**, 2294 (1983).
[243] V. M. F. Choi, J. D. Elliott and W. S. Johnson, Tetrahedron Lett. **25**, 591 (1984).
[244] W. S. Johnson, P. H. Crackett, J. D. Elliott, J. J. Jagodzinski, S. D. Lindell and S. Natarajan, Tetrahedron Lett. **25**, 3951 (1984).
[245] S. D. Lindell, J. D. Elliott and W. S. Johnson, Tetrahedron Lett. **25**, 3947 (1984).
[246] S. Miyano, M. Tobita, M. Nawa, S. Sato and H. Hashimoto, J. C. S. Chem. Comm. 1233 (1980).
[247] S. Miyano, M. Tobita and H. Hashimoto, Bull. Chem. Soc. Jap. **54**, 3522 (1981).
[248] S. Miyano, S. Handa, K. Shimizu, K. Tagami and H. Hashimoto, Bull. Chem. Soc. Jap. **57**, 1943 (1984).
[249] J. M. Wilson and D. J. Cram, J. Am. Chem. Soc. **104**, 881 (1982).
[250] A. I. Meyers and K. A. Lutowski, J. Am. Chem. Soc. **104**, 879 (1982).
[251] K. Suzuki, E. Katayama and G. Tsuchihashi, Tetrahedron Lett. 4997 (1983).
[252] G. Tsuchihashi, K. Tomooka and K. Suzuki, Tetrahedron Lett. **25**, 4253 (1984).
[253] K. Suzuki, E. Katayama and G. Tsuchihashi, Tetrahedron Lett. **25**, 1817 (1984).

Chapter 7

[1] R. B. Woodward and R. Hoffmann, Angew. Chem. **81**, 797 (1969); Int. Ed. Engl. **8**, 781 (1969).
[2] T. L. Gilchrist and R. C. Storr, *Organic Reactions and Orbital Symmetry*. University Press, Cambridge, 1972.
[3] R. Huisgen, Angew. Chem. **80**, 329 (1968); Int. Ed. Engl. **7**, 321 (1968).
[4] H. Wurziger, Kontakte, 3 (1984).
[5] IUPAC Nomenclature for Straightforward Transformation (Provisional). Pure Appl. Chem. **53**, 305 (1981).

[6] K. N. Houk, J. Am. Chem. Soc. **95**, 4092 (1973).
[7] A. Korolev and V. Mur, Dokl. Akad. Nauk SSSR, **59**, 251 (1948); C. A. **42**, 6776 (1948).
[8] H. M. Walborsky, L. Barash and T. C. Davis, Tetrahedron Lett. **19**, 2333 (1963).
[9] J. Sauer and J. Kredel, Tetrahedron Lett. 6359 (1966).
[10] E. J. Corey and H. E. Ensley, J. Am. Chem. Soc. **97**, 6908 (1975).
[11] H. E. Ensley, C. A. Parnell and E. J. Corey, J. Org. Chem. **43**, 1610 (1978).
[12] W. Oppolzer, M. Kurth, D. Reichlin, C. Chapuis, M. Mohnhaupt and F. Moffatt, Helv. Chim. Acta **64**, 2802 (1981).
[13] H. E. Ensley and R. V. C. Carr, Tetrahedron Lett. 513 (1977).
[14] W. Oppolzer, C. Chapuis and M. J. Kelly, Helv. Chim. Acta **66**, 2358 (1983).
[15] W. Oppolzer, *Curr. Trends in Org. Synth.* Pergamon Press, Oxford, 1983, p. 131.
[16] J. D. Morrison and H. S. Mosher, *Asymmetric Organic Reactions*, Prentice-Hall, Englewood Cliffs, N. J., 1971, p. 256.
[17] T. Poll, G. Helmchen and B. Bauer, Tetrahedron Lett. **25**, 2191 (1984).
[18] R. F. Farmer and J. Hamer, J. Org. Chem. **31**, 2418 (1966).
[19] W. Oppolzer, C. Chapuis, M. D. Guo, D. Reichlin and T. Godel, Tetrahedron Lett. 4781 (1982).
[20] W. Oppolzer, C. Chapuis and G. Bernardinelli, Tetrahedron Lett. **25**, 5885 (1984).
[21] D. A. Evans, K. T. Chapman and J. Bisaha, J. Am. Chem. Soc. **106**, 4261 (1984).
[22] W. Oppolzer and C. Chapuis, Tetrahedron Lett. **25**, 5383 (1984).
[23] R. K. Boeckman, Jr., P. C. Naegely and S. D. Arthur, J. Org. Chem. **45**, 752 (1980).
[24] W. Oppolzer and C. Chapuis, Tetrahedron Lett. 4665 (1983).
[25] W. Choy, L. A. Reed and S. Masamune, J. Org. Chem. **48**, 1137 (1983).
[26] S. Masamune, L. A. Reed, J. T. Davis and W. Choy, J. Org. Chem. **48**, 4441 (1983).
[27] J. K. Whitesell, A. Bhattacharya, D. A. Aguilar and K. Henke, J. C. S. Chem. Comm. 989 (1982).
[28] G. Helmchen and R. Schmierer, Angew. Chem. **93**, 208 (1981); Int. Ed. Engl. **20**, 205 (1981).
[29] S. David, A. Lubineau and A. Thieffry, Tetrahedron, **34**, 299 (1978).
[30] P. A. T. W. Porskamp, R. C. Haltiwanger and R. C. Zwanenburg, Tetrahedron Lett. 2035 (1983).
[31] B. M. Trost, S. A. Godieski and J. P. Genét, J. Am. Chem. Soc. **100**, 3930 (1978).
[32] B. M. Trost, D. O'Krongly and J. L. Belletire, J. Am. Chem. Soc. **102**, 7595 (1980).
[33] M. M. Guseinov, I. M. Akhmedov and E. G. Mamedov, Azerb. Khim. Zh. 46 (1976); C. A. **85**, 176925 z (1976).
[34] S. Hashimoto, N. Komeshima and K. Koga, J. C. S. Chem. Comm. 437 (1979).
[35] L.-F. Tietze and G. Kiedrowski, Tetrahedron Lett. 219 (1981).
[36] L.-F. Tietze and K.-H. Glusenkamp, Angew. Chem. **95**, 901 (1983); Int. Ed. Engl. **22**, 887 (1983).
[37] T. Mukaiyama and N. Iwasawa, Chem. Lett. 29 (1981).
[38] D. A. Evans, K. T. Chapman and J. Bisaha, Tetrahedron Lett. **25**, 4071 (1984).
[39] W. Oppolzer, Synthesis 793 (1978).
[40] W. Oppolzer, Pure Appl. Chem. **53**, 1181 (1981).
[41] W. Oppolzer and D. A. Roberts, Helv. Chim. Acta **63**, 1703 (1980).
[42] C. Belzecki and I. Panfil, J. Org. Chem. **44**, 1212 (1979).
[43] K. Koizumi, H. Hirai and E. Yoshii, J. Org. Chem. **47**, 4004 (1982).
[44] A. Vasella and R. Voeffray, Helv. Chim. Acta **65**, 1953 (1982).
[45] A. Vasella and R. Voeffray, J. C. S. Chem. Comm. 97 (1981).
[46] B. S. Green, A. T. Hagler, Y. Rabinson and M. Rejtő, Isr. J. Chem. **15**, 124 (1976/77).
[47] L. M. Tolbert and M. B. Ali, J. Am. Chem. Soc. **104**, 1742 (1982).
[48] H. Gotthardt and W. Lenz, Angew. Chem. **91**, 926 (1979); Int. Ed. Engl. **18**, 868 (1979).
[49] S. Jarosz and A. Zamojski, Tetrahedron **38**, 1447 (1982).
[50] H. Koch, J. Runsink and H.-D. Scharf, Tetrahedron Lett. 3217 (1983).
[51] I. Ojima and S. Inaba, Tetrahedron Lett. 2077 (1980).
[52] I. Ojima and S. Inaba, Tetrahedron Lett. 2081 (1980).

[53] C. Gluchowski, L. Cooper, D. E. Bergbreiter and M. Newcomb. J. Org. Chem. **45**, 3413 (1980).
[54] H. Wynberg and E. G. J. Staring, J. Am. Chem. Soc. **104**, 166 (1982).
[55] H. Nozaki, H. Takaya, S. Moriuti and R. Noyori, Tetrahedron **24**, 3655 (1968).
[56] P. E. Krieger and J. A. Landgrebe, J. Org. Chem. **43**, 4447 (1978).
[57] D. Holland, D. A. Laidler and D. J. Milner, J. Mol. Catal. **11**, 119 (1981).
[58] T. Aratani, T. Yoneyoshi and T. Nagase, Tetrahedron Lett. 1707 (1975).
[59] T. Aratani, Y. Yoneyoshi and T. Nagase, Tetrahedron Lett. 2599 (1977).
[60] T. Aratani, Y. Yoneyoshi and T. Nagase, Tetrahedron Lett. 685 (1982).
[61] Y. Tatsuno, A. Konishi, A. Nakamura and S. Otsuka, J. C. S. Chem. Comm. 588 (1974).
[62] A. Nakamura, A. Konishi, Y. Tatsuno and S. Otsuka, J. Am. Chem. Soc. **100**, 3443 (1978).
[63] A. Nakamura, A. Konishi, R. Tsujitani, M. Kudo and S. Otsuka, J. Am. Chem. Soc. **100**, 3449 (1978).
[64] H. Abdallah, R. Grée and R. Carrié, Tetrahedron Lett. 503 (1982).
[65] A. Monpert, J. Martelli, R. Grée and R. Carrié, Tetrahedron Lett. **22**, 1961 (1981).
[66] G. Quinkert, U. Schwartz, H. Stark, W.-D. Weber, F. Adam, H. Baier, G. Frank and G. Dürner, Liebigs Ann. Chem. 1999 (1982).
[67] M. J. De Vos and A. Krief, Tetrahedron Lett. 103 (1983).
[68] S. J. Rhoads, in *Molecular Rearrangements* Part 1, (P. de Mayo, ed.). Interscience, New York, 1963, p. 655.
[69] W. von E. Doering and W. R. Roth, Angew. Chem. Int. Ed. Engl. **2**, 115 (1963).
[70] R. K. Hill and N. W. Gilman, J. C. S. Chem. Comm. 619 (1967).
[71] R. K. Hill and A. G. Edwards, Tetrahedron Lett. 3239 (1964).
[72] G. B. Bennett, Synthesis 589 (1977).
[73] F. E. Ziegler, Acc. Chem. Res. **10**, 227 (1977).
[74] R. K. Hill, R. Soman and S. Sawada, J. Org. Chem. **37**, 3737 (1972).
[75] K.-K. Chan, N. Cohen, J. P. De Noble, A. C. Specian, Jr. and G. Saucy, J. Org. Chem. **41**, 3505 (1976).
[76] G. Saucy and N. Cohen, in *New Synthetic Methodology and Biologically Active Substances*, (Z. Yoshida, ed.). Elsevier, New York, 1981, p. 155.
[77] N. Cohen, W. F. Eichel, R. J. Lopresti, C. Neukom and G. Saucy, J. Org. Chem. **41**, 3512 (1976).
[78] K.-K. Chan, A. C. Specian Jr. and G. Saucy, J. Org. Chem. **43**, 3435 (1978).
[79] K.-K. Chan, N. Cohen, J. P. De Noble, A. C. Specian and G. Saucy, J. Org. Chem. **41**, 3497 (1976).
[80] P. A. Bartlett and W. F. Hahne, J. Org. Chem. **44**, 882 (1979).
[81] R. E. Ireland, R. M. Mueller and A. K. Willard, J. Am. Chem. Soc. **98**, 2868 (1976).
[82] T. Sato, K. Tajima and T. Fujisawa, Tetrahedron Lett. 729 (1983).
[83] R. E. Ireland and M. D. Varney, J. Am. Chem. Soc. **106**, 3668 (1984).
[84] S. R. Wilson and R. S. Myers, J. Org. Chem. **40**, 3309 (1975).
[85] M. Nagatsuma, F. Shirai, N. Sayo and T. Nakai, Chem. Lett. 1393 (1984).
[86] T. Fujisawa, K. Tajima and T. Sato, Chem. Lett. 1669 (1984).
[87] S. E. Denmark and M. A. Harmata, J. Org. Chem. **48**, 3369 (1983).
[88] Y. Tamaru, Y. Furukawa, M. Mizutani, O. Kitao and Z. Yoshida, J. Org. Chem. **48**, 3631 (1983).
[89] Y. Tamaru, T. Harada and Z. Yoshida, J. Am. Chem. Soc. **100**, 1923 (1978).
[90] M. J. Kurth and O. H. W. Decker, Tetrahedron Lett. 4535 (1983).
[91] T. Nakai, K. Mikami and S. Taya, J. Am. Chem. Soc. **103**, 6492 (1981).
[92] K. Mikami, Y. Kimura, N. Kishi and T. Nakai, J. Org. Chem. **48**, 279 (1983).
[93] K. Mikami, K. Azuma and T. Nakai, Tetrahedron **40**, 2303 (1984).
[94] K. Mikami, K. Fujimoto and T. Nakai, Tetrahedron Lett. 513 (1983).
[95] J. A. Marshall and T. M. Jenson, J. Org. Chem. **49**, 1707 (1984).
[96] K. Hiroi and K. Nakazawa, Chem. Lett. 1077 (1980).
[97] K.-K. Chan and G. Saucy, J. Org. Chem. **42**, 3828 (1977).
[98] M. Moriwaki, Y. Yamamoto, J. Oda and Y. Inouye, J. Org. Chem. **41**, 300 (1976).

[99] M. Moriwaki, S. Sawada and Y. Inouye, J. C. S. Chem. Comm. 419 (1970).
[100] Y. Yamamoto, J. Oda and Y. Inouye, J. Org. Chem. **41**, 303 (1976).
[101] R. W. Hoffmann, R. Gerlach and S. Goldmann, Tetrahedron Lett. 2599 (1978).
[102] P. Bickart, F. W. Carson, J. Jacobus, E. G. Müller and K. Mislow, J. Am. Chem. Soc. **90**, 4869 (1968).
[103] R. W. Hoffmann, S. Goldmann, R. Gerlach and N. Maak, Chem. Ber. **113**, 845 (1980).
[104] S. Goldmann, R. W. Hoffmann, N. Maak and K.-J. Geueke, Chem. Ber. **113**, 831 (1980).
[105] K. Hiroi, P. Kitayama and S. Sato, J. C. S. Chem. Comm. 1470 (1983).
[106] R. K. Hill and M. Rabinovitz, J. Am. Chem. Soc. **86**, 965 (1964).
[107] G. B. Gill and B. Wallace, J. C. S. Chem. Comm. 382 (1977).
[108] W. G. Dauben and T. Brookhart, J. Am. Chem. Soc. **103**, 237 (1981).
[109] W. G. Dauben and T. Brookhart, J. Org. Chem. **47**, 3921 (1982).
[110] J. V. Duncia, P. T. Lansbury, Jr., T. Miller and B. B. Snider, J. Am. Chem. Soc. **104**, 1930 (1982).
[111] C. A. Townsend, T. Scholl and D. Arigoni, J. C. S. Chem. Comm. 921 (1975).
[112] W. Oppolzer, K. K. Mahanalabis and K. Bättig, Helv. Chim. Acta **60**, 2388 (1977).
[113] W. Oppolzer, K. Bättig and T. Hudlicky, Helv. Chim. Acta **62**, 1493 (1979).
[114] P. D. Kennewell, S. S. Matharu, J. B. Taylor and P. G. Sammes, J. C. S. Perkin 1 2563 (1982).
[115] W. Oppolzer and C. Robbiani, Helv. Chim. Acta **63**, 2010 (1980).
[116] W. Oppolzer, C. Robbiani and K. Bättig, Helv. Chim. Acta **63**, 2015 (1980).
[117] W. Oppolzer, C. Robbiani and K. Bättig, Tetrahedron **40**, 1391 (1984).
[118] W. Oppolzer and K. Thirring, J. Am. Chem. Soc. **104**, 4978 (1982).
[119] H. Felkin, L. D. Kwart, G. Swierczewski and J. D. Umpleby, J. C. S. Chem. Comm. 242 (1975).
[120] W. Oppolzer, R. Pitteloud and H. F. Strauss, J. Am. Chem. Soc. **104**, 6476 (1982).
[121] W. S. Johnson, Angew. Chem. **88**, 33 (1976); Int. Ed. Engl. **15**, 9 (1976).
[122] W. S. Johnson, C. A. Harbert, B. E. Ratcliffe and R. D. Stpanovic, J. Am. Chem. Soc. **98**, 6188 (1976).
[123] W. S. Johnson, J. D. Elliott and G. J. Hanson, J. Am. Chem. Soc. **106**, 1138 (1984).
[124] W. S. Johnson and G. E. Dubois, J. Am. Chem. Soc. **98**, 1038 (1976).
[125] W. S. Jonson, S. Escher and B. W. Metcalf, J. Am. Chem. Soc. **98**, 1039 (1976).
[126] W. S. Johnson, R. S. Brinkmeyer, V. M. Kapoor and T. M. Yarnell, J. Am. Chem. Soc. **99**, 8342 (1977).
[127] M. B. Groen and F. J. Zeelen, J. Org. Chem. **43**, 1961 (1978).
[128] J. A. M. Peters, T. A. P. Posthumus, N. P. van Vliet and F. J. Zeelen, J. Org. Chem. **45**, 2208 (1980).
[129] W. S. Johnson, D. Berner, D. J. Dumas, P. J. R. Nederlof and J. Welch, J. Am. Chem. Soc. **104**, 3508 (1982).
[130] W. R. Roush, H. R. Gillis and A. I. Ko, J. Am. Chem. Soc. **104**, 2269 (1982).

Chapter 8

[1] D. J. Abbott, S. Colonna and J. M. Stirling, J. C. S. Perkin 1 492 (1976).
[2] M. Furukawa, T. Okawara and Y. Terawaki, Chem. Pharm. Bull. **25**, 1319 (1977).
[3] R. K. Hill and L. A. Renbaum, Tetrahedron **38**, 1959 (1982).
[4] T. Wakabayashi, Y. Kato and K. Watanabe, Chem. Lett. 1283 (1976).
[5] J. E. Bäckvall and E. E. Björkman, J. Org. Chem. **45**, 2893 (1980).

[6] J.-E. Bäckvall, E. E. Björkmann, S. E. Byström and A. Solladié-Cavallo, Tetrahedron Lett. 943 (1982).
[7] B. M. Trost and E. Keinan, J. Am. Chem. Soc. **100**, 7779 (1978).
[8] R. Urban and I. Ugi, Angew. Chem. **87**, 67 (1975); Int. Ed. Engl. **14**, 61 (1975).
[9] G. Eberle, I. Lagerlund, I. Ugi and R. Urban, Tetrahedron **35**, 977 (1979).
[10] Y. Kawakami, J. Hiratake, Y. Yamamoto and J. Oda, J. C. S. Chem. Comm. 779 (1984).
[11] M. Mikolajczyk, Pure Appl. Chem. **52**, 959 (1980).
[12] S. J. Abbott, S. R. Jones, S. A. Weinman and J. R. Knowles, J. Am. Chem. Soc. **100**, 2558 (1978).
[13] C. R. Hall and T. D. Inch, J. C. S. Perkin 1 1104 (1979).
[14] T. Glowiak and W. Sawka-Dobrowolska, Tetrahedron Lett. **42**, 3965 (1977).
[15] T. Sato, H. Ueda, K. Nakagawa and N. Bodor, J. Org. Chem. **48**, 98 (1983).
[16] H. C. Brown and G. Zweifel, J. Am. Chem. Soc. **83**, 486 (1961).
[17] H. C. Brown, P. K. Jadhav and A. K. Mandal, Tetrahedron **37**, 3547 (1981).
[18] H. C. Brown, N. R. Ayyangar and G. Zweifel, J. Am. Chem. Soc. **86**, 1071 (1964).
[19] H. C. Brown and N. M. Yoon, Isr. J. Chem. **15**, 12 (1976/77).
[20] J. J. Partridge, N. K. Chadha and M. R. Uskokovic, J. Am. Chem. Soc. **95**, 7171 (1973).
[21] G. Grethe, J. Sereno, T. H. Williams and M. R. Uskokovic, J. Org. Chem. **48**, 5315 (1983).
[22] D. J. Sandman, K. Mislow, W. P. Giddings, J. Dirlam and G. C. Hanson, J. Am. Chem. Soc. **90**, 4877 (1968).
[23] H. C. Brown, M. C. Desai and P. K. Jadhav, J. Org. Chem. **47**, 5065 (1982).
[24] A. K. Mandal and N. M. Yoon, J. Organomet. Chem. **156**, 183 (1978).
[25] A. K. Mandal, P. K. Jadhav and H. C. Brown, J. Org. Chem. **45**, 3543 (1980).
[26] H. C. Brown and P. K. Jadhav, J. Org. Chem. **46**, 5047 (1981).
[27] H. C. Brown, P. K. Jadhav and A. K. Mandal, J. Org. Chem. **47**, 5074 (1982).
[28] P. K. Jadhav and H. C. Brown, J. Org. Chem. **46**, 2988 (1981).
[29] H. C. Brown, P. K. Jadhav and M. C. Desai, Tetrahedron **40**, 1325 (1984).
[30] H. C. Brown, P. K. Jadhav and M. C. Desai, J. Am. Chem. Soc. **104**, 4303 (1982).
[31] C. Zioudrou, I. Moustakali-Mavridis, P. Chrysochou and G. J. Karabatsos, Tetrahedron **34**, 3181 (1978).
[32] W. C. Still and J. C. Barrish, J. Am. Chem. Soc. **105**, 2487 (1983).
[33] W. C. Still and K. R. Shaw, Tetrahedron Lett. 3725 (1981).
[34] D. J. Morgens, Jr., Tetrahedron Lett. 3721 (1981).
[35] W. C. Still and K. P. Darst, J. Am. Chem. Soc. **102**, 7385 (1980).
[36] K. N. Houk, N. G. Rondan, Y.-D. Wu, J. T. Metz and M. N. Paddon-Row, Tetrahedron **40**, 2257 (1984).
[37] S. S. Jew, S. Terashima and K. Koga, Tetrahedron **35**, 2337 (1979).
[38] S. S. Jew, S. Terashima and K. Koga, Tetrahedron **35**, 2345 (1979).
[39] M. Hayashi, S. Terashima and K. Koga, Tetrahedron **37**, 2797 (1981).
[40] A. R. Chamberlin, M. Dezube and P. Dussault, Tetrahedron Lett. **22**, 4611 (1981).
[41] P. A. Bartlett, D. P. Richardson and J. Myerson, Tetrahedron **40**, 2317 (1984).
[42] Y. Tamaru, M. Mizutani, Y. Furukawa, S. Kuwamura, Z. Yoshida, K. Yanagi and M. Minobe, J. Am. Chem. Soc. **106**, 1079 (1984).
[43] A. R. Chamberlin and R. L. Mulholland, Jr., Tetrahedron **40**, 2297 (1984).
[44] T. Yamashita, H. Yasueda and N. Nakamura, Bull. Chem. Soc. Jap. **52**, 2165 (1979).
[45] J. Ichikawa, M. Asami and T. Mukaiyama, Chem. Lett. 949 (1984).
[46] T. Mukaiyama, I. Tomioka and M. Shimizu, Chem. Lett. 49 (1984).
[47] J. Gawronski, K. Gawronska and H. Wynberg, J. C. S. Chem. Comm. 307 (1981).
[48] H. Hiemstra and H. Wynberg, J. Am. Chem. Soc. **103**, 417 (1981).
[49] K. Suzuki, A. Ikegawa and T. Mukaiyama, Bull. Chem. Soc. Jap. **55**, 3277 (1982).
[50] N. Kobayashi and K. Iwai, Macromolecules **13**, 31 (1980).
[51] N. Kobayashi and K. Iwai, Tetrahedron Lett. 2167 (1980).
[52] G. Solladié, Synthesis 185 (1981).

[53] C. Mioskowski and G. Solladié, Tetrahedron **36**, 227 (1980).
[54] C. Mioskowski and G. Solladié, Tetrahedron Lett. 3341 (1975).
[55] F. Wudle and T. B. K. Lee, J. Am. Chem. Soc. **95**, 6349 (1973).
[56] K. Hiroi, S. Sato and R. Kitayama, Chem. Lett. 1595 (1980).
[57] L. Duhamel and J.-C. Plaquevent, Tetrahedron Lett. **42**, 2285 (1977).
[58] L. Duhamel and J.-C. Plaquevent, Bull. Soc. Chim. France, 69 (1982).
[59] L. Duhamel and J.-C. Plaquevent, Bull. Soc. Chim. France, 75 (1982).
[60] M. Yamaguchi, S. Yamamatsu, H. Oikawa, M. Saburi and S. Yoshikawa, Inorg. Chem. **20**, 3179 (1981).
[61] J. Hine, W.-S. Li and J. P. Zeigler, J. Am. Chem. Soc. **102**, 4403 (1980).
[62] E. J. Corey and T. A. Engler, Tetrahedron Lett. **25**, 149 (1984).
[63] S. R. Wilson and M. F. Price, J. Am. Chem. Soc. **104**, 1124 (1982).
[64] S. R. Wilson and M. F. Price, Tetrahedron Lett. 569 (1983).
[65] H. Kumobayashi and S. Akutagawa, J. Am. Chem. Soc. **100**, 3949 (1978).
[66] H. J. Hansen, R. Schmid, M. Schmid, Eur. Pat. Appl. EP. 104.376; C. A. **101**, 38124 (1984).
[67] H. J. Hansen, R. Schmid and M. Schmid, Eur. Pat. EP. Appl. 104.375; C. A. **101**, 111.119 (1984).
[68] K. Tani, T. Yamagata, S. Akutagawa, H. Kumobayashi, T. Taketomi, H. Takaya, A. Miyashita, R. Noyori and S. Otsuka, J. Am. Chem. Soc. **106**, 5209 (1984).

Subject Index

Adjectives closely connected with the subject of this bock, such as chiral, stereoselective, enantioselective, etc. were omitted.